$$X'' + \frac{\beta}{m} X' + \frac{\lambda}{m} X = 0 \qquad w = mg$$

$$x = e^{\lambda t}(c_1 \cos \kappa t$$

$$\text{period} = \frac{2\pi}{k}$$

ELEMENTARY DIFFERENTIAL EQUATIONS WITH APPLICATIONS

WILLIAM R. DERRICK
STANLEY I. GROSSMAN
University of Montana

ADDISON-WESLEY PUBLISHING COMPANY
Reading, Massachusetts · Menlo Park, California
London · Amsterdam · Don Mills, Ontario · Sydney

For Judith and Arlene

This book is in the
Addison-Wesley Series in Mathematics

Lynn H. Loomis,
Consulting Editor

PREFACE

Since the time of Newton, differential equations have been of fundamental importance in the study and application of mathematical analysis. The subject is crucial for obtaining an understanding of the physical sciences, and lately has gained increasing importance in the biological and social sciences. It is a truism that nothing is permanent except change. New information concerning the theory of differential equations and its application to all sciences appears at an increasing rate, leading to an increased interest in its study by college students.

Our goal in writing this book has been threefold:

1. To help the student grasp the nature and significance of differential and difference equations;

2. To provide the numerical tools necessary to solve problems not amenable to direct solution;

3. To provide a wealth of examples and problems in the biological and social sciences as well as in the physical sciences (applications are taken from chemistry, chemical engineering, air pollution, genetics, population dynamics, and gambling, as well as the more standard models of electrical circuits and mechanics).

Generally, students enrolled in an elementary differential equations course are poorly prepared for a rigorous treatment of the subject. We have tried to alleviate this problem by isolating the material that requires greater sophistication than that normally acquired in the first year of calculus. The emphasis throughout is on making the text readable both by frequent examples and by the inclusion of enough steps in the working of problems so that the student does not get bogged down in complicated calculations.

Almost every scientist has available the resources provided by a modern high-speed automatic computer. Problems which in the past would have been intractable are now easily approximated by a number of basic numerical techniques. We believe these methods should be made available to the student at an early stage in his career. It is also important that he should be aware of the limitations of these procedures, and not place an inordinate reliance on the computer. To this end, we include:

a) A primer on the two most common computer languages, FORTRAN and BASIC, which provides enough background to enable a beginner to program in these languages.

b) A chapter dedicated to the numerical solution of ordinary differential equations, which includes the development of each method, its limitations, and sample programs for their use on a computer.

Differential and difference equations are having a vast influence in present applications in the biological and social sciences. The role of difference equations in these areas is often more important than that of differential equations, since in many situations only discrete values of the parameters are known. The complexity of the problems posed by biological processes or economic interdependencies leads naturally to macroscopic models, often analyzed as compartmental systems. We have included a great number of problems in these areas and a short discussion of compartmental analysis.

The first three chapters of the text discuss methods for finding solutions to elementary differential equations, while Section 2.6 and Chapter 4 consist of a parallel development for difference equations. If time is a critical factor, the latter material can be deleted with no loss of continuity. Chapters 5 and 6 give a thorough discussion of power series and Laplace transform methods for solving ordinary differential equations. Included in Chapter 6 is a brief discussion of Laplace transform methods for difference equations.

In Chapter 7, we discuss systems of differential equations. The major part of this chapter (the first nine sections) does not require any knowledge of matrix theory. However, for those students who have had a course in matrices, we have included four additional sections which discuss systems of differential equations from the matrix theory point of view.

Chapter 8 contains a discussion of various numerical methods (single step and multistep) that can be used to approximate solutions of differential equations. Included is an analysis of round-off error, numerical instability, and speed of convergence for these methods.

Chapters 9 and 10 contain an introductory treatment of boundary value problems, Fourier series, and partial differential equations. The connection between generalized Fourier series and solutions to nonhomogeneous problems is included.

Chapter 11 provides the theoretical justification for much of the material which has preceded it. In particular, the Picard existence-uniqueness theorems for initial value problems are proved. The continuous dependence on initial conditions is also discussed. Chapter 12 provides a fairly complete discussion of Lyapunov stability theory for differential equations.

There are several different courses which can be taught from this book. The table below shows chapter interdependence. The sections in parentheses depend on the respective chapters in parentheses.

Chapter Number	Depends On
1	—
2	1
3	2
4	2, 3
5	2, 3
6 (6.5)	2, 3, (4)
7 (7.8)	2, 3, (6)
8	2, 3
9	2, 3
10	2, 3, 9
11 (11.2)	2, 3, (7)
12	2, 3, 7, 11

A typical one-semester course for non-math majors could include Chapters 1, 2, 3, 5, 6, and parts of Chapters 4 and 7 as time permits. In a one-year course for students with no matrix background, the 12 chapters could be covered with the deletion of the matrix material in Chapter 7 and the second section of Chapter 11. In a one-year course for non-math majors, it might be necessary to completely delete the last two chapters, depending on the mathematical training of the students. Throughout, starred sections may be omitted without loss of continuity.

We are grateful to Carolyn Chase and Joyce Schlieter, who so skillfully typed the manuscript of this text. We are especially grateful to the editorial staff at Addison-Wesley for the help and encouragement they provided us in the final writing stages. Also, we wish to thank Professors Stacy G. Langton, the University of Minnesota; Jerry D. Schuur, Michigan State University; and Anthony L. Peressini, the University of Illinois, for their helpful reviews of the manuscript.

Missoula, Montana   W.R.D.
January 1976   S.I.G.

CONTENTS

Chapter **4**
**Linear Second Order
Difference Equations**

Chapter **5**
**Power Series Solutions
of Differential Equations**

Chapter **6** **Laplace Transforms**

INTRODUCTION

1

1.1 DIFFERENTIAL EQUATIONS

Many of the basic laws of the physical sciences and, more recently, of the biological and social sciences, are formulated in terms of mathematical relations involving certain known and unknown quantities and their derivatives. Such relations are called *differential* equations.

In the next section we will see how the various types of differential equations are classified. In the succeeding chapters we will discuss methods of solution or, when that is not possible, of obtaining information about the solutions (when they exist) of several classes of equations. In this section we will show how some simple differential equations can arise.

Before citing any examples, however, we should emphasize that in the study of differential equations the most difficult problem is often to describe a real situation quantitatively. To do this, it is usually necessary to make simplifying assumptions that can be expressed in mathematical terms. Thus, for example, we initially describe the motion of a mass in space by assuming that (a) it is a point mass and (b) that there is no friction or air resistance. These assumptions are not realistic, but the scientist can often glean valuable information from even highly idealized models which, once understood, can be modified to take other observable factors into account.

Example 1.1.1 A ball is dropped from the top of a building 44.1 meters high (1 meter = 3.28 feet). When will the ball hit the ground?

Assuming that the ball is a point mass and that there is no air resistance, the acceleration of the ball will be due entirely to the force of gravity. If we denote the height of the ball at any time t by $h(t)$, then $h'(t)$ is the velocity of the ball at time t, since velocity is the instantaneous rate of change of height with respect to time. Similarly, $h''(t)$ is the upward acceleration of the ball at time t, since acceleration is the rate of change of velocity with respect to time.

It has been found experimentally that the gravitational acceleration g is approximately 980 cm/sec^2, 980 centimeters per second per second ($= 32.2$ ft/sec^2) at the surface of the earth. Thus the acceleration of the ball is constant:

$$h''(t) = -980, \tag{1.1}$$

where the minus sign on the right-side of the equation indicates that the force of gravity acts downward. Integrating both sides of Eq. (1.1) with respect to t, we obtain

$$h'(t) = -980t + C_1, \tag{1.2}$$

where C_1 is a constant that we must determine. To find C_1, we set $t = 0$ in Eq. (1.2). Then $C_1 = h'(0) = 0$, since the initial velocity $y'(0)$ of the ball is zero as the ball is dropped from rest. Finally, integrating (1.2) once more with respect to t, we obtain

$$h(t) = -490t^2 + C_2. \tag{1.3}$$

The constant C_2 can also be determined by setting $t = 0$; $C_2 = h(0) = 4410$, the initial height (in centimeters). Since we wish to find the length of time it takes the ball to strike the ground, we set the left-hand side of Eq. (1.3) equal to zero and solve for t:

$$490t^2 = 4410 \quad \text{or} \quad t^2 = 9.$$

Thus $t = \pm 3$, and since $t = -3$ has no physical significance, the answer is $t = 3$ sec.

Example 1.1.2 *Newton's law of cooling* states that the rate of change of the temperature difference between an object and its surrounding medium is proportional to the temperature difference. Let $\Delta(t)$ denote this temperature difference at any time t. Since a rate of change is expressed mathematically by a derivative, we may translate Newton's law of cooling into the equation

$$\frac{d\Delta}{dt} = k\Delta, \tag{1.4}$$

where k is the constant of proportionality. Recalling from elementary calculus that $(ce^{kt})' = k(ce^{kt})$, we are led to a solution

$$\Delta(t) = \Delta_0 e^{kt} \tag{1.5}$$

of the differential equation (1.4), where Δ_0 is constant. To verify that (1.5) is a solution of equation (1.4), we simply substitute (1.5) in (1.4) and perform the indicated operations:

$$(\Delta_0 e^{kt})' = k\Delta_0 e^{kt} = k(\Delta_0 e^{kt}).$$

If we set $t = 0$ in (1.5), we obtain

$$\Delta(0) = \Delta_0,$$

indicating that Δ_0 is the initial temperature difference between the object and its surrounding medium.

Since the temperature difference $\Delta(t)$ will approach zero as time increases without bound, the constant k in Eq. (1.5) must be negative.

Example 1.1.3 Consider a bacteria population that is changing at a rate proportional to its size. If $P(t)$ represents the population at time t, then the equation of population growth is

$$\frac{dP}{dt} = \alpha P,$$

where α can be positive or negative depending on whether the population is growing or declining. As in the last example, a solution is given by

$$P(t) = P(0)e^{\alpha t},$$

where $P(0)$ is the initial population. If $\alpha > 0$, the population will *grow exponentially*, while if $\alpha < 0$, the population will *decline exponentially*. In the special case of $\alpha = 0$ the population will remain constant at its equilibrium point $P(0)$.

Example 1.1.4 The growth rate per individual in a population is the difference between the average birth rate and the average death rate. Suppose that in a given population the average birth rate is a positive constant β, but the average death rate, due to the effects of crowding and increased competition for the available food, is proportional to the size of the population. Suppose that this constant of proportionality is $\delta > 0$. Since dP/dt is the growth rate of the population, the growth rate per individual of the population is

$$\frac{1}{P}\frac{dP}{dt}.$$

Then the differential equation that governs the growth of this population is

$$\frac{1}{P}\frac{dP}{dt} = \beta - \delta P.$$

Multiplying both sides of this equation by P, we have

$$\frac{dP}{dt} = P(\beta - \delta P), \tag{1.6}$$

which is called the *logistic equation*. The growth exhibited by this equation is called *logistic growth*. In Example 2.1.3 we shall find the solution

$$P(t) = \frac{\beta}{\delta + [\beta P(0)^{-1} - \delta]e^{-\beta t}} \tag{1.7}$$

for Eq. (1.6). Observe that as t gets larger, the term $e^{-\beta t}$ approaches zero since $\beta > 0$. Thus the population approaches a limiting value of β/δ beyond which it cannot increase, since setting $P = \beta/\delta$ in Eq. (1.6) yields $dP/dt = 0$.

Example 1.1.5 Assume that an ecosystem contains a predator species that feeds exclusively on a prey species and that the prey population has an ample food supply at all times. Let $y(t)$ and $x(t)$ denote the populations of the predator and prey species, respectively. Since food is readily available, the birth rate of the prey species is very likely to be a constant independent of time. However the death rate will certainly depend on the number of predators.

On the other hand, the birth rate of the predator species will be affected by the uncertain food supply, while its death rate may well be constant. As in Example 1.1.4, we can write the growth rates per individual for the two species as

$$\frac{1}{x}\frac{dx}{dt} = a - by, \qquad \frac{1}{y}\frac{dy}{dt} = \alpha x - \beta, \tag{1.8}$$

where a, b, α, β are positive constants of proportionality. Observe that neither equation is independent of the other, which explains the fact that the equations are called a *system* of first order equations. It is possible to eliminate one of the variables entirely: Solving the second equation for x, we have

$$x = \frac{dy/dt + \beta y}{\alpha y}. \tag{1.9}$$

Differentiating (1.9) yields

$$\frac{dx}{dt} = \frac{y(d^2y/dt^2) - (dy/dt)^2}{\alpha y^2}, \tag{1.10}$$

which when introduced into the first equation of (1.8) gives

$$y\frac{d^2y}{dt^2} - \left(\frac{dy}{dt}\right)^2 = y(a - by)\left(\beta y + \frac{dy}{dt}\right). \tag{1.11}$$

We have no explicit solution for this predator-prey system, but it is possible to establish relation between x and y that disclose some properties of the solution. (See Example 7.1.4.)

Example 1.1.6 Consider a cylindrical rod composed of a uniform heat-conducting material of length L and radius R (so that the cross-sectional area is $A = \pi R^2$). We assume that heat can enter and leave the rod only through its ends; that is, the lateral surface of the rod is completely insulated. Let x measure the distance along the rod (see Fig. 1.1) and let $T(x,t)$ denote the absolute temperature at time t at a point x units along the rod (we assume that the temperature is uniform across any cross-sectional area of the rod). Let ρ be the *density* of the rod (mass per unit volume) which is assumed to be constant. The *specific heat* c of the rod is defined to be the amount of heat (calories, BTU's) that must be supplied to raise the temperature of one unit mass (grams, pounds) of the rod by one degree (centigrade, fahrenheit).

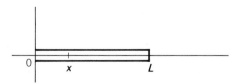

Figure 1.1

Consider a section of the rod between x and $x + dx$. Since mass = volume × density, the mass between these two points is $\rho A \, dx$. In order to change the temperature of the rod between these two points from 0 to $T(x,t)$ degrees, we must supply $T(x,t) \cdot c\rho A \cdot dx$ units of heat. Thus between any two points x_0 and x_1, the heat energy contained in the rod at time t is

$$Q(t) = \int_{x_0}^{x_1} T(x,t)c\rho A \, dx. \tag{1.12}$$

By the law of conservation of energy, if there are no heat sources within the rod, the heat energy in any part of the rod can increase or decrease only because of a lateral flow of heat through the boundaries of that part of the rod. This *heat flux*, written as $Q_F(x,t)$, is the quantity of heat energy per second passing through a unit area in the cross section x units from the left-hand end in a positive (rightward) direction. Thus, to find the rate of change of the heat energy between the x_0 and x_1 cross sections, we need only take the difference

$$\frac{dQ}{dt} = AQ_F(x_0,t) - AQ_F(x_1,t), \tag{1.13}$$

where the first term on the right-hand side is the heat energy flowing in and the second is that flowing out (see Fig. 1.2).

$Q_F(x_0, t) \longrightarrow \qquad \longrightarrow Q_F(x_1, t)$

0 $\qquad x_0 \quad x_1 \qquad L$

Figure 1.2

It is an empirical law of physics (and a commonsense observation) that the heat flux at any point is proportional to the *temperature gradient* $\partial T(x,t)/\partial x$ at that point. The constant of proportionality is called the *heat conductivity* of the rod and is denoted by κ. Since heat flows in the direction of decreasing temperature, we have

$$Q_F(x,t) = -\kappa \frac{\partial T}{\partial x}(x,t), \tag{1.14}$$

where $\kappa > 0$. Since no heat sources are in this part of the rod, we have

$$\begin{aligned}
AQ_F(x_0,t) - AQ_F(x_1,t) &= A\left[\kappa \frac{\partial T}{\partial x}(x_1,t) - \kappa \frac{\partial T}{\partial x}(x_0,t)\right] \\
&= \int_{x_0}^{x_1} \frac{\partial}{\partial x}\left[\kappa \frac{\partial T}{\partial x}(x,t)\right] A \, dx.
\end{aligned} \tag{1.15}$$

Equating the results obtained in (1.13) and (1.15) and differentiating (1.12), we obtain

$$\int_{x_0}^{x_1} \frac{\partial}{\partial x}\left(\kappa \frac{\partial T}{\partial x}\right) A \, dx = \frac{dQ}{dt} = \int_{x_0}^{x_1} c\rho \frac{\partial T}{\partial t} A \, dx.$$

Subtracting the left-hand side of this equation from the right-hand side yields

$$\int_{x_0}^{x_1}\left[c\rho \frac{\partial T}{\partial t} - \frac{\partial}{\partial x}\left(\kappa \frac{\partial T}{\partial x}\right)\right] A \, dx = 0. \tag{1.16}$$

Since (1.16) must hold for any interval $x_0 \leqslant x \leqslant x_1$, no matter how small, and all functions involved are assumed to be continuous, the expression in the brackets must vanish. To see this, we note that if it were positive at some point, then by continuity it would need to be positive on some interval, in which case (1.16) would not hold. Similarly, the expression cannot be negative. Therefore,

$$c\rho \frac{\partial T}{\partial t}(x,t) = \frac{\partial}{\partial x}\left[\kappa \frac{\partial T}{\partial x}(x,t)\right], \tag{1.17}$$

for all x in the interval $[0,L]$. Finally, if we assume that κ is a constant, then

$$\frac{\partial T}{\partial t}(x,t) = \delta \frac{\partial^2 T}{\partial x^2}(x,t), \tag{1.18}$$

where $\delta = \kappa/c\rho$ is the *diffusivity* of the material of the rod.

Equations of this type involving partial derivatives are called *partial differential equations*. This particular one (1.18) is called the *heat equation*. Of course, we could make our description more complicated (and more realistic) by assuming successively that (a) the density ρ is not constant, (b) the specific heat c is not constant, (c) there are internal sources of heat so that the integral in (1.16) is not zero, and (d) the conductivity is not constant. A solution of the heat equation will be obtained in Chapter 10.

EXERCISES 1.1

1. Solve the following differential equations by integrating both sides of the equation and evaluating y at the given x value.

   a) $\dfrac{dy}{dx} = x + 3$, $y(0) = 2$

   b) $\dfrac{dy}{dx} = \dfrac{2x}{x^2 + 5}$, $y(1) = 4$

   c) $\dfrac{dy}{dx} = \tan^2 x$, $y(0) = 1$

d) $\dfrac{d^2y}{dx^2} = x^2 - 9$, $y(0) = 1$, $y'(0) = 3$

e) $\dfrac{d^2y}{dx^2} = \sin x - \cos x$, $y(\pi/2) = 0$, $y'(\pi/2) = 1$

2. An object is falling in a vacuum with constant acceleration g. Express its velocity as a function of its height.

3. A drag racer has a 2000-lb. race car. If he wishes to travel $\frac{1}{2}$ mile in 20 sec, how fast must he accelerate?

4. The half-life of a radioactive substance is defined as the time required to decompose 50 percent of the substance. If $r(t)$ denotes the amount of the radioactive substance present after t years, $r(0) = r_0$, and the half-life is H years, what is a differential equation for $r(t)$ taking all side conditions into account?

5. A rocket is launched from an initial position (x_0, y_0) with an initial speed v_0 and an angle θ $(0 \leqslant \theta \leqslant \pi/2)$. Find its horizontal and vertical coordinates $x(t)$ and $y(t)$ as functions of time. Assume that there is no air resistance, and the force of gravity g is constant.

6. A bacteria population is known to double every three hours. If the population initially consists of 1000 bacteria, when will the population be 10,000?

1.2 CLASSIFICATION OF DIFFERENTIAL EQUATIONS

It should be apparent, if only from reading the examples in the previous section, that a great variety of types of differential equations can arise in the study of familiar phenomena. It is clearly necessary (and expedient) to study independently more restricted classes of these equations.

The most obvious classification is based on the nature of the derivative(s) in the equation. A differential equation involving only ordinary derivatives (derivatives of functions of one variable) is called an *ordinary differential equation*, while one containing partial derivatives is called a *partial differential equation*. Examples 1.1.1 through 1.1.5 led us to ordinary differential equations, while in Example 1.1.6 we obtained a partial differential equation. We will postpone the further classification of partial differential equations until Chapter 10.

Since this section and the next two chapters are concerned with ordinary differential equations, we will drop the word ordinary and refer simply to differential equations. The *order* of a differential equation is defined to be the order of the highest derivative appearing in the equation.

Example 1.2.1 The following are examples of differential equations with indicated orders:

a) $\dfrac{dy}{dx} = ay$ (first order),

b) $x''(t) - 3x'(t) + x(t) = \cos t$ (second order),

c) $(y')^9 - 2 \tan y = 6e^x$ (first order),

d) $(y^{iv})^{3/5} - 2y'' = \cos x$ (fourth order).

In Example 1.1.2 we obtained the differential equation (1.4):

$$\frac{d\Delta}{dt} = k\Delta, \qquad \Delta(0) = \Delta_0.$$

A solution was found to be

$$\Delta(t) = \Delta_0 e^{kt}.$$

This solution (which, as we shall see later, is the *only* solution) depends on the *initial* value of the unknown function $\Delta(t)$. Problems of this type are called *initial value problems*. In general, an *initial value problem* is a differential equation in which the values of the function and some of its derivatives at one point, called the *initial point*, are specified. Specifying the initial populations $x(0)$ and $y(0)$ in Example 1.1.5 would yield an initial value problem for this system of differential equations. Note that by using these values in the second equation of (1.8), we could obtain the value $y'(0)$. The values $y(0)$ and $y'(0)$ together with Eq. (1.11) constitute a second order initial value problem.

We define a *solution* for an nth order initial value problem, as a function which is n times differentiable, satisfies the given differential equation, and satisfies the given initial conditions.

In Chapter 11 we will prove that in a large class of problems, *there is a unique solution to an nth order differential equation if the value at a given point of the unknown function and all its derivatives up to the $(n - 1)$st are specified.*

On the other hand, we may not know the population size of the prey species but have adequate information concerning the population size of the predators at two points in time $y(0)$ and $y(a)$. These values and Eq. (1.11) constitute a boundary value problem. For a differential equation together with certain values of the function and its derivatives to qualify as a *boundary value problem*, it is necessary only that values be given for at least two different points. Knowing the initial temperatures $T(0,0)$ and $T(L,0)$ at both ends of the rod in Example 1.1.6 will transform the problem into a boundary value problem whose governing equation is a partial differential equation. Boundary value problems will be discussed at greater length in Chapter 9.

The physical examples given in the previous section are such that in each case we know a solution exists in nature. However, there is an inherent danger in confusing physical reality with the mathematical model given by the differential equation we are using to represent the real problem. It may well be that our reasoning was faulty, in which case, the equations obtained might bear no connection with reality. Then solutions to the equations need not exist. We should also note that not all differential equations have solutions. For example, the equation

$$\left(\frac{dy}{dx}\right)^2 + 3 = 0 \tag{1.19}$$

has no real-valued solutions, since dy/dx is imaginary. On the other hand, the equation

$$\left(\frac{dy}{dx}\right)^2 + y^2 = 0 \tag{1.20}$$

has zero as its only solution, while the equation

$$\frac{dy}{dx} + y = 0 \tag{1.21}$$

has an infinite set of solutions $y = ce^{-x}$ for any constant c.

EXERCISES 1.2

1. State the order of each of the following differential equations:

   a) $y' + ay = \sin^2 x$

   b) $\left(\dfrac{d^2x}{dt^2}\right)^3 - 3x\dfrac{dx}{dt} = 4\cos t$

   c) $s'''(t) - s''(t) = 0$

   d) $\dfrac{d^5y}{dx^5} = 0$

   e) $y'' + y = 0$

   f) $\left(\dfrac{dx}{dt}\right)^3 = x^5$

   g) $x' - x^2 = 3x'''$

2. State whether each of the following differential equations is an initial value or boundary value problem.

   a) $y'' + \omega^2 y = 0$, $y(0) = 0$, $y(1) = 1$

   b) $y'' + \omega^2 y = 0$, $y(0) = 0$, $y'(0) = 1$

   c) $y'' + \omega^2 y = 0$, $y(0) = 0$, $y'(1) = 1$

   d) $\left(\dfrac{dx}{dt}\right)^3 - 4x^2 = \sin t$, $x(0) = 3$

   e) $y''' + 3y'' - (y')^2 + e^y = \sin x$, $y(0) = 0$, $y'(0) = 3$, $y''(0) = 5$

   f) $y''' + 3y'' - (y')^2 + e^y = \sin x$, $y(0) = y(1) = 0$, $y'(0) = 2$

3. For each of the following differential equations verify that the given function or functions are solutions.

a) $y'' + y = 0$; $y_1 = 2 \sin x$, $y_2 = -5 \cos x$

b) $y''' - y'' + y' - y = x$; $y_1 = e^x - x - 1$, $y_2 = 3 \cos x - x - 1$, $y_3 = \cos x + \sin x + e^x - x - 1$

c) $x^2 y'' - 2xy' + 2y = 0$; $y_1 = x$, $y_2 = x^2$, $y_3 = 2x - 3x^2$

d) $y'' - y = e^x$; $y_1 = \dfrac{x}{2} e^x$, $y_2 = \left(4 + \dfrac{x}{2}\right) e^x + 3e^{-x}$

e) $y' = 2xy + 1$; $y_1 = e^{x^2} + e^{x^2} \int_0^x e^{-s^2} ds$

f) $x^2 y'' + 5xy' + 4y = 0 \ (x > 0)$; $y_1 = \dfrac{4 \ln x}{x^2}$, $y_2 = \dfrac{-6}{x^2}$

4. Verify that $v(x,t) = f_1(x + at) + f_2(x - at)$ is a solution of the partial differential equation

$$a^2 \frac{\partial^2 v}{\partial x^2} = \frac{\partial^2 v}{\partial t^2},$$

where f_1 and f_2 are any two twice differentiable functions. This equation is called the *wave equation* (see Chapter 10).

5. By "guessing" that there is a solution to $y'' - 3y' - 4y = 0$ of the form $y = e^{ax}$ for some constant a, find two solutions of the equation.

6. Given that $y_1(x)$ and $y_2(x)$ are two solutions of the equation in Exercise 5, check to see that $y_3(x) = c_1 y_1(x) + c_2 y_2(x)$ is also a solution, where c_1 and c_2 are arbitrary constants, by substituting y_3 into the differential equation.

7. Determine $\varphi(x)$ so that the functions $\sin \ln x$ and $\cos \ln x \ (x > 0)$ are solutions of the differential equation

$$[\varphi(x)y']' + \frac{y}{x} = 0.$$

8. Show that $\sin (1/x)$ and $\cos (1/x)$ are solutions of the differential equation

$$\frac{d}{dx}\left(x^2 \frac{dy}{dx}\right) + \frac{y}{x^2} = 0.$$

9. Verify that $y_1 = \sinh x$ and $y_2 = \cosh x$ are solutions of the differential equation $y'' - y = 0$. [*Hint*: $\cosh x = \frac{1}{2}(e^x + e^{-x})$ and $\sinh x = \frac{1}{2}(e^x - e^{-x})$.]

10. Suppose that $\varphi(x)$ is a solution of the initial value problem $y'' + yy' = x^3$, with $y(-1) = 1$, $y'(-1) = 2$. Find $\varphi''(-1)$ and $\varphi'''(-1)$.

11. Let $\varphi(x)$ be a solution to $y' = x^2 + y^2$ with $y(1) = 2$. Find $\varphi'(1)$, $\varphi''(1)$, and $\varphi'''(1)$.

1.3 DIFFERENCE EQUATIONS

Difference equations are the discrete analogs of differential equations. Differential equations arise in physical and biological systems where we are interested in studying the instantaneous rate of change (derivative) of one variable with respect to another. Thus we may be interested in the velocity at a given instant of a particle moving in a trajectory or the rate of growth at a particular moment of a bacteria population. On the other hand, there are many systems where it is of prime interest to study changes (differences) in the systems, rather than rates of change.

While the theory of differential equations has been with us for a long time, it is only recently that difference equations have received the kind of attention they deserve. This recent attention is due, in large part, to the advent of the computer with which many types of equations (including ordinary and partial differential equations) can be solved numerically by making use of difference equation formulations. In a sense this is an ironic development, since many differential equations are obtained by starting with difference equations and, by taking a limit, deriving an approximating differential equation. (Most population growth models are of this type.) This used to be the procedure because until recently the differential equations were easier to solve. Unfortunately, differential equations that arose originally from difference equations are sometimes approximated by new difference equations, often very different from the original ones. In Chapter 8 we will study numerical methods and illustrate the process by which a differential equation can be approximated by a difference equation.

Example 1.3.1 A patient in a hospital is suddenly switched to oxygen. Let V be the volume of gas contained in the lungs after inspiration, and V_D the amount present at the end of expiration (commonly called the *dead space*). Assuming uniform and complete mixing of the gases in the lungs, what is the concentration of nitrogen in the lungs at the end of the nth inspiration?

The amount of nitrogen in the lungs at the end of the nth inspiration must equal the amount in the dead space at the end of the $(n - 1)$st expiration. If x_n is the concentration of nitrogen in the lungs at the end of the nth inspiration, then

$$Vx_n = V_D x_{n-1}. \qquad (1.22)$$

Subtracting Vx_{n-1} from both sides of Eq. (1.22), we have

$$V(x_n - x_{n-1}) = (V_D - V)x_{n-1}. \qquad (1.23)$$

The difference $x_n - x_{n-1}$ is the discrete analog of a derivative, since it measures the change in concentration of nitrogen in the lungs. Thus (1.23) is a discrete version of a first order differential equation.

If we can find an expression (in n) for the concentrations x_n which satisfies (1.22) for all values of $n = 0, 1, 2, \ldots$, then we say we have a solution for the difference equation (1.22). It is easy to solve (1.22), since

$$x_n = \frac{V_D}{V} x_{n-1} = \left(\frac{V_D}{V}\right)^2 x_{n-2} = \cdots = \left(\frac{V_D}{V}\right)^n x_0,$$

where x_0 is the concentration of nitrogen in the air.

Example 1.3.2 A fish population grows in such a way that the growth in any year is twice the growth in the preceding year. To analyze this situation, we let P_n denote the population of the fish after n years. The growth during the period between the nth year and the $(n + 1)$st year is $P_{n+1} - P_n$. The equation governing population growth is therefore

$$P_{n+1} - P_n = 2(P_n - P_{n-1}), \tag{1.24}$$

which can be rewritten as

$$P_{n+1} = 3P_n - 2P_{n-1}. \tag{1.25}$$

If we specify the initial population P_0 and the population after one year, P_1, then Eq. (1.25) becomes an *initial value problem* that clearly has a unique solution. For example, if $P_0 = 50$ and $P_1 = 70$, then $P_2 = 3P_1 - 2P_0 = 110$, $P_3 = 3P_2 - 2P_1 = 190$, $P_4 = 3P_3 - 2P_2$ and so on. Of course, this method of obtaining P_n is cumbersome. In Chapters 2 and 4 we will indicate how certain difference equations can be solved in a more efficient manner.

Before citing further examples, we can formally define a *difference equation* as an equation that relates the values of x_n for different values of n. If N_1 and N_2 are respectively the largest and smallest values of n that occur in the equation, then the *order* of the difference equation is $N_1 - N_2$.

Equation (1.25) is a difference equation of order $(n + 1) - (n - 1) = 2$, while (1.22) is of the first order.

Example 1.3.3 Suppose that transverse forces $F_1, \ldots, F_n$ are applied to a string at n points spaced equally h units apart on the x-axis (see Fig. 1.3). To analyze this situation, we assume that between any two successive points at which forces are applied, the tension is constant (of course this constant may vary with the interval). We also assume that the weight of the string is negligible. If y_k denotes the deflection of the string in the y (transverse) direction, and if θ_k denotes the angle between the direction of tension and the positive x-axis, then we have (see Fig. 1.3)

$$\tan \theta_k = \frac{y_{k+1} - y_k}{h}$$

or

$$y_{k+1} - y_k = h \tan \theta_k. \tag{1.26}$$

Note that between any two successive points, since the tension is constant, the direction of the tension vector is constant. If we assume that the string is stationary ($y' = 0$), then the resultant of all the forces at the point (x_k, y_k) must be zero. The component of T_k in the y-direction is $T_k \sin \theta_k$, and the component of $-T_{k-1}$ in that direction is $-T_{k-1} \sin \theta_{k-1}$. Adding the applied force F_k at that point,

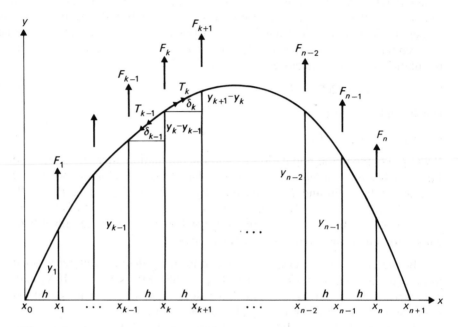

Figure 1.3

we obtain

$$T_k \sin \theta_k - T_{k-1} \sin \theta_{k-1} + F_k = 0. \tag{1.27}$$

Similarly, in the x-direction we obtain

$$T_k \cos \theta_k - T_{k-1} \cos \theta_{k-1} = 0. \tag{1.28}$$

Equation (1.28) indicates that the x-component of the tension does not vary, so we denote $T_k \cos \theta_k$ by T^*, a constant. Substituting this into (1.27), we obtain

$$T^* \frac{\sin \theta_k}{\cos \theta_k} - T^* \frac{\sin \theta_{k-1}}{\cos \theta_{k-1}} + F_k = 0$$

or

$$T^*(\tan \theta_k - \tan \theta_{k-1}) + F_k = 0. \tag{1.29}$$

Combining (1.26) and (1.29), we obtain the second order difference equation

$$\frac{T^*}{h}[(y_{k+1} - y_k) - (y_k - y_{k-1})] + F_k = 0$$

or

$$y_{k+1} - 2y_k + y_{k-1} + \frac{h}{T^*} F_k = 0. \tag{1.30}$$

However, unlike the previous example, it is clearly inappropriate to specify initial values for y_0 and y_1. Rather, we have the obvious boundary conditions

$$y_0 = 0, \qquad y_{n+1} = 0. \tag{1.31}$$

Equations (1.30) and (1.31) together form a second order *boundary value problem*. This problem so far has not been completely formulated since T^* is not known. However, if y_1 and the initial tension T_0 are known, then $\tan \theta_0 = y_1/h$, and we find that

$$T^* = T_0 \cos \theta_0 = \frac{T_0}{\sqrt{1 + (y_1/h)^2}}.$$

Example 1.3.4 Consider a population which is growing from generation to generation. Let P_n represent the population in the nth generation. Clearly, the population in the nth generation depends on the population in the previous generation and it may also depend on the populations in the preceding generations. For example, the preceding generations may have exhausted most of the available resources, thereby making reproduction difficult or impossible. The equation

$$x_{n+2} = rx_{n+1} + sx_n \tag{1.32}$$

can be thought of as a model of population growth where the population in the $(n + 2)$nd generation is made up of contributions from the $(n + 1)$st generation and the nth generation. The constants r and s measure the relative importance of these two terms.

Example 1.3.5 Suppose that two men are betting in some game where the probability of the first winning (at each turn) is p and the probability of the second winning is q, where $p + q = 1$. Suppose that they start with N dollars between them and they bet one dollar on each play of the game. Let x_k denote the probability that the first player eventually will win all of the other players' money where k is the number of dollars the first player has, $0 \leqslant k \leqslant N$. If the first player has k dollars now, then the probability that he will have $k + 1$ dollars at the next turn is p, while it is q that he will have $k - 1$ dollars. Hence,

$$x_k = px_{k+1} + qx_{k-1}, \tag{1.33}$$

and the boundary conditions must be

$$x_0 = 0, \qquad x_N = 1. \tag{1.34}$$

When we return to this problem in Section 4.8, we shall see that if p is only slightly greater than $\frac{1}{2}$, then the first player has a very high probability of winning all the money even though he may start with much less than the second player. This is the situation that holds in a gambling casino where the house always has the odds in its favor.

Difference equations are also useful in the determination of integrals.

Example 1.3.6 The following integral is related to the *complementary error function* erfc(x) that arises frequently in physical problems:

$$I_k(x) = \int_x^\infty \frac{(t-x)^k}{k!} e^{-t^2} \, dt, \tag{1.35}$$

where $k! = k(k-1)(k-2)\cdots 3\cdot 2\cdot 1$, k a positive integer, and $0! = 1$. We may rewrite $I_k(x)$ in the form

$$\begin{aligned} I_k(x) &= \frac{1}{k} \int_x^\infty (t-x) \frac{(t-x)^{k-1}}{(k-1)!} e^{-t^2} \, dt \\ &= -\frac{x}{k} I_{k-1}(x) + \frac{1}{k} \int_x^\infty \frac{(t-x)^{k-1}}{(k-1)!} t e^{-t^2} \, dt. \end{aligned} \tag{1.36}$$

Integrating the last term in Eq. (1.36) by parts yields

$$I_k(x) = -\frac{x}{k} I_{k-1}(x) + \frac{1}{2k} I_{k-2}(x). \tag{1.37}$$

Such difference equations are often called *recurrence relations*. Observe that

$$I_0(x) = \int_x^\infty e^{-t^2} \, dt,$$

and integrating by parts, we have

$$I_1(x) = \int_x^\infty (t-x)e^{-t^2} \, dt = \frac{e^{-x^2}}{2} - x I_0(x).$$

Knowing these values for a given x allows the computation of $I_k(x)$ for all k. In particular, if $x = 0$, we can use formula 58 in Appendix 1 to obtain $I_0(0) = \sqrt{\pi/2}$, and $I_1(0) = \frac{1}{2}$. Then by (1.37), we have

$$I_k(0) = \frac{1}{2k} I_{k-2}(0),$$

so that

$$I_{2n}(0) = \frac{\sqrt{\pi}}{2^{2n+1} n!},$$

and

$$I_{2n+1}(0) = \frac{(n+1)!}{(2n+2)!}.$$

EXERCISES 1.3

1. State the order of each of the following difference equations.

   a) $y_{n+1} = (y_n)^2$

   b) $y_{n+3} = (y_n)^2 - 3n$

   c) $y_{n+3} = (y_{n+1})^3$

   d) $(y_{n+3})^5 = \sqrt{y_{n+2}}$

   e) $y_{n+5} - 2y_{n+3} + y_n^2 = n^2 + 6$

   f) $y_{n+1} - n + y_n - \sqrt{n-2} = 0$

   g) $y_n y_{n+1} y_{n+2} - y_{n+3} y_{n-1} = 2n^2$ ⬥

2. Use graphs to illustrate the discrete growth processes, from the initial time $n = 0$, given by the following formulas. Does x_n approach a limiting (or equilibrium) value as n approaches infinity?

   a) $x_n = 100 + 100(2)^{-n}$

   b) $x_n = 100 + 100\left(\frac{1}{1+n^2}\right)$

   c) $x_n = 100 + 5n^2$

   d) $x_n = \sqrt{9+n}$

3. Verify that $x_n = n^2 + n$ is a solution of the difference equation $x_{n+1} - x_n = 2(n+1)$ and that $y_n = n^2 + n + c$ is also a solution for any constant c.

4. Verify that, for any constant k, $x_n = k2^{n(n+1)/2}$ is a solution of the difference equation $x_n = 2^n x_{n-1}$.

5. The growth of a bacteria culture in a nutrient medium is observed every two hours, and every time it is found that the bacteria population is thirty percent larger than at the previous measurement.

   a) Describe this growth process by means of a difference equation for P_n, the population after n hours.

   b) What is the order of the difference equation?

   c) Given that the initial population is 1000, determine P_2 and P_4.

6. The per capita production of garbage in the United States is estimated to be approximately five pounds per day and is increasing at the rate of four percent per year. Let g_n be the average daily production of garbage per capita n years from now.

   a) Describe the growth of garbage production by means of a difference equation.

   b) Determine g_2 and g_4.

7. By "guessing" that there are solutions to $y_{n+2} - 3y_{n+1} - 4y_n = 0$ of the form $y_n = \lambda^n$ for some constant λ, find two solutions of the equation.

8. Given that y_n and z_n are solutions of the difference equation

   $$y_{n+2} + a_n y_{n+1} + b_n y_n = 0,$$

   where a_n and b_n are functions of n, show that for any two constants c_1 and c_2, $w_n = c_1 y_n + c_2 z_n$ is also a solution.

9. Show that for any two constants c_1 and c_2,

$$y_n = \tfrac{1}{14}(n - \tfrac{9}{14})^2 + c_1 + c_2(-6)^n$$

is a solution to the difference equation

$$y_{n+2} + 5y_{n+1} - 6y_n = n.$$

10. Verify that $y_n = n2^{n-1}$ is a solution of the difference equation

$$y_{n+2} - 3y_{n+1} + 2y_n = 2^n.$$

11. Verify that $x_n = 2^n$ and $y_n = n2^n$ are solutions to the difference equation $y_{n+2} - 4y_{n+1} + 4y_n = 0$.

12. Verify that for any constants c_1 and c_2,

$$z_n = c_1 2^n + c_2 n2^n + \frac{n^2}{8} 2^n$$

is a solution of the difference equation

$$x_{n+2} - 4x_{n+1} + 4x_n = 2^n.$$

13. Verify that $x_n = c_1 \sin(2n\pi/3) + c_2 \cos(2n\pi/3)$ is a solution of the difference equation $y_{n+2} + y_{n+1} + y_n = 0$, for any two constants c_1 and c_2. Find a solution that satisfies the initial conditions $y_0 = 1$ and $y_1 = 2$.

14. The gamma function is defined by

$$\Gamma(n) = \int_0^\infty e^{-t} t^{n-1} \, dt.$$

Show that it satisfies the difference equation $\Gamma(n) = (n - 1)\Gamma(n - 1)$, and prove that $\Gamma(n) = (n - 1)!$, where n is a positive integer.

15. Let

$$I_k(x) = \int_0^\pi \frac{\cos kt - \cos kx}{\cos t - \cos x} \, dt.$$

Show that I_k satisfies the difference equation

$$I_{k+1}(x) - 2I_k(x) \cos x + I_{k-1}(x) = 0.$$

FIRST
ORDER
EQUATIONS

2

Solving a first order differential or difference equation can be very difficult, since there is no general method that can be used for all cases. In this chapter we discuss a few of the most useful methods for solving first order equations. We shall discover in this and the subsequent chapters that our ability to solve linear differential equations is limited only by our capacity to perform integrations based upon a well-developed general theory. On the other hand, as we shall also see, a multitude of methods are required to attack nonlinear equations and there is no guarantee that any of these techniques will succeed. For this reason, the procedures we discuss will appear to be merely a collection of tricks. They are based, however, on three underlying principles: substitution, separation of variables, and multiplication by a suitable function.

2.1 DIRECT METHODS

Consider the first order differential equation

$$\frac{dy}{dx} = f(x,y). \tag{2.1}$$

If the function $f(x,y)$ can be written in such a way that it does not involve the dependent variable y, then the equation can be solved by integrating both sides with respect to x.

Example 2.1.1 $\dfrac{dy}{dx} = \dfrac{x + xy}{1 + y}$. The right-hand side of the equation equals x after cancellation of the common term $(1 + y)$ in the numerator and denominator. Thus

$$y = \int x \, dx + C = \frac{x^2}{2} + C.$$

Clearly this method also will work for higher-order equations of the form $y^{(n)} = f(x)$.

Almost as simple is the method of *separation of variables*, which may be used whenever the function $f(x,y)$ can be factored into the form

$$f(x,y) = \frac{g(x)}{h(y)},$$

where $g(x)$ and $h(y)$ are each functions of only *one* variable. Then Eq. (2.1) can be written in the form

$$h(y)\frac{dy}{dx} = g(x).$$

Integrating both sides of this equation with respect to x and changing variables on the left-hand side, we have

$$\int h(y)\, dy = \int h(y) \frac{dy}{dx}\, dx = \int g(x)\, dx + C.$$

Example 2.1.2 $\dfrac{dy}{dx} = 2xy$. We rewrite this equation as

$$\frac{1}{y}\, dy = 2x\, dx$$

so that we may automatically perform the change of variables. Integrating both sides yields

$$\ln y = x^2 + C,$$

and exponentiating both sides of the equation, we have

$$y = e^{x^2+C} = e^C e^{x^2} = C_1 e^{x^2}.$$

This is the *general solution* of the equation and involves an unspecified real constant C_1. Thus the differential equation $y' = 2xy$ has an infinite number of solutions, depending on the particular choice of C_1.

In Section 1.2 we mentioned the fundamental existence and uniqueness theorem for initial value problems. If we specify an initial condition in addition to the differential equation of Example 2.1.2, then the specified condition will completely determine the solution of the initial value problem. For example, if the initial condition is $y(1) = 2$, then substituting $x = 1$ into the general solution yields

$$2 = y(1) = C_1 e \qquad \text{or} \qquad C_1 = 2/e.$$

Thus the unique solution of the initial value problem

$$\frac{dy}{dx} = 2xy, \qquad y(1) = 2,$$

is

$$y = 2e^{x^2-1}.$$

To obtain the unique solution of an initial value problem, it is necessary to specify as many initial conditions as the order of the equation. This fact will be proved in Chapter 11.

Example 2.1.3 In Example 1.1.4 we derived the *logistic equation*

$$\frac{dP}{dt} = P(\beta - \delta P), \tag{2.2}$$

where β and δ were given constants. Separating the variables, we have

$$\frac{dP}{P(\beta - \delta P)} = dt. \tag{2.3}$$

It is easy to verify that

$$\frac{1}{P(\beta - \delta P)} = \frac{1}{\beta P} + \frac{\delta}{\beta(\beta - \delta P)}.$$

Substituting the right-hand side of this equation into (2.3) and integrating, we obtain

$$\frac{1}{\beta} \ln P - \frac{1}{\beta} \ln (\beta - \delta P) = t + C$$

or

$$\ln \left(\frac{P}{\beta - \delta P} \right)^{1/\beta} = t + C.$$

Exponentiating both sides of this equation and denoting the arbitrary constant e^C by C, we have

$$\frac{P}{\beta - \delta P} = Ce^{\beta t}. \tag{2.4}$$

[*Warning: We shall frequently make such changes of constant without further notice.*]
Setting $t = 0$, we find that

$$\frac{P(0)}{\beta - \delta P(0)} = C, \tag{2.5}$$

and substituting this value of C into Eq. (2.4) yields

$$\frac{P(t)}{\beta - \delta P(t)} = \frac{P(0)}{\beta - \delta P(0)} e^{\beta t}.$$

Cross multiplying and solving for $P(t)$, we obtain (after some algebra) the equation

$$P(t) = \frac{\beta}{\delta + [\beta P(0)^{-1} - \delta]e^{-\beta t}}, \tag{2.6}$$

which is the solution of the logistic equation (1.7).

It is often useful to write Eq. (2.1) in the form

$$dy - f(x,y)\, dx = 0.$$

It may then be possible to multiply the equation by some function to get the differential of another known function.

Example 2.1.4 $\dfrac{dy}{dx} = \dfrac{y}{x + x^2 y^2}$. Rewriting this equation, we obtain

$$(x + x^2 y^2)\, dy - y\, dx = 0,$$

which may be rearranged in the form

$$x^2 y^2\, dy + (x\, dy - y\, dx) = 0. \tag{2.7}$$

The quantity in parentheses should remind the reader of the differential formula

$$d\left(\frac{y}{x}\right) = \frac{x\, dy - y\, dx}{x^2}. \tag{2.8}$$

Dividing Eq. (2.7) by x^2 yields

$$d\left(\frac{y^3}{3}\right) + d\left(\frac{y}{x}\right) = 0,$$

and integrating, we find that

$$\frac{y^3}{3} + \frac{y}{x} = C.$$

The following differential formulas are often useful in similar situations:

$$d\left(\frac{x}{y}\right) = \frac{y\, dx - x\, dy}{y^2}, \tag{2.9}$$

$$d(xy) = x\, dy + y\, dx, \tag{2.10}$$

$$d(x^2 + y^2) = 2(x\, dx + y\, dy), \tag{2.11}$$

$$d(\sqrt{x^2 + y^2}) = \frac{x\, dy + y\, dy}{\sqrt{x^2 + y^2}}, \tag{2.12}$$

$$d\left(\tan^{-1}\frac{x}{y}\right) = \frac{y\, dx - x\, dy}{x^2 + y^2}, \tag{2.13}$$

$$d\left(\ln\frac{x}{y}\right) = \frac{y\, dx - x\, dy}{xy}. \tag{2.14}$$

Example 2.1.5. Free fall According to Newton's second law of motion, the acceleration a of a body of mass m equals the total force F acting on it divided by the mass. Thus $F = ma$. If a body of mass m falls freely under the influence of gravity alone, then the total force acting on it is mg, where g is the acceleration due to gravity (which may be considered constant on the surface of the earth and approximately 32 feet per second per second). Let y be the distance of the body above the surface of the earth; then the upward acceleration of the body is $d^2 y/dt^2$ and

$$m \frac{d^2 y}{dt^2} = -mg, \tag{2.15}$$

where the negative sign indicates that gravity pulls downward. Canceling the m on both sides and integrating, we find that

$$\frac{dy}{dt} = -gt + v_0, \tag{2.16}$$

where v_0 is the velocity of the body at time $t = 0$. Integrating once more, we obtain

$$y = -\frac{gt^2}{2} + v_0 t + y_0, \tag{2.17}$$

where y_0 is the height of the body at time $t = 0$.

Retarded fall If we assume in addition that air exerts a *resisting* force proportional to the velocity of the body, then Eq. (2.15) becomes

$$\frac{d^2 y}{dt^2} = -g - c \frac{dy}{dt}, \qquad c > 0. \tag{2.18}$$

(The minus sign indicates that the air resistance causes a deceleration.) Letting $v = dy/dt$, we have

$$\frac{dv}{dt} = -g - cv, \tag{2.19}$$

and separating variables yields

$$\int \frac{dv}{cv + g} = -\int dt + c_1,$$

so that

$$\frac{1}{c} \ln (cv + g) = -t + c_1,$$

which after exponentiation becomes

$$v = -\frac{g}{c} + c_2 e^{-ct}. \tag{2.20}$$

Since $c > 0$, $v \to -g/c$ as $t \to \infty$. This limiting value is called the *terminal velocity*. Observe by setting $t = 0$ in (2.20) that $c_2 = v_0 + g/c$, and if $v_0 = 0$, we obtain

$$v = \frac{g}{c} (e^{-ct} - 1).$$

To find the height y at any time t, we must perform an additional integration on Eq. (2.20).

Example 2.1.6 Find the shape of a curved mirror such that light from a source at the origin will be reflected in a beam parallel to the x-axis.

By symmetry, the mirror has the shape of a surface of revolution obtained by revolving a curve about the x-axis. Let (x, y) be any point on the cross section in the xy-plane (see Fig. 2.1). The law of reflection states that the angle of incidence α must equal the angle of reflection β; thus $\alpha = \beta = \theta$. Since the interior angles of a triangle add up to $180°$, we have $\psi = \alpha + \theta = 2\theta$. What is important about the angles θ and ψ is that

$$y' = \tan \theta \qquad \text{and} \qquad \frac{y}{x} = \tan \psi.$$

Using the trigonometric formula for the tangent of a double angle, we have

$$\frac{y}{x} = \tan \psi = \tan 2\theta = \frac{2 \tan \theta}{1 - \tan^2 \theta} = \frac{y'}{1 - (y')^2}.$$

Solving for y', we obtain the quadratic equation

$$y(y')^2 + xy' - y = 0,$$

and by the quadratic formula we find that

$$y' = \frac{-x \pm \sqrt{x^2 + y^2}}{y}$$

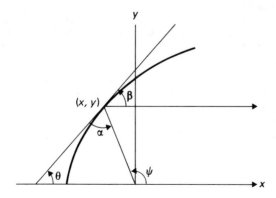

Figure 2.1

which may be written as

$$x \, dx + y \, dy = \pm \sqrt{x^2 + y^2} \, dx.$$

It follows, according to Eq. (2.12), that

$$\sqrt{x^2 + y^2} = \pm x + c,$$

which yields, by squaring both sides,

$$y^2 = \pm 2cx + c^2.$$

This is the equation of the family of all parabolas with focus at the origin which are symmetric with respect to the x-axis.

EXERCISES 2.1

In Exercises 1 through 20, find the general solution explicitly, if possible. Otherwise find a relation that defines the solution implicitly. When an initial condition is given, find the particular solution that satisfies the condition.

1. $\dfrac{dy}{dx} = \dfrac{e^x}{2y}$

2. $xy' = 3y, \ y(2) = 5$

3. $\dfrac{dy}{dx} = \dfrac{e^y x}{e^y + x^2 e^y}$

4. $\dfrac{dx}{dy} = x \cos y, \ x\left(\dfrac{\pi}{2}\right) = 1$

5. $\dfrac{dz}{dr} = r^2(1 + z^2)$

6. $\dfrac{dy}{dx} + y = y(xe^{x^2} + 1), \ y(0) = 1$

7. $\dfrac{dP}{dQ} = P(\cos Q + \sin Q)$

8. $\dfrac{dy}{dx} = y^2(1 + x^2), \ y(0) = 1$

9. $\dfrac{ds}{dt} + 2s = st^2, \ s(0) = 1$

10. $\dfrac{dy}{dx} = \sqrt{1 - y^2}$

11. $(1 + x) \, dy + 3y \, dx = 0, \ y(6) = 7$

12. $\dfrac{dx}{dt} + (\cos t)e^x = 0$

13. $(y + 3) \, dx + \cot x \, dy = 0$

14. $\dfrac{dx}{dt} = x(1 - \sin 2t), \ x(0) = 1$

15. $\sqrt{1 - x^2} \, dy + \sqrt{1 - y^2} \, dx = 0$

16. $(\sin x \cos y) \, dx + (\cos x \sin y) \, dy = 0, \ y(0) = 0$

17. $x^2 \, dy + y^2 \, dx = 0, \ y(1) = 3$

18. $\dfrac{dy}{dx} = \dfrac{y^3 + 2y}{x^2 + 3x}, \ y(1) = 1$

19. $e^x\left(\dfrac{dx}{dt} + 1\right) = 1, \ x(0) = 1$

20. $(r^2 - 2r - 8) \, ds = (s^2 + s - 2) \, dr, \ s(0) = 0$

21. Consider a population $P(t)$ that is growing according to the equation $dP/dt = P(\alpha - \beta P)$ of logistic growth. Prove that the growth rate is at a maximum when the population is equal to half its equilibrium size.

22. Bacteria are supplied as food to a protozoan population at a constant rate μ. It is observed that the bacteria are consumed at a rate that is proportional to the square of their numbers. The concentration $c(t)$ of the bacteria therefore satisfies the differential equation $dc/dt = \mu - \lambda c^2$, where λ is a positive constant.

    a) Determine $c(t)$ in terms of $c(0)$.
    b) What is the equilibrium concentration of the bacteria?

23. In some chemical reactions certain products catalyze their own formation. If $x(t)$ is the amount of such a product at time t, a possible model for the reaction is given by the differential equation $dx/dt = \alpha(\beta - x)$, where α and β are positive constants. According to this model, the reaction is completed when $x = \beta$, since this condition indicates that one of the chemicals has been depleted.

    a) Solve the equation in terms of the constants α, β, and $x(0)$.
    b) For $\alpha = 1$, $\beta = 200$, and $x(0) = 20$, draw a graph of $x(t)$ for $t > 0$.

24. On a certain day it began to snow early in the morning and the snow continued to fall at a constant rate. The velocity at which a snowplow is able to clear a road is inversely proportional to the height of the accumulated snow. The snowplow started at 11 a.m. and has cleared four miles by 2 p.m. By 5 p.m. it has cleared another two miles. When did it start snowing?

25. A large open cistern filled with water has the shape of a hemisphere with radius 25 ft. The bowl has a circular hole of radius 1 ft in the bottom. By Torricelli's law*, water will flow out of the hole with the same speed it would attain in falling freely from the level of the water to the hole. How long will it take for all the water to flow from the cistern?

26. In Exercise 25, find the shape of the cistern that would ensure that the water level drops at a constant rate.†

27. The king and queen of Transylvania order coffee. The king adds a teaspoon of cool cream to the coffee at once but does not drink it immediately. The queen waits ten minutes and then adds the cream (at the same temperature). They then drink their coffee. Who drinks the hotter coffee? [*Hint*: Use Newton's law of cooling and assume that the temperature of the cream is less than the temperature of the air.]

2.2 SUBSTITUTION TECHNIQUES

In this section we present three special substitutions that may be used occasionally to solve a differential equation. Suppose that we are given a first order differential equation of the form

$$\frac{dy}{dx} = F\left(\frac{y}{x}\right);$$ (2.21)

---

* Evangelista Torricelli (1608–1647) was an Italian physicist.

† The ancient Egyptians (1380 B.C.) used water clocks based on this principle to tell time.

that is, the right-hand side of the equation can be written as a function of the variable y/x. It is then natural to try the substitution $z = y/x$. Since the function y depends on x, so does the function z. Differentiating $y = xz$ with respect to x, we have

$$\frac{dy}{dx} = z + x\frac{dz}{dx}. \tag{2.22}$$

Substituting $z = y/x$ and replacing the left-hand side of (2.21) by the right-hand side of (2.22), we obtain

$$z + x\frac{dz}{dx} = F(z).$$

The variables in this equation can be separated, since

$$x\frac{dz}{dx} = F(z) - z,$$

so that

$$\frac{dz}{F(z) - z} = \frac{dx}{x}.$$

A complete solution can now be obtained by integrating both sides of this equation and replacing each z by y/x. We illustrate this procedure in the next example.

Example 2.2.1 $\dfrac{dy}{dx} = \dfrac{x - y}{x + y}$. Dividing the numerator and denominator of the right-hand side of this equation by x yields

$$\frac{dy}{dx} = \frac{1 - (y/x)}{1 + (y/x)} = \frac{1 - z}{1 + z} = F(z).$$

Replacing the left-hand side by $z + x(dz/dx)$ and separating variables, we have (after some algebra)

$$\left(\frac{1 + z}{1 - 2z - z^2}\right) dz = \frac{dx}{x}.$$

After integrating, we obtain

$$\ln(1 - 2z - z^2) = -2\ln x + c = \ln(cx^{-2}),$$

so that exponentiating and replacing z by y/x leads to the implicit solution

$$1 - \frac{2y}{x} - \frac{y^2}{x^2} = \frac{c}{x^2}.$$

Finally, multiplying both sides by x^2 yields $x^2 - 2xy - y^2 = c$.

Another useful technique applies to equations of the form

$$\frac{dy}{dx} = F(ax + by + c), \tag{2.23}$$

where a, b, and c are any real constants. If the substitution $z = ax + by + c$ is made in Eq. (2.23), we will obtain

$$\frac{z' - a}{b} = F(z),$$

since $z' = a + by'$. This equation has separable variables.

Example 2.2.2 $\dfrac{dy}{dx} = (x + y + 1)^2 - 2$. Letting $z = x + y + 1$, we have $z' = 1 + y'$, so $z' - 1 = z^2 - 2$ or $z' = z^2 - 1$, which may be written as

$$dx = \frac{dz}{z^2 - 1} = \frac{1}{2}\left(\frac{1}{z - 1} - \frac{1}{z + 1}\right)dz.$$

Integrating, we find that $\ln\left[(z - 1)/(z + 1)\right] = 2x + c$, so that $(z - 1)/(z + 1) = ce^{2x}$ or, after some algebra,

$$x + y + 1 = z = \frac{1 + ce^{2x}}{1 - ce^{2x}}.$$

A third method is useful when we have an equation of the form

$$\frac{dy}{dx} = F\left(\frac{ax + by + c}{\alpha x + \beta y + \gamma}\right), \qquad a\beta \neq \alpha b. \tag{2.24}$$

We note that if the constants c and γ are zero, the right-hand side of (2.24) can be written as

$$F\left(\frac{ax + by}{\alpha x + \beta y}\right) = F\left(\frac{a + b(y/x)}{\alpha + \beta(y/x)}\right) = G\left(\frac{y}{x}\right),$$

which is a function of the variable y/x and therefore the equation can be solved in the same way that we solved (2.21). Suppose that we let $x = u + h$ and $y = v + k$, where h and k are so chosen that the right-hand side of (2.24) equals

$$F\left(\frac{au + bv}{\alpha u + \beta v}\right).$$

Such a selection can always be done uniquely by solving the equations

$$ah + bk = -c,$$
$$\alpha h + \beta k = -\gamma,$$

simultaneously for h and k, since the determinant $a\beta - \alpha b$ of the system is non-zero (see Appendix 4). Now

$$\frac{dy}{dx} = \frac{d(v + k)}{du}\frac{du}{dx} = \frac{dv}{du},$$

and Eq. (2.24) becomes

$$\frac{dv}{du} = F\left(\frac{au + bv}{\alpha u + \beta v}\right) = F\left(\frac{a + b(v/u)}{\alpha + \beta(v/u)}\right),$$

which is an equation to which the first method applies.

Example 2.2.3 $\dfrac{dy}{dx} = \dfrac{x - y - 5}{x + y - 1}$. Solving $h - k = 5$ and $h + k = 1$ simultaneously yields $h = 3$ and $k = -2$ so the substitution $x = u + 3$, $y = v - 2$ yields

$$\frac{dv}{du} = \frac{u - v}{u + v}.$$

We saw in Example 2.2.1 that the solution of this equation is $u^2 - 2uv - v^2 = c$. We therefore obtain as a solution of the original problem

$$(x - 3)^2 - 2(x - 3)(y + 2) - (y + 2)^2 = c.$$

In the event that $a\beta = \alpha b$ in Eq. (2.24), we can set $z = ax + by$ so that $a\beta/b = \alpha$ and

$$\frac{\beta}{b}z = \frac{a\beta}{b}x + \beta y = \alpha x + \beta y.$$

Replacing $ax + by$ and $\alpha x + \beta y$ in (2.24) by z and $(\beta/b)z$, respectively, we obtain

$$\frac{dz}{dx} = bF\left(\frac{z + c}{(\beta/b)z + \gamma}\right) + a,$$

since $z' = a + by'$. The variables in this equation can be separated.

Example 2.2.4 $\dfrac{dy}{dx} = \dfrac{x + y + 1}{2x + 2y - 1}$. Letting $z = x + y$, we have $z' = 1 + y'$, and, after some algebra, the equation becomes

$$\frac{dz}{dx} = \frac{3z}{2z - 1}.$$

By separation we then obtain $2z - \ln z = 3x + c$ or $x + y = c \exp(2y - x)$.

The results of this section are summarized in the table below.

Form of Equation	Substitution	New Equation
$\dfrac{dy}{dx} = F\left(\dfrac{y}{x}\right)$	$z = \dfrac{y}{x}$	$\dfrac{dz}{F(z) - z} = \dfrac{dx}{x}$
$\dfrac{dy}{dx} = F(ax + by + c)$	$z = ax + by + c$	$\dfrac{dz}{dx} = bF(z) + a$
$\dfrac{dy}{dx} = F\left(\dfrac{ax + by + c}{\alpha x + \beta y + \gamma}\right)$		
$\quad\begin{cases} \alpha\beta = \alpha b \\[6pt] \alpha\beta \neq \alpha b \end{cases}$	$z = ax + by$ $x = u + h*$ $y = v + k$	$\dfrac{dz}{dx} = bF\left(\dfrac{z + c}{(\beta/b)z + \gamma}\right) + a$ $\dfrac{dv}{du} = F\left(\dfrac{a + b(v/u)}{\alpha + \beta(v/u)}\right)$

* h, k chosen so that $ah + bk = -c$ and $\alpha h + \beta k = -\gamma$

EXERCISES 2.2

If possible, find the general solution, for each equation in Exercises 1 through 12. Otherwise, find a relation that defines the solution implicitly. When an initial condition is given, find the particular solution that satisfies the condition.

1. $x\, dy - y\, dx = \sqrt{xy}\, dx$

2. $\dfrac{dx}{dt} = \dfrac{x}{t} + \cosh \dfrac{x}{t}$

3. $(xe^{y/x} + y)\, dx = x\, dy,\ y(1) = 0$

4. $(x^2 + y^2)\, dx = 2xy\, dy,\ y(-1) = 0$

5. $t\dfrac{ds}{dt} + s = 2\sqrt{st}$

6. $(x + v)\, dx = x\, dv$

7. $(y^2 - 2xy)\, dx + (2xy - x^2)\, dy = 0,\ y(1) = 2$

8. $(x^2 + y^2)\, dx + 3xy\, dy = 0$

9. $\sqrt{x^2 + y^2}\, dx = x\, dy - y\, dx$

10. $(x^2y + 2xy^2 - y^3)\, dx - (2y^3 - xy^2 + x^3)\, dy = 0$

11. $y' = \dfrac{y^2 + xy}{x^2},\ y(1) = 1$

12. $x' = \dfrac{x - t\cos^2(x/t)}{t},\ x(1) = \pi/4$

13. Solve the following equations:

   a) $2\dfrac{dy}{dx} = x^2 + 4xy + 4y^2 + 3$

   b) $(x + y - 1)^2\, dx + 9\, dy = 0$

   c) $(x + y)\, dy = (2x + 2y - 3)\, dx$

14. Use the results of this section to solve:

   a) $(x + 2y + 2)\, dx + (2x - y)\, dy = 0$

   b) $(x - y + 1)\, dx + (x + y)\, dy = 0$

   c) $(x + y + 4)\, dx = (2x + 2y - 1)\, dy,\ y(0) = 0$

15. Solve the equation

$$\frac{dy}{dx} = \frac{1 - xy^2}{2x^2 y}$$

by making the substitution $v = y/x^n$ for an appropriate value of n.

16. Use the method of Exercise 15 to solve

$$\frac{dx}{dt} = \frac{x - tx^2}{t + xt^2}.$$

2.3 LINEAR EQUATIONS

An nth order differential equation is *linear* if it can be written in the form

$$\frac{d^n y}{dx^n} + a_{n-1}(x)\frac{d^{n-1}y}{dx^{n-1}} + \cdots + a_1(x)\frac{dy}{dx} + a_0(x)y = f(x).$$

Hence a first order linear equation has the form

$$\frac{dy}{dx} + a(x)y = f(x),$$

while a second order linear equation can be written as

$$\frac{d^2 y}{dx^2} + a(x)\frac{dy}{dx} + b(x)y = f(x).$$

The notation indicates that $a(x)$, $b(x)$, $f(x)$, etc., are functions of x alone.
Before dealing with the general first order linear equation

$$\frac{dy}{dx} + a(x)y = f(x), \tag{2.25}$$

we will illustrate some special cases. If $f(x) \equiv 0$ and $a(x) \equiv a$ is constant, then (2.25) becomes

$$\frac{dy}{dx} + ay = 0. \tag{2.26}$$

Separation of variables yields $dy/y = -a\,dx$. Upon integration we obtain $\ln y = -ax + c$ or

$$y = ce^{-ax}. \tag{2.27}$$

Equation (2.27) is the equation of *exponential decay* (if a is positive) or *exponential growth* (if a is negative), as shown in Examples 1.1.2 and 1.1.3 and illustrated in Fig. 2.2(a) and (b).

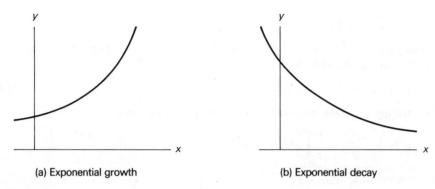

(a) Exponential growth   (b) Exponential decay

Figure 2.2

Example 2.3.1 Consider a population growing at a rate of 10 percent of the population size per unit time period and with an initial size of 1000 individuals. If $P(t)$ is the population at time t, then the growth equation is

$$\frac{dP}{dt} = 0.1P \quad \text{or} \quad \frac{dP}{dt} - 0.1P = 0.$$

This is Eq. (2.26) with $a = -0.1$ and it has the general solution $P(t) = ce^{0.1t}$. Setting $t = 0$ yields the equation $1000 = P(0) = c$. Thus the solution to the problem is $P(t) = 1000e^{0.1t}$. For example, if each unit of time t is measured in weeks, the population after ten weeks is $P(10) = 1000e \approx 2718$.

Let us now consider the more general example

$$\frac{dy}{dx} + ay \doteq f(x). \tag{2.28}$$

It is now impossible to separate the variables. However, there is a simple method of solving (2.28) by multiplying both sides of the equation by an *integrating factor*. Since

$$\frac{d}{dx}(e^{ax}y) = e^{ax}\frac{dy}{dx} + e^{ax}ay = e^{ax}\left(\frac{dy}{dx} + ay\right),$$

we can multiply both sides of (2.28) by e^{ax} to obtain

$$\frac{d}{dx}(e^{ax}y) = e^{ax}\left(\frac{dy}{dx} + ay\right) = e^{ax}f(x). \tag{2.29}$$

Integrating both ends of Eq. (2.29), we obtain

$$e^{ax}y = \int e^{ax}f(x)\,dx + c \tag{2.30}$$

or

$$y = e^{-ax}[\int f(x)e^{ax} \, dx + c]. \tag{2.31}$$

Example 2.3.2 Consider the equation $dy/dx + 2y = x$. The integrating factor is e^{2x}, and we obtain by Eq. (2.30)

$$e^{2x}y = \int xe^{2x} \, dx + c. \tag{2.32}$$

Integrating the right-hand side of (2.32) by parts, we have

$$e^{2x}y = \frac{xe^{2x}}{2} - \frac{1}{2}\int e^{2x} \, dx + c,$$

from which it follows that

$$y = e^{-2x}\left[\frac{xe^{2x}}{2} - \frac{e^{2x}}{4} + c\right]$$

or

$$y = \frac{x}{2} - \frac{1}{4} + ce^{-2x}.$$

We now return to the general linear first order equation (2.25). Since differentiation and integration are inverse operations,

$$\frac{d}{dx}\int a(x) \, dx = a(x).$$

Also we have

$$\frac{d}{dx}(e^{g(x)}y) = e^{g(x)}\left(\frac{dy}{dx} + y\frac{dg}{dx}\right).$$

Then

$$\frac{d}{dx}(e^{\int a(x) \, dx}y) = e^{\int a(x) \, dx}\left[\frac{dy}{dx} + a(x)y\right],$$

so it is clear that $e^{\int a(x) \, dx}$ is a suitable integrating factor for Eq. (2.25). Multiplying both sides of (2.25) by this integrating factor, we find that

$$\frac{d}{dx}(e^{\int a(x) \, dx}y) = e^{\int a(x) \, dx}\left[\frac{dy}{dx} + a(x)y\right] = e^{\int a(x) \, dx}f(x),$$

and integrating both sides yields

$$e^{\int a(x) \, dx}y = \int f(x)e^{\int a(x) \, dx} \, dx + c, \tag{2.33}$$

which can be rewritten in the abbreviated form

$$y = e^{-\int a\, dx}[\int f e^{\int a\, dx}\, dx + c]. \tag{2.34}$$

Equation (2.34) verifies the claim we made at the beginning of this chapter, since our ability to solve the first order equation (2.25) depends entirely on our capacity to perform the integrations in (2.34).

Example 2.3.3 Solve

$$\frac{dy}{dx} + \frac{2}{x}y = 5x^2. \tag{2.35}$$

We observe that Eq. (2.35) has the same form as (2.25) with $a(x) = 2/x$, and

$$\int a(x)\, dx = 2\int \frac{dx}{x} = 2\ln x = \ln x^2.$$

Multiplying both sides of the equation by $\exp(\ln x^2) = x^2$ and integrating yields, according to Eq. (2.33),

$$x^2 y = \int 5x^2 \cdot x^2\, dx + c = x^5 + c.$$

Thus $y = x^3 + cx^{-2}$ is the general solution of (2.35).

Example 2.3.4 Consider the equation $dy/dx = x^3 - 2xy$, where $y = 1$ when $x = 1$. Rewriting the equation as $dy/dx + 2xy = x^3$, we see that $a(x) = 2x$ and the integrating factor is $\exp\left[\int a(t)\, dt\right] = \exp(x^2)$. Thus multiplying both sides by $\exp(x^2)$ and integrating, we have

$$e^{x^2}y = \int x^3 e^{x^2}\, dx + c,$$

so that

$$y = e^{-x^2}[\int x^3 e^{x^2}\, dx + c].$$

We can integrate the integral by parts as follows:

$$\int x^3 e^{x^2}\, dx = \int x^2(x e^{x^2})\, dx = \frac{x^2 e^{x^2}}{2} - \int x e^{x^2}\, dx$$

$$= e^{x^2}\left(\frac{x^2 - 1}{2}\right).$$

Thus replacing this term for the integral above, we have

$$y = e^{-x^2}\left[e^{x^2}\left(\frac{x^2 - 1}{2}\right) + c\right] = \frac{x^2 - 1}{2} + ce^{-x^2}.$$

Setting $x = 1$ yields

$$1 = y(1) = ce^{-1},$$

so $c = e$ and the solution to the problem is

$$y = \tfrac{1}{2}(x^2 - 1) + e^{1-x^2}.$$

Example 2.3.5. Intravenous feeding of glucose Infusion of glucose into the bloodstream is an important medical technique. To study this process, we define $G(t)$ to be the amount of glucose in the bloodstream of a patient at time t. Suppose that glucose is infused into the bloodstream at the constant rate of k grams per minute. At the same time the glucose is converted and removed from the bloodstream at a rate proportional to the amount of glucose present. Then the function $G(t)$ satisfies the first order differential equation

$$\frac{dG}{dt} = k - aG,$$

where a is a positive constant. To solve this equation, we write $dG/dt + aG = k$ and multiply both sides by the integrating factor e^{at}. The solution is $G(t) = ce^{-at} + k/a$. When $t = 0$, $c = G(0) - k/a$, so the solution can be written as

$$G(t) = \frac{k}{a} + \left[G(0) - \frac{k}{a} \right] e^{-at}.$$

As $t \to \infty$, the concentration of glucose approaches the equilibrium value k/a.

EXERCISES 2.3

In Exercises 1 through 11 find the general solution for each equation. When an initial condition is given, find the particular solution that satisfies the condition.

1. $\dfrac{dx}{dt} = 3x$

2. $\dfrac{dy}{dx} + 22y = 0,\ y(1) = 2$

3. $\dfrac{dx}{dt} = x + 1,\ x(0) = 1$

4. $\dfrac{dy}{dx} + y = \sin x,\ y(0) = 0$

5. $\dfrac{dx}{dy} - x \ln y = y^y$

6. $\dfrac{dy}{dx} + y = \dfrac{1}{1 + e^{2x}}$

7. $\dfrac{dy}{dx} - \dfrac{3}{x} y = x^3,\ y(1) = 4$

8. $\dfrac{dx}{dt} + x \cot t = 2t \csc t$

9. $x' - 2x = t^2 e^{2t}$

10. $y' + \dfrac{2}{x} y = \dfrac{\cos x}{x^2},\ y(\pi) = 0$

11. $\dfrac{ds}{du} + s = ue^{-u} + 1$

12. Solve the equation

$$y - x\frac{dy}{dx} = \frac{dy}{dx} y^2 e^y$$

by reversing the roles of x and y (i.e., treat x as the dependent variable).

13. Use the method of Exercise 12 to solve

$$\frac{dy}{dx} = \frac{1}{e^{-y} - x}.$$

14. Find the solution of $dy/dx = 2(2x - y)$ that passes through the point $(0, -1)$.

15. Suppose that $T(t)$ is the temperature difference at time t between an object and its surrounding medium. By Newton's law of cooling, $dT/dt = -kT$, where $k > 0$. In terms of k, calculate the length of time it takes the temperature difference to decrease to

   a) one-half its original value,
   b) one-fourth its original value.

16. A chemical substance S is produced at the rate of r moles per minute in a chemical reaction. At the same time it is consumed at a rate of c moles per minute per mole of S. Let $S(t)$ be the number of moles of the chemical present at time t.

   a) Obtain the differential equation satisfied by $S(t)$.
   b) Determine $S(t)$ in terms of $S(0)$.
   c) Find the equilibrium amount of the chemical.

17. An infectious disease is introduced to a large population. The proportion of people who have been exposed to the disease increases with time. Suppose that $P(t)$ is the proportion of people who have been exposed to the disease within t years of its introduction. If $P'(t) = [1 - P(t)]/3$ and $P(0) = 0$, after how many years will the proportion have increased to 90 percent?

2.4 SOME SPECIAL NONLINEAR EQUATIONS

Certain nonlinear first order equations can be reduced to linear equations by a suitable change of variables. The equation

$$\frac{dy}{dx} + a(x)y = f(x)y^n, \tag{2.36}$$

which is known as *Bernoulli's equation,* is of this type. Set $z = y^{1-n}$. Then $z' = (1 - n)y^{-n}y'$, so if we multiply both sides of (2.36) by $(1 - n)y^{-n}$, we obtain

$$(1 - n)y^{-n}y' + (1 - n)ay^{1-n} = (1 - n)f$$

or

$$\frac{dz}{dx} + (1 - n)a(x)z = (1 - n)f(x).$$

The equation is now linear and may be solved as before.

Example 2.4.1 Solve

$$\frac{dy}{dx} - \frac{y}{x} = -\frac{5}{2}x^2y^3. \tag{2.37}$$

Here $n = 3$, so we let $z = y^{-2}$, $z' = -2y^{-3}y'$, and multiply Eq. (2.37) by $-2y^{-3}$ to obtain

$$z' + \frac{2}{x}z = 5x^2,$$

which, we saw in Example 2.3.3, has the solution

$$y^{-2} = z = x^3 + cx^{-2}.$$

Thus

$$y = (x^3 + cx^{-2})^{-1/2}.$$

A similar procedure is used in the next problem.

Example 2.4.2 Solve

$$\frac{dy}{dx} + a(x)y = f(x)y \ln y. \tag{2.38}$$

We let $z = \ln y$. Then $z' = y'/y$, so that dividing (2.38) by y, we obtain the linear equation

$$\frac{dz}{dx} + a(x) = f(x)z.$$

The nonlinear equation

$$\frac{dy}{dx} + y^2 + a(x)y + b(x) = 0 \tag{2.39}$$

is called the *Riccati equation*. It occurs frequently in physical applications* and may be solved by a simple substitution that reduces it to a linear equation. Let $y = z'/z$. Then $z' = zy$ and $z'' = zy' + yz'$. But using the original substitution, we see that

$$yz' = \frac{(z')^2}{z} = z\left(\frac{z'}{z}\right)^2 = zy^2.$$

Hence $z'' = z(y' + y^2)$. It is now apparent that multiplying Eq. (2.39) by z may produce some interesting consequences:

$$z'' + a(x)z' + b(x)z = zy' + zy^2 + azy + bz = 0.$$

Thus by means of this substitution we have transformed the Riccati equation into a second order linear equation

$$z'' + a(x)z' + b(x)z = 0,$$

---

* See, for example, R. Bellman, *Introduction to the Mathematical Theory of Control Process*, Vol. 1, Academic Press, New York, 1968.

which we will solve in Chapter 3 for the special case when $a(x)$ and $b(x)$ are constants. (See Exercise 19, Section 3.4.)

Another first order equation of considerable importance is the *Clairaut equation*:

$$y = xy' + f(y'). \tag{2.40}$$

We can differentiate both sides with respect to x to obtain

$$y' = y' + xy'' + f'(y')y'',$$

where the last term is a result of the chain rule. Canceling like terms, we get

$$[x + f'(y')]y'' = 0. \tag{2.41}$$

Since one of the factors must vanish, there are two cases to consider.

i. If $y'' = 0$, then $y' = c$, and substituting this value into Eq. (2.40) produces the *general solution*

$$y = cx + f(c), \tag{2.42}$$

which is a collection of straight lines.

ii. If $x + f'(y') = 0$, then $x = -f'(y')$, and Eq. (2.40) may be rewritten as

$$y = f(y') - y'f'(y'). \tag{2.43}$$

Here x and y are both expressed in terms of functions of y', so that if we rename $y' = t$, we will obtain the parametrized curve

$$x = -f'(t), \qquad y = f(t) - tf'(t). \tag{2.44}$$

We must check that points on this curve satisfy Eq. (2.40). Since

$$\frac{dy}{dx} = \frac{dy/dt}{dx/dt} = \frac{f'(t) - f'(t) - tf''(t)}{-f''(t)} = t,$$

the slope of the curve at each point is given by the parameter t, provided $f''(t) \neq 0$. Replacing t by y' and $-f'(t)$ by x in (2.44) yields Eq. (2.40), provided that $f''(t) \neq 0$. Furthermore, (2.44) is not a special case of the general solution (2.42), since for that equation y' is constant while in (2.44) y' depends on the parameter t. The special solution (2.44) is called a *singular solution* of (2.40).

Example 2.4.3 Solve

$$y = xy' + (y')^3. \tag{2.45}$$

According to (2.42), the general solution is $y = cx + f(c) = cx + c^3$. Since $f(t) = t^3$, the parametric equations of the singular solution are given by

$$x = -3t^2, \qquad y = t^3 - t(3t^2) = -2t^3, \qquad t \neq 0. \tag{2.46}$$

[At $t = 0, f''(0) = 0$.] Eliminating the parameter t, we obtain

$$4x^3 = -108t^6 = -27y^2, \tag{2.47}$$

so $27y^2 = -4x^3$ is a singular solution of (2.45) except at the point (0,0), where y' does not exist. (See Fig. 2.3.) Note that the general solution does not include this singular solution.

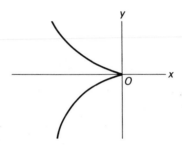

Figure 2.3

Example 2.4.4 Consider the equation

$$xy' - e^{y'} - y = 0,$$

which can be written as

$$y = xy' - e^{y'}. \tag{2.48}$$

The general solution is $y = cx - e^c$, while the singular solution is

$$x = e^t, \qquad y = e^t(t - 1). \tag{2.49}$$

Therefore, $t = \ln x$, which is defined only when $x > 0$, so the singular solution is given by $y = x(\ln x - 1), x > 0$.

EXERCISES 2.4

1. Reduce the following nonlinear equations to linear second order equations:

   a) $y' + y^2 - 1 = 0$

   b) $\dfrac{dx}{dt} + x^2 + 1 = 0$

   c) $y' + y^2 + y - 1 = 0$

2. For arbitrary constants a, b, and c, rewrite the nonlinear equation $y' + ay^2 + by + c = 0$ as a second order linear equation.

In Exercises 3 through 11 find the general solution for each equation and a particular solution when an initial condition is given.

3. $\dfrac{dy}{dx} = -\dfrac{(6y^2 - x - 1)y}{2x}$

4. $x' = 1 + 6te^{t-x}$

5. $y' = -y^3xe^{-2x} + y$

6. $\dfrac{dy}{dx} = \dfrac{\sin^2 x + \cos^2 y}{2 \tan x \sin y \cos y}$, $y\left(\dfrac{\pi}{2}\right) = 0$

7. $(2y^3 - x^3)\,dx + 3xy^2\,dy = 0$, $y(1) = 1$

8. $x\dfrac{dy}{dx} + y = x^4y^3$, $y(1) = 1$

9. $tx^2\dfrac{dx}{dt} + x^3 = t \cos t$

10. $\dfrac{dy}{dx} + \dfrac{3}{x}y = x^2y^2$, $y(1) = 2$

11. $xyy' - y^2 + x^2 = 0$

In Exercises 12 through 18 find the general solution and the singular solution for each equation.

12. $y = x\dfrac{dy}{dx} + \dfrac{1}{4}\left(\dfrac{dy}{dx}\right)^4$

13. $y = x\dfrac{dy}{dx} - \dfrac{1}{2}\left(\dfrac{dy}{dx} - 4\right)^2$

14. $y = x\dfrac{dy}{dx} - e^{dy/dx}$

15. $y = x\dfrac{dy}{dx} + \left(\dfrac{dy}{dx}\right)^{-1}$

16. $(y - xy')^2 - (y')^2 = 1$

17. $y = xy' - \sqrt{y'}$

18. $y = xy' + \ln y'$

2.5 EXACT EQUATIONS

We shall now use partial derivatives to solve ordinary differential equations. Suppose that we take the total differential of the equation $g(x,y) = c$:

$$dg = \frac{\partial g}{\partial x}\,dx + \frac{\partial g}{\partial y}\,dy = 0. \tag{2.50}$$

For example, the equation $xy = c$ has the total differential $y\,dx + x\,dy = 0$, which may be rewritten as the differential equation $y' = -y/x$. Reversing the situation, suppose that we start with the differential equation

$$M(x,y)\,dx + N(x,y)\,dy = 0. \tag{2.51}$$

If we can find a function $g(x,y)$ such that

$$\frac{\partial g}{\partial x} = M \quad \text{and} \quad \frac{\partial g}{\partial y} = N,$$

then (2.51) becomes $dg = 0$, so that $g(x,y) = c$ is the general solution of (2.51). In this case $M\,dx + N\,dy$ is said to be an *exact differential*, and (2.51) is called an exact differential equation.

How can we tell if a differential equation is exact? We may recall from basic calculus that

$$\frac{\partial^2 g}{\partial y\, \partial x} = \frac{\partial^2 g}{\partial x\, \partial y} \tag{2.52}$$

if both sides of the equation exist and are continuous. In terms of our functions M and N, Eq. (2.52) becomes

$$\frac{\partial M}{\partial y} = \frac{\partial N}{\partial x}, \tag{2.53}$$

so (2.53) is a necessary condition for (2.51) to be exact. We shall show that (2.53) is also sufficient by showing how the function g can be obtained so as to satisfy (2.51). The method proceeds in two steps.

1. Integrate the function $M = \partial g/\partial x$ with respect to x:

$$\int M\, dx = \int \frac{\partial g}{\partial x}\, dx = g + h(y). \tag{2.54}$$

The "constant of integration" $h(y)$ occurring in (2.54) is an arbitrary function of y since we must introduce the most general term that vanishes under *partial* differentiation with respect to x. So now the task is to discover what the function $h(y)$ is.

2. Take the partial derivative with respect to y of Eq. (2.54):

$$\frac{\partial}{\partial y} \int M\, dx = \frac{\partial g}{\partial y} + h'(y) = N + h'(y).$$

Thus

$$h'(y) = \frac{\partial}{\partial y} \int M\, dx - N.$$

Taking the partial derivative inside the integral and replacing $\partial M/\partial y$ by $\partial N/\partial x$, we have

$$h'(y) = \int \frac{\partial M}{\partial y}\, dx - N = \int \frac{\partial N}{\partial x}\, dx - N. \tag{2.55}$$

The right-hand side of (2.55) is a function of y only, since its partial derivative with respect to x vanishes. Integrating both sides of (2.55), we obtain

$$h(y) = \int \left(\int \frac{\partial N}{\partial x}\, dx - N \right) dy + c,$$

which, when substituted into (2.54), gives the general solution of Eq. (2.51).

Example 2.5.1 $(1 - \sin x \tan y)\, dx + (\cos x \sec^2 y)\, dy = 0$. Here

$$M(x, y) = 1 - \sin x \tan y \qquad \text{and} \qquad N(x, y) = \cos x \sec^2 y$$

so that

$$\frac{\partial M}{\partial y} = -\sin x \sec^2 y = \frac{\partial N}{\partial x},$$

and the equation is exact. To find the solution, we note that (2.54) implies:

$$g(x,y) = \int M\, dx - h(y) = x + \cos x \tan y - h(y).$$

Taking the partial derivative with respect to y of both sides yields

$$\cos x \sec^2 y = N = \frac{\partial g}{\partial y} = \cos x \sec^2 y - h'(y).$$

Thus $h'(y) = 0$, so that $h(y)$ is constant and the general solution is

$$g(x,y) = x + \cos x \tan y + c = 0.$$

It should be apparent that exact equations are comparatively rare, since condition (2.53) requires a precise balance of the functions M and N. For example,

$$(3x + 2y)\, dx + x\, dy = 0$$

is clearly not exact. However, if we multiply the equation by x, then the new equation

$$(3x^2 + 2xy)\, dx + x^2\, dy = 0$$

is exact. The question we now must ask is: if

$$M(x,y)\, dx + N(x,y)\, dy = 0 \tag{2.56}$$

is not exact, under what conditions does an *integrating factor* $\mu(x,y)$ exist such that

$$\mu M\, dx + \mu N\, dy = 0$$

is exact? Surprisingly, the answer is, whenever (2.56) has a general solution $g(x,y) = c$. To see this, we solve Eq. (2.56) for dy/dx:

$$\frac{dy}{dx} = -\frac{M}{N} = -\frac{\partial g/\partial x}{\partial g/\partial y},$$

from which it follows that

$$\frac{\partial g/\partial x}{M} = \frac{\partial g/\partial y}{N}.$$

Denote either side of the equation above by $\mu(x,y)$. Then

$$\frac{\partial g}{\partial x} = \mu M, \qquad \frac{\partial g}{\partial y} = \mu N, \tag{2.57}$$

and Eq. (2.56) has at least one integrating factor μ. However, finding integrating factors is in general very difficult. There is one procedure which is sometimes successful. Since Eq. (2.57) indicates that $\mu M \, dx + \mu N \, dy = 0$ is exact, by (2.53) we have

$$\mu \frac{\partial M}{\partial y} + M \frac{\partial \mu}{\partial y} = \frac{\partial}{\partial y}(\mu M) = \frac{\partial}{\partial x}(\mu N) = \mu \frac{\partial N}{\partial x} + N \frac{\partial \mu}{\partial x},$$

so that

$$\frac{1}{\mu}\left(N \frac{\partial \mu}{\partial x} - M \frac{\partial \mu}{\partial y}\right) = \frac{\partial M}{\partial y} - \frac{\partial N}{\partial x}. \tag{2.58}$$

In case the integrating factor μ depends only on x, Eq. (2.58) becomes

$$\frac{1}{\mu}\frac{d\mu}{dx} = \frac{\partial M/\partial y - \partial N/\partial x}{N} = k \tag{2.59}$$

Since the left-hand side of this equation consists only of functions of x, k *must* also be a function of x. If this is indeed true, then μ can be found by separating the variables: $\mu(x) = \exp\left[\int k(x) \, dx\right]$. A similar result holds if μ is a function of y alone, in which case

$$K = \frac{\partial M/\partial y - \partial N/\partial x}{-M}$$

is also a function of y. In this case $\mu(y) = \exp\left[\int K(y) \, dy\right]$ is the integrating factor.

Example 2.5.2 $(3x^2 - y^2) \, dy - 2xy \, dx = 0$. In this problem $M = -2xy$, $N = 3x^2 - y^2$, so that

$$\frac{\partial M}{\partial y} = -2x, \qquad \frac{\partial N}{\partial x} = 6x.$$

Then

$$K = \frac{\partial M/\partial y - \partial N/\partial x}{-M} = \frac{-4}{y},$$

so that

$$\mu = \exp\left[\int -\frac{4}{y} \, dy\right] = \exp\left(-4 \ln y\right) = y^{-4}.$$

Now we obtain

$$\left(\frac{3x^2 - y^2}{y^4}\right) dy - \frac{2x}{y^3} dx = 0,$$

which is exact. Thus the general solution is

$$g = \int M \, dx + h(y) = -\frac{x^2}{y^3} + h(y);$$

and differentiating this equation with respect to y, we have

$$\frac{3x^2 - y^2}{y^4} = N = \frac{3x^2}{y^4} + h'(y).$$

Hence $h'(y) = -y^{-2}$, so $h(y) = y^{-1} + c$, yielding

$$g(x,y) = -\frac{x^2}{y^3} + \frac{1}{y} + c = 0,$$

or

$$cy^3 + y^2 - x^2 = 0.$$

EXERCISES 2.5

In Exercises 1 through 11 verify that each given differential equation is exact and find the general solution. Find a particular solution when an initial condition is given.

1. $2xy \, dx + (x^2 + 1) \, dy = 0$

2. $[x \cos (x + y) + \sin (x + y)] \, dx + x \cos (x + y) \, dy = 0, \; y(1) = \pi/2 - 1$

3. $\left(4x^3y^3 + \frac{1}{x}\right) dx + \left(3x^4y^2 - \frac{1}{y}\right) dy = 0, \; x(e) = 1$

4. $\left[\frac{\ln (\ln y)}{x} + \frac{2}{3}xy^3\right] dx + \left[\frac{\ln x}{y \ln y} + x^2y^2\right] dy = 0$

5. $(x - y \cos x) \, dx - \sin x \, dy = 0, \; y(\pi/2) = 1$

6. $\cosh 2x \cosh 2y \, dx + \sinh 2x \sinh 2y \, dy = 0$

7. $(ye^{xy} + 4y^3) \, dx + (xe^{xy} + 12xy^2 - 2y) \, dy = 0, \; y(0) = 2$

8. $(3x^2 \ln x + x^2 - y) \, dx - x \, dy = 0, \; y(1) = 5$

9. $(2xy + e^y) \, dx + (x^2 + xe^y) \, dy = 0$

10. $(x^2 + y^2) \, dx + 2xy \, dy = 0, \; y(1) = 1$

11. $\left(\frac{1}{x} - \frac{y}{x^2 + y^2}\right) dx + \left(\frac{x}{x^2 + y^2} - \frac{1}{y}\right) dy = 0$

In Exercises 12 through 16 find the integrating factor for each differential equation and obtain the general solution.

12. $y \, dx + (y - x) \, dy = 0$

13. $(x^2 + y^2 + x) \, dx + y \, dy = 0$

14. $2y^2 \, dx + (2x + 3xy) \, dy = 0$

15. $(x^2 + 2y) \, dx - x \, dy = 0$

16. $(x^2 + y^2) \, dx + (3xy) \, dy = 0$

17. Solve $xy \, dx + (x^2 + 2y^2 + 2) \, dy = 0$.

18. Let $M = ya(xy)$ and $N = xb(xy)$. Show that $1/(xM - yN)$ is an integrating factor for $M \, dx + N \, dy = 0$.

19. Use the result of Exercise 18 to solve the equation

$$2x^2y^3 \, dx + x^3y^2 \, dy = 0.$$

20. Solve $(x^2 + y^2 + 1) \, dx - (xy + y) \, dy = 0$. [*Hint*: Try an integrating factor of the form $\mu(x, y) = x^n y^m$.]

2.6 LINEAR FIRST ORDER DIFFERENCE EQUATIONS

The general first order linear difference equation can be written in the form

$$y_{n+1} = a_n y_n + f_n, \tag{2.60}$$

where a_n and f_n are known for all n. Before proceeding to the general formula for the solution of (2.60), it is important for us to develop an understanding for the analogies between linear differential and difference equations. Consider the much simpler equation

$$y_{n+1} = a y_n, \tag{2.61}$$

where a is a given constant. Observe that proceeding inductively, we have $y_1 = a y_0$, $y_2 = a y_1 = a^2 y_0$, ..., and in general

$$y_{n+1} = a y_n = a(a y_{n-1}) = \cdots = a^{n+1} y_0, \tag{2.62}$$

which is the general solution of Eq. (2.61). If we compare (2.61) and (2.62) with the differential equation $y' = ay$, which has the general solution $y(x) = ce^{ax}$, it is easy to see that $n + 1$, a, and y_0 correspond to x, e^a, and c, respectively.

Suppose that our equation is

$$y_{n+1} = a_n y_n. \tag{2.63}$$

Using the same procedure as above, we see that $y_1 = a_0 y_0$, $y_2 = a_1 y_1 = a_1 a_0 y_0$, ..., and in general

$$y_{n+1} = a_n a_{n-1} \cdots a_1 a_0 y_0 = \left(\prod_{k=0}^{n} a_k \right) y_0. \tag{2.64}$$

The comparable differential equation $y' = a(x)y$ has the general solution $y(x) = ce^{\int a(x)\,dx}$, so $\prod_{k=0}^{n} a_k$ corresponds to $e^{\int a(x)\,dx}$.

We now consider

$$y_{n+1} = y_n + f_n. \tag{2.65}$$

Our iterative procedure yields $y_1 = y_0 + f_0$, $y_2 = y_1 + f_1 = y_0 + f_0 + f_1, \ldots$, so that

$$y_{n+1} = y_0 + \sum_{k=0}^{n} f_k. \tag{2.66}$$

Since $y' = y + f(x)$ has the general solution $y = ce^x + e^x\int f(x)e^{-x}\,dx$, the correspondence is clear (recall that $\prod_{k=0}^{n} 1 = 1$ corresponds to e^0).

Finally, we look at the general first order linear equation

$$y_{n+1} = a_n y_n + f_n.$$

Proceeding inductively, we obtain

$$y_1 = a_0 y_0 + f_0, \qquad y_2 = a_1 y_1 + f_1 = a_1 a_0 y_0 + a_1 f_0 + f_1, \ldots,$$

and in general

$$
\begin{aligned}
y_{n+1} &= (a_n a_{n-1} \cdots a_1 a_0)y_0 + (a_n \cdots a_1)f_0 + (a_n \cdots a_2)f_1 + \cdots + a_n f_{n-1} + f_n \\
&= \left(\prod_{k=0}^{n} a_k\right) y_0 + \sum_{k=0}^{n} \left(\prod_{j=k+1}^{n} a_j\right) f_k. \tag{2.67}
\end{aligned}
$$

(*Note.* We define the "empty" product $\prod_{j=n+1}^{n} a_j = 1$.) This equation should be compared with Eq. (2.34) of Section 2.3, since it is the discrete analog of that equation.

Example 2.6.1 $y_{n+1} = (n + 1)y_n$. Here $a_n = n + 1$ and $f_n = 0$. Using Eq. (2.64) or proceeding inductively, we find that

$$y_{n+1} = \left(\prod_{k=0}^{n} (k + 1)\right) y_0 = (n + 1)! y_0.$$

Example 2.6.2   An amoeba population has an initial size of 1000. It is observed that on the average one out of every ten amoebas reproduces by cell division every hour. Approximately how many amoebas will there be after twenty hours?

We let y_n be the number of amoebas present after n hours. Then the population growth over the next hour is given by

$$y_{n+1} - y_n = \tfrac{1}{10}y_n, \tag{2.68}$$

which implies that $y_{n+1} = (1.1)y_n$. The general solution given by (2.62) is

$$y_{n+1} = (1.1)^{n+1}y_0, \qquad y_0 = 1000.$$

Thus $y_{20} = (1.1)^{20}(1000) \approx 6727$.

Example 2.6.3 Suppose in the previous example that a leak from another container is introducing 30 additional amoebas into the population every hour. How many will there be after 20 hours?

The equation becomes

$$y_{n+1} - y_n = \tfrac{1}{10}y_n + 30,$$

or $y_{n+1} = (1.1)y_n + 30$, which by (2.67) has the solution

$$y_{n+1} = (1.1)^{n+1}(1000) + \sum_{k=0}^{n} (1.1)^{n-k}(30).$$

The sum of the first $n + 1$ terms of a geometrical progression satisfies the identity $1 + a + a^2 + \cdots + a^n = (a^{n+1} - 1)/(a - 1)$, so that the solution of Example 2.6.3 is given by

$$y_{n+1} = 1000(1.1)^{n+1} + 30\left(\frac{(1.1)^{n+1} - 1}{1.1 - 1}\right),$$

and $y_{20} \approx 8445$.

It is possible to solve some nonlinear difference equations that are similar to the equations in Section 2.4. We consider two of them here.

The equation

$$y_{n+1} = \frac{a_n y_n}{1 - f_n y_n}, \tag{2.69}$$

can be written in the form

$$y_{n+1} = a_n y_n + f_n y_n y_{n+1}. \tag{2.70}$$

This is comparable to Bernoulli's equation (2.36). Proceeding in a similar manner, we try the substitution

$$z_n = \frac{1}{y_n}. \tag{2.71}$$

After dividing both sides of (2.70) by $y_n y_{n+1}$, we obtain

$$z_n = a_n z_{n+1} + f_n \tag{2.72}$$

or

$$z_{n+1} = \frac{z_n - f_n}{a_n}. \tag{2.73}$$

Using the methods of this section, we find that

$$z_{n+1} = \left(\prod_{k=0}^{n} a_k^{-1}\right) z_0 - \sum_{k=0}^{n}\left(\prod_{j=k+1}^{n} a_j^{-1}\right)\frac{f_k}{a_k}. \tag{2.74}$$

Substitution of (2.71) into (2.74) then yields the solution

$$y_{n+1}^{-1} = \left(\prod_{k=0}^{n} a_k^{-1}\right) y_0^{-1} - \sum_{k=0}^{n}\left(\prod_{j=k}^{n} a_j^{-1}\right) f_k.$$

Similarly, we have the nonlinear *Riccati* (difference) *equation*

$$y_n y_{n-1} + a_n y_n + b_n y_{n-1} = c_n. \tag{2.75}$$

To find the resemblance to the Riccati differential equation (2.39), we rewrite (2.75) in the form

$$(y_n - y_{n-1}) + y_n y_{n+1} + [(a_n - 1)y_n + (b_n + 1)y_{n-1}] - c_n = 0.$$

In order to obtain a linear equation, we substitute

$$y_n = \frac{x_n}{x_{n+1}} - b_n,$$

into Eq. (2.75). After simplification Eq. (2.75) becomes

$$(a_n b_n + c_n)x_{n+1} - (a_n - b_{n-1})x_n - x_{n-1} = 0, \tag{2.76}$$

which is a second order linear equation. Methods for solving such equations when a_n, b_n, and c_n are constants will be presented in Chapter 4.

EXERCISES 2.6

In Exercises 1 through 9 find the general solution of each difference equation and a particular solution when an initial condition is specified.

1. $y_{n+1} - y_n = 2^{-n}$

2. $y_{n+1} = \dfrac{n+5}{n+3} y_n$

3. $y_{n+1} - 3y_n = 3y_{n+1} - y_n$

4. $2y_{n+1} = y_n, y_0 = 1$

5. $(n+1)y_{n+1} = (n+2)y_n, y_0 = 1$

6. $y_{n+1} = ny_n$

7. $y_{n+1} - 5y_n = 2, y_0 = 2$

8. $y_{n+1} - ny_n = n!, y_0 = 5$

9. $y_{n+1} - e^{-2n}y_n = e^{-n^2}$

10. Radium transmutes at a rate of one percent every twenty-five years. Consider a sample of r_0 grams of radium. If r_n is the amount of radium remaining in the sample after $25n$ years, obtain a difference equation for r_n and find its solution. How much radium will be left after one hundred years?

11. A fair coin is marked 1 on one side and 2 on the other side. The coin is tossed repeatedly and a cumulative score of the outcomes is recorded. Define P_n to be the probability that at some time the cumulative score will take on the value n. Prove that $P_n = 1 - \frac{1}{2}P_{n-1}$. Assuming that $P_0 = 1$, derive the formula for P_n.

12. In constructing a mathematical model of a population, it is assumed that the probability P_n that a couple produces exactly n offspring satisfies the equation $P_n = 0.7P_{n-1}$. Find P_n in terms of P_0 and determine P_0 from the fact that $P_0 + P_1 + P_2 + \cdots = 1$.

13. An alternative model to Exercise 12 is given by $P_n = (1/n)P_{n-1}$. For this model, find P_n in terms of P_0 and prove that $P_0 = 1/e$.

14. Let x_k denote the number of permutations of n objects taken k at a time. For every permutation of k objects, we can get a total of $n - k$ permutations of $k + 1$ objects by taking one of the remaining $n - k$ objects and placing it at the end. Thus $x_{k+1} = (n - k)x_k$. Prove that the number of permutations of n objects taken k at a time is $n!/(n - k)!$

15. If in Exercise 14 we let x_n denote the number of combinations of the n objects (order does not count), then every permutation of $k + 1$ objects occurs (in different orders) $k + 1$ times. Thus

$$x_{k+1} = \frac{n - k}{k + 1} x_n.$$

Prove that the number of combinations of n objects taken k at a time is

$$\frac{n!}{(n - k)!k!} \equiv \binom{n}{k}.$$

The expression on the right-hand side is called the *binomial coefficient*.

16. Reduce the equation $y_n(1 + ay_{n-1}) = 1$ to a linear second order equation by making an appropriate substitution.

17. Reduce each of the following nonlinear Riccati equations to linear second-order equations by making an appropriate substitution.

   a) $y_{n+1}y_n + 2y_{n+1} + 4y_n = n$
   b) $y_{n+1}y_n - 3^n y_{n+1} + (3^{n+1} - 2)y_n + 3^n(3^{n+1} - 2) = 0$

18. Solve the equation $y_{n+1} = 2y_n^2 - 1$ by means of the substitution $y_n = \cos x_n$.

19. The *Clairaut difference equation* is of the form

$$y_n = n(y_{n+1} - y_n) + f(y_{n+1} - y_n).$$

Note that the *first difference* $y_{n+1} - y_n$ (denoted by Δy_n) plays the role of the derivative in differential equations. For the equation $y_n = n \Delta y_n + (\Delta y_n)^2$, first take the first differences of both sides and then make the substitution $\Delta y_n = x_n$. Show that

   a) the resulting equation can be written in the form

$$(x_{n+1} - x_n)(x_{n+1} + x_n + n + 1) = 0;$$

   b) the condition $x_{n+1} - x_n = 0$ yields the *general solution* $y_n = nc + c^2$, c arbitrary;

c) the condition $x_{n+1} + x_n + n + 1 = 0$ implies that $y_{n+1} = -x_n x_{n+1}$, and thus we get the *singular solution*

$$y_n = \left[\sqrt{y_0} + \frac{1 - (-1)^n}{4} \right] - \left(\frac{n}{2} \right)^2.$$

2.7 AN APPLICATION OF FIRST ORDER DIFFERENCE EQUATIONS: NEWTON'S METHOD

An important problem in mathematics is finding the roots of a given equation

$$F(x) = 0. \tag{2.77}$$

Using Taylor's theorem centered at the value x_n, we can express this function in the form

$$0 = F(x) = F(x_n) + F'(x_n)(x - x_n) + \frac{F''(x_n)}{2!}(x - x_n)^2 + \cdots. \tag{2.78}$$

Omitting all but the first two terms on the right-hand side of Eq. (2.78) and solving for x yields the equation

$$x = x_n - \frac{F(x_n)}{F'(x_n)}.$$

If we call this new value x_{n+1}, we obtain the first order difference equation

$$x_{n+1} = x_n - \frac{F(x_n)}{F'(x_n)}, \tag{2.79}$$

known as *Newton's formula*. The value x_{n+1} is an approximation to some root of Eq. (2.77). Of course, since we have ignored all but the first two terms of Taylor's series to obtain this value, it is very unlikely that x_{n+1} is actually a root of (2.77). However, if the value x_n is close to a root x, then the quantity $x - x_n$ is near zero, so that the effect of the terms involving higher powers of $x - x_n$, in Taylor's series, may be negligible.

Rewriting (2.79) as

$$F'(x_n) = \frac{F(x_n)}{x_n - x_{n+1}},$$

we obtain a graphic interpretation of the procedure (see Fig. 2.4). Since the derivative $F'(x_n)$ is the slope of the tangent to the curve $y = F(x)$ at $x = x_n$ and this slope is $\tan \alpha$, the tangent line intersects the x-axis at x_{n+1}.

The procedure for finding a root consists in making an initial guess x_0 and repeatedly applying (2.79) to generate a sequence $\{x_n\}$ which we hope converges

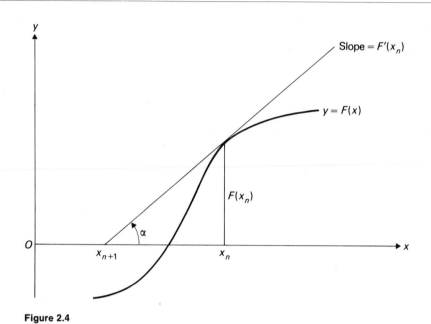

Figure 2.4

to a solution of (2.77). (See Fig. 2.5.) Conditions which guarantee the method will work are given in the following theorem.

Theorem 2.1 Suppose that on some interval $a \leqslant x \leqslant b$, $F(x)$ is defined, twice continuously differentiable, and that:

i) $F(a)$ and $F(b)$ have different signs;
ii) $F'(x) \neq 0$ for every x in $a \leqslant x \leqslant b$;
iii) $F''(x)$ does not change sign on $a \leqslant x \leqslant b$;
iv) if $F'(a) \leqslant F'(b)$, then $|F(a)/F'(a)| \leqslant b - a$; if $F'(b) \leqslant F'(a)$, then $|F(b)/F'(b)| \leqslant b - a$.

Then Newton's method converges to the *unique* solution x^* of $F(x) = 0$ for any initial choice x_0 in the interval $a \leqslant x \leqslant b$.

The proof of this theorem will not be given here, but the reader can find one in the footnoted reference below.†

This method can be applied to a variety of practical problems.

---

† Peter Henrici, *Elements of Numerical Analysis*, Wiley, New York, 1964, p. 79.

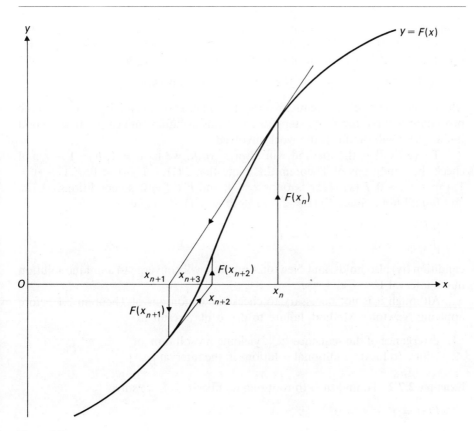

Figure 2.5

Example 2.7.1 Let $r > 1$ and let us formulate an algorithm for calculating the square root of r. Let

$$F(x) = x^2 - r, \qquad x > 0.$$

Then $F'(x) = 2x$ and Newton's method yields the difference equation

$$x_{n+1} = x_n - \frac{(x_n^2 - r)}{2x_n}$$

or

$$x_{n+1} = \frac{1}{2}\left(x_n + \frac{r}{x_n}\right). \tag{2.80}$$

It is easy to use formula (2.80) on a pocket calculator. Let x_0 be any integer in the interval $1 \leqslant x_0 \leqslant 1 + r$. Divide r by x_0, add x_0 to this quotient; then divide the result by 2 to get x_1. Repeat the process with x_n instead of x_0 to get x_{n+1}, for $n = 1, 2, 3, \ldots$. For example, if $r = 15$ and $x_0 = 4$, then (2.80) yields

$$x_1 = 3.875, \qquad x_2 = 3.872984, \qquad x_3 = 3.872983345,$$

which is correct to eight decimal places. If another choice is made for x_0, then the procedure will require more steps to equal this accuracy, indicating that a good initial guess will diminish the work involved.

To verify that the method will always work, we let $a = 1$, $b = 1 + r$ and check the conditions of Theorem 2.1. Note that $F(1) = 1 - r < 0$, $F(1 + r) = 1 + r + r^2 > 0$, $F'(x) = 2x > 0$ for $x > 1$, and $F''(x) = 2$, so conditions (i), (ii), and (iii) all hold. Since $F'(1) = 2 < 2 + 2r = F'(1 + r)$ and

$$\left| \frac{F(1)}{F'(1)} \right| = \frac{r - 1}{2} \leqslant r = (r + 1) - 1,$$

condition (iv) also holds, and Newton's method will converge to a unique solution in the interval $1 \leqslant x \leqslant 1 + r$.

Although it is not necessary to check the conditions of Theorem 2.1 before applying Newton's Method, failure to do so may result in

1. divergence of the sequence $\{x_n\}$, yielding no solution, or
2. failure to locate additional solutions in the interval $[a,b]$.

Example 2.7.2 To find the kth root of a number $r > 1$, we solve

$$F(x) = x^k - r = 0, \qquad x > 0,$$

by Newton's method, obtaining

$$x_{n+1} = x_n - \frac{x_n^k - r}{k x_n^{k-1}} = \left(1 - \frac{1}{k} \right) x_n + \frac{r}{k x_n^{k-1}}.$$

This formula will converge to $r^{1/k}$ for any choice of $x_0 > 1$.

Example 2.7.3 Find the roots of the polynomial

$$F(x) = x^3 + x^2 + 7x - 3. \tag{2.81}$$

We note that $F(0) = -3$ and $F(1) = 6$, so (2.81) has a root between 0 and 1. Letting $a = 0$, $b = 1$, we can easily verify conditions (i), (ii), (iii) of Theorem 2.1; and since $F'(0) = 7 < 12 = F'(1)$ and $3/7 < 1$, condition (iv) also holds. Thus

$$x_{n+1} = x_n - \frac{F(x_n)}{F'(x_n)} = \frac{2x_n^3 + x_n^2 + 3}{3x_n^2 + 2x_n + 7} \tag{2.82}$$

is Newton's formula for (2.81). An initial guess of $x_0 = 0$ yields $x_1 = 0.4286$, $x_2 = 0.3973$, $x_3 = 0.3971$, which is correct to four places.

Example 2.7.4 If we wish to find a reciprocal without dividing,* we can let $F(x) = 1/x - r$. Then $F'(x) = -1/x^2$, and (2.79) yields

$$x_{n+1} = x_n - \frac{x_n^{-1} - r}{x_n^{-2}} = x_n(2 - rx_n).$$

This expression will converge to $1/r$ for any x_0 such that $0 < x_0 < 2/r$. To calculate $1/\pi$ for $\pi = 3.1415926$, let $x_0 = 0.5$. Then we have

$$x_0 = 0.5, \qquad x_3 = 0.3147,$$
$$x_1 = 0.2146, \qquad x_4 = 0.31827,$$
$$x_2 = 0.2845, \qquad x_5 = 0.31830989,$$

the last of which is correct to eight decimal places.

Note that in Examples 2.7.1 and 2.7.4 the convergence is quite rapid. The rate of convergence in Newton's method is proportional to the quadratic $(x - x_n)^2$, which can be factored from each of the terms in (2.78) that we have omitted. For this reason, it is often called *quadratic convergence*.

EXERCISES 2.7

1. Using Newton's method, formulate an algorithm for calculating the cube root of a given positive number and then calculate $\sqrt[3]{6}$ to four decimal places.

2. Calculate $\frac{1}{6}$ without division to eight decimal places. Start the Newton iteration with 0.15.

3. Show by means of a graph that there is only one solution to the equation $x = e^{-x}$ and determine this solution to four decimal places by means of Newton's method.

4. For a suitable choice of x_0, formulate an algorithm for finding the smallest positive root of $4 \cos x = e^x$ and calculate the root to four decimal places. [*Hint*: Draw graphs of $4 \cos x$ and e^x to see what is happening.]

5. Find to four decimal places the smallest positive root of $\frac{1}{2} \tan x = x$.

6. Write a computer program in BASIC or FORTRAN to carry out the computations called for in Exercises 3, 4, and 5.

7. The derivative $F'(x_n)$ can be approximated by the difference quotient

$$\frac{F(x_n) - F(x_{n-1})}{x_n - x_{n-1}}.$$

Clearly, this quotient converges to the derivative as the difference between successive iterations approaches zero. Using the difference quotient instead of the derivative in

---

* Some computers use this algorithm to do divisions.

Newton's method, derive a second order difference equation that defines successive iterates. (*Note.* This defines a method, known as *regula falsi*, for the numerical solution of equations. The method is useful when calculation of derivatives is undesirable.)

8. Using the method indicated in Exercise 7, formulate algorithms for calculating square roots and cube roots and calculate $\sqrt{15}$ and $\sqrt[3]{6}$. Compare these computations to those of Newton's method. (Note that two initial choices must be made.)

2.8 SIMPLE ELECTRIC CIRCUITS

In this section we shall consider simple electric circuits containing a resistor and an inductor or capacitor in series with a source of electromotive force (emf). Such circuits are shown in Fig. 2.6(a) and (b), and their action can be understood very easily without any special knowledge of electricity.

(a)   (b)

Figure 2.6

1. An electromotive force (emf) E (volts), usually a battery or generator, drives an electric charge Q (coulombs) and produces a current I (amperes). The current is defined as the rate of flow of the charge, and we can write

$$I = \frac{dQ}{dt}. \tag{2.83}$$

2. A resistor of resistance R (ohms) is a component of the circuit that opposes the current, dissipating the energy in the form of heat. It produces a drop in voltage given by *Ohm's law*

$$E_R = RI. \tag{2.84}$$

3. An inductor of inductance L (henries) opposes any change in current by producing a voltage drop of

$$E_L = L\frac{dI}{dt}. \tag{2.85}$$

4. A capacitor of capacitance C (farads) stores charge. In so doing, it resists the flow of further charge, causing a drop in the voltage amounting to

$$E_C = \frac{Q}{C}. \tag{2.86}$$

The quantities R, L, C are usually constants associated with the particular component in the circuit. E may be a constant or a function of time. The fundamental principle guiding such circuits is given by *Kirchhoff's voltage law*:

The algebraic sum of all voltage drops around a closed circuit is zero.

In the circuit of Fig. 2.6(a) the resistor and the inductor cause voltage drops of E_R and E_L, respectively. The emf, however, *provides* a voltage of E (that is, a voltage drop of $-E$). Thus Kirchhoff's voltage law yields

$$E_R + E_L - E = 0.$$

Transposing E to the other side of the equation and using Eq. (2.84) and (2.85) to replace E_R and E_L, we have

$$L\frac{dI}{dt} + RI = E, \tag{2.87}$$

which is a linear differential equation. Dividing by L and employing Eq. (2.34), we obtain

$$I(t) = e^{-Rt/L}\left[\frac{1}{L}\int E(t)e^{Rt/L}\,dt + c\right]. \tag{2.88}$$

If E is constant, this equation reduces to

$$I(t) = \frac{E}{R} + ce^{-Rt/L},$$

and setting $t = 0$, we obtain

$$I(t) = \frac{E}{R} + \left[I(0) - \frac{E}{R}\right]e^{-Rt/L}. \tag{2.89}$$

As t increases, the last term, called the *transient* part of the current, approaches zero, leaving us with the *steady-state* part of the current E/R.

Example 2.8.1 An inductance of 2 henries (h) and a resistance of 10 ohms (Ω) are connected in series with an emf of 100 volts (v). If the current is zero when $t = 0$, what is the current at the end of 0.1 second?

Applying Eq. (2.89) with $E = 100$, $R = 10$, $L = 2$, $I(0) = 0$, and $t = 0.1$, we have

$$I(0.1) = 10(1 - e^{-0.5}) = 3.93 \text{ amp.}$$

Example 2.8.2 Suppose that in the previous example the emf $E = 100 \sin 60t$

volts. Then using Eq. (2.88) and formula 50 of Appendix 1, we obtain

$$I(t) = e^{-5t} \left[50 \int (\sin 60t)e^{5t} \, dt + c \right]$$

$$= e^{-5t} \left[50e^{5t} \left(\frac{5 \sin 60t - 60 \cos 60t}{3625} \right) + c \right]$$

$$= \frac{2 \sin 60t - 24 \cos 60t}{29} + ce^{-5t}.$$

Thus setting $t = 0$, we find that $c = 24/29$ and

$$I(0.1) = \frac{2 \sin 6 - 24 \cos 6}{29} + \frac{24}{29} e^{-0.5}. = -.31 \text{ amp.}$$

For the circuit in Fig. 2.4(b) we have $E_R + E_C - E = 0$ or

$$R \frac{dQ}{dt} + \frac{Q}{C} = E, \tag{2.90}$$

since $I = dQ/dt$. This equation is also linear, and so

$$Q(t) = e^{-t/CR} \left[\frac{1}{R} \int E(t)e^{t/CR} \, dt + c \right]. \tag{2.91}$$

If E is a constant, we obtain

$$Q(t) = EC + [Q(0) - EC]e^{-t/CR}. \tag{2.92}$$

Example 2.8.3 If a resistance of 2000 ohms and a capacitance of 5×10^{-6} farad (f) are connected in series with an emf of 100 volts, what is the current at $t = 0.1$ second if $I(0) = 0.01$ ampere?

Setting $R = 2000, C = 5 \times 10^{-6}, E = 100$ in Eq. (2.92), we have the charge at any time t in terms of the charge at time $t = 0$:

$$Q(t) = 5 \times 10^{-4} + [Q(0) - 5 \times 10^{-4}]e^{-100t}. \tag{2.93}$$

Since $I = dQ/dt$, we find that

$$I(t) = (-100)[Q(0) - 5 \times 10^{-4}]e^{-100t}.$$

Letting $t = 0$ yields $Q(0) = 4 \times 10^{-4}$ coulombs. Thus

$$I(0.1) = 10^{-2}e^{-10} \text{ amp.}$$

EXERCISES 2.8

1. Solve the problem in Example 2.8.3 with an emf of $100 \sin 120\pi t$ volts.

2. An inductance of 1 henry and a resistance of 2 ohms are connected in series with a battery of $6e^{-0.001t}$ volt. No current is flowing initially. When will the current measure 0.5 ampere?

3. A variable resistance $R = 1/(5 + t)$ ohms and a capacitance of 5×10^{-6} farad are connected in series with an emf of 100 volts. If $Q(0) = 0$, what is the charge on the capacitor after one minute?

4. In the RC-circuit [Fig. 2.4(b)] with constant voltage E, how long will it take the current to decrease to one-half its original value?

5. Suppose that the voltage in an RC-circuit is $E(t) = E_0 \cos \omega t$ where $2\pi/\omega$ is the period of the cycle. Assuming that the initial charge is zero, what are the charge and current as functions of R, C, ω, and t?

6. Show that the current in Exercise 5 consists of two parts: a steady-state term that has a period of $2\pi/\omega$ and a transient term that tends to zero as t increases.

7. In Exercise 6 show that if R is small, then the transient term can be quite large for small values of t. (This is why fuses can blow when a switch is flipped.)

8. Find the steady-state current, given that a resistance of 2000 ohms and a capacitance of 3×10^{-6} farad are connected in series with an alternating emf of $120 \cos 2t$ volts.

2.9 CURVES OF PURSUIT *

Many interesting differential equations arise in studying the path of a pursuer in tracking his prey. One of the earliest such problems is stated below.

Example 2.9.1 A heavy weight P located at the point $(a,0)$ is attached to a man Q by a chain of length a. We wish to find the path that P will follow if Q starts at the origin and moves up the y-axis (see Fig. 2.7). This path is called a *tractrix*.

Observe that the line PQ is tangent to the path that P will follow. Hence its slope is given by

$$\frac{dy}{dx} = -\frac{\sqrt{a^2 - x^2}}{x},\tag{2.94}$$

since the length of PQ is a. Integrating both sides of (2.94) and using formula 16 of Appendix 1, we have

$$y = -\int \frac{\sqrt{a^2 - x^2}}{x}\,dx + c$$

$$= a \ln\left(\frac{a + \sqrt{a^2 - x^2}}{x}\right) - \sqrt{a^2 - x^2} + c.\tag{2.95}$$

Since $y = 0$ when $x = a$, we see that $c = 0$ and thus the equation of the tractrix is

$$y = a \ln\left(\frac{a + \sqrt{a^2 - x^2}}{x}\right) - \sqrt{a^2 - x^2}.$$

---

* An interesting discussion of this topic is contained in A. Bernhart, "Curves of pursuit II," *Scripta Mathematica*, vol. 23 (1957), pp. 49–66.

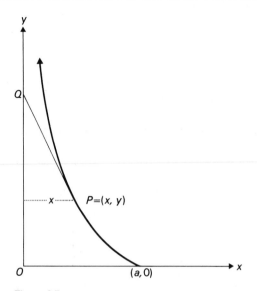

Figure 2.7

Example 2.9.2 Suppose that a hawk P at the point $(a,0)$ spots a pigeon Q at the origin flying along the y-axis at a speed v. The hawk immediately flies towards the pigeon at a speed w. What will be the flight path of the hawk?

Let time $t = 0$ at the instant the hawk starts flying toward the pigeon. After t seconds the pigeon will be at the point $Q = (0,vt)$ and the hawk at $P = (x,y)$. Since the line PQ is again tangent to the path (see Fig. 2.7), we find that its slope is given by $y' = (y - vt)/x$, so that

$$xy' - y = -vt. \tag{2.96}$$

On the other hand, the length of the path traveled by the hawk can be computed by the formula for arc length of basic calculus

$$wt = \int ds = \int_x^a \sqrt{1 + (y')^2}\, dx. \tag{2.97}$$

Solving Eqs. (2.96) and (2.97) for t and equating them, we have

$$\frac{y - xy'}{v} = \frac{1}{w} \int_x^a \sqrt{1 + (y')^2}\, dx. \tag{2.98}$$

Differentiating both sides of (2.98) with respect to x yields

$$xy'' = \frac{v}{w} \sqrt{1 + (y')^2}. \tag{2.99}$$

Setting $p = y'$, we find that Eq. (2.99) becomes

$$xp' = \frac{v}{w}\sqrt{1 + p^2},$$

and we can separate the variables to obtain

$$\frac{dp}{\sqrt{1 + p^2}} = \frac{v}{w}\frac{dx}{x}.$$

Integrating both sides of this equation (see formula 9 of Appendix 1), we have

$$\ln (p + \sqrt{1 + p^2}) = \frac{v}{w}\ln x - c. \tag{2.100}$$

Since $p = y' = 0$ when $x = a$ (the slope of the line PQ at $t = 0$ is zero), it follows that $c = (v/w)\ln a$. Exponentiating both sides of this equation yields

$$p + \sqrt{1 + p^2} = \left(\frac{x}{a}\right)^{v/w},$$

which, after some algebra, yields

$$\frac{dy}{dx} = p = \frac{1}{2}\left[\left(\frac{x}{a}\right)^{v/w} - \left(\frac{x}{a}\right)^{-v/w}\right]. \tag{2.101}$$

If we assume that the hawk flies faster than the pigeon ($w > v$), we may integrate (2.101) to obtain

$$y = \frac{a}{2}\left[\frac{(x/a)^{1 + v/w}}{1 + v/w} - \frac{(x/a)^{1 - v/w}}{1 - v/w} + c\right]. \tag{2.102}$$

Since $y = 0$ when $x = a$, we have

$$c = -\frac{a}{2}\left[\frac{1}{1 + v/w} - \frac{1}{1 - v/w}\right] = \frac{avw}{w^2 - v^2}.$$

The hawk will catch the pigeon at $x = 0$ and $y = c = avw/(w^2 - v^2)$. The situation where the hawk flies no faster than the pigeon ($w \leqslant v$) is discussed in Exercises 1 and 2.

Example 2.9.3 A destroyer is in a dense fog, which lifts for an instant, disclosing an enemy submarine on the surface 4 miles away. Suppose that the submarine dives immediately and proceeds at full speed in an unknown direction. What path should the destroyer select to be certain of passing directly over the submarine, if its velocity v is three times that of the submarine?

Suppose that the destroyer has traveled 3 miles toward the place where the submarine was spotted. Then the submarine lies on the circle of radius 1 mile centered at where it was when spotted (see Fig. 2.8), since its velocity is one-third

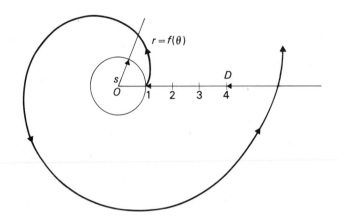

$r = f(\theta)$

Figure 2.8

that of the destroyer. Since the location of the submarine can be easily described in polar coordinates, we make use of polar coordinates and assume that $r = f(\theta)$ is the path the destroyer must follow to be certain of passing over the submarine, regardless of the direction the latter chooses. Then the distance traveled by the submarine to the point where the paths will cross is $r - 1$, while that of the destroyer (which is three times longer) is given by the arc length formula in polar coordinates:

$$3(r - 1) = \int_0^\theta ds = \int_0^\theta \sqrt{(dr)^2 + (r\,d\theta)^2}$$

$$= \int_0^\theta \sqrt{(dr/d\theta)^2 + r^2}\, d\theta. \tag{2.103}$$

Differentiating both sides of (2.103) with respect to θ yields the differential equation

$$3r' = \sqrt{(r')^2 + r^2},$$

which simplifies to $8(r')^2 = r^2$. Taking the square roots of both sides and separating the variables, we have

$$\frac{dr}{r} = \frac{d\theta}{\sqrt{8}},$$

from which it follows that $\ln r = \theta/\sqrt{8} + c$ or

$$r = ce^{\theta/\sqrt{8}}. \tag{2.104}$$

Since $r = 1$ when $\theta = 0$, it follows that $c = 1$ and the path that the destroyer should follow is the spiral $r = e^{\theta/\sqrt{8}}$ after proceeding 3 miles towards where the submarine was spotted.

It should be noted that this path is not the only curve the destroyer could follow. For example, suppose that the destroyer has gone 6 miles toward where the submarine was spotted (see Fig. 2.9). At this point, we can again follow a path $r = g(\theta)$. Since by now the submarine is 2 miles from the origin, the distance traveled by the submarine to where the paths will cross is $r - 2$, while the destroyer must go a distance

$$3(r - 2) = \int_{-\pi}^{\theta} \sqrt{(dr/d\theta)^2 + r^2} \, d\theta. \tag{2.105}$$

Equation (2.105) again leads to the general solution (2.104), but in this case $r = 2$ when $\theta = -\pi$, so that $c = 2e^{\pi/\sqrt{8}}$. Thus the spiral the destroyer must follow is

$$r = 2e^{(\theta - \pi)/\sqrt{8}}.$$

Of course, the submarine captain can evade detection by not going at full speed or by following a curved trajectory.

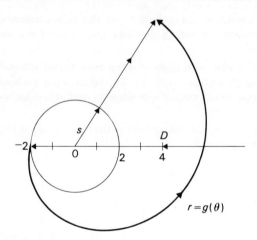

Figure 2.9

EXERCISES 2.9

1. Suppose that $v = w$ in Example 2.9.2. Prove that

$$y = \frac{a}{2} \left\{ \frac{1}{2} \left[\left(\frac{x}{a}\right)^2 - 1 \right] - \ln \frac{x}{a} \right\},$$

so that the hawk will never catch the pigeon. Using Eqs. (2.97) and (2.101), show that the distance between the hawk and the pigeon is $(x^2 + a^2)/2a$ whenever the hawk is at the point (x,y) on the path. Thus the hawk will not come as close as $a/2$ to the pigeon.

2. Suppose that $v > w$ in Example 2.9.2. Then show that

$$y = \frac{a}{2}\left[\frac{(x/a)^{1+v/w} - 1}{1 + v/w} + \frac{(a/x)^{v/w-1} - 1}{(v/w) - 1}\right],$$

so that the hawk will never catch the pigeon. Find the distance between the hawk and the pigeon in terms of the variable x.

3. Let the y-axis and the line $x = b$ be the banks of a river whose current has velocity v (in the negative y-direction). A man is at the origin, and his dog is at the point $(b,0)$. When the man calls the dog enters the river, swimming towards the man at a constant velocity $w \, (> v)$. What will be the path of the dog?

4. Where will the dog of Exercise 3 land if $w = v$?

5. Show that the dog of Exercise 3 will never land if $w < v$. Suppose that the man walks down river at the velocity v while calling his dog. Will the dog now be able to land?

6. In Example 2.9.3 suppose the destroyer proceeds to where the submarine was sighted, then turns $90°$ left and proceeds 2 miles before beginning the spiral search pattern. What is the equation of the path the destroyer should now follow?

7. Suppose the destroyer in Example 2.9.3 is only twice as fast as the submarine and the submarine was spotted when it was 3 miles away. Find a path that will guarantee the destroyer's passing over the submarine, assuming that both ships execute the same maneuvers as given in the example.

8. Three snails at the corners of an equilateral triangle of side a begin to move with the same velocity, each toward the snail to its right. Centering the triangle at the origin with one vertex along the positive x-axis, find an equation for the slime path left by the snail that started on the x-axis.

9. Consider the same problem with four snails at the corners of the square $[0,a] \times [0,a]$. How far will the snails travel before they meet?

2.10 COMPARTMENTAL ANALYSIS

A complicated physical or biological process can often be divided into several distinct stages. The entire process can then be described by the interactions between the individual stages. Each such stage is called a *compartment* or pool, and the contents of each compartment are assumed to be well mixed. Material from one compartment is transferred to another and is immediately incorporated into the latter. On account of the name we have given to the stages, the entire process is called a *compartmental system.** An *open* system is one in which there are inputs to or outputs from the system through one or more compartments. A system which is not open is said to be *closed*.

---

* This name is frequently used in mathematical biology. Engineers refer to such systems as *block diagrams*.

In this section we will investigate only the simplest such systems: the one-compartment system. Additional work on more complicated systems will be found in later chapters.

Figure 2.10 illustrates a one-compartment system consisting of a quantity $x(t)$ of material in the compartment, an input rate $i(t)$ at which material is being introduced to the system, and a *fractional transfer coefficient k* indicating the fraction of the material in the compartment that is being removed from the system per unit time. It is clear that the rate at which the quantity x is changing depends on the difference between the input and output at any time t, leading to the differential equation

$$\frac{dx}{dt} = i(t) - kx(t). \tag{2.106}$$

As we saw in Section 2.3, this linear equation has the solution

$$x(t) = e^{-kt}[\int i(t)e^{kt}\, dt + c]. \tag{2.107}$$

This simple model applies to many different problems, as we shall illustrate below.

Figure 2.10

Example 2.10.1 Strontium 90 (Sr^{90}) has a half-life of twenty-five years. If 10 grams of Sr^{90} are initially placed in a sealed container, how many grams will remain after ten years?

Let $x(t)$ be the number of grams of Sr^{90} at time t (years). Since the number of atoms present is very large, the number decaying per unit time is directly proportional to the number present at that time. The constant of proportionality k is the fractional transfer coefficient. Since there is no input, the equation involved is

$$\frac{dx}{dt} = -kx(t). \tag{2.108}$$

Equation (2.108) has the solution $x(t) = x_0 e^{-kt}$, where $x_0 = 10$ grams. To find k, we set $t = 25$ to obtain

$$5 = 10e^{-25t},$$

from which we find, after taking logarithms, that $k = (\ln 2)/25$. Thus

$$x(10) = 10e^{-(10 \ln 2)/25} = 10(2)^{-2/5} \approx 7.578 \text{ g}.$$

Example 2.10.2 Consider a tank holding 100 gallons of water in which are dissolved 50 pounds of salt. Suppose that 2 gallons of brine, each containing 1 pound of dissolved salt, run into the tank per minute, and the mixture, kept uniform by high-speed stirring, runs out of the tank at the rate of 2 gallons per minute. Find the amount of salt in the tank at any time t.

Let $x(t)$ be the number of pounds of salt at the end of t minutes. Since each gallon of brine input to the compartment (tank) contains 1 pound of salt we have that $i(t) = 2$. On the other hand, $k = \frac{2}{100}$ since two of the 100 gallons in the tank are being removed each minute. Thus Eq. (2.106) becomes

$$\frac{dx}{dt} = 2 - \frac{2}{100} x,$$

which has the solution

$$x(t) = e^{-t/50} [2 \int e^{t/50} \, dt + c],$$
$$= 100 + ce^{-t/50}.$$

At $t = 0$ we have

$$50 = x(0) = 100 + c,$$

so that

$$x(t) = 100 - 50e^{-t/50}.$$

Observe that x increases and approaches the ratio of salt to water in the input stream as time increases.

The fractional transfer coefficient k may be a function of time, as we shall see in the next example.

Example 2.10.3 Suppose that in Example 2.10.2, 3 gallons of brine, each containing 1 pound of salt, run into the tank each minute, and all other facts are as before. Now $i(t) = 3$, but since the quantity of brine in the tank increases with time, the fraction that is being transferred is $k = 2/(100 + t)$. The numerator of k is the number of gallons being removed, and $100 + t$ is the number of gallons in the tank at time t. The equation describing the system is

$$\frac{dx}{dt} = 3 - \frac{2x}{100 + t}. \tag{2.109}$$

Using Eq. (2.107), we see that (2.109) has the solution

$$x(t) = \exp\left(-\int \frac{2 \, dt}{100 + t}\right) \left[3 \int \exp\left(\int \frac{2 \, dt}{100 + t}\right) dt + c\right]$$
$$= (100 + t) + c(100 + t)^{-2}.$$

Setting $t = 0$, we find that $c = -50(100)^2$, so that

$$x(t) = 100 + t - 50(1 + t/100)^{-2}.$$

After 100 minutes, we have

$$x(100) = 200 - 50/4 = 187.5 \text{ lb}.$$

of salt in the tank.

The input function $i(t)$ may depend not only on time but also on the quantity present. Examples 1.1.3 and 1.1.4 are situations where the input function depends on the quantity present, and in Example 1.1.4 the fractional transfer coefficient $k = \delta P$ also depends on the quantity present.

Systems with periodic inputs and fractional transfer coefficients often occur in biological processes due to the diurnal period of activity. For example, ACTH (adrenocorticotropic hormone) secretion by the anterior pituitary follows a twenty-four-hour cycle which drives the secretion of adrenal steroids in such a way that the levels of these steroids in the blood plasma peaks near 8 a.m. and is at a minimum near 8 p.m.

Example 2.10.4 Let $k(t) = A + B \sin \omega t$, with $A > B$, in (2.106), leading to the equation

$$\frac{dx}{dt} = i(t) - (A + B \sin \omega t)x. \tag{2.110}$$

Since

$$\int (A + B \sin \omega t)\, dt = At - \frac{B}{\omega} \cos \omega t + c,$$

we may use the integrating factor $\exp\left[At + (B/\omega)(1 - \cos \omega t)\right]$ on both sides of (2.110):

$$\frac{d}{dt}(e^{At + (B/\omega)(1 - \cos \omega t)}x(t)) = e^{At + (B/\omega)(1 - \cos \omega t)}[x' + (A + B \sin \omega t)x]$$
$$= e^{At + (B/\omega)(1 - \cos \omega t)}i(t). \tag{2.111}$$

Integrating both sides of (2.111) from 0 to t, we have

$$e^{At + (B/\omega)(1 - \cos \omega t)}x(t)\Big|_0^t = \int_0^t i(t)e^{At + (B/\omega)(1 - \cos \omega t)}\, dt$$

or

$$x(t) = e^{-At - (B/\omega)(1 - \cos \omega t)}\left[x(0) + \int_0^t i(t)e^{At + (B/\omega)(1 - \cos \omega t)}\, dt\right]. \tag{2.112}$$

Since $1 - \cos \omega t = 2 \sin^2 (\omega t/2)$, we can write (2.112) as

$$x(t) = e^{-At - 2B \sin^2 (\omega t/2)/\omega}\left[x(0) + \int_0^t i(t)e^{At + 2B \sin^2 (\omega t/2)/\omega}\, dt\right]. \tag{2.113}$$

If $i(t) = 0$, then $x(t)$ behaves as shown in Fig. 2.11, where $x(0)e^{-At}$ is an upper bound, and the factor $\exp\left[-2B\sin^2(\omega t/2)/\omega\right]$ oscillates between $\exp(-2B/\omega)$ and 1.

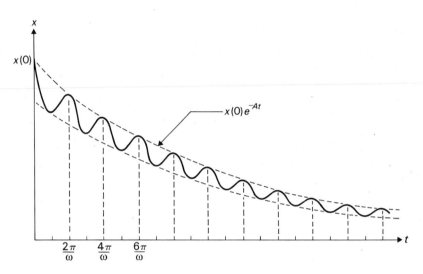

Figure 2.11

EXERCISES 2.10

1. Carbon 14 (C^{14}) has a half-life of 5700 years and is uniformly distributed in the atmosphere in the form of carbon dioxide. Living plants absorb carbon dioxide and maintain a fixed ratio of C^{14} to the stable element C^{12}. At death, the disintegration of C^{14} changes this ratio. Compare the concentrations of C^{14} in two identical pieces of wood, one of them freshly cut, the other 2000-years old.

2. Radioactive iodine I^{131} is often used as a tracer in medicine. Suppose that a given dose Q_0 is injected into the blood stream at time $t = 0$ and is evenly distributed in the entire blood stream before any loss occurs. If the daily removal rate of the iodine by the kidney is k_1 percent, and k_2 percent by the thyroid gland, what percentage of the initial amount will still be in the blood after one day?

3. Suppose that an infected individual is introduced into a population of size N, all of whom are susceptible to the disease. If we assume that the rate of infection is proportional to the product of the numbers of infectives and susceptibles present, what will be the number of infections at any time t? Let k be the *specific infection rate*.

4. A tank initially contains 100 liters of fresh water. Brine containing 20 grams per liter of salt flows into the tank at the rate of 4 liters per minute, and the mixture, kept uniform

by stirring, runs out at the same rate. How long will it take for the quantity of salt in the tank to become 1 kilogram?

5. Given the same data as in Exercise 4, how long will it take for the quantity of salt in the tank to increase from 1 kilogram to $1\frac{1}{2}$ kilograms?

6. A tank contains 100 gallons of fresh water. Brine containing 2 pounds per gallon of salt runs into the tank at the rate of 4 gallons per minute, and the mixture, kept uniform by stirring, runs out at the rate of 2 gallons per minute. Find:

   a) the amount of salt present when the tank has 120 gallons of brine,
   b) the concentration of salt in the tank at the end of 20 minutes.

7. A tank contains 50 liters of water. Brine containing x grams per liter of salt enters the tank at the rate of 1.5 liters per minute. The mixture, thoroughly stirred, leaves the tank at the rate of 1 liter per minute. If the concentration is to be 20 grams per liter at the end of 20 minutes, what is the value of x?

8. A tank holds 500 gallons of brine. Brine containing 2 pounds per gallon of salt flows into the tank at the rate of 5 gallons per minute, and the mixture, kept uniform, flows out at the rate of 10 gallons per minute. If the maximum amount of salt is found in the tank at the end of 20 minutes, what was the initial salt content of the tank?

9. Phosphate excretion is at a minimum at 6 a.m. and rises to a peak at 6 p.m. If the rate of excretion is

$$\frac{1}{6} + \frac{1}{3}\cos\frac{\pi}{12}(t - 6)$$

grams per hour at time t hours ($0 \leqslant t \leqslant 24$), the body contains 400 grams of phosphate, and the patient is only allowed to drink water, what is the amount of phosphate in the patient's body at all times t?

10. Suppose in Exercise 9 that the patient is allowed three meals during the day in such a way that the body takes in phosphate at a rate given by the formula

$$i(t) = \begin{cases} 1/3 \text{ g/hr}, & 8 \leqslant t \leqslant 16, \\ 0 \text{ g/hr}, & \text{otherwise}. \end{cases}$$

Obtain a formula for the amount of phosphate in the patient's body at all times t. When is it at a maximum?

11. Given a one-compartment system with k constant and $i(t) = A + B \sin \omega t$, $A > B$, find a solution of the system. How does it differ from that of the system in which the input is constant and the fractional transfer coefficient is periodic [see Eq. (2.113)]?

LINEAR
SECOND
ORDER
DIFFERENTIAL
EQUATIONS

3

3.1 INTRODUCTION

The most general second order differential equation is $F(x,y,y',y'') = 0$, that is, an equation involving x, y, y', and y''. There is no procedure for solving explicitly arbitrary equations in this form, and in fact, many of these equations do not have solutions. In this chapter we shall study a class of second order equations for which there are always unique solutions and present some methods for calculating the solutions.

We recall that a differential equation is linear if it does not involve nonlinear functions (squares, exponentials, etc.) or products of the unknown function and its derivatives. Thus $y'' + (x^3 \sin x)^5 y' + y = \cos x^3$ is linear, while $y'' + (y')^2 + y = 0$ is nonlinear. The most general second order linear equation is

$$y''(x) + a(x)y'(x) + b(x)y(x) = f(x). \tag{3.1}$$

The following theorem will be proved in Chapter 11.

Theorem 3.1 Let $a(x)$, $b(x)$, and $f(x)$ be continuous on the closed interval $[x_0,x_1]$ and let two constants c_0 and c_1 be given. Then there exists a unique function $y(x)$ which is twice continuously differentiable and satisfies (3.1) on $[x_0,x_1]$, and for which $y(x_0) = c_0$ and $y'(x_0) = c_1$.

The theorem states that if a, b, and f are continuous, then the initial value problem has a unique solution. However, in some important applications the continuity conditions do not hold. Thus in the case of the equation

$$(x - 1)y'' + x^2 y' + y = 0, \qquad y(0) = 1, \qquad y'(0) = 0,$$

the functions $a(x)$ and $b(x)$ are $x^2/(x - 1)$ and $1/(x - 1)$, respectively, which are discontinuous at $x = 1$. There are special techniques for handling some problems of this sort that will be discussed in Chapter 5. In the remainder of this chapter we will assume, unless otherwise stated, that the functions $a(x)$, $b(x)$, and $f(x)$ are continuous for all real values of x.

If the function $f(x)$ is identically zero, we say that Eq. (3.1) is *homogeneous*. Otherwise, it is *nonhomogeneous*. If the coefficient functions $a(x)$ and $b(x)$ are constants $[a(x) \equiv a, b(x) \equiv b]$, then the equation is said to have *constant coefficients*. As we shall see, linear differential equations with constant coefficients are the easiest to solve.

EXERCISES 3.1

In each of Exercises 1 through 10 determine whether the given equation is linear or nonlinear. If it is linear, state whether it is homogeneous or nonhomogeneous with constant or variable coefficients.

1. $y'' + 2x^3 y' + y = 0$
2. $y'' + 2y' + y^2 = x$
3. $y'' + 3y' + yy' = 0$
4. $y'' + 3y' + 4y = 0$
5. $y'' + 3y' + 4y = \sin x$
6. $y'' + y(2 + 3y) = e^x$
7. $y'' + 4xy' + 2x^3 y = e^{2x}$
8. $y'' + [\sin (xe^x)]y' + 4xy = 0$
9. $3y'' + 16y' + 2y = 0$
10. $yy'y'' = 1$

11. Let $y_1(x)$ be a solution of the homogeneous equation

$$y'' + a(x)y' + b(x)y = 0,$$

on the interval $\alpha \leqslant x \leqslant \beta$. Suppose that the curve y_1 is tangent to the x-axis at some point of this interval. Prove that y_1 must be identically zero.

12. Let $y_1(x)$ and $y_2(x)$ be two solutions of the homogeneous equation

$$y'' + a(x)y' + b(x)y = 0$$

on the interval $\alpha \leqslant x \leqslant \beta$. Suppose $y_1(x_0) = y_2(x_0) = 0$ for some point $\alpha \leqslant x_0 \leqslant \beta$. Show that y_2 is a constant multiple of y_1.

3.2 PROPERTIES OF SOLUTIONS TO THE LINEAR EQUATION

In this section we shall lay the theoretical groundwork for much of the material in this chapter. We begin by analyzing the homogeneous equation

$$y'' + a(x)y' + b(x)y = 0. \tag{3.2}$$

We shall assume throughout this section that $a(x)$ and $b(x)$ are continuous so that our basic existence-uniqueness theorem (Theorem 3.1) applies. The definitions and terminology we will introduce are applicable to higher order equations as well as to the second order equation (3.2) in question. Higher order equations are considered in Exercises 8 through 14 and, more generally, in Section 3.12.

Let y_1 and y_2 be any two functions. By a *linear combination* of y_1 and y_2 we mean a function $y(x)$ that can be written in the form

$$y(x) = c_1 y_1(x) + c_2 y_2(x)$$

for some constants c_1 and c_2. Two functions are *linearly independent* on an interval $[x_0, x_1]$ whenever the relation $c_1 y_1(x) + c_2 y_2(x) = 0$ for all x in $[x_0, x_1]$ implies that $c_1 = c_2 = 0$. Otherwise, they are *linearly dependent*. There is, however, an easier way to see that two functions y_1 and y_2 are linearly dependent. If $c_1 y_1(x) + c_2 y_2(x) = 0$ (where not both c_1 and c_2 are zero), we may suppose that $c_1 \neq 0$. Then, dividing the above expression by c_1, we obtain

$$y_1(x) + \frac{c_2}{c_1} y_2(x) = 0$$

or

$$y_1(x) = -\frac{c_2}{c_1} y_2(x) = cy_2(x).$$

Therefore, *two functions are linearly dependent on the interval* $[x_0, x_1]$ *if and only if one of the functions is a constant multiple of the other.*

The notions of linear combination and linear independence are central to the theory of linear homogeneous equations, as illustrated by the results that follow.

Theorem 3.2 Let $y_1(x)$ and $y_2(x)$ be any two solutions of Eq. (3.2). Then any linear combination of them is also a solution of (3.2).

Proof. Let $y(x) = c_1 y_1(x) + c_2 y_2(x)$. Then

$$y'' + ay' + by = c_1 y_1'' + c_2 y_2'' + c_1 a y_1' + c_2 a y_2' + c_1 b y_1 + c_2 b y_2$$
$$= c_1(y_1'' + ay_1' + by_1) + c_2(y_2'' + ay_2' + by_2) = 0,$$

since y_1 and y_2 are solutions.

Let y_1 and y_2 be two linearly independent solutions of Eq. (3.2). We shall show that *any other solution of* (3.2) *can be written as a linear combination of* y_1 *and* y_2. This remarkable fact means that once we have found two linearly independent solutions of (3.2), we have "essentially" found *all* solutions of (3.2). However, before proving this claim, we introduce a new function which will be very useful in our calculations.

Let $y_1(x)$ and $y_2(x)$ be any two solutions to Eq. (3.2). The *Wronskian* of y_1 and y_2, $W(y_1, y_2)$, is defined to be

$$W(y_1, y_2)(x) = y_1(x)y_2'(x) - y_1'(x)y_2(x). \tag{3.3}$$

Using the product rule of differentiation on (3.3), we see that

$$W'(y_1, y_2) = y_1 y_2'' + y_1' y_2' - y_1' y_2' - y_1'' y_2$$
$$= y_1 y_2'' - y_1'' y_2.$$

Since y_1 and y_2 are solutions,

$$y_1'' + ay_1' + by_1 = 0, \qquad y_2'' + ay_2' + by_2 = 0.$$

Multiplying the first of these equations by y_2 and the second by y_1 and subtracting, we obtain

$$y_1 y_2'' - y_2 y_1'' + a(y_1 y_2' - y_2 y_1') = 0, \tag{3.4}$$

which is just

$$W' + aW = 0. \tag{3.5}$$

By the methods given in the previous chapter, we see that the solution of (3.5) can be written as

$$W(y_1,y_2) = ce^{-\int a(x)\,dx} \tag{3.6}$$

for some arbitrary constant c. Formula (3.6) is known as *Abel's formula*. Since an exponential is never zero, we see that $W(y_1,y_2)$ is either always zero (when $c = 0$) or never zero (when $c \neq 0$). The importance of this fact is given by the following lemma.

Lemma 3.3 The solutions $y_1(x)$ and $y_2(x)$ of Eq. (3.2) are linearly independent on $[x_0,x_1]$ if and only if $W(y_1,y_2) \neq 0$.

Proof: We first show that if $W(y_1,y_2) = 0$, then y_1 and y_2 are linearly dependent. Let x_2 be a point in the interval $x_0 \leqslant x \leqslant x_1$. Consider the system of equations

$$c_1 y_1(x_2) + c_2 y_2(x_2) = 0,$$
$$c_1 y_1'(x_2) + c_2 y_2'(x_2) = 0. \tag{3.7}$$

The determinant of the system (3.7) is

$$y_1(x_2)y_2'(x_2) - y_2(x_2)y_1'(x_2) = W(y_1,y_2)(x_2) = 0.$$

Thus according to the theory of determinants (see Appendix 4), there exists a solution c_1, c_2 for (3.7) where c_1 and c_2 are not both equal to zero. Define $y(x) = c_1 y_1(x) + c_2 y_2(x)$. By Theorem 3.2, $y(x)$ is a solution of (3.2). Then since c_1 and c_2 solve (3.7),

$$y(x_2) = c_1 y_1(x_2) + c_2 y_2(x_2) = 0$$

and

$$y'(x_2) = c_1 y_1'(x_2) + c_2 y_2'(x_2) = 0.$$

Thus $y(x)$ solves the initial value problem

$$y'' + a(x)y' + b(x)y = 0, \qquad y(x_2) = y'(x_2) = 0.$$

But this initial value problem also has the solution $y_3(x) \equiv 0$ for all values of x in $x_0 \leqslant x \leqslant x_1$. By Theorem 3.1 the solution of this initial value problem is unique so that necessarily $y(x) = y_3(x) = 0$. Thus

$$y(x) = c_1 y_1(x) + c_2 y_2(x) = 0,$$

for all values of x in $x_0 \leqslant x \leqslant x_1$, which proves that y_1 and y_2 are linearly dependent.

We now assume that $W(y_1,y_2) \neq 0$ in $[x_0,x_1]$ and shall prove that y_1 and y_2 are linearly independent. If y_1 and y_2 are not linearly independent, then there is a constant c such that $y_2 = cy_1$ or $y_1 = cy_2$. Assume that $y_2 = cy_1$. Then $y_2' = cy_1'$ and

$$W(y_1,y_2) = y_1 y_2' - y_1' y_2 = y_1(cy_1') - y_1'(cy_1) = 0.$$

But this contradicts the assumption that $W \neq 0$. Hence the solutions y_1 and y_2 must be independent.

Example 3.2.1 Consider the equation $x'' + x = 0$. It is easy to verify that $x_1 = \cos t$ and $x_2 = \sin t$ are solutions. The Wronskian $W(x_1, x_2) = \cos t \cos t - \sin t(-\sin t) = \cos^2 t + \sin^2 t = 1$, so the solutions are linearly independent.

Using Lemma 3.3, we can prove the basic theorem mentioned above.

Theorem 3.4 Let $y_1(x)$ and $y_2(x)$ be two linearly independent solutions to Eq. (3.2) on the interval $[x_0, x_1]$ and let $y(x)$ be any other solution. Then $y(x)$ can be written as a linear combination of y_1 and y_2.

Proof. Let $y(x_0) = a$ and $y'(x_0) = b$. Consider the linear system of equations in two unknowns c_1 and c_2:

$$y_1(x_0)c_1 + y_2(x_0)c_2 = a,$$
$$y_1'(x_0)c_1 + y_2'(x_0)c_2 = b. \tag{3.8}$$

As we saw earlier, the determinant of this system is $W(y_1, y_2)(x_0)$, which is nonzero since the solutions are linearly independent. Thus there is a unique solution (c_1, c_2) to Eqs. (3.8) and a solution $y^*(x) = c_1 y_1(x) + c_2 y_2(x)$ which satisfies the conditions $y^*(x_0) = a$ and $y^{*'}(x_0) = b$. Since every initial value problem has a unique solution (see Theorem 3.1), it must follow that $y(x) = y^*(x)$ on the interval $x_0 \leq x \leq x_1$, and so the proof is complete.

In the above theorem, we showed that *every* solution can be written as a linear combination of linearly independent solutions. Therefore, we can talk about the general solution to Eq. (3.2). Let y_1 and y_2 be linearly independent solutions of (3.2). Then the *general solution* of (3.2) is given by

$$y(x) = c_1 y_1(x) + c_2 y_2(x), \tag{3.9}$$

where c_1 and c_2 are arbitrary constants.

We now turn briefly to the nonhomogeneous equation (3.1). Let $y_p(t)$ be any solution to (3.1). If we know the general solution to (3.2), we can find all solutions to (3.1).

Theorem 3.5 Let $y_p(x)$ be a solution of Eq. (3.1) and let $y^*(x)$ be any other solution. Then $y^*(x) - y_p(x)$ is a solution of (3.2); that is, $y^*(t) = c_1 y_1(t) + c_2 y_2(t) + y_p(t)$ for some constants c_1 and c_2 where y_1, y_2 are two linearly independent solutions of (3.2).

Proof. We have

$$(y^* - y_p)'' + a(y^* - y_p)' + b(y^* - y_p)$$
$$= (y^{*''} + ay^{*'} + by^*) - (y_p'' + ay_p' + by_p) = f - f = 0.$$

Thus *in order to find all solutions to the nonhomogeneous problem, we need only find one solution to the nonhomogeneous problem and the general solution of the homogeneous problem.*

Example 3.2.2 It is easy to verify that $\frac{1}{4}e^t(2t - 1)$ is a particular solution of

$$x'' - x = e^t. \tag{3.10}$$

Two linearly independent solutions of $x'' - x = 0$ are given by $x_1 = e^t$ and $x_2 = e^{-t}$. The general solution to Eq. (3.10) is therefore $\frac{1}{4}e^t(2t - 1) + c_1 e^t + c_2 e^{-t}$. Note that x_1 and x_2 are independent since $W(x_1, x_2) = e^t(-e^{-t}) - e^{-t}(e^t) = -2$.

In the next three sections we will present two methods for finding the general solution of the homogeneous problem, and in Sections 3.6 and 3.7 techniques for obtaining the solution of the nonhomogeneous problem will be developed.

EXERCISES 3.2

1. a) Show that $y_1(x) = \sin x^2$ and $y_2(x) = \cos x^2$ are linearly independent solutions of $xy'' - y' + 4x^3 y = 0$.

   b) Calculate $W(y_1, y_2)$ and show that it is zero when $x = 0$. Does this result contradict Lemma 3.3? [*Hint*: In Lemma 3.3, as elsewhere in this section, it is assumed that $a(x)$ and $b(x)$ are continuous.]

2. Show that two solutions $y_1(x)$ and $y_2(x)$ to Eq. (3.2) are linearly dependent if and only if one is a constant multiple of the other.

3. Show that $y_1(x) = \sin x$ and $y_2(x) = 4 \sin x - 2 \cos x$ are linearly independent solutions of $y'' + y = 0$. Write the solution $y_3(x) = \cos x$ as a linear combination of y_1 and y_2.

4. Prove that $e^x \sin x$ and $e^x \cos x$ are linearly independent solutions of the equation

   $$y'' - 2y' + 2y = 0.$$

   a) Find a solution that satisfies the conditions $y(0) = 1$, $y'(0) = 4$.

   b) Find another pair of linearly independent solutions.

5. Assume that some nonzero solution of

   $$y'' + a(x)y' + b(x)y = 0, \qquad y(0) = 0,$$

   vanishes at some point x_1, where $x_1 > 0$. Prove that any other solution vanishes at $x = x_1$.

6. Define the function $s(x)$ to be the unique solution of the initial value problem

   $$y'' + y = 0; \qquad y(0) = 0, \qquad y'(0) = 1,$$

   and the function $c(x)$ as the solution of

   $$y'' + y = 0; \qquad y(0) = 1, \qquad y'(0) = 0.$$

Without using trigonometry, prove that:

a) $\dfrac{ds}{dx} = c(x)$, b) $\dfrac{dc}{dx} = -s(x)$,

c) $s^2 + c^2 = 1$.

7. a) Show that $y_1 = \sin \ln x^2$ and $y_2 = \cos \ln x^2$ are linearly independent solutions to

$$y'' + \frac{1}{x}y' + \frac{4}{x^2}y = 0 \qquad (y > 0).$$

b) Calculate $W(y_1, y_2)$.

8. Consider the third order equation

$$y''' + a(x)y'' + b(x)y' + c(x)y = 0,$$

where a, b, and c are continuous functions of x in some interval I. Prove that if $y_1(x)$, $y_2(x)$, and $y_3(x)$ are solutions to the equation, then so is any linear combination of them.

9. In Exercise 8 define the Wronskian $W(y_1, y_2, y_3)$ by the determinant

$$W(y_1, y_2, y_3)(x) = \begin{vmatrix} y_1 & y_2 & y_3 \\ y_1' & y_2' & y_3' \\ y_1'' & y_2'' & y_3'' \end{vmatrix}.$$

a) Show that W satisfies the differential equation

$$W'(x) = -a(x)W.$$

b) Prove that $W(y_1, y_2, y_3)(x)$ is either always zero or never zero.

10. a) Prove that the solutions $y_1(x)$, $y_2(x)$, $y_3(x)$ of the equation in Exercise 8 are linearly independent on $[x_0, x_1]$ if and only if $W(y_1, y_2, y_3) \neq 0$. (This is the analogue of Lemma 3.3 for third order equations.)

b) Show that $\sin t$, $\cos t$, and e^t are linearly independent solutions of

$$y''' - y'' + y' - y = 0$$

on any interval (a, b), where $-\infty < a < b < \infty$.

11. Using the result of Exercise 10, find the unique solution to $y''' - y'' + y' - y = 0$ that satisfies the initial conditions $y(0) = 1$, $y'(0) = 0$, $y''(0) = 4$.

12. Show that if $y_1(x)$, $y_2(x)$, and $y_3(x)$ are three linearly independent solutions to the homogeneous equation of Exercise 8, then any other solution can be written as a linear combination of them. [*Hint*: Assume the third order analogue of Theorem 3.1.]

13. Assume that $y_1(x)$ and $y_2(x)$ are two solutions to

$$y''' + a(x)y'' + b(x)y' + c(x)y = f(x).$$

Prove that $y_3(x) = y_1(x) - y_2(x)$ is a solution of the associated homogeneous equation.

*14. Generalize the results of Exercises 8, 9, 10(a), 12, and 13 for the nth order linear equation

$$y^{(n)} + a_1(x)y^{(n-1)} + a_2(x)y^{(n-2)} + \cdots + a_{n-1}(x)y' + a_n(x)y = f(x),$$

where the functions $a_i(x)$, $i = 1, 2, \ldots, n$, are continuous on some interval I.

15. Let $y_1(x)$ be a solution of the equation

$$y'' + a(x)y' + b(x)y = f_1(x),$$

and let $y_2(x)$ be a solution of the equation

$$y'' + a(x)y' + b(x)y = f_2(x).$$

Show that $y_3(x) = y_1(x) + y_2(x)$ is a solution of the equation

$$y'' + a(x)y' + b(x)y = f_1(x) + f_2(x).$$

This important relation is called the *principle of superposition*.

3.3 USING ONE SOLUTION TO FIND ANOTHER

As we saw in Theorem 3.4, it is easy to write down the general solution of the homogeneous equation

$$y'' + a(x)y' + b(x)y = 0, \tag{3.11}$$

provided that we know two linearly independent solutions y_1 and y_2 of Eq. (3.11). The general solution is then given by

$$y = c_1 y_1 + c_2 y_2, \tag{3.12}$$

where c_1 and c_2 are arbitrary constants. Unfortunately, there is no general procedure for determining y_1 and y_2. However, a standard procedure does exist for finding y_2 when y_1 is known. This method is of considerable importance, since it is often possible to find one solution by inspecting the equation or by trial and error.

We assume that y_1 is a nonzero solution of Eq. (3.11) and seek another solution y_2 such that y_1 and y_2 are linearly independent. If it can be found, then

$$\frac{y_2}{y_1} = v(x)$$

must be a nonconstant function of x, and $y_2 = vy_1$ must satisfy Eq. (3.11). Thus

$$(vy_1)'' + a(vy_1)' + b(vy_1) = 0$$

or, after differentiations and factoring,

$$v(y_1'' + ay_1' + by_1) + v'(2y_1' + ay_1) + v''y_1 = 0. \tag{3.13}$$

The first term in parentheses in Eq. (3.13) vanishes since y_1 is a solution of (3.11), so we obtain the equation

$$v''y_1 + v'(2y_1' + ay_1) = 0.$$

Dividing by $v'y_1$, we can rewrite this equation in the form

$$\frac{v''}{v'} = -2\frac{y_1'}{y_1} - a. \tag{3.14}$$

Integrating Eq. (3.14), we find that

$$\ln v' = -2\ln y_1 - \int a(x)\,dx,$$

so that

$$v' = \frac{1}{y_1^2}\,e^{-\int a(x)\,dx}. \tag{3.15}$$

Since the exponential is never zero, v is nonconstant. To find v, we perform another integration and obtain

$$y_2 = vy_1 = y_1(x)\int \frac{e^{-\int a(x)\,dx}}{y_1^2(x)}\,dx. \tag{3.16}$$

It remains to be shown that y_2 is a solution of Eq. (3.11). But this is obvious since v is a solution of (3.14) and vy_1 is a solution of (3.13). We shall make use of formula (3.16) in later chapters; the following examples are illustrations of how it can be used.

Example 3.3.1 Note that $y_1 = x$ is a solution of

$$x^2 y'' - xy' + y = 0, \qquad x > 0. \tag{3.17}$$

To find y_2, we rewrite (3.17) as

$$y'' - \left(\frac{1}{x}\right)y' + \left(\frac{1}{x^2}\right)y = 0.$$

Then $\int a(x)\,dx = -\ln x$, so that using formula (3.16), we have

$$y_2 = x\int \frac{x}{x^2}\,dx = x\ln x.$$

Thus the general solution of Eq. (3.17) is $y = c_1 x + c_2 x\ln x,\ x > 0$.

Example 3.3.2 Consider the *Legendre equation of order one*:

$$(1 - x^2)y'' - 2xy' + 2y = 0, \qquad -1 < x < 1, \tag{3.18}$$

or

$$y'' - \frac{2x}{1 - x^2}y' + \frac{2y}{1 - x^2} = 0.$$

It is easy to verify that $y_1 = x$ is a solution. To find y_2, we note that $\int a(x)\,dx = \ln(1 - x^2)$, so that by (3.16)

$$
\begin{aligned}
y_2 &= x \int \frac{e^{-\ln(1-x^2)}}{x^2}\,dx = x \int \frac{dx}{x^2(1 - x^2)} \\
&= x \int \left(\frac{1}{x^2} + \frac{1}{1 - x^2} \right) dx = x \left[-\frac{1}{x} + \frac{1}{2} \ln \left(\frac{1 + x}{1 - x} \right) \right] \\
&= \frac{x}{2} \ln \left(\frac{1 + x}{1 - x} \right) - 1.
\end{aligned}
$$

Note that in this example y_2 is defined in $-1 < x < 1$ even though $v(0)$ is undefined.

EXERCISES 3.3

In each of Exercises 1 through 5 a second order differential equation and one solution $y_1(x)$ are given. Verify that $y_1(x)$ is indeed a solution and find a second linearly independent solution.

1. $y'' - 2y' + y = 0,\ y_1(x) = e^x$
2. $y'' - 2xy' + 2y = 0,\ y_1(x) = x \qquad (x > 0)$
3. $y'' - \left(\dfrac{2x}{1 - x^2} \right) y' + \left(\dfrac{6}{1 - x^2} \right) y = 0,\ y_1(x) = \dfrac{3x^2 - 1}{2} \qquad (|x| < 1)$

   (This equation is called the *Legendre differential equation of order two*.)
4. $y'' + \left(\dfrac{3}{x} \right) y' = 0,\ y_1(x) = 1$
5. $x^2 y'' + xy' - 4y = 0,\ y_1(x) = x^2$
6. The *Bessel* differential equation is given by

   $$x^2 y'' + xy' + (x^2 - p^2)y = 0.$$

   For $p = \frac{1}{2}$, verify that $y_1(x) = (\sin x)/\sqrt{x}$ is a solution for $x > 0$. Find a second, linearly independent solution.
7. Letting $p = 0$ in the equation of Exercise 6, we obtain the *Bessel* differential equation of index zero that we will study in Chapter 5. One solution is the *Bessel function of order zero* denoted by $J_0(x)$. In terms of $J_0(x)$, find a second, linearly independent solution.
8. Consider the third order equation

   $$y''' + a(x)y'' + b(x)y' + c(x)y = 0$$

   and let $y_1(x)$ and $y_2(x)$ be two linearly independent solutions. Define $y_3(x) = v(x)y_1(x)$ and assume that $y_3(x)$ is a solution to the equation.

   a) Find a second order differential equation that is satisfied by v'.
   b) Show that $(y_2/y_1)'$ is a solution of this equation.

    c) Use the result of part (b) to find a second, linearly independent solution of the equation derived in part (a).

9. Consider the equation

$$y''' - \left(\frac{3}{x^2}\right)y' + \left(\frac{3}{x^3}\right)y = 0 \qquad (x > 0).$$

    a) Show that $y_1(x) = x$ and $y_2(x) = x^3$ are two linearly independent solutions.
    b) Use the results of Exercise 8 to get a third linearly independent solution.

3.4 HOMOGENEOUS EQUATIONS WITH CONSTANT COEFFICIENTS: THE CASE OF REAL ROOTS

In this section we shall present a simple procedure for finding the general solution to the linear homogeneous equation with constant coefficients

$$y'' + ay' + by = 0. \tag{3.19}$$

We recall that for the comparable first order equation $y' + ay = 0$ the general solution is $y(x) = ce^{-ax}$. It is then not implausible to "guess" that there may be a solution to Eq. (3.19) of the form $y(x) = e^{\lambda x}$ for some number λ (real or complex). Setting $y(x) = e^{\lambda x}$ in Eq. (3.19), we obtain

$$\lambda^2 e^{\lambda x} + a\lambda e^{\lambda x} + be^{\lambda x} = 0. \tag{3.20}$$

Since $e^{\lambda x} \neq 0$, we can divide this equation by $e^{\lambda x}$ to obtain

$$\lambda^2 + a\lambda + b = 0, \tag{3.21}$$

where a and b are real numbers. Equation (3.21) is called the *auxiliary equation* of the differential equation (3.19). It is clear that if λ satisfies Eq. (3.21), then $y(x) = e^{\lambda x}$ is a solution to Eq. (3.19). As we saw in Section 3.2, we need only obtain two linearly independent solutions. Equation (3.21) has the roots

$$\lambda_1 = \frac{-a + \sqrt{a^2 - 4b}}{2}, \qquad \lambda_2 = \frac{-a - \sqrt{a^2 - 4b}}{2}. \tag{3.22}$$

There are three possibilities: $a^2 - 4b > 0$, $a^2 - 4b = 0$, $a^2 - 4b < 0$.

CASE 1. If $a^2 - 4b > 0$, then λ_1 and λ_2 are distinct real numbers [given by (3.22)] and $y_1(x) = e^{\lambda_1 x}$ and $y_2 = e^{\lambda_2 x}$ are distinct solutions. We can now show that they are linearly independent:

$$W(y_1, y_2) = e^{\lambda_1 x} \cdot \lambda_2 e^{\lambda_2 x} - \lambda_1 e^{\lambda_1 x} e^{\lambda_2 x}$$
$$= (\lambda_2 - \lambda_1)e^{(\lambda_2 + \lambda_1)x} \neq 0,$$

since $\lambda_1 \neq \lambda_2$. According to Theorem 3.4, the solutions are independent and thus we have proved:

Theorem 3.6 If $a^2 - 4b > 0$, then the general solution to Eq. (3.19) is given by

$$y(x) = c_1 e^{\lambda_1 x} + c_2 e^{\lambda_2 x}, \tag{3.23}$$

where c_1 and c_2 are arbitrary constants and λ_1 and λ_2 are the real roots of Eq. (3.21).

Example 3.4.1 Consider the equation

$$y'' + 3y' - 10y = 0.$$

The auxiliary equation is $\lambda^2 + 3\lambda - 10 = 0$, $a^2 - 4b = 49$, and the roots are $\lambda_1 = 2$ and $\lambda_2 = -5$ (the order in which the roots are taken is irrelevant). The general solution is

$$y(x) = c_1 e^{2x} + c_2 e^{-5x}.$$

If we specify the initial conditions $y(0) = 1$ and $y'(0) = 3$, for example, then we obtain the simultaneous equations

$$c_1 + c_2 = 1,$$
$$2c_1 - 5c_2 = 3,$$

which have the unique solution $c_1 = \frac{8}{7}$ and $c_2 = -\frac{1}{7}$. The unique solution to the initial value problem is therefore

$$y(x) = \tfrac{1}{7}(8e^{2x} - e^{-5x}).$$

CASE 2. $a^2 - 4b = 0$. In this case (3.21) has the double root $\lambda_1 = \lambda_2 = -a/2$. Thus $y_1(x) = e^{-ax/2}$ is a solution of Eq. (3.19). To find the second solution y_2, we make use of Eq. (3.16), since one solution is known:

$$y_2 = y_1 \int \frac{e^{-ax}}{y_1^2} \, dx = e^{-ax/2} \int dx = xe^{-ax/2}.$$

We saw in Section 3.3 that y_1 and y_2 are linearly independent. Hence we have the following result.

Theorem 3.7 If $a^2 - 4b = 0$, then the general solution to Eq. (3.19) is given by

$$y(x) = c_1 e^{-(a/2)x} + c_2 x e^{-(a/2)x}, \tag{3.24}$$

where c_1 and c_2 are arbitrary constants.

Example 3.4.2 Consider the equation

$$y'' - 6y' + 9 = 0.$$

The auxiliary equation is $\lambda^2 - 6\lambda + 9 = 0$, and $a^2 - 4b = 0$, yielding the unique double root $\lambda_1 = -a/2 = 3$. The general solution is

$$y(x) = c_1 e^{3x} + c_2 x e^{3x}.$$

If we use the initial conditions $y(0) = 1$, $y'(0) = 3$, we obtain the simultaneous equations

$$c_1 = 1,$$
$$3c_1 + c_2 = 3,$$

which yield the unique solution

$$y(x) = e^{3x}.$$

We will deal with the more complicated case of complex roots ($a^2 - 4b < 0$) in the next section.

EXERCISES 3.4

In Exercises 1 through 12, find the general solution of each equation. When initial conditions are specified, give the particular solution that satisfies them.

1. $y'' - 4y = 0$
2. $x'' + x' - 3x = 0$, $x(0) = 0$, $x'(0) = 1$
3. $y'' - 3y' + 2y = 0$
4. $y'' + 5y' + 6y = 0$, $y(0) = 1$, $y'(0) = 2$
5. $4x'' + 20x' + 25x = 0$, $x(0) = 1$, $x'(0) = 2$
6. $y'' + 6y' + 9y = 0$
7. $x'' - x' - 6x = 0$, $x(0) = -1$, $x(1) = 1$
8. $y'' - 8y' + 16y = 0$, $y(0) = 2$, $y'(0) = -1$
9. $y'' - 5y' = 0$
10. $y'' + 17y' = 0$, $y(0) = 1$, $y'(0) = 0$
11. $y'' + 2\pi y' + \pi^2 y = 0$, $y(1) = 1$, $y'(1) = 1/\pi$
12. $y'' - 13y' + 42y = 0$
13. Consider the linear third order homogeneous equation with constant coefficients

    $$y''' + ay'' + by' + cy = 0.$$

    a) Show that $e^{\lambda x}$ is a solution if it satisfies the auxiliary equation

    $$\lambda^3 + a\lambda^2 + b\lambda + c = 0.$$

    b) Show that if the roots $\lambda_1, \lambda_2, \lambda_3$ of the auxiliary equation are distinct, then $y_1 = e^{\lambda_1 x}$, $y_2 = e^{\lambda_2 x}$, and $y_3 = e^{\lambda_3 x}$ are linearly independent solutions of the equation (see

Exercise 3.2.10). According to the result of Exercise 3.2.12, what is the general solution to the equation in this case?

c) Given that λ_1 is a double root, prove that $y_1(x) = e^{\lambda_1 x}$, $y_2(x) = xe^{\lambda_1 x}$, and $y_3(x) = e^{\lambda_2 x}$ are linearly independent solutions. What is the general solution?

d) Given that λ_1 is a triple root [i.e., $\lambda^3 + a\lambda^2 + b\lambda + c = (\lambda - \lambda_1)^3$], show that $y_1(x) = e^{\lambda_1 x}$, $y_2(x) = xe^{\lambda_1 x}$, $y_3(x) = x^2 e^{\lambda_1 x}$ are linearly independent solutions. What is the general solution? (Note that the solution $x^2 e^{\lambda_1 x}$ can be derived from the solutions $e^{\lambda_1 x}$ and $xe^{\lambda_1 x}$ by using the technique outlined in Exercise 3.3.8.)

In Exercises 14 through 18 use the technique outlined in Exercise 13 to find the general solution for each of the given differential equations. Find a particular solution when indicated.

14. $y''' - y'' - y' + y = 0$

15. $y''' - 3y'' + 3y' - y = 0$, $y(0) = 1$, $y'(0) = 2$, $y''(0) = 3$

16. $x''' + 5x'' - x' - 5x = 0$

17. $y''' - 9y' = 0$, $y(0) = 1$, $y'(0) = 0$, $y''(0) = 2$

18. $y''' - 6y'' + 3y' + 10y = 0$

19. In Section 2.4 we discussed the nonlinear Riccati equation

$$y' + y^2 + a(x)y + b(x) = 0.$$

Using the techniques of that section and the present one, solve the equation

$$y' + y^2 - 1 = 0.$$

20. Find the general solution of $y' + y^2 + y - 1 = 0$.

21. Find the general solution of $y' + ay^2 + by + c = 0$, where a, b, and c are constants.

22. For any functions $a(x)$ and $b(x)$ find the solution to the initial value problem

$$y'' + a(x)y' + b(x)y = 0, \qquad y(x_0) = y'(x_0) = 0,$$

at any point x_0 at which both $a(x)$ and $b(x)$ are continuous.

3.5 HOMOGENEOUS EQUATIONS WITH CONSTANT COEFFICIENTS: COMPLEX ROOTS

Before dealing with the remaining case, let us briefly review some facts about complex numbers. A *complex number* z is any expression of the form $z = \alpha + i\beta$, where α and β are real and $i^2 = -1$. If $\beta = 0$, then z is a real number, while if $\alpha = 0$ and $\beta \neq 0$, then z is a *pure imaginary* number. Any complex number z can be represented as a vector in the xy-plane where α equals the x-coordinate and β equals the y-coordinate (see Fig. 3.1).

Any complex number $z = \alpha + i\beta$ can be described in terms of *polar coordinates*. Let r represent the distance from z to the origin and let θ represent the angle the

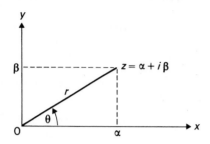

Figure 3.1

vector z makes with the positive x-axis (see Fig. 3.1). Then it is evident that

$$r = \sqrt{\alpha^2 + \beta^2}, \qquad \theta = \tan^{-1}\frac{\beta}{\alpha},$$
$$\beta = r\sin\theta, \qquad \alpha = r\cos\theta. \tag{3.25}$$

Using the last two relations in (3.25), we have

$$z = r(\cos\theta + i\sin\theta). \tag{3.26}$$

As can be proved by the ratio test, the following power series are convergent for all complex values of θ:

$$e^\theta = 1 + \theta + \frac{\theta^2}{2!} + \cdots + \frac{\theta^n}{n!} + \cdots = \sum_{n=0}^{\infty} \frac{\theta^n}{n!},$$

$$\cos\theta = 1 - \frac{\theta^2}{2!} + \frac{\theta^4}{4!} - \cdots = \sum_{n=0}^{\infty} (-1)^n \frac{\theta^{2n}}{(2n)!}, \tag{3.27}$$

$$\sin\theta = \theta - \frac{\theta^3}{3!} + \frac{\theta^5}{5!} - \cdots = \sum_{n=0}^{\infty} (-1)^n \frac{\theta^{2n+1}}{(2n+1)!}.$$

Replacing θ by $i\theta$ in the power series e^θ and using the identity $i^2 = -1$, we have *Euler's formula*:

$$e^{i\theta} = 1 + i\theta - \frac{\theta^2}{2!} - \frac{i\theta^3}{3!} + \frac{\theta^4}{4!} + \frac{i\theta^5}{5!} - \frac{\theta^6}{6!} - \frac{i\theta^7}{7!} + \cdots,$$

$$= \cos\theta + i\sin\theta. \tag{3.28}$$

Thus any complex number $z = \alpha + i\beta$ can be represented in *polar form* by the expression

$$z = re^{i\theta}, \tag{3.29}$$

where r and θ are as in (3.25). Although Euler's Formula was obtained by means of a power series, we shall make no further use of power series in this chapter.

Now we can return to our examination of the situation where $a^2 - 4b < 0$.

CASE 3. $a^2 - 4b < 0$. The roots of the auxiliary equation (3.21) are

$$\lambda_1 = \alpha + i\beta, \qquad \lambda_2 = \alpha - i\beta, \tag{3.30}$$

where $\alpha = -a/2$ and $\beta = \sqrt{4b - a^2}/2$. Thus $y_1 = e^{\lambda_1 x}$ and $y_2 = e^{\lambda_2 x}$ are solutions to Eq. (3.19). However, knowing that any linear combination of solutions is also a solution, we let

$$y_1^* = \frac{1}{2}(y_1 + y_2), \qquad y_2^* = \frac{1}{2i}(y_1 - y_2). \tag{3.31}$$

Then y_1^* and y_2^* are solutions and

$$y_1^* = \frac{e^{\lambda_1 x} + e^{\lambda_2 x}}{2} = \frac{e^{\alpha x}}{2}(e^{i\beta x} + e^{-i\beta x})$$

$$= \frac{e^{\alpha x}}{2}[\cos \beta x + i \sin \beta x + \cos(-\beta x) + i \sin(-\beta x)]$$

$$= e^{\alpha x} \cos \beta x,$$

since $\cos(-\theta) = \cos \theta$ and $\sin(-\theta) = -\sin \theta$. Similarly,

$$y_2^* = e^{\alpha x} \sin \beta x.$$

That y_1^* and y_2^* are linearly independent follows easily since

$$\frac{y_1^*}{y_2^*} = \cot \beta x, \qquad \beta \neq 0,$$

which is not a constant. Alternatively, it is also easy to compute $W(y_1^*, y_2^*) = \beta e^{\alpha x} \neq 0$. Thus we have proved:

Theorem 3.8 If $a^2 - 4b < 0$, then the general solution to Eq. (3.19) is given by

$$y(x) = e^{\alpha x}(c_1 \cos \beta x + c_2 \sin \beta x), \tag{3.32}$$

where c_1 and c_2 are arbitrary constants and

$$\alpha = -\frac{a}{2}, \qquad \beta = \frac{\sqrt{4b - a^2}}{2}.$$

Example 3.5.1 Let $y'' + y = 0$. Then the auxiliary equation is $\lambda^2 + 1 = 0$ with roots $\pm i$. We have $\alpha = 0$ and $\beta = 1$ so that the general solution is

$$y(x) = c_1 \cos x + c_2 \sin x. \tag{3.33}$$

This is the equation of *harmonic motion*.

Example 3.5.2 Consider the equation $y'' + y' + y = 0$, $y(0) = 1$, $y'(0) = 3$. We have $\lambda^2 + \lambda + 1 = 0$ with roots $\lambda_1 = (-1 + i\sqrt{3})/2$ and $\lambda_2 = (-1 - i\sqrt{3})/2$. Then $\alpha = -\frac{1}{2}$ and $\beta = \sqrt{3}/2$, so that the general solution is

$$y(x) = e^{-x/2}\left(c_1 \cos \frac{\sqrt{3}}{2}x + c_2 \sin \frac{\sqrt{3}}{2}x\right).$$

To solve the initial value problem, we solve the simultaneous equations

$$c_1 = 1,$$

$$\frac{\sqrt{3}}{2}c_2 - \frac{1}{2}c_1 = 3.$$

Thus $c_1 = 1$, $c_2 = 7/\sqrt{3}$, and

$$y(x) = e^{-x/2}\left(\cos \frac{\sqrt{3}}{2}x + \frac{7}{\sqrt{3}} \sin \frac{\sqrt{3}}{2}x\right).$$

EXERCISES 3.5

In the following exercises find the general solution of each equation. When initial conditions are specified, give the particular solution that satisfies them.

1. $y'' + 2y' + 2y = 0$
2. $8y'' + 4y' + y = 0$, $y(0) = 0$, $y'(0) = 1$
3. $x'' + x' + 7x = 0$
4. $y'' + y' + 2y = 0$
5. $\dfrac{d^2x}{d\theta^2} + 4x = 0$, $x\left(\dfrac{\pi}{4}\right) = 1$, $x'\left(\dfrac{\pi}{4}\right) = 3$
6. $y''' - y'' + y' - y = 0$ [*Hint:* See Exercise 3.2.10.]
7. $y''' - 3y'' + 4y' - 2y = 0$, $y(0) = 1$, $y'(0) = 2$, $y''(0) = 3$

3.6 NONHOMOGENEOUS EQUATIONS: THE METHOD OF UNDETERMINED COEFFICIENTS

In this and the next sections we shall present methods for finding a particular solution to the nonhomogeneous linear equations with constant coefficients

$$y'' + ay' + by = f(x). \tag{3.34}$$

Let $P_n(x)$ denote a polynomial of degree n in x. In this section we assume that $f(x)$ takes one of the special forms:

1. $P_n(x)$,
2. $P_n(x)e^{ax}$,
3. $P_n(x)e^{ax} \sin bx$ or $P_n(x)e^{ax} \cos bx$.

The technique we will employ is to "guess" that there is a solution to Eq. (3.34) in the same basic "form" as $f(x)$ and then substitute this "guessed" solution back into (3.34) and "determine" the unknown coefficients. This method is best illustrated by a number of examples.

Example 3.6.1 Consider the equation

$$y'' - y = x^2. \tag{3.35}$$

Since $f(x) = x^2$ is a polynomial of degree two, we "guess" that Eq. (3.35) has a solution $y_p(x)$ that is a polynomial of degree two, that is,

$$y_p(x) = a + bx + cx^2.$$

After noting that $y_p'' = 2c$, we substitute $y_p(x)$ into Eq. (3.35) to obtain

$$2c - (a + bx + cx^2) = x^2.$$

Equating coefficients, we have

$$2c - a = 0, \qquad -b = 0, \qquad -c = 1,$$

which immediately yields $a = -2, b = 0, c = -1$, and the solution

$$y_p(x) = -2 - x^2.$$

This solution is easily verified by substitution into Eq. (3.35). Finally, since the general solution of the homogeneous equation $y'' - y = 0$ is given by

$$y = c_1 e^x + c_2 e^{-x},$$

the general solution of Eq. (3.35) is

$$y = c_1 e^x + c_2 e^{-x} - 2 - x^2.$$

Example 3.6.2 Let

$$y'' - 3y' + 2y = e^x \sin x. \tag{3.36}$$

Since $f(x)$ is of form 3 with $P_n(x) = 1$, or a polynomial of degree zero, we "guess" that there is a solution to Eq. (3.36) of the form

$$y_p(x) = ae^x \sin x + be^x \cos x.$$

Then

$$y_p'(x) = ae^x(\sin x + \cos x) + be^x(\cos x - \sin x)$$

and

$$y_p''(x) = 2ae^x \cos x - 2be^x \sin x.$$

Thus substitution into Eq. (3.37) yields

$$e^x(2a \cos x - 2b \sin x) - 3e^x[(a - b) \sin x + (a + b) \cos x]$$
$$+ e^x(a \sin x + b \cos x) = e^x \sin x.$$

Dividing both sides by e^x and equating coefficients, we have

$$2a - 3(a + b) + b = 0,$$
$$-2b - 3(a - b) + a = 1,$$

which yield $a = -\frac{2}{5}$ and $b = \frac{1}{5}$ so that

$$y_p = \frac{e^x}{5}(\cos x - 2 \sin x).$$

Again this result is easily verified by substitution. Finally, the general solution of Eq. (3.36) is

$$y = c_1 e^{2x} + c_2 e^x + \frac{e^x}{5}(\cos x - 2 \sin x).$$

Example 3.6.3 $y'' + y = xe^{2x}$. Here $f(x)$ is of form 2, where $P_n(x)$ is a polynomial of degree one, so we try a solution of the form

$$y_p(x) = e^{2x}(a + bx).$$

Then

$$y_p'(x) = e^{2x}(2a + b + 2bx), \qquad y_p''(x) = e^{2x}(4a + 4b + 4bx),$$

and substitution yields

$$e^{2x}(4a + 4b + 4bx) + e^{2x}(a + bx) = xe^{2x}.$$

Dividing both sides by e^{2x} and equating like powers of x, we obtain the equations

$$5a + 4b = 0, \qquad 5b = 1.$$

Thus $a = -\frac{4}{25}$, $b = \frac{1}{5}$, and a particular solution is

$$y_p(x) = \frac{e^{2x}}{25}(5x - 4).$$

Therefore, the general solution of this example is

$$y(x) = c_1 \sin x + c_2 \cos x + \frac{e^{2x}}{25}(5x - 4).$$

Difficulties arise in connection with problems of this type whenever any term of $f(x)$ is a solution of the homogeneous equation

$$y'' + ay' + by = 0.$$

Example 3.6.4 Consider the equation

$$y'' - y = 2e^x.$$

Since the general solution of $y'' - y = 0$ is

$$y(x) = c_1 e^x + c_2 e^{-x},$$

we see that $f(x)$ is indeed a solution to the homogeneous equation. If we try to find a solution of the form $y_p(x) = ae^x$, we obtain $y_p''(x) = ae^x$, and substitution into the equation $y'' - y = 2e^x$ yields $0 = 2e^x$, which is impossible. We must therefore look for a different solution. A similar problem arose in the case of a double root of the auxiliary equation that came up in the solution of a homogeneous differential equation. We employ the same "solution" here and "guess" that there is a solution of the form

$$y_p(x) = axe^x.$$

Then $y_p'(x) = ae^x(x + 1)$, $y_p''(x) = ae^x(x + 2)$, and

$$y_p'' - y_p = ae^x(x + 2) - axe^x = 2ae^x = 2e^x.$$

Hence $a = 1$ and $y_p = xe^x$. Thus the general solution is

$$y(x) = c_1 e^x + c_2 e^{-x} + xe^x.$$

Finally, we note that if $f(x) = f_1(x) + f_2(x)$, where f_1 and f_2 are both of one of the three forms above, then we may use the principle of superposition (see Exercise 3.2.15). That is, we may separately solve the equations

$$y'' + ay' + by = f_1(x),$$

with particular solution $y_{p_1}(x)$, and

$$y'' + ay' + by = f_2(x),$$

with particular solution $y_{p_2}(x)$. Then the particular solution to

$$y'' + ay' + by = f(x) = f_1(x) + f_2(x),$$

is given by $y_p(x) = y_{p_1}(x) + y_{p_2}(x)$.

Example 3.6.5 Consider the equation

$$y'' - y = x^2 + 2e^x.$$

Using the results of Examples 3.6.1 and 3.6.4 and the principle of superposition, we find immediately that a particular solution is given by

$$y_p(x) = -2 - x^2 + xe^x.$$

Let us now summarize the results of this section as follows: Consider the nonhomogeneous equation

$$y'' + ay' + by = f(x) \tag{3.37}$$

and the homogeneous equation

$$y'' + ay' + by = 0. \tag{3.38}$$

CASE 1. *No term in* $y_p(x)$ *is a solution of* Eq. (3.38). A particular solution of Eq. (3.37) will have the form $y_p(x)$ according to the table below:

$f(x)$	$y_p(x)$
$P_n(x)$	$a_0 + a_1x + a_2x^2 + \cdots + a_nx^n$
$P_n(x)e^{ax}$	$(a_0 + a_1x + a_2x^2 + \cdots + a_nx^n)e^{ax}$
$P_n(x)e^{ax} \sin bx$ or $P_n(x)e^{ax} \cos bx$	$(a_0 + a_1x + a_2x^2 + \cdots + a_nx^n)e^{ax} \sin bx +$ $(c_0 + c_1x + c_2x^2 + \cdots + c_nx^n)e^{ax} \cos bx$

CASE 2. If any term of $y_p(x)$ is a solution of Eq. (3.38), then multiply the appropriate function $y_p(x)$ of Case 1 by x^k, where k is the smallest integer such that no term in $x^k y_p(x)$ is a solution of Eq. (3.38).

We note that the method of undetermined coefficients should be used only when $f(x)$ is in a "correct" form. The more general situation is dealt with in the next section.

EXERCISES 3.6

In Exercises 1 through 13 find the general solution of each given differential equation. If initial conditions are given, then find the particular solution that satisfies them.

1. $y'' + 4y = 3 \sin x$
2. $y'' - y' - 6y = 20e^{-2x}$, $y(0) = 0$, $y'(0) = 6$
3. $y'' - 3y' + 2y = 6e^{3x}$
4. $y'' + y' = 3x^2$, $y(0) = 4$, $y'(0) = 0$
5. $y'' - 2y' + y = -4e^x$
6. $y'' - 4y' + 4y = 6xe^{2x}$, $y(0) = 0$, $y'(0) = 3$
7. $y'' - 7y' + 10y = 100x$, $y(0) = 0$, $y'(0) = 5$
8. $y'' + y = 1 + x + x^2$
9. $y'' + y' = x^3 - x^2$
10. $y'' + 4y = 16x \sin 2x$
11. $y'' - 4y' + 5y = 20 \cosh 2x \cos x$
12. $y'' - y' - 2y = x^2 + \cos x$
13. $y'' + 6y' + 9y = 10e^{-3x}$

Use the principle of superposition to find the general solution of each of the equations in Exercises 14 through 17.

14. $y'' + y = 1 + 2 \sin x$ 15. $y'' - 2y' - 3y = x - x^2 + e^x$

16. $y'' + 4y = 3 \cos 2x - 7x^2$ 17. $y'' + 4y' + 4y = xe^x + \sin x$

18. Show by the methods of this section that a particular solution of

$$y'' + 2ay' + b^2 y = A \sin \omega x \qquad (a, \omega > 0)$$

is given by

$$y = \frac{A \sin (\omega x - \alpha)}{\sqrt{(b^2 - \omega^2)^2 + 4\omega^2 a^2}},$$

where

$$\alpha = \tan^{-1} \frac{2a\omega}{(b^2 - \omega^2)}, \qquad (0 < \alpha < \pi).$$

19. Let $f(x)$ be a polynomial of degree n. Show that there is always a solution that is a polynomial of degree n for the equation $y'' + ay' + by = f(x)$.

20. Use the method indicated in Exercise 19 to find a particular solution of $y'' + 3y' + 17y = 1 + x - x^2$.

3.7 NONHOMOGENEOUS EQUATIONS: VARIATIONS OF CONSTANTS

Here we consider again the equation

$$y'' + ay' + by = f(x) \tag{3.39}$$

and assume that we have found two linearly independent solutions y_1 and y_2 of the homogeneous equation $y'' + ay' + by = 0$, so that its general solution is

$$c_1 y_1(x) + c_2 y_2(x). \tag{3.40}$$

Thus any particular solution y^* of Eq. (3.39) must have the property that y^*/y_1 and y^*/y_2 are not constants, suggesting that we should replace the constants c_1 and c_2 in Eq. (3.40) by two functions $c_1(x)$ and $c_2(x)$ and look for a solution of (3.39) of the form

$$y(x) = c_1(x)y_1(x) + c_2(x)y_2(x). \tag{3.41}$$

This replacement of constants by variables gives the method its name. Differentiating (3.41), we obtain

$$y'(x) = c_1(x)y_1'(x) + c_2(x)y_2'(x) + c_1'(x)y_1(x) + c_2'(x)y_2(x).$$

To simplify this expression, it is convenient to set

$$c_1'(x)y_1(x) + c_2'(x)y_2(x) = 0. \tag{3.42}$$

Then

$$y'(x) = c_1(x)y_1'(x) + c_2(x)y_2'(x).$$

Differentiating once again, we obtain

$$y''(x) = c_1(x)y_1''(x) + c_2(x)y_2''(x) + c_1'(x)y_1'(x) + c_2'(x)y_2'(x).$$

Substitution of the expressions for $y(t)$, $y'(t)$ and $y''(t)$ in Eq. (3.39) yields

$$y''(x) + ay'(x) + by(x) = c_1(x)(y_1'' + ay_1' + by_1) + c_2(x)(y_2'' + ay_2' + by_2)$$
$$+ c_1'y_1' + c_2'y_2'$$
$$= f(x).$$

But y_1 and y_2 are solutions to the homogeneous equation so that the equation above reduces to

$$c_1'y_1' + c_2'y_2' = f(x). \tag{3.43}$$

This gives a second equation relating $c_1'(x)$ and $c_2'(x)$, and we have the simultaneous equations

$$y_1c_1' + y_2c_2' = 0,$$
$$y_1'c_1' + y_2'c_2' = f(x). \tag{3.44}$$

Before continuing with the discussion, we note that the problem has essentially been solved. Equations (3.44) uniquely determine c_1' and c_2', so that c_1 and c_2 can be obtained by integration.

To complete the derivation, we multiply the first of these equations by y_2', the second by y_2, and subtract to obtain $c_1'(x)$; c_2' can be determined in a similar manner. The solutions are thus

$$c_1'(x) = \frac{-f(x)y_2(x)}{y_1(x)y_2'(x) - y_1'(x)y_2(x)},$$
$$c_2'(x) = \frac{f(x)y_1(x)}{y_1(x)y_2'(x) - y_1'(x)y_2(x)}. \tag{3.45}$$

Note that the denominators here are not zero since they are both equal to $W(y_1,y_2)$, and y_1 and y_2 are linearly independent.

Finally, we can integrate c_1' and c_2' and then substitute c_1 and c_2 into Eq. (3.41) to obtain a particular solution to the nonhomogeneous equation.

Example 3.7.1 Solve $y'' - y = e^x$ by the variation of constants method. The solutions to the homogeneous equation are $y_1 = e^{-x}$ and $y_2 = e^x$. $W(y_1,y_2) = 2$, so that (3.45) becomes

$$c_1'(x) = \frac{-e^x e^x}{2} = \frac{-e^{2x}}{2}, \qquad c_2'(x) = \frac{e^x e^{-x}}{2} = \frac{1}{2}.$$

Integrating these functions, we obtain $c_1(x) = -e^{2x}/4$ and $c_2(x) = x/2$. A particular solution is therefore

$$c_1(x)y_1(x) + c_2(x)y_2(x) = -\frac{e^x}{4} + \frac{x}{2}e^x,$$

and the general solution is

$$y(x) = c_1e^x + c_2e^{-x} + \frac{x}{2}e^x - \frac{e^x}{4} = c_1^*e^x + c_2^*e^{-x} + \frac{x}{2}e^x.$$

Example 3.7.2 Determine the solution of $y'' + y = 2\cos x$ that satisfies $y(0) = 5$ and $y'(0) = 2$. Here $y_1(x) = \cos x$, $y_2(x) = \sin x$, and $W(y_1,y_2) = 1$, so that $c_1' = -2\cos x \sin x$ and $c_2' = 2\cos^2 x$. Immediately we have $c_1(x) = \cos^2 x$ and $c_2(x) = x + \sin x \cos x$. Thus the general solution is

$$y = c_1 \cos x + c_2 \sin x + \cos^3 x + x \sin x + \sin^2 x \cos x,$$
$$= c_1 \cos x + c_2 \sin x + x \sin x.$$

Since $5 = y(0) = c_1$ and $2 = y'(0) = c_2$, the required solution is

$$y(x) = (x + 2) \sin x + 5 \cos x.$$

Example 3.7.3 Solve $y'' + y = \tan x$. The solutions to the homogeneous equation are $y_1 = \cos x$ and $y_2 = \sin x$. Also $W(y_1,y_2) = 1$, so that Eqs. (3.45) become

$$c_1'(x) = -\tan x \sin x = -\frac{\sin^2 x}{\cos x} = \frac{\cos^2 x - 1}{\cos x} = \cos x - \sec x,$$

$$c_2'(x) = \tan x \cos x = \sin x.$$

Hence using the integrals 17, 22, and 33 in Appendix 1, we have

$$c_1(x) = \sin x - \ln |\sec x + \tan x|$$

and

$$c_2(x) = -\cos x.$$

Thus the particular solution is

$$y_p(x) = c_1(x)y_1(x) + c_2(x)y_2(x)$$
$$= \cos x \sin x - \cos x \ln |\sec x + \tan x| - \sin x \cos x$$
$$= -\cos x \ln |\sec x + \tan x|,$$

and the general solution is

$$y(x) = c_1 \cos x + c_2 \sin x - \cos x \ln |\sec x + \tan x|.$$

Example 3.7.3 illustrates that there are instances in which we cannot apply the method of undetermined coefficients. (Try to "guess" a solution in this case.) As a rule, the method of undetermined coefficients is easier to use if the function

$f(x)$ is in the right form. However, the method of variation of parameters is far more general, since it will yield a solution whenever the functions c_1' and c_2' have a known indefinite integral.

EXERCISES 3.7

In Exercises 1 through 10 find the general solution of each equation by the variation of constants method.

1. $y'' + y = \cos 2x$
2. $y'' + y = \cot x$
3. $y'' + 4y = \sec 2x$
4. $y'' + 4y = \sec x \tan x$
5. $y'' - 2y' + y = \dfrac{e^x}{(1 - x)^2}$
6. $y'' - y = \sin^2 x$
7. $y'' + y' = \cosh x$
8. $y'' - 3y' - 4y = e^x + e^{2x}$ [*Hint*: Use the principle of superposition.]
9. $y'' - 4y' + 4y = e^{2x}$
10. $y'' + 2y' + y = e^{-x} \ln |x|$

11. Find a particular solution of

$$y'' + \frac{1}{x} y' - \frac{y}{x^2} = \frac{1}{x^2 + x^3} \qquad (x > 0),$$

given that two solutions of the associated homogeneous equation are $y_1 = x$ and $y_2 = 1/x$.

12. Find a particular solution of

$$y'' - \frac{2}{x} y' + \frac{2}{x^2} y = \frac{\ln |x|}{x} \qquad (x > 0),$$

given the two homogeneous solutions $y_1 = x$ and $y_2 = x^2$.

13. Verify that

$$y = \frac{1}{\omega} \int_0^x f(t) \sin \omega(x - t)\, dt$$

is a particular solution of $y'' + \omega^2 y = f(x)$.

14. Consider the third order equation

$$y''' + ay'' + by' + cy = f(x). \qquad (*)$$

Let $y_1(x)$, $y_2(x)$, and $y_3(x)$ be three linearly independent solutions (see Exercises 3.2.8, 3.2.9, 3.2.10). Assume that there is a solution of (*) of the form $y(x) = c_1(x)y_1(x) + c_2(x)y_2(x) + c_3(x)y_3(x)$.

a) Following the steps used in deriving the variation of constants procedure for the second order equations, derive the same procedure for the third order equation.

b) Find a particular solution of the equation

$$y''' - 2y' - 4y = e^{-x} \sin x.$$

15. Use the method derived in Exercise 14 to find a particular solution of

$$y''' + 5y'' + 9y' + 5y = 3e^{2x}.$$

3.8 VARIABLE COEFFICIENTS: THE EULER EQUATION '

In this section we shall consider a special equation with variable coefficients for which methods of solution are known.

The *Euler equation* has the form

$$x^2 y'' + ax y' + by = 0, \tag{3.46}$$

where a and b are constants. It arises in the solution of the problem of steady-state two-dimensional heat flow on a disk, as we shall see in Chapter 10. We can solve it by seeking a solution of the form $y = x^\lambda$ for some constant λ. Making this substitution into Eq. (3.46), we obtain

$$x^2 [\lambda(\lambda - 1)x^{\lambda - 2}] + ax\lambda x^{\lambda - 1} + bx^\lambda = 0.$$

Divided by x^λ, this expression becomes

$$\lambda^2 + (a - 1)\lambda + b = 0. \tag{3.47}$$

If r_1 and r_2 are distinct roots of Eq. (3.47), then we have the linearly independent solutions $y_1 = x^{r_1}$ and $y_2 = x^{r_2}$ and the general solution $y(x) = c_1 x^{r_1} + c_2 x^{r_2}$. The expression x^{r_1} makes sense even when r_1 is complex, as we shall see in Example 3.8.4, where the roots are complex conjugates.

In the case of a double root $r = r_1 = r_2$ we use the method of Section 3.3. Clearly $y(x) = x^r$ is a solution. To find another linearly independent solution, we use Eq. (3.16) to obtain

$$y_2(x) = x^r \int \frac{\exp\left[-\int(a/x)\,dx\right]}{x^{2r}}\,dx = x^r \int x^{-a-2r}\,dx.$$

Since r is a double root of Eq. (3.47), the quadratic formula yields $r = (1 - a)/2$ so that $-a - 2r = -1$. Hence

$$y_2(x) = x^r \int x^{-1}\,dx = x^r \ln x. \tag{3.48}$$

The general solution to Eq. (3.46) in the case of a double root is therefore given by

$$y(x) = c_1 x^r + c_2 x^r \ln x. \tag{3.49}$$

We should point out that this method applies equally well to the Euler equation of higher order

$$x^n y^{(n)} + a_{n-1} x^{n-1} y^{(n-1)} + \cdots + a_1 x y' + a_0 y = 0. \tag{3.50}$$

In the nth order case it is of course necessary to find the roots of the nth degree polynomial obtained after substituting $y = x^\lambda$ into (3.49) and dividing by x^λ.

Example 3.8.1 $x^2 y'' + xy' - y = 0$. Setting $y = x^\lambda$ yields $\lambda(\lambda - 1)x^\lambda + \lambda x^\lambda - x^\lambda = 0$ or, after dividing by x^λ and simplifying,

$$\lambda^2 - 1 = 0.$$

Since the roots of this equation are ± 1, the general solution is

$$y = c_1 x + c_2 x^{-1}.$$

Example 3.8.2 $x^2 y'' - 5xy' + 9y = 0$. Setting $y = x^\lambda$, we obtain, after dividing by x^λ,

$$\lambda(\lambda - 1) - 5\lambda + 9 = \lambda^2 - 6\lambda + 9 = 0.$$

The only root is $r = 3$. Hence by (3.49), the general solution is

$$y(x) = c_1 x^3 + c_2 x^3 \ln x.$$

Example 3.8.3 $x^3 y''' + x^2 y'' - 2xy' + 2y = 0$. This is a third order Euler equation of the type (3.50). Setting $y = x^\lambda$, differentiating, and simplifying, we obtain the equation

$$(\lambda^3 - 2\lambda^2 - \lambda + 2)x^\lambda = 0.$$

Dividing by x^λ, we obtain the cubic equation

$$\lambda^3 - 2\lambda^2 - \lambda + 2 = 0$$

with the roots $\lambda_1 = 1$, $\lambda_2 = -1$, and $\lambda_3 = 2$. The general solution is therefore given by

$$y(x) = c_1 x + c_2 x^{-1} + c_3 x^2.$$

Example 3.8.4 Let $x^2 y'' - xy' + 2y = 0$, $x > 0$. Setting $y = x^\lambda$, we obtain after simplification $\lambda^2 - 2\lambda + 2 = 0$ with the roots $\lambda_1 = 1 + i$ and $\lambda_2 = 1 - i$. Therefore, $y_1 = x^{1+i}$ and $y_2 = x^{1-i}$ are two linearly independent solutions. To make these into more usable form, we recall that $x = e^{\ln x}$ and make use of Euler's formula $e^{i\theta} = \cos \theta + i \sin \theta$ as follows:

$$y_1 = x^{1+i} = x \cdot x^i = xe^{i \ln x} = x(\cos \ln x + i \sin \ln x).$$

Similarly, $y_2 = x(\cos \ln x - i \sin \ln x)$. Using the technique of Eq. (3.31), if we let

$$y_1^* = \frac{1}{2}(y_1 + y_2), \qquad y_2^* = \frac{1}{2i}(y_1 - y_2),$$

then we have the linearly independent solutions

$$y_1^* = x \cos \ln x, \qquad y_2^* = x \sin \ln x.$$

Of course, these solutions are defined only when $x > 0$.

The above method can be used whenever complex roots are involved. All the results of this section are summarized in the following theorem.

Theorem 3.9 Given the Euler equation (3.46), let λ_1 and λ_2 be the roots of (3.47).
1. If λ_1 and λ_2 are distinct real numbers, then the general solution is

$$y = c_1 |x|^{\lambda_1} + c_2 |x|^{\lambda_2}.$$

2. If $\lambda_1 = \lambda_2$, then the general solution is

$$y = c_1|x|^{\lambda_1} + c_2|x|^{\lambda_2} \ln |x|.$$

3. If λ_1 and λ_2 are complex conjugates ($\lambda_1 = a + ib$, $\lambda_2 = a - ib$), then the general solution is

$$y = |x|^a[c_1 \cos (\ln |x|^b) + c_2 \sin (\ln |x|^b)].$$

EXERCISES 3.8

In Exercises 1 through 7 find the general solution of each given differential equation. Find a particular solution when initial conditions are given.

1. $x^2y'' - 2y = 0$, $y(1) = 3$, $y'(1) = 1$
2. $4x^2y'' - 4xy' + 3y = 0$, $x > 0$, $y(1) = 0$, $y'(1) = 1$
3. $x^2y'' + 3xy' + 2y = 0$
4. $x^2y'' - 3xy' + 3y = 0$
5. $x^2y'' + 5xy' + 4y = 0$, $x > 0$, $y(1) = 1$, $y'(1) = 3$
6. $x^2y'' + 5xy' + 5y = 0$
7. $x^3y''' + 4x^2y'' - 8xy' + 8y = 0$
8. Let us develop an alternative method for solving the Euler equation (3.46).

    a) Let $x = e^t$ and use the chain rule to show that

    $$\frac{dy}{dt} = x\frac{dy}{dx}, \qquad \frac{d^2y}{dt^2} = x^2\frac{d^2y}{dx^2} + x\frac{dy}{dx}.$$

    b) Making the above substitutions, convert (3.46) into a homogeneous linear equation with constant coefficients.

9. Solve $x^2y'' + 2xy' - 12y = 0$ by the method developed in Exercise 8. First solve the equation in terms of t and then convert the solution back to one in terms of x.

10. Solve $x^2y'' + xy' + y = 0$ by the method of Exercises 8 and 9.

In the next five exercises use the method of Exercise 8 to convert each equation to a non-homogeneous linear equation with constant coefficients and then find the general solution by the methods of Section 3.6 or 3.7.

11. $x^2y'' + 7xy' + 5y = x$
12. $x^2y'' + 3xy' - 3y = 5x^2$
13. $x^2y'' - 2y = \ln x$, $x > 0$
14. $4x^2y'' - 4xy' + 3y = \sin \ln (-x)$, $x < 0$
15. Calculate the Wronskian of the two solutions in each of the three cases in Theorem 3.9 and prove that it is never zero.

3.9 VIBRATORY MOTION

Differential equations were first studied in attempts to describe the motion of particles. As a simple example, consider a mass m attached to a coiled spring of length l_0, the upper end of which is securely fastened (see Fig. 3.2).

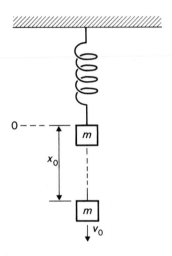

Figure 3.2

We have denoted by the number zero the equilibrium position of the mass on the spring, that is the point where the mass remains at rest. Suppose that the mass is given an initial displacement x_0, and an initial velocity v_0. Can we describe the future movement of the mass? To do so, we make the following assumptions about the force exerted by the spring on the mass:

a) It moves along a vertical line through the center of gravity of the mass (which is then treated as if it were a point mass), and its direction is always from the mass toward the point of equilibrium.

b) At any time t the magnitude of the force exerted on the mass is proportional to the difference between the length l of the spring and its equilibrium length l_0. The positive constant of proportionality λ is called the *spring constant*, and the principle above is known as Hooke's Law.

Newton's second law of motion states that the force F acting on this particle moving with varying velocity v is equal to the time rate of change of the momentum

mv:

$$F = \frac{d(mv)}{dt} = ma.$$

Equating the two forces and applying Hooke's law, we have

$$\frac{d^2x}{dt^2} = -\lambda x, \tag{3.51}$$

where $x(t)$ denotes the displacement from equilibrium of the spring and is positive when the spring is stretched. The negative sign in Eq. (3.51) is due to the fact that the force always acts toward the equilibrium position and therefore is in the negative direction when x is positive.

Note that we have assumed that all other forces acting on the spring (such as friction, air resistance, etc.) can be ignored. Equation (3.51) yields the initial value problem

$$\frac{d^2x}{dt^2} + \frac{\lambda}{m}x = 0, \qquad x(0) = x_0, \qquad x'(0) = v_0. \tag{3.52}$$

To find the solution of (3.52), we note that the auxiliary equation has the complex roots $\pm i\sqrt{\lambda/m}$, leading to the general solution

$$x(t) = A \cos \sqrt{\lambda/m}\, t + B \sin \sqrt{\lambda/m}\, t,$$

which has the form of Eq. (3.33). The motion of the mass is called *simple harmonic motion*. Using the initial conditions, we find that $A = x_0$ and $B = v_0/\sqrt{\lambda/m}$. Thus the solution of (3.52) is given by

$$x(t) = x_0 \cos \sqrt{\lambda/m}\, t + (v_0/\sqrt{\lambda/m}) \sin \sqrt{\lambda/m}\, t. \tag{3.53}$$

Using the formula $\sin (A + B) = \sin A \cos B + \cos A \sin B$ and letting $a = \sqrt{x_0^2 + mv_0^2/\lambda}$, $\tan \alpha = x_0\sqrt{\lambda/m}/v_0$, we can rewrite Eq. (3.53) as

$$x(t) = a \sin (\sqrt{\lambda/m}\, t + \alpha). \tag{3.54}$$

From this equation it is clear that the mass oscillates between the extreme positions $\pm a$; a is called the *amplitude* of the motion. Since the sine term has period $2\pi\sqrt{m/\lambda}$, this is the time required for each complete oscillation. The *frequency* f of the motion is the number of complete oscillations per unit time:

$$f = \frac{\sqrt{\lambda/m}}{2\pi}. \tag{3.55}$$

Note that while the amplitude depends on the initial conditions, the frequency does not.

Example 3.9.1 Consider a spring fixed at its upper end and supporting a weight of 10 pounds at its lower end. Suppose the 10-pound weight stretches the spring by 6 inches. Find the equation of motion of the weight if it is drawn to a position 4 inches below its equilibrium position and released.

*Solution.** By Hooke's law, since a force of 10 lb. stretches the spring by $\frac{1}{2}$ ft, $10 = \lambda(\frac{1}{2})$ or $\lambda = 20$ (lb./ft). We are given the initial values $x_0 = \frac{1}{3}$ (ft) and $v_0 = 0$, so by (3.53) and the identity $\lambda/m = g\lambda/\omega = 64$ we obtain

$$x(t) = \tfrac{1}{3} \cos 8t.$$

Thus the amplitude is $\frac{1}{3}$ ft ($= 4$ in.), and the frequency is $f = 4/\pi$ oscillations per second.

Damped Vibrations Throughout the above discussion we made the assumption that there were no external forces acting on the spring. This assumption, however, is not very realistic. To take care of such things as friction in the spring and air resistance, we now assume that there is a *damping* force (that tends to slow things down) which can be thought of as the resultant of all external forces (except gravity) acting on the spring. It is reasonable to assume that the magnitude of the damping force is proportional to the velocity of the particle (for example, the slower the movement, the smaller the air resistance). Therefore, to Eq. (3.51) we add the term $\mu(dx/dt)$, where μ is the damping constant that depends on all external factors. This constant could be determined experimentally. The equation of motion then becomes

$$\frac{d^2x}{dt^2} = -\frac{\lambda}{m}x - \frac{\mu}{m}\frac{dx}{dt}, \qquad x(0) = x_0, \qquad x'(0) = v_0. \qquad (3.56)$$

[Of course, since μ depends on external factors, it may very well not be a constant at all but may vary with time and position. In that case μ is really $\mu(t,x)$, and the equation becomes much harder to analyze than the constant coefficient case.]

To study Eq. (3.56), we first find the roots of the auxiliary equation:

$$\frac{-\mu \pm \sqrt{\mu^2 - 4m\lambda}}{2m}. \qquad (3.57)$$

The nature of the general solution will depend on the discriminant $\sqrt{\mu^2 - 4m\lambda}$. If $\mu^2 > 4m\lambda$, then both roots will be negative since $\sqrt{\mu^2 - 4m\lambda} < \mu$. So in this case

$$x(t) = c_1 \exp\left(\frac{-\mu + \sqrt{\mu^2 - 4m\lambda}}{2m}t\right)$$

$$+ c_2 \exp\left(\frac{-\mu - \sqrt{\mu^2 - 4m\lambda}}{2m}t\right) \qquad (3.58)$$

---

* The identity $m = w/g$ may be used to convert weight to mass. Keep in mind that pounds are a unit of weight (force) while kilograms are a unit of mass.

will become small as t becomes large whatever the initial conditions may be. Similarly, in the event the discriminant vanishes, then

$$x(t) = e^{(-\mu/2m)t}(c_1 + c_2 t), \tag{3.59}$$

and the solution has a similar behavior. For example, if $\mu = 5$ (lb.-sec/ft) in Example 3.9.1, then the discriminant vanishes and

$$x(t) = e^{-8t}(c_1 + c_2 t).$$

Applying the initial conditions yields

$$x(t) = \tfrac{1}{3} e^{-8t}(1 + 8t),$$

which has the graph shown in Fig. 3.3. We observe that the solution does not oscillate. This type of motion can take place in a viscous medium (such as oil or water).

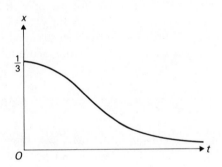

Figure 3.3

If $\mu^2 < 4m\lambda$, then the general solution is

$$x(t) = e^{(-\mu/2m)t}\left(c_1 \cos \frac{\sqrt{4m\lambda - \mu^2}}{2m} t + c_2 \sin \frac{\sqrt{4m\lambda - \mu^2}}{2m} t\right), \tag{3.60}$$

which shows an oscillation with frequency

$$f = \frac{\sqrt{4m\lambda - \mu^2}}{4\pi m}.$$

The factor $e^{(-\mu/2m)t}$ is called the *damping factor*. Letting $\mu = 4$ (lb.-sec/ft) in Example 3.9.1 leads to the general solution

$$x(t) = e^{-32t/5}(c_1 \cos \tfrac{24}{5} t + c_2 \sin \tfrac{24}{5} t).$$

Using the initial values, we find that $c_1 = \frac{1}{3}$, $c_2 = \frac{4}{9}$, and the motion is illustrated in Fig. 3.4.

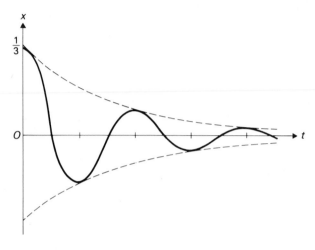

Figure 3.4

Forced Vibrations The motion of the mass considered in the two cases above is determined by the inherent forces of the spring-weight system and the natural forces acting on the system. Accordingly, the vibrations are called *free* or *natural* vibrations. We will now assume that the system is also subject to an external periodic force $k \sin \omega t$, perhaps due to the motion of the object to which the upper end of the spring is attached (see Fig. 3.2). In this case the mass will undergo *forced vibrations*.

Equation (3.56) may be replaced by the nonhomogeneous second order differential equation

$$m \frac{d^2x}{dt^2} = -\lambda x - \mu \frac{dx}{dt} + k \sin \omega t,$$

which we write in the form

$$\frac{d^2x}{dt^2} + \frac{\mu}{m} \frac{dx}{dt} + \frac{\lambda}{m} x = \frac{k}{m} \sin \omega t. \tag{3.61}$$

By the method of undetermined coefficients, we know that $x(t)$ has a particular solution of the form

$$x_p(t) = A_1 \cos \omega t + A_2 \sin \omega t. \tag{3.62}$$

Substituting this function into (3.61) yields the simultaneous equations

$$\left(\frac{\lambda}{m} - \omega^2\right) A_1 + \frac{\mu\omega}{m} A_2 = 0,$$

$$-\frac{\mu\omega}{m} A_1 + \left(\frac{\lambda}{m} - \omega^2\right) A_2 = \frac{k}{m}, \tag{3.63}$$

from which we obtain

$$A_1 = \frac{-k\mu\omega}{(\lambda - m\omega^2)^2 + (\mu\omega)^2}, \qquad A_2 = \frac{k(\lambda - m\omega^2)}{(\lambda - m\omega^2)^2 + (\mu\omega)^2}.$$

Using the same method we used to obtain Eq. (3.54), we have

$$x_p(t) = a \sin(\omega t + \alpha), \tag{3.64}$$

where

$$a = \frac{k}{\sqrt{(\lambda - m\omega^2)^2 + (\mu\omega)^2}}, \qquad \tan \alpha = \frac{\mu\omega}{m\omega^2 - \lambda}.$$

To find the general solution of Eq. (3.61), we add the general solution of the homogeneous equation to the particular solution (3.64). The nature of the general solution again depends on the discriminant $\sqrt{\mu^2 - 4m\lambda}$. If $\mu^2 \geq 4m\lambda$, we are superimposing the periodic function (3.64) on either Eq. (3.58) or (3.59). Since (3.58) and (3.59) damp out as t increases, the motion for large t will be very close to (3.64). On the other hand, if $\mu^2 < 4m\lambda$, we will suppose initially that $\mu = 0$. Then the general solution, for $\sqrt{\lambda/m} \neq \omega$, is

$$x(t) = c_1 \cos \sqrt{\lambda/m}\, t + c_2 \sin \sqrt{\lambda/m}\, t + \frac{k}{m(\lambda/m - \omega^2)} \sin \omega t. \tag{3.65}$$

Note that the effect of the sinusoidal force $k \sin \omega t$ is merely to superimpose itself on the simple harmonic motion of the system. The period of this added motion is the same as that of the external force, with constant amplitude for each fixed ω, but the amplitude approaches infinity as ω approaches $\sqrt{\lambda/m}$. If $\omega = \sqrt{\lambda/m}$, then the particular solution must have the form

$$x_p(t) = A_1 t \cos \omega t + A_2 t \sin \omega t. \tag{3.66}$$

Substituting (3.66) into (3.52) yields $A_1 = -k/2\omega$, $A_2 = 0$, so we have the general solution

$$x(t) = c_1 \cos \omega t + c_2 \sin \omega t - \frac{k}{2\omega} t \cos \omega t. \tag{3.67}$$

We note that as t increases, the vibrations will get larger and larger without bound. Of course, this situation cannot arise in practice, since some resistance to the motion will always be present ($\mu \neq 0$). However, it is evident that if the resistance is small and the external forcing motion agrees with the natural vibration, then the displacement x may become so large that the elastic limit of the spring will be exceeded and a permanent set or distortion will occur. If $\omega = \sqrt{\lambda/m}$, we say the external force is in *resonance* with the vibrating mass. The phenomenon of resonance is of great importance in engineering. It is necessary in some cases, as with our spring problem, to avoid a resonant condition so that no undue stresses will occur; on the other hand, resonance is desirable in many radio circuit problems.

EXERCISES 3.9∗

1. One end of a rubber band is fixed at a point A. A 1-kg mass, attached to the other end, stretches the rubber band vertically to the point B in such a way that the length AB is 16 cm greater than the natural length of the band. If the mass is further drawn to a position 8 cm below B and released, what will be its velocity (if we neglect resistance) as it passes the position B?

2. If in Exercise 1 the mass is released at a position 8 cm above B, what will be its velocity as it passes 1 cm above B?

3. A cylindrical block of wood of radius and height 1 ft and weighing 12.48 lb. floats with its axis vertical in water (62.4 lb. per ft^3). If it is depressed so that the surface of the water is tangent to the block, and is then released, what will be its period of vibration and equation of motion? Neglect resistance. [*Hint*: The upward force on the block is equal to the weight of the water displaced by the block.]

4. A cubical block of wood, 1 ft on a side, is depressed so that its upper face lies along the surface of the water, and is then released. The period of vibration is found to be 1 sec. Neglecting resistance, what is the weight of the block of wood?

5. A 10-g mass suspended from a spring vibrates freely, the resistance being numerically equal to half the velocity (in m/sec) at any instant. If the period of the motion is 8 sec, what is the spring constant (in g/sec^2)?

6. A weight w (lb) is suspended from a spring whose constant is 10 lb./ft. The motion of the weight is subject to a resistance (lb.) numerically equal to half the velocity (ft/sec). If the motion is to have a 1-sec period, what are the possible values of w?

7. A 50-g mass is hanging at rest on a spring which is stretched 10 cm by the weight. The upper end of the spring is given the periodic force $10 \sin 2t$ g·m/sec^2 and the resistance has

---

∗ See footnote on page 102.

a magnitude $100\sqrt{17}$ (g/sec) times the velocity in meters per second. Find the equation of motion of the weight.

8. Suppose in Exercise 7 that the upper end of the spring is given, instead, the force $1 - e^{-\sqrt{g}t}$ gm/sec^2. Find the displacement of the mass after 1 sec.

3.10 ELECTRIC CIRCUITS

We shall make use of the concepts developed in Section 2.8 and the methods of this chapter to study a simple electric circuit containing a resistor, an inductor, and a capacitor in series with an electromotive force (Fig. 3.5). Suppose that R, L, C, and E are constants. Applying Kirchhoff's law, we obtain

$$L\frac{dI}{dt} + RI + \frac{Q}{C} = E. \tag{3.68}$$

Since $dQ/dt = I$, we differentiate Eq. (3.68) to get the second order homogeneous differential equation

$$L\frac{d^2I}{dt^2} + R\frac{dI}{dt} + \frac{I}{C} = 0. \tag{3.69}$$

To solve this equation, we write the auxiliary equation

$$\lambda^2 + \frac{R}{L}\lambda + \frac{1}{CL} = 0, \tag{3.70}$$

with the following roots:

$$\lambda_1 = \frac{-R + \sqrt{R^2 - 4L/C}}{2L}, \qquad \lambda_2 = \frac{-R - \sqrt{R^2 - 4L/C}}{2L}. \tag{3.71}$$

Figure 3.5

Example 3.10.1 Let $L = 1$ henry (h), $R = 100$ ohms (Ω), $C = 10^{-4}$ farads (f), and $E = 1000$ volts (v) in the circuit shown in Fig. 3.5. Suppose that no charges are present and no current is flowing at time $t = 0$ when E is applied. By Eq. (3.71) we see that the auxiliary equation has the roots $\lambda_1 = -50 + 50\sqrt{3}i$ and $\lambda_2 = -50 - 50\sqrt{3}i$. Thus

$$I(t) = e^{-50t}(c_1 \cos 50\sqrt{3}t + c_2 \sin 50\sqrt{3}t).$$

Applying the initial condition $I(0) = 0$, we have $c_1 = 0$. Hence

$$I(t) = c_2 e^{-50t} \sin 50\sqrt{3}t. \tag{3.72}$$

To establish the value of c_2, we must make use of Eq. (3.68) and the initial condition $Q(0) = 0$. Since

$$Q(t) = C\left(E - L\frac{dI}{dt} - RI\right)$$

$$= \frac{1}{10} - \frac{c_2}{200} e^{-50t}(\sin 50\sqrt{3}t + \sqrt{3}\cos 50\sqrt{3}t),$$

it follows that for $t = 0$, $c_2 = 20/\sqrt{3}$. Thus

$$Q(t) = \frac{1}{10} - \frac{1}{10\sqrt{3}} e^{-50t}(\sin 50\sqrt{3}t + \sqrt{3}\cos 50\sqrt{3}t)$$

and

$$I(t) = \frac{20}{\sqrt{3}} e^{-50t} \sin 50\sqrt{3}t.$$

From these equations we observe that the current will rapidly damp out and that the charge will rapidly approach its steady-state value of $\frac{1}{10}$ coulomb (coul).

Example 3.10.2 Let the inductance, resistance, and capacitance in Example 3.10.1 remain the same, but suppose $E = 1000 \sin 60t$. By Eq. (3.68) we have

$$\frac{dI}{dt} + 100I + 10^4 Q = 1000 \sin 60t, \tag{3.73}$$

and converting (3.73) so that all expressions are in terms of $Q(t)$, we obtain

$$\frac{d^2Q}{dt^2} + 100\frac{dQ}{dt} + 10^4 Q = 1000 \sin 60t. \tag{3.74}$$

It is evident that Eq. (3.74) has a particular solution of the form

$$Q_p(t) = A_1 \sin 60t + A_2 \cos 60t. \tag{3.75}$$

To determine the values A_1 and A_2, we substitute (3.75) into Eq. (3.73), obtaining the simultaneous equations

$$6400A_1 - 6000A_2 = 1000,$$
$$6000A_1 + 6400A_2 = 0.$$

Thus $A_1 = \frac{40}{481}$, $A_2 = -\frac{75}{962}$, and since the general solution of the homogeneous equation is the same as that of (3.69), the general solution of Eq. (3.68) is

$$Q(t) = e^{-50t}(c_1 \cos 50\sqrt{3}t + c_2 \sin 50\sqrt{3}t)$$
$$+ \frac{40}{481} \sin 60t - \frac{75}{962} \cos 60t. \qquad (3.76)$$

Differentiating Eq. (3.76), we obtain

$$I(t) = 50e^{-50t}[(\sqrt{3}c_2 - c_1) \cos 50\sqrt{3}t - (c_2 + \sqrt{3}c_1) \sin 50\sqrt{3}t]$$
$$+ \frac{2400}{481} \cos 60t + \frac{2250}{481} \sin 60t. \qquad (3.77)$$

Setting $t = 0$ and using the initial conditions, we obtain the simultaneous equations

$$c_1 = \frac{75}{962},$$
$$50(\sqrt{3}c_2 - c_1) = -\frac{2400}{481}.$$

Therefore, $c_1 = \frac{75}{962}$ and $c_2 = -7\sqrt{3}/962$, so that

$$Q(t) = \frac{e^{-50t}}{962}(75 \cos 50\sqrt{3}t - 7\sqrt{3} \sin 50\sqrt{3}t) + \frac{80 \sin 60t - 75 \cos 60t}{962},$$

$$I(t) = \frac{-e^{-50t}}{481}(2400 \cos 50\sqrt{3}t + 1700\sqrt{3} \sin 50\sqrt{3}t)$$
$$+ \frac{2400 \cos 60t + 2250 \sin 60t}{481}.$$

EXERCISES 3.10

1. In Example 3.10.1, let $L = 2$ h, $R = 50 \, \Omega$, $C = 3 \times 10^{-4}$ f, and $E = 2000$ v. With the same assumptions, calculate the current and charge for all values of $t \geq 0$.

2. In Example 3.10.2, suppose instead that $E = 50 \cos 30t$. Find $Q(t)$ for $t \geq 0$.

3. Given that $L = 1$ h, $R = 1200 \, \Omega$, $C = 10^{-6}$ f, $I(0) = Q(0) = 0$, and $E = 100 \sin 600t$ volts, determine whether the transient current or the steady-state current is numerically the larger when $t = 0.001$ sec.

4. Find the ratio of the current flowing in the circuit of Exercise 3 to that which would be flowing if there were no resistance, at $t = 0.001$ sec.

5. Consider the system governed by Eq. (3.68) for the case where the resistance is zero and $E = E_0 \sin \omega t$. Show that the solution consists of two parts, a general solution with frequency $1/\sqrt{LC}$ and a particular solution with frequency ω. The frequency $1/\sqrt{LC}$ is called the *natural* frequency of the circuit. Note that if $\omega = 1/\sqrt{LC}$, then the particular solution disappears.

6. To allow for different variations of the voltage, let us assume in Eq. (3.68) that $E = E_0 e^{it}$ ($= E_0 \cos t + iE_0 \sin t$). Assume also, as in Exercise 5, that $R = 0$. Finally, for simplicity assume that $E_0 = L = C = 1$. Then $1 = \omega = 1/\sqrt{LC}$.

a) Show that Eq. (3.69) becomes

$$\frac{d^2 I}{dt^2} + I = e^{it}.$$

b) Determine λ such that $I(t) = \lambda t e^{it}$ is a solution.

c) Calculate the general solution and show that the magnitude of the current increases without bound as t increases. This phenomenon will produce *resonance*.

7. Let an inductance of L henries, a resistance of R ohms, and a capacitance of C farads be connected in series with an emf of $E_0 \sin \omega t$ volts. Suppose $Q(0) = I(0) = 0$, and $4L > R^2 C$.

   a) Find the expressions for $Q(t)$ and $I(t)$.

   b) What value of ω will produce resonance?

8. Solve Exercise 7 for $4L = R^2 C$.

9. Solve Exercise 7 for $4L < R^2 C$.

3.11 COMPARTMENTAL ANALYSIS II

We now show how the methods in this chapter may be used to solve some two-compartment systems.

Example 3.11.1 Consider two tanks x and y. Suppose that tank x contains 100 gallons of brine in which 100 pounds of salt are dissolved, while tank y contains 100 gallons of water. Water flows into tank x at the rate of 3 gallons per minute, and the mixture flows from x into y at the rate of 4 gallons per minute. From y, one gallon per minute is pumped back to x, while 3 gallons per minute are removed from the system (see Fig. 3.6). Find the maximum amount of salt in tank y.

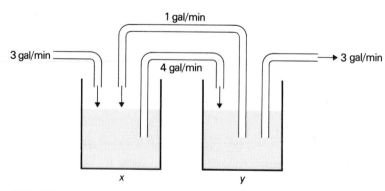

Figure 3.6

This problem can be described by the open two-compartment system in Fig. 3.7, where the fractional transfer coefficients (see Section 2.10) are given by $k_{12} = \frac{4}{100}$, $k_{21} = \frac{1}{100}$, and $k = \frac{3}{100}$. Let $x(t)$ and $y(t)$ denote the amount of salt in tanks x and y, respectively, at time t. Then we can express the rate at which the amount of salt in each tank is changing by the pair of differential equations

$$\frac{dx}{dt} = k_{21}y - k_{12}x, \tag{3.78}$$

$$\frac{dy}{dt} = k_{12}x - (k_{21} + k)y. \tag{3.79}$$

(This is a *system* of equations that will be discussed in much greater generality in Chapter 7.) Note that $i(t) = 0$, since only water is being put in. If we differentiate Eq. (3.79) and replace the resulting dx/dt by Eq. (3.78), we will obtain

$$\frac{d^2y}{dt^2} = k_{12}(k_{21}y - k_{12}x) - (k_{21} + k)\frac{dy}{dt}. \tag{3.80}$$

Replacing the $k_{12}x$ in Eq. (3.80) by its value in Eq. (3.79), we obtain the homogeneous second order equation

$$\frac{d^2y}{dt^2} + (k_{12} + k_{21} + k)\frac{dy}{dt} + kk_{12}y = 0. \tag{3.81}$$

The roots of the auxiliary equation are $-\frac{1}{50}$ and $-\frac{3}{50}$, so that the general solution of Eq. (3.81) is

$$y(t) = c_1 e^{-t/50} + c_2 e^{-3t/50}. \tag{3.82}$$

To find c_1 and c_2, we must obtain two initial conditions. We know that $y(0) = 0$, and by (3.79)

$$y'(0) = k_{12}x(0) - (k_{21} + k)y(0) = 4.$$

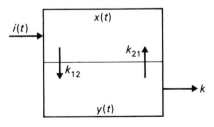

$i(t)$

$x(t)$

k_{21}

k_{12}

k

$y(t)$

Figure 3.7

Using (3.82) and these conditions, we obtain the simultaneous equations

$$c_1 + c_2 = 0,$$

$$-\frac{c_1}{50} - \frac{3c_2}{50} = 4.$$

Thus $c_1 = 100$ and $c_2 = -100$. Hence the amount of salt in tank y at any time t is

$$y(t) = 100(e^{-t/50} - e^{-3t/50}). \tag{3.83}$$

To find the time at which this amount is maximal, we differentiate (3.83):

$$y'(t) = 6e^{-3t/50} - 2e^{-t/50} = 0. \tag{3.84}$$

Solving for t, we find $e^{t/25} = 3$, so that $t = 25 \ln 3$. It is easy to see that a maximum occurs at $t = 25 \ln 3$ (minutes) when tank y contains

$$y(25 \ln 3) = 100(e^{-(\ln 3)/2} - e^{-(3 \ln 3)/2}) = \frac{200}{3\sqrt{3}} \text{ lb.}$$

of salt. It is easy to obtain an expression for $x(t)$ by substituting (3.83) and (3.84) into Eq. (3.79):

$$x(t) = 50(e^{-t/50} + e^{-3t/50}).$$

Example 3.11.2. Chemical kinetics Suppose that a chemical bimolecular reaction is described by the equation

$$A + B \underset{k_2}{\overset{k_1}{\rightleftarrows}} AB. \tag{3.85}$$

Let x_1, x_2, x_3 be the concentrations of the compounds A, B, and AB, respectively. Two basic processes occur in such a system: a forward reaction of rate r_f and a backward reaction of rate r_b. The backward reaction in (3.85) is a *first order reaction*, since it depends only on the concentration of the compound AB; that is, $r_b = k_2 x_3$. The forward reaction requires the collision of the molecules of A with the molecules of B, thus is proportional to the product of the concentrations of A and B. Hence $r_f = k_1 x_1 x_2$, and we say that r_f is a *second order reaction*. The constants k_1 and k_2 are called the *specific reaction rates* and indicate the rate at which the reaction proceeds (assuming constant temperature). Since the concentrations of A and B are changing at the same rate, the rate of change of the concentrations of A, B, and AB can be described by the differential equations

$$\frac{dx_1}{dt} = \frac{dx_2}{dt} = r_b - r_f = k_2 x_3 - k_1 x_1 x_2,$$

$$\frac{dx_3}{dt} = r_f - r_b = k_1 x_1 x_2 - k_2 x_3.$$

We shall study a simpler reaction, namely the dimerization

$$A + A \underset{k_2}{\overset{k_1}{\rightleftharpoons}} A_2. \tag{3.86}$$

Assume that the reaction takes place in a closed container of fixed volume V. Let x be the concentration of A, and y be the concentration of A_2. Then the rates of change of the concentrations are given by

$$\frac{dx}{dt} = 2k_2 y - 2k_1 x^2, \tag{3.87}$$

$$\frac{dy}{dt} = k_1 x^2 - k_2 y, \tag{3.88}$$

where the twos in Eq. (3.87) denote the fact that two As are produced by the decomposition of one A_2. We can describe this situation by the closed three-compartment system below (Fig. 3.8).

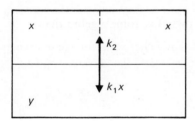

Figure 3.8

Proceeding as we did in Example 3.11.1, we differentiate Eq. (3.87) and apply (3.87) and (3.88) to obtain a second order equation in terms of x alone:

$$x'' = 2k_2(k_1 x^2 - k_2 y) - 4k_1 x x' = -k_2 x' - 4k_1 x x'$$

or

$$\frac{d^2 x}{dt^2} + \frac{dx}{dt}(4k_1 x + k_2) = 0. \tag{3.89}$$

This is a nonlinear second order equation, but the substitutions

$$p = \frac{dx}{dt}, \qquad p\frac{dp}{dx} = \frac{d^2 x}{dt^2}$$

will transform it into the equation

$$p\left(\frac{dp}{dx} + 4k_1x + k_2\right) = 0. \tag{3.90}$$

Either $p = 0$, in which case we also have $dy/dt = 0$, and the reaction is in equilibrium, or $dp = -(4k_1x + k_2)\, dx$, which implies

$$\frac{dx}{dt} = c_1 - k_2x - 2k_1x^2. \tag{3.91}$$

Comparing Eqs. (3.91) and (3.87), we see that

$$2k_2y = -k_2x + c_1, \tag{3.92}$$

which will fix c_1 if the initial concentrations $x(0)$ and $y(0)$ are known. We observe that since the slope of the line (3.92) is negative, if the concentration of A_2 increases, that of A will decrease, and vice versa. Finally, the variables in Eq. (3.91) can be separated to yield

$$\frac{1}{a} \ln \frac{4k_1x + k_2 + a}{4k_1x + k_2 - a} = t + c_0,$$

where $a = \sqrt{k_2^2 + 8c_1k_1}$, from which it follows after some algebra that

$$x(t) = \frac{a}{4k_1}\left[\frac{c_0e^{at} + 1}{c_0e^{at} - 1} - \frac{k_2}{4k_1}\right], \tag{3.93}$$

where

$$c_0 = \frac{4k_1x(0) + k_2 + a}{4k_1x(0) + k_2 - a}.$$

From (3.93) we see that as t increases, x tends to $(a - k_2)/4k_1$. A solution $y(t)$ can be easily obtained by inserting (3.93) into Eq. (3.87).

It should be noted that the predator-prey model of Example 1.1.5 may also be described by a compartmental system.

EXERCISES 3.11

1. Tank x contains 500 gal of brine in which 500 lb. of salt are dissolved. Tank y contains 500 gal of water. Water flows into tank x at the rate of 30 gal/min, and the mixture flows into y at the rate of 40 gal/min. From y the solution is pumped back into x at the rate of 10 gal/min and into a third tank at the rate of 30 gal/min. Find the maximum amount of salt in y. When does this concentration occur?

2. Suppose in Exercise 1 that tank x has a leak which removes 1 gal/min of solution. Solve the problem, given that all other conditions are unchanged.

3. Solve the bimolecular reaction equation (3.85), given that the concentrations of A and B are the same. Describe this situation as a two-compartment system.

4. Describe the predator-prey model of Example 1.1.5 as a two-compartment system.

5. In a study concerning the distribution of radioactive potassium K^{42} between red blood cells and the plasma of the human blood, C. W. Sheppard and W. R. Martin [*J. Gen. Physiol*, **33** (1950), 703–722] added K^{42} to freshly drawn blood. They discovered that although the total amount of potassium (stable and radioactive) in the red cells and in the plasma remained practically constant during the experiment, the radioactivity was gradually transmitted from the plasma to the red cells. Thus the behavior of the radioactivity is that of a linear closed two-compartment system. If the fractional transfer coefficient from the plasma to the cells is $k_{12} = 30.1$ percent per hour, while $k_{21} = 1.7$ percent per hour, and the initial radioactivity was 800 counts per minute in the plasma and 25 counts per minute in the red cells, what is the number of counts per minute in the red cells after 300 minutes?

3.12 HIGHER ORDER LINEAR EQUATIONS

In this section we will show how most of the results of the preceding sections can be immediately generalized for application to equations of order higher than two. We shall begin by considering the nth order linear differential equation

$$y^{(n)} + a_1(x)y^{(n-1)} + a_2(x)y^{(n-2)} + \cdots + a_{n-1}(x)y' + a_n(x)y = f(x) \qquad (3.94)$$

and the associated homogeneous equation

$$y^{(n)} + a_1(x)y^{(n-1)} + a_2(x)y^{(n-2)} + \cdots + a_{n-1}(x)y' + a_n(x)y = 0, \qquad (3.95)$$

where the functions $a_i(x)$, $i = 1, 2, \ldots, n$, and $f(x)$ are assumed to be continuous. The theory of Section 3.2 carries over immediately for both Eqs. (3.94) and (3.95). We say that $y_1, y_2, \ldots, y_n$ are *linearly independent* over an interval $x_0 \leqslant x \leqslant x_1$ if the condition $c_1 y_1(x) + c_2 y_2(x) + \cdots + c_n y_n(x) = 0$, for all x so that $x_0 \leqslant x \leqslant x_1$, implies that $c_1 = c_2 \cdots = c_n = 0$. We define the *Wronskian* $W(y_1, y_2, \ldots, y_n)(x)$ by the determinant

$$W(y_1, y_2, \ldots, y_n)(x) = \begin{vmatrix} y_1 & y_2 & \cdots & y_n \\ y_1' & y_2' & \cdots & y_n' \\ \cdot & & & \\ \cdot & & & \\ \cdot & & & \\ y_1^{(n-1)} & & \cdots & y_n^{(n-1)} \end{vmatrix}.$$

It can be proved, in a manner analogous to the proof of Lemma 3.3, that $W(y_1, y_2, \ldots, y_n)(x) = 0$ if and only if the functions $y_1, y_2, \ldots, y_n$ are linearly dependent solutions of Eq. (3.95). We can also prove the generalization of Eq. (3.6), known as *Abel's formula*:

$$W(y_1, \ldots, y_n)(x) = ce^{-\int a_1(x)\,dx}. \tag{3.96}$$

Furthermore, as in Theorem 3.4, we can show that if $y_1, y_2, \ldots, y_n$ are n linearly independent solutions of the homogeneous equation (3.95), then any other solution $y(x)$ of (3.95) can be written in the form

$$y(x) = c_1 y_1(x) + c_2 y_2(x) + \cdots + c_n y_n(x), \tag{3.97}$$

where $c_1, c_2, \ldots, c_n$ are constants. Thus (3.97) may be called the *general solution* of Eq. (3.95), and the problem is reduced to finding n linearly independent solutions to (3.95). Finally, if y_{p_1} and y_{p_2} are two solutions of the nonhomogeneous equation (3.94), then by the principle of superposition, it follows that $y(x) = y_{p_1}(x) - y_{p_2}(x)$ is a solution of the homogeneous equation (3.95). Hence, as in the second order case, it is necessary to find only one particular solution of Eq. (3.94). The proofs of all these statements are immediate generalizations of the results in Section 3.2 and are left as exercises. The remainder of this section is concerned with finding solutions of higher order equations.

As with the case of second order equations, there is no general method for obtaining a closed-form solution of Eq. (3.94) or (3.95) unless the functions $a_i(x)$, $i = 1, 2, \ldots, n$, are all constants. For this reason we shall concentrate on finding the solution of nth order linear equations with constant (real) coefficients

$$y^{(n)} + a_1 y^{(n-1)} + \cdots + a_{n-1} y' + a_n y = 0. \tag{3.98}$$

As in the second order case, we "guess" that there is a solution of the form

$$y = e^{\lambda x}. \tag{3.99}$$

Substituting (3.99) into Eq. (3.98) and noting that the kth derivative of $e^{\lambda x}$ is $\lambda^k e^{\lambda x}$, we obtain

$$e^{\lambda x}(\lambda^n + a_1 \lambda^{n-1} + \cdots + a_{n-1}\lambda + a_n) = 0$$

or, after division by $e^{\lambda x}$,

$$P(\lambda) = \lambda^n + a_1 \lambda^{n-1} + \cdots + a_{n-1}\lambda + a_n = 0. \tag{3.100}$$

The polynomial (3.100) is called, as expected, the *auxiliary equation* for the homogeneous equation (3.98), and it is evident that if λ is a root of (3.100), then $e^{\lambda x}$ is a solution to Eq. (3.98).

The polynomial $P(\lambda)$ can be factored:

$$P(\lambda) = (\lambda - \lambda_1)^{n_1}(\lambda - \lambda_2)^{n_2} \cdots (\lambda - \lambda_k)^{n_k},$$

where n_j is the *multiplicity* of the root λ_j and $n_1 + n_2 + \cdots + n_k = n$. If $n_j = 1$ for a given j, then the root λ_j is called a *simple root* of (3.100). Since we are assuming that the coefficients a_j, $j = 1, \ldots, n$, are real, any complex roots of (3.100) will appear in conjugate pairs.

Let us now summarize the nature of the solutions of Eq. (3.98) corresponding to different types of roots of (3.100).

i) Let $\lambda_1, \lambda_2, \ldots, \lambda_m$ be m real simple roots of (3.100). Then

$$e^{\lambda_1 x}, e^{\lambda_2 x}, \ldots, e^{\lambda_m x}$$

are m linearly independent solutions of Eq. (3.98).

ii) If λ_j is a root of (3.100) of multiplicity $n_j > 1$, then

$$e^{\lambda_j x}, xe^{\lambda_j x}, x^2 e^{\lambda_j x}, \ldots, x^{n_j - 1} e^{\lambda_j x}$$

are n_j linearly independent solutions of Eq. (3.98).

iii) If $\lambda_j = \alpha + i\beta$ and $\lambda_k = \alpha - i\beta$ are a pair of simple complex conjugate roots of (3.100), then

$$e^{\alpha x} \cos \beta x \qquad \text{and} \qquad e^{\alpha x} \sin \beta x$$

are two linearly independent solutions of Eq. (3.98).

iv) If $\lambda_j = \alpha + i\beta$ and $\lambda_k = \alpha - i\beta$ are a pair of complex conjugate roots of multiplicity $n_j = n_k > 1$, then

$$e^{\alpha x} \cos \beta x, \ e^{\alpha x} \sin \beta x, \ xe^{\alpha x} \cos \beta x, \ xe^{\alpha x} \sin \beta x,$$
$$x^2 e^{\alpha x} \cos \beta x, x^2 e^{\alpha x} \sin \beta x, \ldots, x^{n_j - 1} e^{\alpha x} \cos \beta x, x^{n_j - 1} e^{\alpha x} \sin \beta x$$

are $2n_j$ linearly independent solutions of Eq. (3.98).

The proofs of these statements are obvious generalizations of the results of Sections 3.4 and 3.5. We shall illustrate these results with a number of examples and leave the proofs as exercises.

Example 3.12.1 Let

$$y''' + 4y'' + y' - 6y = 0.$$

The auxiliary equation is

$$\lambda^3 + 4\lambda^2 + \lambda - 6 = 0,$$

which has three real distinct roots $\lambda_1 = 1$, $\lambda_2 = -2$, $\lambda_3 = -3$. The general solution is therefore

$$y(x) = c_1 e^x + c_2 e^{-2x} + c_3 e^{-3x}.$$

Before continuing with our discussion, we should mention that finding roots of polynomials of degree greater than two is generally a tedious task. The examples given here all have easily calculable roots, but the reader should not be left with the impression that this is usually the case. In the case of most higher order polynomials the roots can only be approximated, and many methods for performing such approximations are known. The interested reader should consult any of a number of books on numerical analysis.*

Example 3.12.2 Consider the equation

$$y''' - 3y' - 2y = 0,$$

with the auxiliary equation

$$\lambda^3 - 3\lambda - 2 = (\lambda + 1)^2(\lambda - 2) = 0.$$

We have a double root $\lambda = -1$ and a simple root $\lambda = 2$. The general solution is thus

$$y(x) = c_1 e^{-x} + c_2 x e^{-x} + c_3 e^{2x}.$$

Example 3.12.3 Consider the equation

$$y^{iv} + 16y''' + 96y'' + 256y' + 256y = 0,$$

with the auxiliary equation

$$\lambda^4 + 16\lambda^3 + 96\lambda^2 + 256\lambda + 256 = (\lambda + 4)^4 = 0.$$

Here $\lambda = -4$ is a root of multiplicity four, and the general solution is

$$y(x) = c_1 e^{-4x} + c_2 x e^{-4x} + c_3 x^2 e^{-4x} + c_4 x^3 e^{-4x}.$$

Example 3.12.4 Let $y''' + y' = 0$. The auxiliary equation is $\lambda^3 + \lambda = 0$, with the roots 0 and $\pm i$. The general solution is, then,

$$y(x) = c_1 + c_2 \cos x + c_3 \sin x.$$

Example 3.12.5 Consider the equation

$$y^{iv} - 16y''' + 42y'' - 104y' + 169y = 0.$$

The auxiliary equation is

$$\lambda^4 - 16\lambda^3 + 42\lambda^2 - 104\lambda + 169 = (\lambda^2 - 4\lambda + 13)^2 = 0,$$

with two double roots, $2 \pm 3i$. The general solution is thus

$$y(x) = c_1 e^{2x} \cos 3x + c_2 e^{2x} \sin 3x + c_3 x e^{2x} \cos 3x + c_4 x e^{2x} \sin 3x.$$

Example 3.12.6 Consider the equation

$$y^{iv} + y = 0.$$

---

* See, for example, Peter Henrici, *Elements of Numerical Analysis*, Wiley, New York, 1964.

The auxiliary equation is $\lambda^4 + 1 = 0$. The four roots here are the four fourth roots of -1, which may be obtained in the following way by means of Euler's formula (3.28). We observe that

$$e^{\pi i} = \cos \pi + i \sin \pi = -1$$

and

$$e^{2\pi i} = \cos 2\pi + i \sin 2\pi = 1.$$

Thus

$$-1 = e^{\pi i} = e^{\pi i}e^{2\pi i} = e^{3\pi i} = e^{5\pi i} = e^{7\pi i},$$

and taking the fourth root of these terms, we have

$$e^{\pi i/4} = \cos \frac{\pi}{4} + i \sin \frac{\pi}{4} = \frac{(1 + i)}{\sqrt{2}},$$

$$e^{3\pi i/4} = \cos \frac{3\pi}{4} + i \sin \frac{3\pi}{4} = \frac{(-1 + i)}{\sqrt{2}},$$

$$e^{5\pi i/4} = \cos \frac{5\pi}{4} + i \sin \frac{5\pi}{4} = \frac{(-1 - i)}{\sqrt{2}},$$

$$e^{7\pi i/4} = \cos \frac{7\pi}{4} + i \sin \frac{7\pi}{4} = \frac{(1 - i)}{\sqrt{2}}.$$

Using the identities (3.31), we may again express the general solution in terms of sines and cosines:

$$y(x) = c_1 e^{x/\sqrt{2}} \cos \frac{x}{\sqrt{2}} + c_2 e^{x/\sqrt{2}} \sin \frac{x}{\sqrt{2}}$$

$$+ c_3 e^{-x/\sqrt{2}} \cos \frac{x}{\sqrt{2}} + c_4 e^{-x/\sqrt{2}} \sin \frac{x}{\sqrt{2}}.$$

This procedure can always be used to find the nth roots of any real (or complex) number.

We can now consider the nonhomogeneous equation with constant coefficients

$$y^{(n)} + a_1 y^{(n-1)} + \cdots + a_{n-1} y' + a_n y = f(x). \tag{3.101}$$

As in Section 3.6, this equation can be solved by the method of undetermined coefficients if $f(x)$ is in the "right form." In addition, it can be solved for any function $f(x)$ by the method of variation of parameters. The method of undetermined coefficients works in exactly the same way as it did for second order equations. See the following examples.

Example 3.12.7 Consider the equation

$$y''' + 4y'' + y' - 6y = x^2. \tag{3.102}$$

As in Section 3.6, we "guess" that since x^2 is a polynomial of degree two, there is a solution to Eq. (3.102) which is also a polynomial of degree two:

$$y_p(x) = ax^2 + bx + c.$$

Then $y_p' = 2ax + b$, $y_p'' = 2a$, and $y_p''' = 0$. Substituting these expressions into (3.102) yields

$$8a + (2ax + b) - 6(ax^2 + bx + c) = x^2.$$

Equating the coefficients of the various powers of x, we obtain the system of equations

$$8a + b - 6c = 0,$$
$$2a - 6b = 0,$$
$$-6a = 1.$$

Hence $a = -\frac{1}{6}, b = -\frac{1}{18}, c = -\frac{25}{108}$, and the particular solution of Eq. (3.102) is

$$y_p(x) = -\frac{1}{6}x^2 - \frac{1}{18}x - \frac{25}{108},$$

which can be checked by substitution.

Example 3.12.8 Let

$$y''' - y'' + 4y' - 4y = \sin x. \tag{3.103}$$

The general solution of the related homogeneous equation is

$$y(x) = c_1 e^x + c_2 \cos 2x + c_3 \sin 2x.$$

To find a particular solution of Eq. (3.103), we set

$$y_p(x) = A \sin x + B \cos x.$$

Then $y_p' = A \cos x - B \sin x$, $y_p'' = -A \sin x - B \cos x$, and $y_p''' = -A \cos x + B \sin x$. Substituting these expressions into Eq. (3.103), we obtain

$$-A \cos x + B \sin x + A \sin x + B \cos x + 4A \cos x - 4B \sin x$$
$$- 4A \sin x - 4B \cos x = \sin x.$$

Combining like terms and equating coefficients yields

$$3A - 3B = 0,$$
$$-3A - 3B = 1,$$

or $A = B = -\frac{1}{6}$ and

$$y_p(x) = -\frac{1}{6}(\sin x + \cos x).$$

Example 3.12.9 Consider the equation

$$y''' - 3y' - 2y = 3e^{-x}. \tag{3.104}$$

In Example 3.12.2 we found that the general solution to the related homogeneous equation is

$$y(x) = c_1 e^{-x} + c_2 x e^{-x} + c_3 e^{2x}.$$

Therefore, both e^{-x} and xe^{-x} are solutions of the homogeneous equation. Following the rule given in Section 3.6, we seek a solution of the form

$$y_p(x) = Ax^2 e^{-x}.$$

Then $y_p' = Ae^{-x}(-x^2 + 2x)$, $y_p'' = Ae^{-x}(x^2 - 4x + 2)$, and $y_p''' = Ae^{-x}(-x^2 + 6x - 6)$. Substitution into Eq. (3.104) yields

$$Ae^{-x}(-x^2 + 6x - 6 + 3x^2 - 6x - 2x^2) = -6Ae^{-x} = 3e^{-x},$$

so that $A = -\frac{1}{2}$. Therefore, $y_p = -\frac{1}{2}x^2 e^{-x}$.

The method of variation of parameters for nth order equations is a tedious but direct generalization of the method for second order equations. Let $y_1, y_2, \ldots, y_n$ be n linearly independent solutions to the homogeneous equation (3.98). We seek a solution to Eq. (3.101) of the form

$$y_p(x) = c_1(x)y_1(x) + c_2(x)y_2(x) + \cdots + c_n(x)y_n(x). \tag{3.105}$$

Differentiating (3.105), we obtain

$$y_p'(x) = (c_1 y_1' + c_2 y_2' + \cdots + c_n y_n') + (c_1' y_1 + c_2' y_2 + \cdots + c_n' y_n).$$

It is convenient (as in the second order case) to set

$$c_1' y_1 + c_2' y_2 + \cdots + c_n' y_n = 0.$$

Then $y_p' = c_1 y_1' + c_2 y_2' + \cdots + c_n y_n'$, and another differentiation yields

$$y_p''(x) = (c_1 y_1'' + c_2 y_2'' + \cdots + c_n y_n'') + (c_1' y_1' + c_2' y_2' + \cdots + c_n' y_n').$$

Now we set

$$c_1' y_1' + c_2' y_2' + \cdots + c_n' y_n' = 0,$$

so that $y_p'' = c_1 y_1'' + c_2 y_2'' + \cdots + c_n y_n''$. Continuing in a like manner, we set

$$c_1' y_1^{(k)} + c_2' y_2^{(k)} + \cdots + c_n' y_n^{(k)} = 0 \qquad \text{for} \qquad k = 0, 1, \ldots, n - 2$$

and obtain

$$y_p^{(k)} = c_1 y_1^{(k)} + c_2 y_2^{(k)} + \cdots + c_n y_n^{(k)} \qquad \text{for} \qquad k = 0, 1, \ldots, n - 1.$$

Finally, since

$$y_p^{(n-1)} = c_1 y_1^{(n-1)} + c_2 y_2^{(n-1)} + \cdots + c_n y_n^{(n-1)},$$

a last differentiation yields

$$y_p^{(n)} = [c_1 y_1^{(n)} + c_2 y_2^{(n)} + \cdots + c_n y_n^{(n)}] + [c_1' y_1^{(n-1)} + c_2' y_2^{(n-1)} + \cdots + c_n' y_n^{(n-1)}].$$

We have now obtained all the derivatives up to the nth derivative of y_p and we may substitute these into Eq. (3.101). Recalling that $y_1, y_2, \ldots, y_n$ are solutions of the homogeneous equation (3.98) and that y_p solves Eq. (3.101), we find that

$$c_1' y_1^{(n-1)} + c_2' y_2^{(n-1)} + \cdots + c_n' y_n^{(n-1)} = f(x).$$

We have thus found n equations in the n unknown functions $c_1', c_2', \ldots, c_n'$:

$$
\begin{aligned}
c_1' y_1 \quad &+ c_2' y_2 \quad &+ \cdots + c_n' y_n \quad &= 0, \\
c_1' y_1' \quad &+ c_2' y_2' \quad &+ \cdots + c_n' y_n' \quad &= 0, \\
&\vdots & & \\
c_1' y_1^{(n-2)} &+ c_2' y_2^{(n-2)} + \cdots + c_n' y_n^{(n-2)} &= 0, \\
c_1' y_1^{(n-1)} &+ c_2' y_2^{(n-1)} + \cdots + c_n' y_n^{(n-1)} &= f.
\end{aligned}
\tag{3.106}
$$

The determinant of the system (3.106) is $W = W(y_1, y_2, \ldots, y_n)$, which is nonzero since $y_1, y_2, \ldots, y_n$ are linearly independent. Thus there is a unique solution given by Cramer's rule (see Appendix 4):

$$c_k' = \frac{W_k}{W}, \qquad k = 1, 2, \ldots, n, \tag{3.107}$$

where W_k is the determinant obtained by replacing the kth column of W by

$$
\begin{bmatrix}
0 \\
0 \\
0 \\
\vdots \\
0 \\
f
\end{bmatrix}.
$$

Finally, the functions $c_1, c_2, \ldots, c_n$ may be obtained by integration (if possible).

Example 3.12.9 (continued) We will solve

$$y''' - 3y' - 2y = 3e^{-x}$$

by the method of variation of parameters. Here $y_1 = e^{-x}$, $y_2 = xe^{-x}$, and $y_3 = e^{2x}$.

According to Abel's formula (3.96),

$$W(x) = \begin{vmatrix} e^{-x} & xe^{-x} & e^{2x} \\ -e^{-x} & (1-x)e^{-x} & 2e^{2x} \\ e^{-x} & (x-2)e^{-x} & 4e^{2x} \end{vmatrix} = ce^{\int 0 \, dx} = c,$$

which is a constant. Therefore

$$W(x) = W(0) = \begin{vmatrix} 1 & 0 & 1 \\ -1 & 1 & 2 \\ 1 & -2 & 4 \end{vmatrix} = 9.$$

By Eq. (3.107) we have

$$W_1 = \begin{vmatrix} 0 & xe^{-x} & e^{2x} \\ 0 & (1-x)e^{-x} & 2e^{2x} \\ 3e^{-x} & (x-2)e^{-x} & 4e^{2x} \end{vmatrix} = 9x - 3,$$

$$W_2 = \begin{vmatrix} e^{-x} & 0 & e^{2x} \\ -e^{-x} & 0 & 2e^{2x} \\ e^{-x} & 3e^{-x} & 4e^{2x} \end{vmatrix} = -9,$$

$$W_3 = \begin{vmatrix} e^{-x} & xe^{-x} & 0 \\ -e^{-x} & (1-x)e^{-x} & 0 \\ e^{-x} & (x-2)e^{-x} & 3e^{-x} \end{vmatrix} = 3e^{-3x}.$$

Then

$$c_1'(x) = \frac{W_1}{W} = x - \frac{1}{3}, \qquad c_2'(x) = -1, \qquad c_3'(x) = \frac{e^{-3x}}{3},$$

and

$$c_1(x) = \frac{x^2}{2} - \frac{x}{3}, \qquad c_2(x) = -x, \qquad c_3(x) = -\frac{e^{-3x}}{9}.$$

Finally, we obtain

$$y_p(x) = c_1(x)y_1(x) + c_2(x)y_2(x) + c_3(x)y_3(x)$$
$$= \left(\frac{x^2}{2} - \frac{x}{3}\right)e^{-x} - x(xe^{-x}) - \left(\frac{e^{-3x}}{9}\right)e^{2x}$$
$$= -\frac{x^2}{2}e^{-x} - \left(\frac{x}{3} + \frac{1}{9}\right)e^{-x}.$$

The last term of this equation is part of the general solution of the homogeneous equation. Thus

$$y_p(x) = -\frac{x^2}{2} e^{-x}.$$

As we saw in the previous example, the method of variation of parameters is more tedious to apply than the method of undetermined coefficients. However, as in the case of second order equations, it has the advantage of working for any functions $f(x)$ (so long as the functions c_k' as determined by Eq. (3.107) can be integrated).

EXERCISES 3.12

1. Prove that if $y_1, y_2, \ldots, y_n$ are solutions of Eq. (3.95), then $W(y_1, y_2 \ldots, y_n) = 0$ if and only if $y_1, y_2, \ldots, y_n$ are linearly dependent.

2. Prove Abel's formula for $n = 4$. A generalization of this proof may be used to prove the formula (3.96) for every n.

3. Prove that if $y_1, y_2, \ldots, y_n$ are n linearly independent solutions to Eq. (3.96) and if $y(x)$ is another solution, then there exist constants $c_1, c_2, \ldots, c_n$ such that $y(x) = c_1 y_1(x) + \cdots + c_n y_n(x)$. [*Hint*: See Theorem 3.4.]

4. Given that y_p and y_q are solutions to the nonhomogeneous equation (3.94), show that $y_h = y_p - y_q$ is a solution to the homogeneous equation (3.95).

5. Show that if y_1 is a solution to $y^{(n)} + a_1 y^{(n-1)} + \cdots + a_n y = f_1$ and y_2 is a solution to $y^{(n)} + a_1 y^{(n-1)} + \cdots + a_n y = f_2$, then $y_3 = y_1 + y_2$ is a solution of $y^{(n)} + a_1 y^{(n-1)} + \cdots + a_n y = f_1 + f_2$. This is a generalization to nth order equations of the *principle of super-position* (see Exercise 3.2.15 and Section 3.6).

In Exercises 6 through 14 find the general solution for each given differential equation.

6. $y''' - 5y'' - 2y' + 24y = 0$
7. $y''' - 3y'' + 3y' - y = 0$
8. $y''' - 27y = 0$
9. $y^{iv} + 2y''' - 11y'' - 12y' + 36y = 0$
10. $y^{iv} = 0$
11. $y^{v} + 2y''' + y' = 0$
12. $y^{iv} - 5y'' + 4y = 0$
13. $y^{iv} + 16y = 0$
14. $y^{v} + 10y^{iv} + 40y''' + 80y'' + 80y' + 32y = 0$

15. Find the particular solution for Exercise 6 that satisfies the initial conditions $y(0) = 1$, $y'(0) = 0$, $y''(0) = -1$.

16. Find the particular solution for Exercise 11 that satisfies the conditions $y(\pi/2) = 0$, $y'(\pi/2) = 1$, $y''(\pi/2) = 0$, $y'''(\pi/2) = -3$, $y^{(iv)}(\pi/2) = 0$.

In Exercises 17 through 30 find a particular solution of each given equation by the method of undetermined coefficients (if possible) *and* by the method of variation of constants. Compare these solutions.

17. $y''' - 5y'' - 2y' + 24y = x$
18. $y''' - 5y'' - 2y' + 24y = e^x$
19. $y''' - 5y'' - 2y' + 24y = e^{-2x}$
20. $y''' - 5y'' - 2y' + 24y = e^x \cos 2x$

21. $y''' - 5y'' - 2y' + 24y = 3 + x.$
22. $y^{iv} = 1$
23. $y^{iv} = x + x^2 + x^4$
24. $y''' - 3y'' + 3y' - y = e^x$
25. $y''' - 3y'' + 3y' - y = \cos x$
26. $y''' - 3y'' + 3y' - y = \ln x$
27. $y^v + 2y''' + y' = 1 - e^x$ [*Hint*: Use the principle of superposition.]
28. $y^v + 10y^{iv} + 40y''' + 80y'' + 80y' + 32y = x + \sin x + e^x$
29. $y''' + 3y'' + 2y' = x^3 - e^x$
30. $y''' + y' = \tan x$

LINEAR
SECOND
ORDER
DIFFERENCE
EQUATIONS

4

4.1 INTRODUCTION

The most general second order difference equation is of the form

$$F(n, y_n, y_{n+1}, y_{n+2}) = 0. \tag{4.1}$$

Unlike the general second order differential equation, Eq. (4.1) is *usually* solvable by successive iteration.

Example 4.1.1 Consider the equation

$$e^{y_{n+2}} - [6(n + 1) \sin^2 y_{n+1}]\sqrt{y_n} = 0. \tag{4.2}$$

If we let y_0 and y_1 be the initial values for the unknown function y_n, then

$$y_2 = \ln (6\sqrt{y_0} \sin^2 y_1),$$
$$y_3 = \ln (12\sqrt{y_1} \sin^2 y_2) = \ln [12\sqrt{y_1} \sin^2 (\ln 6\sqrt{y_0} \sin^2 y_1)],$$

and so on. This iteration is defined so long as

$$6(n + 1)(\sin^2 y_{n+1})\sqrt{y_n} \geqslant 1,$$

for if the left-hand side of this expression is less than 1, then y_{n+2} is negative and the expression for y_{n+4} contains the term $\sqrt{y_{n+2}}$ and thus has no real solution. A search for a solution in terms of a simple combination of elementary functions will, in this case, not be fruitful.

Another example will illustrate further difficulties with the iterative method.

Example 4.1.2 Consider the equation

$$\exp \{\sin [(y_{n+2} y_{n+1})^{1/3} y_n^5]\} = \cos (y_{n+2} y_{n+1})^{1/2} y_n.$$

Even if y_0 and y_1 are known, there is no way to derive an *explicit* expression for y_2 in terms of y_0 and y_1.

In this chapter we shall investigate a class of difference equations for which the difficulties inherent in the two examples above do not arise.

A second order difference equation is *linear* if it involves no nonlinear functions or products of the unknown terms y_{n+2}, y_{n+1}, and y_n. Thus $y_{n+2} + (3n^3 \sin n)^5 y_{n+1} + y = \cos n^3$, for example, is linear, while $y_{n+2} + y_{n+1} y_n = 0$ is not. The most general second order linear equation can be written as

$$y_{n+2} + a_n y_{n+1} + b_n y_n = f_n.$$

If a_n, b_n, and f_n are defined for every integer $n \geqslant 0$, and if y_0 and y_1 are given, then

$$y_2 = f_0 - a_0 y_1 - b_0 y_0,$$
$$y_3 = f_1 - a_1 y_2 - b_1 y_1, \ldots$$

and the iteration thus defined will not terminate. Therefore, for linear equations we have the following theorem.

Theorem 4.1 Let c_0 and c_1 be given constants and suppose that a_n, b_n, and f_n are defined for every integer $n \geqslant 0$. Then there is a unique solution y_n to

$$y_{n+2} + a_n y_{n+1} + b_n y_n = f_n \qquad (4.3)$$

which satisfies the conditions $y_0 = c_0$ and $y_1 = c_1$.

In analogy to the terminology of Chapter 3, Eq. (4.3) is said to be *homogeneous* if $f_n = 0$ for every n. Otherwise, it is said to be *nonhomogeneous*. If a_n and b_n are constants independent of n, then Eq. (4.3) is said to have *constant coefficients*.

Although Theorem 4.1 guarantees a unique solution to the linear initial value problem, we shall seek in the next few sections methods for obtaining the general solution to (4.3) in a way that does not involve continued iteration. In the process we will discover important links between the solutions of second order difference equations and those of comparable second order differential equations.

EXERCISES 4.1

In Exercises 1 through 10 determine whether each given difference equation is nonlinear or linear; if the latter, state whether the equation is homogeneous or nonhomogeneous, with constant or variable coefficients.

1. $y_{n+2} - n + y_n = 0$
2. $y_{n+2} - n^2 + y_n^2 = 0$
3. $y_{n+2} - y_{n+1} y_n = \sin n$
4. $6y_{n+2} - (n+1)^2 y_{n+1} + e^{n^2} y_n = \sqrt{6n}$
5. $y_{n+1} - 3y_{n+4} + y_n = 0$
6. $y_{n+2} - \sqrt{y_n} = \sqrt{n}$
7. $y_{n+2} y_n = 3$
8. $2y_{n+2} - 5y_{n+1} + 17y_n = n^{7/2}$
9. $\sqrt{y_{n+2} + y_{n+1}} - y_n = n$
10. $y_{n+2} - \sqrt{y_{n+1}} - y_n = 3$

11.   Show that if two consecutive values of a solution y_n of the homogeneous linear equation

$$y_{n+2} + a_n y_{n+1} + b_n y_n = 0$$

are zero, then $y_n = 0$ for all n.

12.   Suppose that x_n and y_n are solutions of the homogeneous linear equation

$$y_{n+2} + a_n y_{n+1} + b_n y_n = 0,$$

and $x_k = y_k = 0$ for some integer k. Show that there is a constant c such that $y_n = cx_n$ for all n.

4.2 PROPERTIES OF SOLUTIONS TO THE LINEAR EQUATION

We shall begin with the homogeneous equation

$$y_{n+2} + a_n y_{n+1} + b_n y_n = 0. \qquad (4.4)$$

A *solution* to Eq. (4.4) consists of a sequence of real numbers $y_0, y_1, y_2, \ldots, y_n, \ldots$ that satisfy Eq. (4.4) for all integers $n \geqslant 0$. To avoid the cumbersome use of sequences, we shall denote a solution merely by the term y_n.

A *linear combination* of the solutions x_n and y_n of Eq. (4.4) is a sequence $Ax_0 + By_0, Ax_1 + By_1, \ldots, Ax_n + By_n, \ldots$, where A and B are real numbers. Two solutions x_n and y_n of Eq. (4.4) are said to be *linearly independent* if $Ax_n + By_n = 0$ only for $A = B = 0$. This is equivalent to saying that there is no constant c such that $x_n = cy_n$ for all n. These two definitions can be extended to apply to k solutions of Eq. (4.4) in the obvious way.

Theorem 4.2 Any linear combination of solutions of Eq. (4.4) is also a solution of Eq. (4.4).

Proof. We need only show that if x_n and y_n are solutions of Eq. (4.4), then so is $z_n = Ax_n + By_n$:

$$z_{n+2} + a_n z_{n+1} + b_n z_n$$
$$= Ax_{n+2} + By_{n+2} + a_n(Ax_{n+1} + By_{n+1}) + b_n(Ax_n + By_n)$$
$$= A(x_{n+2} + a_n x_{n+1} + b_n x_n) + B(y_{n+2} + a_n y_{n+1} + b_n y_n) = 0.$$

We can now define the analog of the Wronskian for difference equations. Let x_n and y_n be two solutions of Eq. (4.4). Then the *Casoratian* of x_n and y_n is given by

$$C_n(x,y) = x_n y_{n+1} - x_{n+1} y_n. \tag{4.5}$$

In Section 3.2 we proved that homogeneous second order linear differential equations have two linearly independent solutions. This is also true for homogeneous second order linear difference equations, provided that none of the coefficients b_n in Eq. (4.4) is zero. The need for this condition is obvious in the case when all the coefficients b_n vanish, because then Eq. (4.4) is really a first order difference equation and cannot have two linearly independent solutions. The next theorem, whose proof is left as an exercise, shows why every b_n must be nonzero.

Theorem 4.3 Suppose that some coefficient b_N in Eq. (4.4) is zero. Then Eq. (4.4) does not have two linearly independent solutions.

The next result is similar to Lemma 3.3.

Theorem 4.4 Let x_n and y_n be two solutions of Eq. (4.4) and suppose that $b_n \neq 0$ for every $n = 0, 1, 2, \ldots$. Then x_n and y_n are linearly independent if and only if $C_n(x,y) \neq 0$ for some integer n.

Proof. We shall begin by showing that $C_n(x,y)$ is never zero or $C_n(x,y)$ vanishes for all n. Using Eq. (4.5), we see that

$$C_{n+1} = x_{n+1} y_{n+2} - x_{n+2} y_{n+1}. \tag{4.6}$$

Since x_n and y_n are solutions of Eq. (4.4), we may replace x_{n+2} and y_{n+2} in (4.6) by

$$x_{n+2} = -a_n x_{n+1} - b_n x_n, \qquad y_{n+2} = -a_n y_{n+1} - b_n y_n \tag{4.7}$$

to obtain

$$C_{n+1} = x_{n+1}(-a_n y_{n+1} - b_n y_n) - y_{n+1}(-a_n x_{n+1} - b_n x_n)$$
$$= b_n[x_n y_{n+1} - x_{n+1} y_n] \tag{4.8}$$

or

$$C_{n+1} = b_n C_n. \tag{4.9}$$

[Note the similarity between Eq. (4.9) and Eq. (3.5) for the Wronskian.] As we saw in Section 2.6, Eq. (4.9) has the solution

$$C_{n+1} = \left(\prod_{i=0}^{n} b_i\right) C_0.$$

Since $b_i \neq 0$ for every i by assumption, C_n is either always zero or never zero depending on whether C_0 is zero or not.

We can now prove the theorem by showing that the solutions x_n and y_n of Eq. (4.4) are linearly dependent if and only if $C_n(x,y) = 0$ (for all n).

If x_n and y_n are linearly dependent, then there is a constant c such that $y_n = cx_n$ for all n. Hence

$$C_n(x,y) = x_n y_{n+1} - y_n x_{n+1} = x_n(cx_{n+1}) - (cx_n)x_{n+1} = 0.$$

On the other hand, if $C_n(x,y) = 0$, then $C_0(x,y) = 0$ so that

$$x_0 y_1 = x_1 y_0. \tag{4.10}$$

There are two cases to consider, depending on whether $x_0 \neq 0$ or $x_0 = 0$.

CASE 1. If $x_0 \neq 0$, then Eq. (4.10) implies that

$$y_1 = \left(\frac{y_0}{x_0}\right) x_1.$$

Let $c = y_0/x_0$. Then

$$y_0 = \left(\frac{y_0}{x_0}\right) x_0 = cx_0.$$

By Eq. (4.4) it follows that

$$y_2 = -a_0 y_1 - b_0 y_0 = -a_0(cx_1) - b_0(cx_0)$$
$$= c(-a_0 x_1 - b_0 x_0) = cx_2.$$

Similarly, it follows that $y_3 = cx_3$, $y_4 = cx_4$, and so on, for every n. Thus $y_n = cx_n$, and x_n and y_n are linearly dependent.

CASE 2. If $x_0 = 0$, then either $x_1 = 0$ or $x_1 \neq 0$. If $x_1 = 0$, then Eq. (4.4) implies that $x_2 = 0$, $x_3 = 0$, and so on, for every n. Thus $x_n = 0 \cdot y_n$, and the solutions are dependent. If $x_1 \neq 0$, then Eq. (4.10) implies that

$$y_0 = \left(\frac{y_1}{x_1}\right) x_0.$$

Let $c = y_1/x_1$. We then note that

$$y_1 = \left(\frac{y_1}{x_1}\right) x_1 = cx_1.$$

The remaining proof follows as in Case 1.

To complete the discussion of the homogeneous case, we shall prove:

Theorem 4.5 Let x_n and y_n be two linearly independent solutions to Eq. (4.4) and let z_n be another solution. Then z_n can be written as a linear combination of x_n and y_n.

Proof. Consider the linear system

$$x_0 c_1 + y_0 c_2 = z_0,$$
$$x_1 c_1 + y_1 c_2 = z_1, \tag{4.11}$$

where c_1 and c_2 are not known. The determinant of this system is $C_0(x,y)$, which is nonzero since x_n and y_n are linearly independent. Hence the system has a unique solution c_1 and c_2, and the sequence $z_n^* = c_1 x_n + c_2 y_n$ satisfies the conditions $z_0^* = z_0$ and $z_1^* = z_1$. Since the initial value problem has a unique solution, it follows that $z_n^* = z_n$, and the proof is complete.

Once we have the above result, we can speak of the *general solution* of Eq. (4.4) which is given by

$$z_n = c_1 x_n + c_2 y_n, \tag{4.12}$$

where x_n and y_n are any two linearly independent solutions of (4.4).

Let us now turn briefly to the nonhomogeneous equation

$$y_{n+2} + a_n y_{n+1} + b_n y_n = f_n. \tag{4.13}$$

Theorem 4.6 Let y_n be a solution of Eq. (4.13) and let x_n be any other solution. Then $z_n = x_n - y_n$ is a solution of (4.4).

Proof. We have

$$\begin{aligned}
z_{n+2} + a_n z_{n+1} + b_n z_n &= (x_{n+2} - y_{n+2}) + a_n(x_{n+1} - y_{n+1}) + b_n(x_n - y_n) \\
&= (x_{n+2} + a_n x_{n+1} + b_n x_n) - (y_{n+2} + a_n y_{n+1} + b_n y_n) \\
&= f_n - f_n = 0.
\end{aligned}$$

Given this theorem, it is evident that to obtain all solutions of the non-homogeneous equation (4.13), it is necessary only to obtain one solution to Eq. (4.13) and the general solution to Eq. (4.4), as in the case of the corresponding differential equations.

EXERCISES 4.2

In Exercises 1 through 5 verify that x_n and y_n are solutions to each given difference equation and calculate their Casoratian.

1. $y_{n+2} - 3y_{n+1} + 2y_n = 0$; $x_n = 2^n$, $y_n = 1$

2. $y_{n+2} - 4y_{n+1} + 4y_n = 0$; $x_n = 2^n$, $y_n = n2^n$

3. $y_{n+2} + y_n = 0$; $x_n = \cos \dfrac{n\pi}{2}$, $y_n = \sin \dfrac{n\pi}{2}$

4. $y_{n+2} - (n + 2)(n + 1)y_n = 0$; $x_n = n!$, $y_n = (-1)^n n!$

5. $y_{n+2} - 7y_{n+1} + 12y_n = 0$; $x_n = 3^n$, $y_n = 4^n$

6. Let x_n be a solution to $y_{n+2} + a_n y_{n+1} + b_n y_n = f_n$ and let z_n be a solution to $y_{n+2} + a_n y_{n+1} + b_n y_n = g_n$. Show that $x_n + z_n$ is a solution to $y_{n+2} + a_n y_{n+1} + b_n y_n = f_n + g_n$. This is the *principle of superposition* for difference equations (see Exercise 3.2.15).

7. Consider the homogeneous third order difference equation

$$y_{n+3} + a_n y_{n+2} + b_n y_{n+1} + c_n y_n = 0.$$

Show that if x_n, y_n, and z_n are three solutions to the equation, then any linear combination of them is also a solution.

8. Given the third order equation of Exercise 7, again let x_n, y_n, and z_n be three solutions. Define the *Casoratian* $C_n(x,y,z)$ by

$$C_n(x,y,z) = \det \begin{vmatrix} x_n & y_n & z_n \\ x_{n+1} & y_{n+1} & z_{n+1} \\ x_{n+2} & y_{n+2} & z_{n+2} \end{vmatrix}.$$

Prove that $C_{n+1}(x,y,z) = c_n C_n(x,y,z)$ and conclude that if $c_n \neq 0$ for $n = 0, 1, 2, \ldots$, then $C_n(x,y,z)$ is either always zero or never zero.

9. Show that if, in Exercise 8, $c_n \neq 0$ for every n and if none of the solutions x_n, y_n, z_n are identically zero, then $C_n(x,y,z) \neq 0$ if and only if the three solutions are linearly independent.

10. Given that in Exercise 9 the three solutions are linearly independent and w_n is another solution, show that there exist constants α, β, and γ such that $w_n = \alpha x_n + \beta y_n + \gamma z_n$ for every n.

11. Show that if x_n and y_n are solutions to the third order nonhomogeneous equation $y_{n+3} + a_n y_{n+2} + b_n y_{n+1} + c_n y_n = f_n$, then $z_n = x_n - y_n$ is a solution to the associated homogeneous equation.

12. Assume that some nonzero solution of

$$y_{n+2} + a_n y_{n+1} + b_n y_n = 0, \qquad y_0 = 0 \tag{*}$$

satisfies the condition $y_k = 0$ for some $k > 0$. Prove that $x_k = 0$ for any other solution x_n of (*)

* 13. Prove Theorem 4.3.

4.3 USING ONE SOLUTION TO FIND ANOTHER

We shall consider the linear homogeneous equation

$$y_{n+2} + a_n y_{n+1} + b_n y_n = 0, \qquad b_n \neq 0. \tag{4.14}$$

If the sequences a_n and b_n are not constants, then there is no general method for obtaining solutions. However, as in the case of second order homogeneous differential equations, if one solution is known, then another linearly independent solution can be found (see Section 3.3).

Suppose that x_n is a known nonzero solution to Eq. (4.14). We shall seek a second solution of the form

$$y_n = v_n x_n, \tag{4.15}$$

where v_n is a nonconstant sequence still to be determined. Define the sequence u_n by

$$u_n = v_{n+1} - v_n. \tag{4.16}$$

Recalling that the Casoratian $C_n(x,y)$ has the form $C_n(x,y) = x_n y_{n+1} - x_{n+1} y_n$ for any two solutions x_n and y_n of Eq. (4.14) and assuming that y_n is indeed a solution, we obtain, after some simple algebraic manipulations,

$$C_n(x,y) = x_{n+1} x_n \left(\frac{y_{n+1}}{x_{n+1}} - \frac{y_n}{x_n} \right)$$

$$= x_{n+1} x_n (v_{n+1} - v_n) = x_{n+1} x_n u_n \tag{4.17}$$

or

$$u_n = \frac{C_n(x,y)}{x_{n+1} x_n}. \tag{4.18}$$

Using the formula following Eq. (4.8) for the Casoratian, we can obtain

$$u_n = \frac{\prod_{i=0}^{n-1} b_i}{x_{n+1} x_n}, \tag{4.19}$$

where we have assumed that $C_0 = 1$. The last assumption can be made since the unknown solution y_n is independent of x_n (implying $C_0 \neq 0$) and defined up to an arbitrary constant. Finally, from (4.19) we can obtain the first order difference equation

$$v_{n+1} - v_n = \frac{\prod_{i=0}^{n-1} b_i}{x_{n+1} x_n} = d_n, \tag{4.20}$$

which by Eq. (2.66) has the solution

$$v_n = v_0 + \sum_{i=0}^{n-1} d_i. \tag{4.21}$$

Since neither x_n nor b_n is zero, d_n is not zero for any n and v_n is therefore not a constant. Hence $y_n = v_n x_n$ is a second linearly independent solution to Eq. (4.14).

Example 4.3.1 $y_{n+2} - 4y_{n+1} + 4y_n = 0$. One solution is easy to see: $x_n = 2^n$. (More will be said about constant coefficient equations in the next two sections.) Let $y_n = v_n x_n$ be another solution. Then by (4.20) we see that

$$d_n = \frac{\prod_{i=0}^{n-1} b_i}{x_n x_{n+1}} = \frac{4^n}{2^n 2^{n+1}} = \frac{1}{2}$$

and thus

$$v_n = v_0 + \frac{n}{2}.$$

The choice of v_0 is arbitrary; so selecting $v_0 = 0$, we have

$$v_n = \frac{n}{2} \quad \text{and} \quad y_n = \frac{n}{2} x_n = \frac{n 2^n}{2}.$$

The general solution is therefore

$$y_n = c_1 2^n + c_2 n 2^n,$$

where the number 2 has been absorbed into the arbitrary constant c_2.

Example 4.3.2 Let $y_{n+2} - (n + 2)(n + 1)y_n = 0$. It is easy to verify that $x_n = n!$ is a solution. Since $b_n = -(n + 2)(n + 1)$,

$$d_n = \frac{\prod_{i=0}^{n-1} b_i}{x_n x_{n+1}} = \frac{(-1)^n (n + 1)! n!}{(n + 1)! n!} = (-1)^n.$$

Thus

$$v_n = v_0 + \sum_{i=0}^{n-1} d_i = v_0 + \frac{1 - (-1)^n}{2},$$

since the geometric progression $1 + a + a^2 + \cdots + a^{n-1} = (1 - a^n)/(1 - a)$. And since v_0 is arbitrary, we may set $v_0 = -\frac{1}{2}$ to obtain

$$y_n = v_n x_n = \frac{-(-1)^n}{2} n!,$$

and the general solution is given by

$$y_n = c_1 n! + c_2(-1)^n n!.$$

Note that the solutions $n!$ and $(-1)^n n!$ are linearly independent, since $(-1)^n$ is not constant.

EXERCISES 4.3

In each of the following exercises a difference equation is given with one solution. Find a second linearly independent solution.

1. $y_{n+2} - 2y_{n+1} + y_n = 0;\ x_n = 7$
2. $y_{n+2} + 7y_{n+1} + 10y_n = 0;\ x_n = (-2)^n$
3. $(n + 2)y_{n+2} - (n + 1)y_{n+1} - y_n = 0;\ x_n = 3$
4. $y_{n+2} + 16y_n = 0;\ x_n = 4^n \cos \dfrac{n\pi}{2}$
5. $y_{n+2} - 6y_{n+1} + 8y_n = 0;\ x_n = 4^n$
6. $y_{n+2} + 6y_{n+1} + 9y_n = 0;\ x_n = (5n)3^{n-1}$
7. $y_{n+2} - y_{n+1} - (n + 1)^2 y_n = 0;\ x_n = n!$
8. $y_{n+2} - (n + 1)y_{n+1} - (n + 1)y_n = 0;\ x_n = n!$

4.4 HOMOGENEOUS EQUATIONS WITH CONSTANT COEFFICIENTS: THE CASE OF REAL ROOTS

We shall now consider the linear homogeneous equation with constant coefficients

$$y_{n+2} + ay_{n+1} + by_n = 0, \qquad b \neq 0, \tag{4.22}$$

and give a simple method for obtaining the general solution. In the case of the first-order equation $y_{n+1} = ay_n$ we saw in Section 2.6 that the general solution is $y_n = a^n y_0$. It is then, not implausible to "guess" that there are solutions to Eq. (4.22) of the form $y_n = \lambda^n$ for some λ (real or complex). Substituting $y_n = \lambda^n$ into Eq. (4.22), we obtain

$$\lambda^{n+2} + a\lambda^{n+1} + b\lambda^n = 0.$$

Since this equation is true for all $n \geqslant 0$, it holds for $n = 0$, so that

$$\lambda^2 + a\lambda + b = 0. \tag{4.23}$$

This is the *auxiliary equation* for the difference equation (4.22) and is identical to the auxiliary equation for the second order differential equation derived in Section 3.4. As before, the roots are

$$\lambda_1 = \frac{-a + \sqrt{a^2 - 4b}}{2}, \qquad \lambda_2 = \frac{-a - \sqrt{a^2 - 4b}}{2}. \tag{4.24}$$

Again there are three cases.

CASE 1. $a^2 - 4b > 0$. In this case there are two distinct real roots λ_1 and λ_2, given by (4.24), and $x_n = \lambda_1^n$ and $y_n = \lambda_2^n$ are two solutions of Eq. (4.22). Since

$$C_n(x,y) = x_{n+1}y_n - x_n y_{n+1} = \lambda_1^n \lambda_2^n (\lambda_1 - \lambda_2),$$

$C_n(x,y)$ is not zero except in the special case of λ_1 or λ_2 being 0. This can only occur if $b = 0$, which has been ruled out. Thus we have proved:

Theorem 4.7 If $a^2 - 4b > 0$, then the general solution of Eq. (4.22) is given by

$$y_n = c_1\lambda_1^n + c_2\lambda_2^n, \tag{4.25}$$

where c_1 and c_2 are arbitrary constants and λ_1 and λ_2 are given by (4.24).

Example 4.4.1 Consider the equation

$$x_{n+2} + 5x_{n+1} - 6x_n = 0.$$

The auxiliary equation is $\lambda^2 + 5\lambda - 6 = 0$, with the roots $\lambda_1 = 6$ and $\lambda_2 = -1$. The general solution is given by $y_n = c_1(6)^n + c_2(-1)^n$. If we specify the initial conditions $y_0 = 3$, $y_1 = 11$, for example, then we obtain the system

$$c_1 + c_2 = 3,$$
$$6c_1 - c_2 = 11,$$

which has the unique solution $c_1 = 2$, $c_2 = 1$ and the specific solution $y_n = 2 \cdot 6^n + (-1)^n$.

CASE 2. $a^2 - 4b = 0$. Here the two roots of Eq. (4.23) are equal, and we have the solution $x_n = \lambda^n$ where $\lambda = -a/2$. We can obtain a second linearly independent solution by the method of the previous section. Letting $y_n = v_n x_n$ denote another solution, we have, by (4.20),

$$d_n = \frac{\prod_{i=0}^{n-1} b_i}{x_{n+1}x_n} = \frac{b^n}{\lambda^{n+1}\lambda^n} = \frac{b^n}{\lambda^{2n+1}}.$$

But $b = (a/2)^2$ and $\lambda = (-a/2)$, so that

$$d_n = \frac{(a/2)^{2n}}{(-a/2)^{2n+1}} = \frac{1}{(-a/2)} = \frac{1}{\lambda}.$$

Thus $v_n = v_0 + \sum_{i=0}^{n-1} d_i = n/\lambda$ (letting $v_0 = 0$) and $y_n = n\lambda^{n-1}$. We have thus shown the following result.

Theorem 4.8 Let $a^2 - 4b = 0$. Then the general solution of Eq. (4.22) is

$$y_n = c_1\lambda^n + c_2 n\lambda^n, \tag{4.26}$$

where $\lambda = -a/2$ and c_1 and c_2 are arbitrary constants.

Example 4.4.2 Consider the equation

$$y_{n+2} - 6y_{n+1} + 9y_n = 0, \tag{4.27}$$

with initial conditions $y_0 = 5$, $y_1 = 12$. The characteristic equation is $\lambda^2 - 6\lambda + 9 = 0$ with the double root $\lambda = 3$. The general solution is therefore

$$y_n = c_1 3^n + c_2 n 3^n = 3^n(c_1 + nc_2).$$

Using the initial conditions, we obtain

$$c_1 = 5,$$
$$3c_1 + 3c_2 = 12.$$

The unique solution is $c_1 = 5$, $c_2 = -1$, and we have the specific solution to Eq. (4.27): $y_n = 5 \cdot 3^n - n \cdot 3^n = 3^n(5 - n)$.

We shall consider the third case in the next section.

EXERCISES 4.4

In Exercises 1 through 10 find the general solution to each given difference equation. When initial conditions are specified, find the unique solution that satisfies them.

1. $y_{n+2} - 3y_{n+1} - 4y_n = 0$
2. $y_{n+2} + 7y_{n+1} + 6y_n = 0$, $y_0 = 0$, $y_1 = 1$
3. $6y_{n+2} + 5y_{n+1} + y_n = 0$, $y_0 = 1$, $y_1 = 0$
4. $y_{n+2} + y_{n+1} - 6y_n = 0$, $y_0 = 1$, $y_1 = 2$
5. $y_{n+2} + 2y_{n+1} + y_n = 0$
6. $y_{n+2} + 16y_{n+1} + 64y_n = 0$, $y_0 = 2$, $y_1 = 0$
7. $y_{n+2} - \frac{2}{3}y_{n+1} + \frac{1}{9}y_n = 0$
8. $3y_{n+2} - 2y_{n+1} - y_n = 0$, $y_0 = 0$, $y_1 = 3$

9. $10y_{n+2} - y_{n+1} - y_n = 0$

10. $36y_{n+2} - 60y_{n+1} + 25y_n = 0$

11. In a study of infectious diseases a record is kept of outbreaks of measles in a particular school. It is estimated that the probability of at least one new infection occurring in the nth week after an outbreak is $P_n = P_{n-1} - \frac{1}{5}P_{n-2}$. If $P_0 = 0$ and $P_1 = 1$, what is P_n? After how many weeks will the probability of the occurrence of a new case of measles be less than 10 percent?

12. The Fibonacci numbers are a sequence of numbers such that each one is the sum of its two predecessors. The first few Fibonacci numbers are 1, 1, 2, 3, 5, 8, 13, ...

   a) Formulate an initial value difference equation that will generate the Fibonacci numbers.
   b) Find the solution to this equation.
   c) Show that the ratio of successive Fibonacci numbers tends to $(1 + \sqrt{5})/2$ as $n \to \infty$. This ratio, known as the *golden ratio*, was often used in ancient Greek architecture whenever rectangular structures were constructed. It was believed that when the ratio of the sides of a rectangle was this number, the resulting structure was most pleasing to the eye.

13. Two competing species of drosophila (fruit flies) are growing under favorable conditions. In each generation species A increases its population by 60 percent and species B increases by 40 percent. If initially there are 1000 flies of each species, what is the total population after n generations?

14. Consider the third order difference equation

$$y_{n+3} + ay_{n+2} + by_{n+1} + cy_n = 0.$$

   a) Show that $x_n = \lambda^n$ is a solution of the equation if λ satisfies the auxiliary equation

$$\lambda^3 + a\lambda^2 + b\lambda + c = 0. \qquad (*)$$

   b) Given that λ_1, λ_2, and λ_3 are distinct real roots of $(*)$, show that the general solution is given by

$$y_n = c_1\lambda_1^n + c_2\lambda_2^n + c_3\lambda_3^n.$$

   (See Exercises 4.2.8, 4.2.9, and 4.2.10.)
   c) Given that λ_1 is a double root, show that the general solution is

$$y_n = c_1\lambda_1^n + c_2n\lambda_1^n + c_3\lambda_2^n.$$

   d) Given that λ_1 is a triple root, show that the general solution is

$$y_n = c_1\lambda_1^n + c_2n\lambda_1^n + c_3n^2\lambda_1^n.$$

In the following exercises, use the results of Exercise 14 to obtain the general solution for each given equation.

15. $y_{n+3} - 6y_{n+2} + 11y_{n+1} - 6y_n = 0$

16. $y_{n+3} + 6y_{n+2} - 13y_{n+1} - 42y_n = 0$

17. $y_{n+3} - 3y_{n+1} + 2y_n = 0$

18. $y_{n+3} - 6y_{n+2} + 12y_{n+1} - 8y_n = 0$

19. $y_{n+3} - 3y_{n+2} + 3y_{n+1} - y_n = 0$

20. $y_{n+3} + 6y_{n+2} - 32y_n = 0$

4.5 HOMOGENEOUS EQUATIONS WITH CONSTANT COEFFICIENTS: THE COMPLEX CASE

When $a^2 - 4b < 0$, the roots of Eq. (4.23) are

$$\lambda_1 = \alpha + i\beta \quad \text{and} \quad \lambda_2 = \alpha - i\beta, \tag{4.28}$$

where $\alpha = -a/2$ and $\beta = \sqrt{4b - a^2}/2$. Since solutions are of the form λ^n, we must describe how to calculate powers of complex numbers. This is very easy if the numbers are written in polar form. Let $z = re^{i\theta}$. Then $z^n = (re^{i\theta})^n = r^n e^{in\theta}$. But $e^{in\theta} = \cos n\theta + i \sin n\theta$. Thus we have obtained the *DeMoivre formula*

$$(\cos \theta + i \sin \theta)^n = \cos n\theta + i \sin n\theta. \tag{4.29}$$

Example 4.5.1 Let $z = 1 + i$. Then $r = \sqrt{\alpha^2 + \beta^2} = \sqrt{2}$ and $\theta = \tan^{-1}(\beta/\alpha) = \tan^{-1} 1 = \pi/4$. So $z = \sqrt{2}\, e^{i\pi/4}$. Therefore,

$$z^3 = 2^{3/2} e^{i3\pi/4} = 2\sqrt{2}\left(\cos \frac{3\pi}{4} + i \sin \frac{3\pi}{4}\right) = -2 + 2i.$$

Of course, it is easier to cube $1 + i$ directly, but higher powers of $(1 + i)$ should be calculated this way.

Returning to the case where $a^2 - 4b < 0$, we can write the two roots λ_1 and λ_2 in Eq. (4.28) as

$$\lambda_1 = re^{i\theta}, \qquad \lambda_2 = re^{-i\theta}, \tag{4.30}$$

where $r = \sqrt{\alpha^2 + \beta^2}$ and $\theta = \tan^{-1}(\beta/\alpha)$, $0 < \theta < \pi$. The two linearly independent solutions are then

$$x_n = (re^{i\theta})^n = r^n e^{in\theta}, \qquad y_n = r^n e^{-in\theta}. \tag{4.31}$$

Thus $\frac{1}{2}(x_n + y_n)$ is a solution, as is $(1/2i)(x_n - y_n)$. We saw in Section 3.5 that these solutions can be written as $x_n^* = r^n \cos n\theta$ and $y_n^* = r^n \sin n\theta$, respectively. They are clearly linearly independent, since one is not a constant multiple of the other. Alternatively, we show that the Casoratian $C_n(x^*, y^*)$ is nonzero:

$$C_0(x^*, y^*) = x_1^* y_0^* - x_0^* y_1^*$$
$$= r \cos \theta \cdot \sin 0 - \cos 0 \cdot r \sin \theta$$
$$= -r \sin \theta.$$

Since $\beta = \sqrt{4b - a^2}/2 \neq 0$, it follows that $r = \sqrt{\alpha^2 + \beta^2} \neq 0$, because $\alpha^2 + \beta^2$ must be positive. Also $\theta = \tan^{-1}(\beta/\alpha)$ cannot be zero or π, since $\tan 0 = \tan \pi =$

0 and $\tan \theta = \beta/\alpha \neq 0$. Thus $C_0(x^*,y^*) \neq 0$, which implies that $C_n(x^*,y^*)$ is nonzero and x_n^*, y_n^* are linearly independent. These results are summarized in Theorem 4.9.

Theorem 4.9 Let $a^2 - 4b < 0$. Then the general solution to the homogeneous equation $y_{n+2} + ay_{n+1} + by_n = 0$ is given by

$$y_n = c_1 r^n \cos n\theta + c_2 r^n \sin n\theta, \tag{4.32}$$

where $r = \sqrt{\alpha^2 + \beta^2}$, $\theta = \tan^{-1} \beta/\alpha$, $0 < \theta < \pi$, $\alpha = -a/2$, $\beta = \sqrt{4b - a^2}/2$, and c_1 and c_2 are arbitrary constants.

Example 4.5.2 Let $y_{n+2} + y_n = 0$. Then the auxiliary equation is $\lambda^2 + 1 = 0$ with the roots $\pm i$. Here $\alpha = 0$ and $\beta = 1$, so $r = 1$ and $\theta = \pi/2$. The general solution is given by

$$y_n = c_1 \cos \frac{n\pi}{2} + c_2 \sin \frac{n\pi}{2}. \tag{4.33}$$

This is the equation for *discrete harmonic motion*, which corresponds to ordinary harmonic motion as discussed in Example 3.5.1. If we specify $y_0 = 0$ and $y_1 = 1000$, we will obtain

$$c_1 = 0, \qquad c_1 \cos \frac{\pi}{2} + c_2 \sin \frac{\pi}{2} = 1000,$$

with the solution $c_1 = 0$, $c_2 = 1000$, and the specific solution

$$y_n = 1000 \sin \frac{n\pi}{2}.$$

Example 4.5.3 In Example 1.3.4 we derived a difference equation describing the influence of preceding generations on population growth:

$$P_{n+2} - rP_{n+1} - sP_n = 0, \tag{4.34}$$

where r and s are measures of the relative importance of the preceding two generations.

The auxiliary equation is $\lambda^2 - r\lambda - s = 0$ with the roots

$$\lambda_1 = \frac{r + \sqrt{r^2 + 4s}}{2}, \qquad \lambda_2 = \frac{r - \sqrt{r^2 + 4s}}{2}.$$

If $r^2 > -4s$, the roots are real and distinct, and the general solution is $P_n = c_1 \lambda_1^n + c_2 \lambda_2^n$. If $|\lambda_1| < 1$ and $|\lambda_2| < 1$, then $P_n \to 0$ as $n \to \infty$; but if either $|\lambda_1|$ or $|\lambda_2|$ is greater than 1, then $|P_n| \to \infty$. If $r^2 = -4s$ ($s < 0$), then the general solution is $P_n = c_1(r/2)^n + c_2 n(r/2)^{n-1}$, in which case $P_n \to 0$ if $|r| < 2$, but if $|r| \geq 2$, the solution tends to ∞. In the third case, $r^2 < -4s$, the general solution is

$$P_n = (-s)^{n/2}(c_1 \cos n\theta + c_2 \sin n\theta),$$

where $\theta = \tan^{-1}(\sqrt{-r^2 - 4s}/r)$. When $-s < 1$, the solution P_n tends to zero in an oscillatory manner. This is called *damped discrete harmonic motion* (see Fig. 4.1). If $-s = 1$, we have the harmonic motion of the previous example. Finally, if $-s > 1$, the solution grows in an oscillatory motion, called *forced discrete harmonic motion* (see Fig. 4.2).

Figure 4.1

Figure 4.2

It is interesting to see that even in this very simple model, the behavior of solutions depends critically on the relative importance of the populations in the preceding two generations.

EXERCISES 4.5

In Exercises 1 through 4 find the general solution to each given difference equation. When initial conditions are specified, find the unique solution that satisfies them.

1. $y_{n+2} - \sqrt{2}\, y_{n+1} + y_n = 0$

2. $y_{n+2} - 2y_{n+1} + 2y_n = 0$, $y_0 = 1$, $y_1 = 2$

3. $y_{n+2} + 8y_n = 0$

4. $y_{n+2} - 2y_{n+1} + 4y_n = 0$, $y_0 = 0$, $y_1 = 1$

5. Discuss the properties of the solutions of

   a) $x_{n+2} = x_{n+1} + \frac{1}{2}x_n$, and
   b) $x_{n+2} = x_{n+1} - \frac{1}{2}x_n$,

   as special cases of the model of population growth of Example 4.5.3.

6. Consider the integrals

$$I_n(\theta) = \int_0^\pi \frac{\cos nx - \cos n\theta}{\cos x - \cos \theta} \, dx.$$

   a) Show that $I_{n+2}(\theta) - 2 \cos \theta \, I_{n+1}(\theta) + I_n(\theta) = 0$.
   b) Solve this equation and obtain an explicit expression for $I_n(0)$. [*Hint*: It is first necessary to obtain the initial values $I_0(0)$ and $I_1(0)$.]

7. In Chapter 2 we discussed the nonlinear first order Riccati equation

$$y_n y_{n-1} + a_n y_n + b_n y_{n-1} = c_n.$$

   By making the substitution $y_n = x_n/x_{n+1} - b_n$, show that this equation becomes

$$(a_n b_n + c_n)x_{n+1} - (a_n - b_{n-1})x_n - x_{n-1} = 0.$$

Use the results of Exercise 7 to obtain the general solutions of the difference equations in Exercises 8 through 12.

8. $y_n y_{n-1} + 2y_n - 3y_{n-1} = 2$ 9. $y_n y_{n-1} - y_n - y_{n-1} = 1$
10. $y_n y_{n-1} + y_n + y_{n-1} = 1$ 11. $4y_n y_{n-1} + 6y_n - 3y_{n-1} = 0$
12. $2y_n y_{n-1} + 2y_n + 4y_{n-1} = -5$.

Use the technique of Exercise 4.4.14 and the results of this section to obtain the general solutions of the difference equations in the following exercises.

13. $y_{n+3} - 2y_{n+2} + y_{n+1} - 2y_n = 0$

14. $y_{n+3} - y_{n+1} + \sqrt{2}y_n = 0$

15. $y_{n+3} - y_{n+2} + y_{n+1} - y_n = 0$

16. $y_{n+3} - 8y_{n+2} + 8y_{n+1} - 64y_n = 0$

17. $y_{n+3} + 4y_{n+2} - 8y_{n+1} + 24y_n = 0$

4.6 NONHOMOGENEOUS EQUATIONS: VARIATION OF CONSTANTS

There are two methods of solving nonhomogeneous difference equations—undetermined coefficients and variation of constants—which are analogous to the methods of solving nonhomogeneous differential equations. In this section we shall study the very powerful variation of constants method and discuss the technique of undetermined coefficients in the exercises.

As with differential equations, we assume a solution to

$$y_{n+2} + a_n y_{n+1} + b_n y_n = f_n, \qquad b_n \neq 0, \tag{4.35}$$

of the form

$$z_n = c_n x_n + d_n y_n, \tag{4.36}$$

where x_n and y_n are linearly independent solutions of the homogeneous equation. From Eq. (4.36) we have

$$z_{n+1} = c_{n+1} x_{n+1} + d_{n+1} y_{n+1};$$

and adding and subtracting the quantities $c_n x_{n+1}$ and $d_n y_{n+1}$, we obtain

$$z_{n+1} = c_n x_{n+1} + d_n y_{n+1} + (c_{n+1} - c_n) x_{n+1} + (d_{n+1} - d_n) y_{n+1}. \tag{4.37}$$

As the first of two conditions that we will need to determine both c_n and d_n, it is convenient to assume, for all n, that

$$(c_{n+1} - c_n) x_{n+1} + (d_{n+1} - d_n) y_{n+1} = 0. \tag{4.38}$$

This requirement is analogous to that of Eq. (3.42) for differential equations. Thus

$$z_{n+1} = c_n x_{n+1} + d_n y_{n+1} \tag{4.39}$$

and

$$z_{n+2} = c_{n+1} x_{n+2} + d_{n+1} y_{n+2}. \tag{4.40}$$

Assuming that z_n is a solution of Eq. (4.35), we obtain

$$\begin{aligned} f_n &= z_{n+2} + a_n z_{n+1} + b_n z_n \\ &= (c_{n+1} - c_n) x_{n+2} + (d_{n+1} - d_n) y_{n+2} \\ &\quad + c_n [x_{n+2} + a_n x_{n+1} + b_n x_n] + d_n [y_{n+2} + a_n y_{n+1} + b_n y_n]. \end{aligned} \tag{4.41}$$

The bracketed expressions in Eq. (4.41) vanish, since x_n and y_n are solutions of Eq. (4.35), and thus we have the second condition [analogous to (3.43)] that must hold for all n:

$$(c_{n+1} - c_n) x_{n+2} + (d_{n+1} - d_n) y_{n+2} = f_n. \tag{4.42}$$

Combining (4.42) and (4.38), we obtain the following system of two equations in the two unknowns $(c_{n+1} - c_n)$ and $(d_{n+1} - d_n)$:

$$\begin{aligned} (c_{n+1} - c_n) x_{n+1} + (d_{n+1} - d_n) y_{n+1} &= 0, \\ (c_{n+1} - c_n) x_{n+2} + (d_{n+1} - d_n) y_{n+2} &= f_n. \end{aligned} \tag{4.43}$$

The determinant of this system, $C_{n+1}(x,y)$, is nonzero by Theorem 4.4, so that

$$c_{n+1} - c_n = -\frac{f_n y_{n+1}}{C_{n+1}(x,y)}, \qquad d_{n+1} - d_n = \frac{f_n x_{n+1}}{C_{n+1}(x,y)}. \tag{4.44}$$

Equations (4.44) are first order difference equations in the unknowns c_n and d_n which can be solved by means of Eq. (2.66). Thus

$$c_n = c_0 - \sum_{k=0}^{n-1} \frac{f_k y_{k+1}}{C_{k+1}(x,y)}, \qquad d_n = d_0 + \sum_{k=0}^{n-1} \frac{f_k x_{k+1}}{C_{k+1}(x,y)},$$

and the general solution of Eq. (4.35) is given by

$$z_n = c_0 x_n + d_0 y_n - x_n \sum_{k=0}^{n-2} \frac{f_k y_{k+1}}{C_{k+1}(x,y)} + y_n \sum_{k=0}^{n-2} \frac{f_k x_{k+1}}{C_{k+1}(x,y)}, \qquad (4.45)$$

where c_0 and d_0 are arbitrary constants.

Example 4.6.1 Solve

$$y_{n+2} - 3y_{n+1} + 2y_n = 1 + 5^{n+1}. \qquad (4.46)$$

The auxiliary equation has roots $\lambda = 1, 2$, so that the homogeneous equation has the independent solutions $x_n = 1$ and $y_n = 2^n$. The Casoratian is

$$C_{n+1}(x,y) = x_{n+1} y_{n+2} - x_{n+2} y_{n+1} = 2^{n+2} - 2^{n+1} = 2^{n+1},$$

and $f_n = 1 + 5^{n+1}$. Using these values in the sums of Eq. (4.45), we find that

$$\sum_{k=0}^{n-2} \frac{f_k y_{k+1}}{C_{k+1}(x,y)} = \sum_{k=0}^{n-2} (1 + 5^{k+1}) = n + \frac{5^n}{4} - \frac{9}{4},$$

$$\sum_{k=0}^{n-2} \frac{f_k x_{k+1}}{C_{k+1}(x,y)} = \sum_{k=0}^{n-2} \left[\left(\frac{1}{2}\right)^{k+1} + \left(\frac{5}{2}\right)^{k+1} \right] = \frac{2}{3}\left(\frac{5}{2}\right)^n - \left(\frac{1}{2}\right)^{n-1} - \frac{2}{3}, \qquad (4.47)$$

where we have used the following fact about geometric progressions:

$$a + a^2 + a^3 + \cdots + a^{n-1} = \frac{a - a^n}{1 - a}.$$

Inserting (4.47) into Eq. (4.45), combining like terms, and absorbing all stray coefficients into the arbitrary constants c_0 and d_0, we get the general solution of Eq. (4.46):

$$z_n = c_0 + d_0 2^n - n + \frac{(5)^{n+1}}{12}.$$

Example 4.6.2 Consider the equation

$$y_{n+2} + y_n = n + 1. \qquad (4.48)$$

The roots of the auxiliary equation are $\pm i$, so by Theorem 4.9 the homogeneous equation has the independent solutions $x_n = \cos n\pi/2$ and $y_n = \sin n\pi/2$ (see

Example 4.5.2). The Casoratian is

$$C_{n+1}(x,y) = \cos \frac{(n+1)\pi}{2} \sin \frac{(n+2)\pi}{2} - \sin \frac{(n+1)\pi}{2} \cos \frac{(n+2)\pi}{2}$$

$$= \sin \frac{\pi}{2} = 1.$$

Therefore,

$$C_n = c_0 - \sum_{k=0}^{n-1} (k+1) \sin \frac{(k+1)\pi}{2}$$

$$= c_0 - 1 + 3 - 5 + \cdots - n \sin \frac{n\pi}{2}, \tag{4.49}$$

$$d_n = d_0 + \sum_{k=0}^{n-1} (k+1) \cos \frac{(k+1)\pi}{2}$$

$$= d_0 - 2 + 4 - 6 + \cdots + n \cos \frac{n\pi}{2}. \tag{4.50}$$

Note that $c_{2m} = c_{2m-1} = c_0 + (-1)^m m$, and $d_{2m+1} = d_{2m} = d_0 + (-1)^m(m+1) - 1$, so that the general solution of Eq. (4.48) is given by

$$z_n = c_n \cos \frac{n\pi}{2} + d_n \sin \frac{n\pi}{2},$$

where c_n and d_n are as given above. Note that this expression includes c_0 and d_0 as arbitrary constants.

Example 4.6.3 Solve

$$y_{n+2} - (n+2)(n+1)y_n = (n+3)!. \tag{4.51}$$

We saw in Example 4.3.2 that the homogeneous equation has the independent solutions $x_n = n!$ and $y_n = (-1)^n n!$. An easy computation yields the Casoratian $C_{n+1}(x,y) = 2(-1)^n(n+1)!(n+2)!$, so the sums in Eq. (4.45) are given by

$$\sum_{k=0}^{n-2} \frac{f_k y_{k+1}}{C_{k+1}(x,y)} = -\frac{1}{2}[3 + 4 + \cdots + (n+1)] = \frac{3}{2} - \frac{(n+1)(n+2)}{4},$$

$$\sum_{k=0}^{n-2} \frac{f_k x_{k+1}}{C_{k+1}(x,y)} = \frac{1}{2}[3 - 4 + 5 - 6 + \cdots + (-1)^n(n+1)]$$

$$= \frac{(-1)^n}{2}\left(\left[\frac{n}{2}\right] + 1 + (-1)^n\right),$$

where $\left[\frac{n}{2}\right]$ denotes the integral part of $\frac{n}{2}$. The reader should verify this last equation. Thus absorbing constants into the c_0, d_0 terms and collecting like quantities, we obtain the general solution of (4.51):

$$z_n = c_0 n! + d_0(-1)^n n! + n!\left(\frac{n^2}{4} + \frac{3n}{4} + \frac{1}{2}\left[\frac{n}{2}\right]\right).$$

EXERCISES 4.6

Use the method of variation of constants to find particular solutions of the equations in Exercises 1 through 5.

1. $y_{n+2} - 5y_{n+1} + 6y_n = 5^{n+1}$

2. $y_{n+2} + y_{n+1} - 6y_n = e^{n+1} + 3e^n$

3. $y_{n+2} - 4y_{n+1} + 4y_n = n2^n$

4. $y_{n+2} - 2y_{n+1} + y_n = 2^n$

5. $y_{n+2} + 5y_n = \sin\dfrac{n\pi}{2}$

In the following exercises we shall develop the method of undetermined coefficients for difference equations. This method is easier to use than the variation of constants technique, but it applies only when $a_n = a$ and $b_n = b$ are constants (for all n) and f_n is of one of the following three forms: $z_n = A_0 a^n$, $A_1 \sin cn + A_2 \cos cn$, $A_0 + A_1 n + \cdots + A_k n^k$, or any combination of these terms. After finding the independent solutions x_n, y_n of the homogeneous equation, we proceed as follows: write the expression z_n in the same form as f_n, with undetermined coefficients A_j.

i) If no part of z_n is in the general solution of the homogeneous equation, substitute z_n for y_n in the equation

$$y_{n+2} + ay_{n+1} + by_n = f_n, \tag{4.52}$$

and solve for the coefficients A_j.

ii) Otherwise, multiply z_n by the smallest integral power of n such that no part of the product belongs to the general solution of the homogeneous equation, and proceed as with z_n in step (i). (For example, if the homogeneous solutions are 3^n and $n3^n$ and $f_n = 3^n$, then multiply z_n by n^2 so that $z_n = An^2 3^n$ is *not* a solution to the homogeneous equation.)

6. Use $z_n = A_0 5^n$ to show that the equation in Exercise 1 has the general solution

$$y_n = c_0 2^n + c_1 3^n + \tfrac{1}{6}5^{n+1}.$$

7. Solve the equation in Exercise 2 by the method of undetermined coefficients.

8. Solve the equation in Exercise 4 by the method of undetermined coefficients.

9. Solve the equation in Exercise 5 by the method of undetermined coefficients.

10. Solve $y_{n+2} - 4y_{n+1} + 4y_n = 2^n$ by the method of undetermined coefficients.

In the following exercises, find the general solution of each equation by either of the two available methods. In some cases it may be useful to use the principle of superposition (see Exercise 4.2.6).

11. $y_{n+2} - 3y_{n+1} + 2y_n = 2^n + 2^{-n}$

12. $y_{n+2} - y_{n+1} - 6y_n = n + 3^n$

13. $y_{n+2} + y_n = \sin n$

14. $y_{n+3} - 6y_{n+2} + 11y_{n+1} - 6y_n = 2^n$

15. $y_{n+3} - 2y_{n+2} + 3y_n = \cos \dfrac{n\pi}{2}$

16. $y_{n+3} + y_{n+2} - 2y_n = n + n^2 + n^3$

17. $y_{n+2} - y_{n+1} - y_n = n^2 + 1$

18. $y_{n+2} + 2y_{n+1} + y_n = 2^n + 3^n + 4^n$.

4.7 ELECTRICAL NETWORKS

In an electrical network consisting of several coupled closed circuits, we use *Kirchhoff's current law* to determine the relations between the various circuits. This principle may be stated as follows:

> *The algebraic sum of the currents flowing toward any junction in a network must equal zero.*

Using this law and Kirchhoff's voltage law, which was stated in Section 2.8, we can analyze the most complicated electrical networks. Consider, for example, the network shown in Fig. 4.3, where E_j is the voltage relative to the ground at the jth junction, and the shaded region indicates the ground (i.e., zero voltage). We assume that E_0 is a constant voltage, all resistances equal R, and $E_n = 0$. The object of the following discussion is to derive a difference equation for E_k. Using

Figure 4.3

Kirchhoff's current law, we find that

$$I_k = I_k^* + I_{k+1} \tag{4.53}$$

for all integers k so that $1 \leqslant k \leqslant n - 1$. By Ohm's law (see Section 2.8) we may rewrite Eq. (4.53) as

$$\frac{E_{k-1} - E_k}{R} = \frac{E_k}{R} + \frac{E_k - E_{k+1}}{R}$$

or, after some algebra,

$$E_{k+1} - 3E_k + E_{k-1} = 0. \tag{4.54}$$

The auxiliary equation to (4.54) has the roots $(3 \pm \sqrt{5})/2$, so (4.54) has the general solution

$$E_k = c_0 \left(\frac{3 + \sqrt{5}}{2}\right)^k + c_1 \left(\frac{3 - \sqrt{5}}{2}\right)^k. \tag{4.55}$$

Since E_0 is a known constant and $E_n = 0$, we obtain the simultaneous equations

$$c_0 + c_1 = E_0,$$

$$c_0 \left(\frac{3 + \sqrt{5}}{2}\right)^n + c_1 \left(\frac{3 - \sqrt{5}}{2}\right)^n = 0.$$

Noting that

$$\left(\frac{3 + \sqrt{5}}{2}\right)\left(\frac{3 - \sqrt{5}}{2}\right) = 1,$$

we find that

$$c_1 = -c_0 \left(\frac{3 + \sqrt{5}}{2}\right)^{2n}, \qquad c_0 = \frac{E_0}{1 - [(3 + \sqrt{5})/2]^{2n}}.$$

Thus

$$E_k = E_0 \frac{[(3 + \sqrt{5})/2]^{2n-k} - [(3 + \sqrt{5})/2]^k}{[(3 + \sqrt{5})/2]^{2n} - 1}. \tag{4.56}$$

A similar problem will occur if each resistor in Fig. 4.3 is replaced by a capacitor (Fig. 4.4) and the line carries an alternating current [i.e., $E_0(t)$ is a periodic function]. This type of network is called a *string insulator*. Let the capacitance between successive conducting segments be C_1 and the capacitance relative to the ground be C_2. Then Eq. (4.53) can be written in the form

$$C_1 \frac{d}{dt}(E_{k-1} - E_k) = C_2 \frac{d}{dt}(E_k) + C_1 \frac{d}{dt}(E_k - E_{k+1}), \tag{4.57}$$

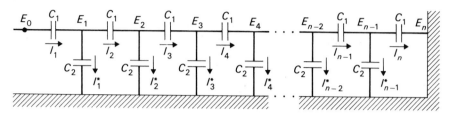

Figure 4.4

since $I = dQ/dt$ and $E_C = Q/C$. Letting $\rho = C_2/2C_1$ and proceeding as in the previous discussion, we find that

$$E_k' = E_0' \frac{(1 + \rho + \sqrt{2\rho + \rho^2})^{2n-k} - (1 + \rho + \sqrt{2\rho + \rho^2})^k}{(1 + \rho + \sqrt{2\rho + \rho^2})^{2n} - 1}$$

$$= E_0' \frac{\sinh a(n - k)}{\sinh an}, \tag{4.58}$$

where $e^a = 1 + \rho + \sqrt{2\rho + \rho^2}$ and $\sinh x = (e^x - e^{-x})/2$. Thus the voltage at the kth junction is given by

$$E_k = E_0 \frac{\sinh a(n - k)}{\sinh an} = A \frac{\sinh a(n - k)}{\sinh an} \cos \omega t, \tag{4.59}$$

if $E_0 = A \cos \omega t$. [We obtained Eq. (4.59) by integrating both sides of Eq. (4.58) and noting that

$$\frac{\sinh a(n - k)}{\sinh an}$$

is a constant relative to t.]

The network in Fig. 4.5 is called a *low-pass filter*, since it damps out all waves above a certain *cut-off* frequency. To see this, we write Eq. (4.53) in the form

$$\frac{E_k - E_{k+1}}{L} = C \frac{d^2}{dt^2} E_{k+1} + \frac{E_{k+1} - E_{k+2}}{L}$$

for $0 \leqslant k \leqslant n - 2$, or

$$E_{k+2} - 2E_{k+1} + E_k = CLE_{k+1}''. \tag{4.60}$$

We now assume that the current in the line is alternating and $E_k = A_k \cos \omega t$ for some fixed frequency ω. Substituting these values into Eq. (4.60) and canceling

Figure 4.5

the common multiple $\cos \omega t$ yield

$$A_{k+2} - (2 - \alpha^2)A_{k+1} + A_k = 0, \tag{4.61}$$

where $\alpha^2 = \omega^2 CL > 0$. The auxiliary equation has the roots

$$1 - \frac{\alpha^2}{2} \pm \sqrt{\left(1 - \frac{\alpha^2}{2}\right)^2 - 1}. \tag{4.62}$$

CASE 1. If $|1 - \alpha^2/2| < 1$, we set $\cos \mu = 1 - \alpha^2/2$ (so $0 < \mu < \pi$). Then Eq. (4.62) becomes

$$\cos \mu \pm i \sin \mu = e^{\pm i\mu}.$$

Hence Eq. (4.61) has the general solution

$$A_k = c_0 \cos k\mu + c_1 \sin k\mu \qquad \text{for} \qquad 0 \leqslant k \leqslant n. \tag{4.63}$$

Now we want to find the constants c_0 and c_1. Two additional conditions hold in the network of Fig. 4.5:

$$\frac{E_0 - E_1}{L} + CE_0'' = 0, \qquad \frac{E_{n-1} - E_n}{L} - CE_n'' = 0$$

at the first and last junctions. These equations yield

$$(1 - \alpha^2)A_0 - A_1 = 0, \qquad A_{n-1} - (1 - \alpha^2)A_n = 0. \tag{4.64}$$

Applying (4.63) to (4.64), we obtain (after numerous steps) the homogeneous system

$$(1 - \cos \mu)c_0 + (\sin \mu)c_1 = 0,$$

$$[\cos n\mu - \cos (n + 1)\mu]c_0 + [\sin n\mu - \sin (n + 1)\mu]c_1 = 0, \tag{4.65}$$

where we used the identities

$$\alpha^2 = 2(1 - \cos \mu),$$

$$2 \cos n\mu \cos \mu = \cos (n + 1)\mu + \cos (n - 1)\mu,$$

$$2 \sin n\mu \cos \mu = \sin (n + 1)\mu + \sin (n - 1)\mu.$$

The equation (4.65) will have a nontrivial solution for c_0 and c_1 if and only if

$$\begin{vmatrix} 1 - \cos \mu & \sin \mu \\ \cos n\mu - \cos (n + 1)\mu & \sin n\mu - \sin (n + 1)\mu \end{vmatrix}$$

$$= \sin (n + 2)\mu - 2 \sin (n + 1)\mu + \sin n\mu$$

$$= -4 \sin^2 \frac{\mu}{2} \sin (n + 1)\mu = 0.$$

Now $\sin (\mu/2) = 0$ only if μ is a multiple of 2π, which is impossible since $0 < \mu < \pi$. Thus $\sin (n + 1)\mu = 0$, which implies that $\mu = N\pi/(n + 1)$ for $N = 1, 2, 3, \ldots, n$. Thus

$$CL\omega^2 = \alpha^2 = 2\left(1 - \cos \frac{N\pi}{n + 1}\right) = 4 \sin^2 \frac{N\pi}{2(n + 1)},$$

implying that the *natural frequencies* of the network are

$$\omega_N = \frac{2}{\sqrt{CL}} \sin \frac{N\pi}{2(n + 1)}, \qquad N = 1, 2, 3, \ldots, n. \tag{4.66}$$

Waves of frequency higher than ω_n $(<2/\sqrt{CL})$ are damped out as t increases. Finally, a short computation shows that the voltages are given by

$$E_k = c_0 \frac{\cos [(2k + 1)N\pi/2(n + 1)]}{\cos [N\pi/2(n + 1)]} \cos \omega_N t \tag{4.67}$$

for $k = 0, 1, \ldots, n$ and $N = 1, 2, 3, \ldots, n$.

CASE 2. If $|1 - \alpha^2/2| \geq 1$, then since $\alpha^2 > 0$, $1 - \alpha^2/2 \leq -1$. If equality holds, $\alpha^2 = 4$, and the roots (4.62) of the auxiliary equation are both -1. This would imply that $A_k = (c_0 + c_1 k)(-1)^k$ would be the general solution of (4.61). Applying this solution to the boundary conditions (4.64), we obtain the homogeneous system

$$2c_0 - c_1 = 0,$$
$$2c_0 + (2n + 1)c_1 = 0,$$

which can have only the trivial solutions $c_0 = c_1 = 0$. Thus $E_k \equiv 0$ for all k, which contradicts our hypothesis.

If $1 - \alpha^2/2 < -1$, then $\alpha^2/2 - 1 > 1$ and there is a $\mu > 0$ so that

$$\frac{\alpha^2}{2} - 1 = \frac{e^\mu + e^{-\mu}}{2}.$$

Then (4.62) becomes $e^{\pm\mu}$, and Eq. (4.61) has the general solution $A_k = c_0 e^{k\mu} + c_1 e^{-k\mu}$. Using this value in the condition of Eqs. (4.64) yields the homogenous system

$$c_0(2e^\mu + 1 + e^{-\mu}) + c_1(e^\mu + 1 + 2e^{-\mu}) = 0,$$
$$c_0 e^{n\mu}(e^\mu + 1 + 2e^{-\mu}) + c_1 e^{-n\mu}(2e^\mu + 1 + e^{-\mu}) = 0.$$

For nontrivial solutions to exist, the determinant must vanish; that is

$$e^{-n\mu}(2e^{\mu} + 1 + e^{-\mu})^2 - e^{n\mu}(e^{\mu} + 1 + 2e^{-\mu})^2 = 0, \tag{4.68}$$

or

$$e^{4\mu}\left(\frac{2 + e^{-\mu} + e^{-2\mu}}{2 + e^{\mu} + e^{2\mu}}\right)^2 = e^{2n\mu}, \tag{4.69}$$

which has no solution for $\mu > 0$. Thus only trivial solutions c_0, c_1 exist, contradicting our assumptions.

EXERCISES 4.7

1. Obtain (4.69) from Eq. (4.68).

2. Investigate the network that is obtained by replacing each resistor in Fig. 4.3 by an inductance L.

3. Consider the network shown in Fig. 4.6, called a *high-pass filter*. Show that it damps out all waves below a certain *cut-off* frequency.

4. The network shown in Fig. 4.7 is called a *band-pass filter*, since frequencies outside a certain band will be damped out. Find the two cut-off frequencies.

Figure 4.6

Figure 4.7

4.8 AN APPLICATION TO GAMES AND QUALITY CONTROL

Suppose that two people, A and B, play a certain unspecified game several times in succession. One of the players must win on each play of the game (ties are ruled out). The probability that A wins is p ($\neq 0$), and the corresponding probability for B is q. Of course, $p + q = 1$. One dollar is bet by each competitor on each play of the game. Suppose that player A starts with k dollars and player B with j dollars. We let P_n denote the probability that A will bankrupt B (i.e., win all his money) when A has n dollars. Clearly $P_0 = 0$ (this is the probability that player A will wipe out player B when he has no money) and $P_{k+j} = 1$ (A has already eliminated B from further competition). To calculate P_n for other values of n between 0 and $k + j$, we have the following difference equation [see Eq. (1.33)]:

$$P_n = qP_{n-1} + pP_{n+1}. \tag{4.70}$$

To understand this equation, we note that if A has n dollars at one turn, his probability of having $n - 1$ dollars at the next turn is q, and similarly for the other term in Eq. (4.70).

Equation (4.70) can be written in our usual format as

$$pP_{n+1} - P_n + qP_{n-1} = 0$$

or

$$P_{n+2} - \frac{1}{p}P_{n+1} + \frac{q}{p}P_n = 0. \tag{4.71}$$

The auxiliary equation is

$$\lambda^2 - \frac{1}{p}\lambda + \frac{q}{p} = 0,$$

which has the roots

$$\frac{1}{2p}(1 \pm \sqrt{1 - 4pq}). \tag{4.72}$$

Since

$$\sqrt{1 - 4pq} = \sqrt{1 - 4p(1 - p)} = \sqrt{4p^2 - 4p + 1} = |1 - 2p|,$$

we discover that the roots of the auxiliary equation are

$$\lambda_1 = \frac{1 - p}{p} = \frac{q}{p}, \qquad \lambda_2 = 1. \tag{4.73}$$

The general solution to Eq. (4.70) is therefore

$$P_n = c_1 \left(\frac{q}{p}\right)^n + c_2 \tag{4.74}$$

when $p \neq q$, and

$$P_n = c_1 + c_2 n \tag{4.75}$$

when $p = q = \frac{1}{2}$ (by the rules for repeated roots). Keeping in mind the fact that $0 \leqslant P_n \leqslant 1$ for all n and using the boundary conditions $P_0 = 0$ and $P_{k+j} = 1$, we can finally obtain

$$P_n = \frac{1 - (q/p)^n}{1 - (q/p)^{k+j}} \quad \text{if} \quad p \neq q, \tag{4.76}$$

and

$$P_n = \frac{n}{k+j} \quad \text{if} \quad p = q = \frac{1}{2}, \tag{4.77}$$

for all n so that $0 \leqslant n \leqslant k + j$.

Example 4.8.1 In a game of roulette a gambler bets against a casino. Suppose that the gambler bets one dollar on red each time. On a typical roulette wheel there are eighteen reds, eighteen blacks, a zero, and a double zero. If the wheel is honest, the probability that the casino will win is $p = \frac{20}{38} \approx 0.5263$. Then $q = 0.4737$ and $q/p = 0.9$. Suppose that both the gambler and the casino start with k dollars. Then the probability that the gambler will be cleaned out is

$$P_k = \frac{1 - (q/p)^k}{1 - (q/p)^{2k}} = \frac{1}{1 + (q/p)^k} = \frac{1}{1 + (0.9)^k}.$$

The second column in Table 4.1 shows the probability that the gambler will lose all his money for different initial amounts k. Note that the more money both start with, the more likely it is that the gambler will lose all his money. It is evident that as the gambler continues to play, the probability that he will be wiped out approaches a certainty. This is why even a very small advantage enables the casino, in almost every case, to break the gambler if he continues to bet.

Table 4.1

k	P_k	P_k^*
1	0.5263	0.5263
5	.6287	.8739
10	.7415	.9491
25	.9331	.9920
50	.9948	.9994
100	.9999	.9999

The third column in Table 4.1 gives the probability that the gambler will lose all his money if he has one dollar and the casino has k dollars:

$$P_k^* = \frac{1 - (q/p)^k}{1 - (q/p)^{k+1}}.$$

Thus, if the gambler has much less money than the casino, he is even more likely to lose it all. The combination of a small advantage, table limits, and huge financial resources makes running a casino a very profitable business indeed.

Example 4.8.2 The previous example can be used as a simple model of competition for resources between two species in a single ecological niche. Suppose that the species, A and B, control a combined "territory" of N resource units. The resource units may be such things as acres of grassland, number of trees, and so on. If species A controls k units, then species B will control the remaining $N - k$ units. Suppose that during each unit of time there is competition for one unit of resource between the two species. Let species A have the probability p of being successful. If we let P_n denote the probability that species B will lose all its resources when species A has n units of resource, then P_n is determined by Eq. (4.70) with initial conditions $P_0 = 0$ and $P_N = 1$. By Eqs. (4.76) and (4.77) the solution to this equation is

$$P_n = \begin{cases} \dfrac{1 - [(1 - p)/p]^n}{1 - [(1 - p)/p]^N} & \text{if } p \neq \dfrac{1}{2}, \\[2ex] \dfrac{n}{N} & \text{if } p = \dfrac{1}{2}. \end{cases}$$

As in the previous example, even a small competitive advantage will virtually ensure the extinction of the weaker species. For example, if $p = 0.55$, and species A has n of a total of 100 units, then we have the following table.

Table 4.2

n	P_n
1	0.1818
2	.3306
3	.4523
4	.5519
5	.6333
10	.8656
25	.9935

Note that even if the species with the competitive advantage starts with as few as four of the 100 units of resource, it has a better than even chance to supplant the weaker species. With one quarter of the resources, it is virtually certain to do so.

A process similar to the one discussed has been used with success to determine whether, in a manufacturing process, a batch of articles is satisfactory. Let us briefly describe this process here. More extensive details can be found in the paper by G. A. Barnard.*

To test whether a batch of articles is satisfactory, we introduce a scoring system. The score is initially set at N. If a randomly sampled item is found to be defective, we subtract k. If it is acceptable, we add 1. The procedure stops when the score reaches either $2N$ or becomes nonpositive. If $2N$, the batch is accepted; if nonpositive, it is rejected. Suppose that the probability of selecting an acceptable item is p, and $q = 1 - p$. Let P_n denote the probability that the batch will be rejected when the score is at n. Then after the next choice, the score will be either increased by 1 with probability p or decreased by k with probability q. Thus

$$P_n = pP_{n+1} + qP_{n-k}, \tag{4.78}$$

which can be written as the $(k + 1)$st order difference equation

$$P_{n+k+1} - \frac{1}{p}P_{n+k} + \frac{q}{p}P_n = 0 \tag{4.79}$$

with boundary conditions

$$P_{1-k} = \cdots = P_0 = 1, \qquad P_{2N} = 0. \tag{4.80}$$

The case of $k = 1$ reduces to Example 4.8.1 with the result that a batch is almost certain to be accepted or rejected depending on whether p is greater or less than $\frac{1}{2}$.

EXERCISES 4.8

1. In Example 4.8.1 assume that player A has a 10-percent competitive advantage. If player A starts with $3, how much money must player B start with to have a better than even chance to win all of A's money? To have an 80-percent chance?

2. Answer the questions in Exercise 1 given that player A has

   a) a 2-percent advantage,
   b) a 20-percent advantage.

3. The games described in Examples 4.8.1 and 4.8.2 can be modified to allow for more than one possibility. Suppose that, on each play of the game player A has the probability p to win one dollar, q to win two dollars, and r to lose one dollar, where $p + q + r = 1$. Let P_n be as before. Write down a difference equation that determines P_N, assuming that each player starts with N dollars.

4. Given that $p = q = \frac{1}{4}$ and $r = \frac{1}{2}$, solve the difference equation of Exercise 3.

---

* G. A. Barnard, "Sequential tests in industrial statistics," Supplement to the *Journal of the Royal Statistical Society*, **8**, No 1, 1946.

5. Given that $p = q = r = \frac{1}{3}$, solve the difference equation of Exercise 3.

6. Suppose that player A has the probability p of winning one dollar, q of losing one dollar, and r of losing two dollars. Let P_n be the probability that A will bankrupt B when A has n dollars, and suppose A and B each have N dollars initially.

   a) Obtain a difference equation with appropriate boundary conditions for this problem.
   b) Solve the equation with $p = \frac{1}{6}, q = \frac{1}{3}, r = \frac{1}{2}$.

7. Analyze a game in which on each play player A has a $\frac{1}{3}$ probability of winning two dollars and $\frac{2}{3}$ probability of losing one dollar.

   a) Assume that A and B initially each have N dollars. Does A have an initial advantage? Suppose that $N = 10$. What is P_{10}?
   b) If A has ten dollars and B has twenty dollars initially, what is the initial probability that A will bankrupt B?

8. Discuss each of the previous exercises in the context of two species competing for resources.

9. Formulate a model comparable to the one in Example 4.8.2 for three species competing for a single resource.

POWER
SERIES
SOLUTIONS
OF
DIFFERENTIAL
EQUATIONS

5

In Chapter 3 we studied several methods for solving second and higher order differential equations. With the exception of the Euler equation and a few equations where one solution was easily guessed, the techniques applied only to linear differential equations with *constant coefficients*. The case of linear differential equations with *variable coefficients* is much more complicated. Unfortunately many of the most important differential equations in applied mathematics, for example Bessel's equation and Legendre's equation, are of this type. In this chapter we will consider a method for obtaining solutions to such equations. Since the solutions so obtained are in the form of power series, the procedure used is known as the *power series method*.

5.1 REVIEW OF POWER SERIES

In this section we will review some of the basic properties of power series. We take it for granted that most readers will have received some background in power series in an earlier course in calculus. A *power series* in $(x - a)$ is an infinite series of the form

$$\sum_{n=0}^{\infty} c_n(x - a)^n = c_0 + c_1(x - a) + c_2(x - a)^2 + \cdots, \tag{5.1}$$

where $c_0, c_1, \ldots$ are constants, called the *coefficients* of the series, a is a constant called the *center* of the series, and x is an independent variable. In particular, a power series centered at zero ($a = 0$) has the form

$$\sum_{n=0}^{\infty} c_n x^n = c_0 + c_1 x + c_2 x^2 + c_3 x^3 + \cdots. \tag{5.2}$$

Note that polynomials are also power series, since they have this form.

A series of the form (5.1) can always be reduced to the form (5.2) by the substitution $X = x - a$. This substitution is merely a translation of the coordinate system. It is easy to see that the behavior of (5.2) near zero is exactly the same as the behavior of (5.1) near a. For this reason we need only study the properties of series of the form (5.2).

The expression

$$s_n(x) = c_0 + c_1 x + c_2 x^2 + \cdots + c_n x^n \tag{5.3}$$

is called the *nth partial sum* of the series (5.2); and the difference between (5.2) and (5.3),

$$R_n(x) = c_{n+1} x^{n+1} + c_{n+2} x^{n+2} + \cdots, \tag{5.4}$$

is called the *remainder after the nth term* of (5.2). Thus, in the case of the series

$$\sum_{n=0}^{\infty} (n + 1)x^n = 1 + 2x + 3x^2 + 4x^3 + \cdots,$$

we have

$$s_1(x) = 1 + 2x, \qquad\qquad R_1(x) = 3x^2 + 4x^3 + \cdots,$$
$$s_2(x) = 1 + 2x + 3x^2, \qquad R_2(x) = 4x^3 + 5x^4 + \cdots, \quad \text{and so on.}$$

In this way we can generate a sequence of partial sums $s_1(x), s_2(x), \ldots$. It often happens that for some particular value $x = x_0$, the sequence of partial sums evaluated at x_0, has a (finite) limit. When that occurs, we say that the series (5.2) *converges* (or *is convergent*) *at* x_0, and write

$$\sum_{n=0}^{\infty} c_n x_0^n = \lim_{n \to \infty} s_n(x_0) = s(x_0). \tag{5.5}$$

The number $s(x_0)$ obtained in (5.5) is called the *sum* of the series (5.2) at x_0. When the series is not convergent at x_0, we say it *diverges* (or *is divergent*) *at* x_0. In case of convergence, the partial sum $s_n(x_0)$ is an approximation of the sum $s(x_0)$, and the error of the approximation

$$|s(x_0) - s_n(x_0)| = |R_n(x_0)|$$

can be made as small as desired by using a sufficiently large n.

If a series (5.2) converges at x_0, then the terms $c_n x_0^n$ tend to zero:

$$\lim_{n \to \infty} c_n x_0^n = \lim_{n \to \infty} s_n(x_0) - s_{n-1}(x_0)$$

$$= \lim_{n \to \infty} s_n(x_0) - \lim_{n \to \infty} s_{n-1}(x_0)$$

$$= s(x_0) - s(x_0) = 0. \tag{5.6}$$

However, even if the terms in the series tend to zero, the series may diverge (see Exercise 5.1.17).

If we let $x = 0$ in (5.2), then the series is reduced to the single term c_0 and thus converges. There may be other values at which the series converges. The values for which the series (5.2) converges lie in an interval centered at 0, called the *interval of convergence*. (This fact is proved in Theorem 5.1.) Hence to each series (5.2) corresponds a number $0 \leq R \leq \infty$, called the *radius of convergence*, with the property that the series converges if $|x| < R$ and diverges if $|x| > R$.

The radius of convergence can often be determined by means of the *root test* formula

$$\frac{1}{R} = \lim_{n \to \infty} \sqrt[n]{|c_n|}, \tag{5.7}$$

provided that the limit exists. (If $\sqrt[n]{|c_n|} \to 0$, we set $R = \infty$.)

Example 5.1.1 Consider the *geometric series*

$$\sum_{n=0}^{\infty} x^n = 1 + x + x^2 + \cdots. \tag{5.8}$$

To find the sum of this series, we note that

$$s_n(x) = 1 + x + \cdots + x^{n-1} + x^n,$$

so that

$$xs_n(x) = x + x^2 + \cdots + x^n + x^{n+1}.$$

Taking the difference between these two expressions, we find that $(1 - x)s_n(x) = 1 - x^{n+1}$ or

$$s_n(x) = 1 + x + \cdots + x^n = \frac{1 - x^{n+1}}{1 - x}. \tag{5.9}$$

If $|x| < 1$, then the numerator in the right-hand term tends to 1 as n tends to infinity. Thus

$$s(x) = \lim_{n \to \infty} s_n(x) = \frac{1}{1 - x}, \tag{5.10}$$

for all x with $|x| < 1$.

If $|x| \geqslant 1$, then the sequence $\{s_n(x)\}$ does not have a finite limit, so the series (5.8) diverges.

Using the root test (5.7), we obtain $R = 1$, confirming the results above. The case $|x| = R$ must always be checked separately.

Theorem 5.1 Let R be determined by the root test formula (5.7). Then the series (5.2) converges for each x in $|x| < R$ and diverges for every x such that $|x| > R$.

Proof. Let $|x| < r < R$ so that $1/R < 1/r$. Since $\sqrt[n]{|c_n|} \to 1/R$, we must have $\sqrt[n]{|c_n|} < 1/r$ for all $n \geqslant n_0$, where n_0 is sufficiently large. Then using the triangle inequality,* we find that the remainder term $R_n(x)$ satisfies the inequality

$$|R_n(x)| \leqslant |c_{n+1}| \, |x|^{n+1} + |c_{n+2}| \, |x|^{n+2} + \cdots < \left|\frac{x}{r}\right|^{n+1} + \left|\frac{x}{r}\right|^{n+2} + \cdots. \tag{5.11}$$

Using (5.10), we can rewrite the last term in (5.11) in the form

$$|R_n(x)| < \left|\frac{x}{r}\right|^{n+1} \left(1 + \left|\frac{x}{r}\right| + \left|\frac{x}{r}\right|^2 + \cdots\right) = \frac{|x/r|^{n+1}}{1 - |x/r|}, \tag{5.12}$$

since $|x/r| < 1$. By (5.12), it is clear that the remainder term vanishes as n tends to infinity. Thus the series converges for each x such that $|x| < R$.

If $|x| > r > R$, we have $\sqrt[n]{|c_n|} > 1/r$, so that

$$|c_n x^n| > \left|\frac{x}{r}\right|^n > 1.$$

---

* The triangle inequality states that $|a + b| \leqslant |a| + |b|$.

Thus the terms in the series do not tend to zero. By (5.6) we see that the series must diverge for all x such that $|x| > R$.

Another useful formula for determining the radius of convergence is given by the *ratio test*:

$$\frac{1}{R} = \lim_{n \to \infty} \left| \frac{c_{n+1}}{c_n} \right| \tag{5.13}$$

whenever the limit exists. The proof that this limit in fact yields the same value of R is left as an exercise (see Exercise 5.1.20).

Example 5.1.2 Find the radius of convergence of the series

$$\sum_{n=0}^{\infty} \frac{x^n}{n!} = 1 + x + \frac{x^2}{2!} + \frac{x^3}{3!} + \cdots.$$

Here $c_n = 1/n!$, so by the ratio test (5.13)

$$\frac{1}{R} = \lim_{n \to \infty} \frac{n!}{(n+1)!} = \lim_{n \to \infty} \frac{1}{n+1} = 0,$$

and $R = \infty$. Thus this series converges for all values of x.

Example 5.1.3 The series

$$\sum_{n=0}^{\infty} n! x^n = 1 + x + 2! x^2 + 3! x^3 + \cdots$$

diverges for all $x \neq 0$, since

$$\frac{1}{R} = \lim_{n \to \infty} \left| \frac{c_{n+1}}{c_n} \right| = \lim_{n \to \infty} \frac{(n+1)!}{n!} = \lim_{n \to \infty} (n+1) = \infty.$$

The most familiar power series are those that are obtained by the use of *Taylor's formula*:

$$f(x) = \sum_{n=0}^{N} \frac{f^{(n)}(x_0)}{n!} (x - x_0)^n + R_N(x - x_0). \tag{5.14}$$

If the function $f(x)$ has derivatives of all orders at the point x_0, and the remainder term $R_N(x - x_0)$ tends to zero as N tends to infinity, we say that $f(x)$ is *analytic at* x_0 and write

$$f(x) = \sum_{n=0}^{\infty} \frac{f^{(n)}(x_0)}{n!} (x - x_0)^n. \tag{5.15}$$

The series (5.15) is called the *Taylor series* of $f(x)$ at the point $x = x_0$. When $x_0 = 0$, (5.15) is often called the *Maclaurin series* of $f(x)$. The following familiar expansions, valid for all x, may all be obtained by this method:

$$e^x = \sum_{n=0}^{\infty} \frac{x^n}{n!} = 1 + x + \frac{x^2}{2!} + \frac{x^3}{3!} + \cdots, \tag{5.16}$$

$$\sin x = \sum_{n=0}^{\infty} \frac{(-1)^n x^{2n+1}}{(2n+1)!} = x - \frac{x^3}{3!} + \frac{x^5}{5!} - \frac{x^7}{7!} + \cdots, \tag{5.17}$$

$$\cos x = \sum_{n=0}^{\infty} \frac{(-1)^n x^{2n}}{(2n)!} = 1 - \frac{x^2}{2!} + \frac{x^4}{4!} - \frac{x^6}{6!} + \cdots, \tag{5.18}$$

$$\sinh x = \sum_{n=0}^{\infty} \frac{x^{2n+1}}{(2n+1)!} = x + \frac{x^3}{3!} + \frac{x^5}{5!} + \frac{x^7}{7!} + \cdots, \tag{5.19}$$

$$\cosh x = \sum_{n=0}^{\infty} \frac{x^{2n}}{(2n)!} = 1 + \frac{x^2}{2!} + \frac{x^4}{4!} + \frac{x^6}{6!} + \cdots. \tag{5.20}$$

There are functions $f(x)$ which have derivatives of all orders at a given point x_0 and yet are not analytic. In these cases, the remainder term $R_N(x - x_0)$ does not tend to zero as N tends to ∞. An example of such a function is given in Exercise 5.1.24.

Let us now consider four operations on power series; all will be used in applying the power series method to differential equations.

1. *Two power series may be added term by term.* To be precise, if the series

$$\sum_{n=0}^{\infty} b_n(x - x_0)^n \quad \text{and} \quad \sum_{n=0}^{\infty} c_n(x - x_0)^n$$

have radii of convergence R_b and R_c and sums $b(x)$ and $c(x)$, respectively, then the series

$$\sum_{n=0}^{\infty} (b_n + c_n)(x - x_0)^n$$

converges to $b(x) + c(x)$ for all x such that $|x - x_0| < R$, where $R = \min(R_b, R_c)$.

2. *Two power series may be multiplied term by term.* Indeed, given the preceding two series, we have

$$b(x)c(x) = \left(\sum_{n=0}^{\infty} b_n(x - x_0)^n \right) \left(\sum_{n=0}^{\infty} c_n(x - x_0)^n \right)$$

$$= \sum_{n=0}^{\infty} (b_0 c_n + b_1 c_{n-1} + \cdots + b_n c_0)(x - x_0)^n \tag{5.21}$$

for all x such that $|x - x_0| < R$, where again $R = \min(R_b, R_c)$.

3. *A power series may be differentiated term by term.* Thus

$$b'(x) = \frac{d}{dx}\left(\sum_{n=0}^{\infty} b_n(x - x_0)^n\right) = \sum_{n=1}^{\infty} nb_n(x - x_0)^{n-1} \tag{5.22}$$

for all x such that $|x - x_0| < R_b$.

4. *A power series may be integrated term by term.* That is,

$$\int b(x)\, dx = \int \sum_{n=0}^{\infty} b_n(x - x_0)^n\, dx = \sum_{n=0}^{\infty} \frac{b_n(x - x_0)^{n+1}}{n + 1} \tag{5.23}$$

for all x such that $|x - x_0| < R_b$.

We note that *the radius of convergence is unaffected by differentiation or integration.*

We will leave the proofs of the first two properties as exercises (see Exercises 5.1.18 and 5.1.19). To show that Eq. (5.22) is indeed true, we observe that

$$\frac{b(x + h) - b(x)}{h} = \sum_{n=0}^{\infty} b_n \frac{(x + h - x_0)^n - (x - x_0)^n}{h}. \tag{5.24}$$

Expanding the term

$$(x - x_0 + h)^n = (x - x_0)^n + nh(x - x_0)^{n-1} + \cdots + h^n,$$

we see that Eq. (5.24) becomes

$$\sum_{n=1}^{\infty} b_n \frac{nh(x - x_0)^{n-1} + \cdots + h^n}{h} = \sum_{n=1}^{\infty} b_n[n(x - x_0)^{n-1} + \cdots + h^{n-1}], \tag{5.25}$$

since the first term drops out. Interchanging the limit and sum* and letting h tend to zero, we note that all but the first term in the brackets vanish in Eq. (5.25). Therefore, we obtain the right-hand side of Eq. (5.22). To see that the series (5.22) has the same radius of convergence, we may use the limit (5.13) (if it exists):

$$\lim_{n \to \infty} \left|\frac{(n + 1)b_{n+1}}{nb_n}\right| = \lim_{n \to \infty} \frac{n + 1}{n} \lim_{n \to \infty} \left|\frac{b_{n+1}}{b_n}\right| = \lim_{n \to \infty} \left|\frac{b_{n+1}}{b_n}\right| = \frac{1}{R_b}.$$

Finally, we prove (5.23) by means of (5.22), noting that

$$\frac{d}{dx} \sum_{n=0}^{\infty} b_n \frac{(x - x_0)^{n+1}}{n + 1} = \sum_{n=0}^{\infty} \frac{d}{dx} b_n \frac{(x - x_0)^{n+1}}{n + 1} = \sum_{n=0}^{\infty} b_n(x - x_0)^n = b(x).$$

Thus (5.23) is, indeed, an integral of $b(x)$.

---

* Interchanging limits and sums requires justification. In this case there is no difficulty, since power series are *uniformly convergent* on closed bounded subsets of $|x| < R_b$. The proof may be found in most advanced calculus texts.

EXERCISES 5.1

Obtain each of the following Taylor series by repeated differentiations.

1. (5.16)   2. (5.17)   3. (5.18)

4. (5.19)   5. (5.20)

6. $\ln (1 + x) = x - \dfrac{x^2}{2} + \dfrac{x^3}{3} - \dfrac{x^4}{4} + \cdots, |x| < 1$

7. $\sin^{-1} x = x + \dfrac{1}{2} \cdot \dfrac{x^3}{3} + \dfrac{1}{2} \cdot \dfrac{3}{4} \cdot \dfrac{x^5}{5} + \dfrac{1}{2} \cdot \dfrac{3}{4} \cdot \dfrac{5}{6} \cdot \dfrac{x^7}{7} + \cdots, |x| < 1$

8. $\ln x = (x - 1) - \dfrac{(x - 1)^2}{2} + \dfrac{(x - 1)^3}{3} - \dfrac{(x - 1)^4}{4} + \cdots, 0 < x < 2$

9. $\dfrac{1}{2 - x} = 1 + (x - 1) + (x - 1)^2 + (x - 1)^3 + \cdots, 0 < x < 2$

10. Use Eq. (5.10) to show that

$$\frac{1}{1 + x} = 1 - x + x^2 - x^3 + \cdots, |x| < 1.$$

Then prove that:

a) $\ln (1 + x) = x - \dfrac{x^2}{2} + \dfrac{x^3}{3} - \dfrac{x^4}{4} + \cdots, |x| < 1,$

b) $\tan^{-1} x = x - \dfrac{x^3}{3} + \dfrac{x^5}{5} - \dfrac{x^7}{7} + \cdots, |x| < 1,$

c) $\dfrac{1}{(1 + x)^2} = 1 - 2x + 3x^2 - 4x^3 + \cdots, |x| < 1.$

Find the radius and interval of convergence of each given series in Exercises 11 through 16.

11. $\displaystyle\sum_{n=1}^{\infty} \frac{x^n}{n^2}$ 12. $\displaystyle\sum_{n=1}^{\infty} \left(\frac{x - 2}{n}\right)^n$ 13. $\displaystyle\sum_{n=1}^{\infty} n^n x^n$

14. $\displaystyle\sum_{n=0}^{\infty} \frac{n^2(x - 1)^n}{2^n}$ 15. $\displaystyle\sum_{n=0}^{\infty} \frac{(2n)! x^n}{(n!)^2}$ 16. $\displaystyle\sum_{n=0}^{\infty} \frac{(k + n)!(x + 1)^n}{k! n!}$

17. Show that the series

$$\sum_{n=1}^{\infty} \frac{x^n}{n} = x + \frac{x^2}{2} + \frac{x^3}{3} + \frac{x^4}{4} + \cdots$$

diverges at $x = 1$ by proving that the partial sums satisfy the inequality

$$s_{2k}(1) \geqslant 1 + k/2.$$

(This exercise shows that even though the terms in a series may tend to zero, the series itself may diverge, in this case at $x = 1$.)

18. Prove that two power series may be added term by term and determine the appropriate interval of convergence.

19. Prove that two power series may be multiplied term by term and determine the appropriate interval of convergence.

20. Prove that formulas (5.13) and (5.7) yield the same result when both limits exist.

21. Prove that the remainder term $R_n(x - x_0)$ of Taylor's formula satisfies the equation

$$R_n(x - x_0) = \frac{f^{(n+1)}(\bar{x})}{(n+1)!}(x - x_0)^{n+1},$$

where $\bar{x}$ is some value between x_0 and x. [*Hint*: Use the mean value theorem of differential calculus.]

22. Use the result of Exercise 21 to verify that e^x, $\sin x$, and $\cos x$ are analytic for all x.

23. Use the result of Exercise 21 to prove that $\ln(1 + x)$ and $\tan^{-1} x$ are analytic for all x such that $|x| < 1$.

24. Consider the function

$$f(x) = \begin{cases} e^{-1/x^2}, & x \neq 0, \\ 0, & x = 0. \end{cases}$$

a) Show that f has derivatives of all orders at $x = 0$ and that

$$f'(0) = f''(0) = \cdots = 0.$$

b) Conclude that $f(x)$ does not have a Taylor series expansion at $x = 0$, even though it is infinitely differentiable there. Thus f is not analytic at $x = 0$.

25. Using Taylor's formula, prove the *binomial formula*

$$(1 + x)^p = 1 + px + \frac{p(p - 1)}{1 \cdot 2}x^2 + \frac{p(p - 1)(p - 2)}{1 \cdot 2 \cdot 3}x^3 + \cdots.$$

5.2 THE POWER SERIES METHOD: EXAMPLES

The fundamental assumption employed in solving a differential equation by the power series method is that the solution of the differential equation can be expressed in the form of a power series, say

$$y = \sum_{n=0}^{\infty} c_n x^n. \tag{5.26}$$

Once this assumption has been made, power series expansions for y', y'', ... can be obtained by differentiating (5.26) term by term,

$$y' = \sum_{n=1}^{\infty} nc_n x^{n-1}, \tag{5.27}$$

$$y'' = \sum_{n=2}^{\infty} n(n - 1)c_n x^{n-2}, \quad \text{etc.} \tag{5.28}$$

and these may then be substituted into the given differential equation. After all the indicated operations have been carried out, and like powers of x have been collected, we obtain an expression of the form

$$k_0 + k_1 x + k_2 x^2 + \cdots = \sum_{n=0}^{\infty} k_n x^n = 0, \tag{5.29}$$

where the coefficients k_0, k_1, k_2, ... are expressions involving the unknown coefficients c_0, c_1, c_2, ... Since Eq. (5.29) must hold for all values of x in some interval, all the coefficients k_0, k_1, k_2, ... must vanish. From the equations

$$k_0 = 0, \quad k_1 = 0, \quad k_2 = 0, \ldots$$

it is then possible to determine successively the coefficients c_0, c_1, c_2, ... In this section we will illustrate this procedure by means of several examples without concerning ourselves with questions of the convergence of the power series under consideration. The next section will contain a discussion of the validity and limitations of the methods illustrated here.

Example 5.2.1 Consider the initial value problem

$$y' = y + x^2, \quad y(0) = 1. \tag{5.30}$$

Inserting (5.26) and (5.27) into the equation, we have

$$c_1 + 2c_2 x + 3c_3 x^2 + 4c_4 x^3 + \cdots = (c_0 + c_1 x + c_2 x^2 + c_3 x^3 + \cdots) + x^2.$$

Collecting like powers of x, we obtain

$$(c_1 - c_0) + (2c_2 - c_1)x + (3c_3 - c_2 - 1)x^2 + (4c_4 - c_3)x^3 + \cdots = 0.$$

Equating each of the coefficients to zero, we obtain the identities

$$c_1 - c_0 = 0, \quad 2c_2 - c_1 = 0, \quad 3c_3 - c_2 - 1 = 0, \quad 4c_4 - c_3 = 0, \ldots,$$

from which we find that

$$c_1 = c_0, \quad c_2 = \frac{c_1}{2} = \frac{c_0}{2!}, \quad c_3 = \frac{c_2 + 1}{3} = \frac{c_0 + 2}{3!}, \quad c_4 = \frac{c_3}{4} = \frac{c_0 + 2}{4!}, \ldots.$$

With these values, (5.26) becomes

$$y = c_0 + c_0 x + \frac{c_0}{2!} x^2 + \frac{c_0 + 2}{3!} x^3 + \frac{c_0 + 2}{4!} x^4 + \frac{c_0 + 2}{5!} x^5 + \cdots$$

$$= (c_0 + 2)\left[1 + x + \frac{x^2}{2!} + \frac{x^3}{3!} + \frac{x^4}{4!} + \cdots\right] - 2\left[1 + x + \frac{x^2}{2!}\right].$$

Using the expansion (5.16), we have the general solution

$$y = (c_0 + 2)e^x - x^2 - 2x - 2.$$

To solve the initial value problem, we set $x = 0$ to obtain

$$1 = y(0) = c_0 + 2 - 2 = c_0.$$

Thus the solution of the initial value problem (5.30) is given by the equation

$$y = 3e^x - x^2 - 2x - 2.$$

Example 5.2.2 Solve

$$y'' + y = 0. \tag{5.31}$$

Using (5.26) and (5.28), we have

$$(2c_2 + 3 \cdot 2c_3 x + 4 \cdot 3c_4 x^2 + \cdots) + (c_0 + c_1 x + c_2 x^2 + \cdots) = 0.$$

Gathering like powers of x yields

$$(2c_2 + c_0) + (3 \cdot 2c_3 + c_1)x + (4 \cdot 3c_4 + c_2)x^2 + \cdots = 0.$$

Setting each of the coefficients to zero, we obtain

$$2c_2 + c_0 = 0, \quad 3 \cdot 2c_3 + c_1 = 0, \quad 4 \cdot 3c_4 + c_2 = 0, \quad 5 \cdot 4c_5 + c_3 = 0, \ldots,$$

and

$$c_2 = -\frac{c_0}{2!}, \quad c_3 = -\frac{c_1}{3!}, \quad c_4 = -\frac{c_2}{4 \cdot 3} = \frac{c_0}{4!}, \quad c_5 = -\frac{c_3}{5 \cdot 4} = \frac{c_1}{5!}, \ldots.$$

Substituting these values into the power series (5.25) for y yields

$$y = c_0 + c_1 x - \frac{c_0}{2!} x^2 - \frac{c_1}{3!} x^3 + \frac{c_0}{4!} x^4 + \frac{c_1}{5!} x^5 + \cdots.$$

Splitting this series into two parts, we have

$$y = c_0 \left(1 - \frac{x^2}{2!} + \frac{x^4}{4!} - \cdots \right) + c_1 \left(x - \frac{x^3}{3!} + \frac{x^5}{5!} - \cdots \right).$$

Using Eqs. (5.17) and (5.18) reveals the familiar general solution

$$y = c_0 \cos x + c_1 \sin x.$$

We observe that in this case the power series method produces two arbitrary constants c_0, c_1 and yields the general solution for Eq. (5.31).

So far we have considered only linear equations with constant coefficients. We turn now to linear equations with variable coefficients.

Example 5.2.3 Consider the initial value problem

$$(1 + x^2)y' = 2pxy, \qquad y(0) = 1, \tag{5.32}$$

where p is a constant. Applying (5.26) and (5.27), we have

$$(1 + x^2) \sum_{n=1}^{\infty} nc_n x^{n-1} = 2px \sum_{n=0}^{\infty} c_n x^n.$$

We use the summation notation in this example in order to develop the skill in manipulating power series that will be required later on. Equation (5.32) may be rewritten in the form

$$\sum_{n=1}^{\infty} nc_n x^{n-1} + \sum_{n=1}^{\infty} nc_n x^{n+1} = \sum_{n=0}^{\infty} 2pc_n x^{n+1}. \tag{5.33}$$

We would like to rewrite each of the sums in Eq. (5.33) so that each general term will contain the same power of x. This can be done by assuming that each general term contains the term x^k. For the first sum, this amounts to substituting $k = n - 1$. Since n ranges from 1 to ∞, $k = n - 1$ will range from 0 to ∞. Substituting $k = n + 1$ with k ranging from 2 to ∞ into the second sum, and $k = n + 1$ with k ranging from 1 to ∞ into the third sum, allows us to rewrite these sums so that the general term will involve the power x^k. We then obtain

$$\sum_{k=0}^{\infty} (k + 1)c_{k+1} x^k + \sum_{k=2}^{\infty} (k - 1)c_{k-1} x^k = \sum_{k=1}^{\infty} 2pc_{k-1} x^k.$$

Now we can gather like terms in x:

$$c_1 + (2c_2 - 2pc_0)x + \sum_{k=2}^{\infty} \{(k + 1)c_{k+1} + [(k - 1) - 2p]c_{k-1}\}x^k = 0. \tag{5.34}$$

Equating each coefficient to zero yields

$$c_1 = 0, \qquad 2c_2 - 2pc_0 = 0, \qquad 3c_3 + (1 - 2p)c_1 = 0,$$

and in general

$$(k + 1)c_{k+1} + [(k - 1) - 2p]c_{k-1} = 0, \qquad k \geqslant 1. \tag{5.35}$$

We note that Eq. (5.35) is a difference equation with variable coefficients. This equation is called a *recursion formula* and may be used to evaluate the constants $c_0, c_1, c_2, \ldots$ successively. We see that

$$c_1 = 0, \qquad c_2 = pc_0, \qquad c_3 = 0,$$

and by (5.35), in general

$$c_{k+1} = \frac{(2p - k + 1)}{k + 1} c_{k-1}, \tag{5.36}$$

so that

$$c_4 = \frac{2p - 2}{4}c_2 = \frac{p(p - 1)}{1 \cdot 2}c_0, \qquad c_5 = 0,$$

$$c_6 = \frac{2p - 4}{6}c_4 = \frac{p(p - 1)(p - 2)}{1 \cdot 2 \cdot 3}c_0, \qquad c_7 = 0, \ldots$$

since $c_3 = 0$. Thus the coefficients with odd-numbered subscripts vanish and the power series for y is given by

$$y = c_0 + \frac{p}{1}c_0 x^2 + \frac{p(p - 1)}{1 \cdot 2}c_0 x^4 + \frac{p(p - 1)(p - 2)}{1 \cdot 2 \cdot 3}c_0 x^6 + \cdots$$

$$= c_0\left(1 + \frac{p}{1}x^2 + \frac{p(p - 1)}{1 \cdot 2}x^4 + \frac{p(p - 1)(p - 2)}{1 \cdot 2 \cdot 3}x^6 + \cdots\right).$$

Replacing x by x^2 in the binomial formula (see Exercise 5.1.25), we have

$$(1 + x)^p = 1 + \frac{p}{1}x + \frac{p(p - 1)}{1 \cdot 2}x^2 + \frac{p(p - 1)(p - 2)}{1 \cdot 2 \cdot 3}x^3 + \cdots, \qquad (5.37)$$

which yields the general solution of the differential equation:

$$y = c_0(1 + x^2)^p.$$

Since $y(0) = 1$, it follows that $c_0 = 1$ and $y = (1 + x^2)^p$.

Example 5.2.4 Consider the differential equation

$$y'' + xy' + y = 0. \qquad (5.38)$$

Using (5.26), (5.27), and (5.28), we obtain the equation

$$\sum_{n=2}^{\infty} n(n - 1)c_n x^{n-2} + x\sum_{n=1}^{\infty} nc_n x^{n-1} + \sum_{n=0}^{\infty} c_n x^n = 0.$$

Reindexing to obtain equal powers of x, we have

$$\sum_{k=0}^{\infty} (k + 2)(k + 1)c_{k+2}x^k + \sum_{k=1}^{\infty} kc_k x^k + \sum_{k=0}^{\infty} c_k x^k = 0.$$

Note that the second sum can also be allowed to range from 0 to ∞. Gathering like terms in x produces the equation

$$\sum_{k=0}^{\infty} [(k + 2)(k + 1)c_{k+2} + (k + 1)c_k]x^k = 0. \qquad (5.39)$$

Setting the coefficients equal to zero, we obtain the general recursion formula

$$(k + 2)(k + 1)c_{k+2} + (k + 1)c_k = 0. \tag{5.40}$$

Therefore $(k + 2)c_{k+2} = -c_k$, and

$$c_2 = -\frac{c_0}{2}, \qquad c_3 = -\frac{c_1}{3}, \qquad c_4 = -\frac{c_2}{4} = \frac{c_0}{2 \cdot 4},$$

$$c_5 = -\frac{c_3}{5} = \frac{c_1}{3 \cdot 5}, \qquad c_6 = -\frac{c_4}{6} = -\frac{c_0}{2 \cdot 4 \cdot 6}, \quad \text{etc.}$$

Hence the power series for y can be written in the form

$$
\begin{aligned}
y &= c_0 + c_1 x - \frac{c_0}{2} x^2 - \frac{c_1}{3} x^3 + \frac{c_0}{2 \cdot 4} x^4 + \frac{c_1}{3 \cdot 5} x^5 - \cdots \\
&= c_0 \left(1 - \frac{x^2}{2} + \frac{x^4}{2 \cdot 4} - \frac{x^6}{2 \cdot 4 \cdot 6} + \cdots \right) \\
&\quad + c_1 \left(x - \frac{x^3}{3} + \frac{x^5}{3 \cdot 5} - \frac{x^7}{3 \cdot 5 \cdot 7} + \cdots \right)
\end{aligned}
\tag{5.41}
$$

by separating the terms that involve c_0 and c_1. At this point we try to see whether we recognize the two series that have been obtained by the power series method. Very frequently this is an unproductive task, but in this instance we are fortunate:

$$
\begin{aligned}
1 - \frac{x^2}{2} + \frac{x^4}{2 \cdot 4} - \frac{x^6}{2 \cdot 4 \cdot 6} + \cdots \\
= 1 + \left(-\frac{x^2}{2} \right) + \frac{1}{2!} \left(-\frac{x^2}{2} \right)^2 + \frac{1}{3!} \left(-\frac{x^2}{2} \right)^3 + \cdots \\
= e^{-x^2/2}.
\end{aligned}
$$

We can't recognize the second series, so we use the method of Section 3.3 of finding one solution when another is known. By formula (3.16) we have

$$
\begin{aligned}
y_2 &= y_1 \int \frac{e^{-\int x \, dx}}{y_1^2} \, dx = e^{-x^2/2} \int \frac{e^{-x^2/2}}{(e^{-x^2/2})^2} \, dx \\
&= e^{-x^2/2} \int e^{x^2/2} \, dx.
\end{aligned}
\tag{5.42}
$$

The integral in (5.42) does not have a closed-form solution. That this is indeed the second series in (5.41) can be verified by integrating the series for $e^{x^2/2}$ term by term and multiplying the result by the series for $e^{-x^2/2}$. Hence the general solution of (5.38) is given by

$$y = c_0 e^{-x^2/2} + c_1 e^{-x^2/2} \int e^{x^2/2} \, dx.$$

Example 5.2.5 Solve the equation

$$xy'' + y' + xy = 0. \tag{5.43}$$

Using the power series (5.26), (5.27), and (5.28) for Eq. (5.43) yields the equation

$$\sum_{n=2}^{\infty} n(n-1)c_n x^{n-1} + \sum_{n=1}^{\infty} nc_n x^{n-1} + \sum_{n=0}^{\infty} c_n x^{n+1} = 0.$$

Reindexing the series to obtain like powers of x, we have

$$\sum_{k=1}^{\infty} (k+1)kc_{k+1} x^k + \sum_{k=0}^{\infty} (k+1)c_{k+1} x^k + \sum_{k=1}^{\infty} c_{k-1} x^k = 0.$$

Condensing the three series in one yields, after some algebra,

$$c_1 + \sum_{k=1}^{\infty} [(k+1)^2 c_{k+1} + c_{k-1}] x^k = 0. \tag{5.44}$$

Setting the coefficients to zero, we have $c_1 = 0$ and

$$(k+1)^2 c_{k+1} = -c_{k-1}, \qquad k = 1, 2, 3, \ldots. \tag{5.45}$$

The recursion formula (5.45) together with $c_1 = 0$ imply that all coefficients with odd-numbered subscripts vanish, and

$$c_2 = -\frac{c_0}{2^2}, \quad c_4 = -\frac{c_2}{4^2} = \frac{c_0}{2^2 4^2}, \quad c_6 = -\frac{c_4}{6^2} = -\frac{c_0}{2^2 4^2 6^2}, \ldots.$$

Hence

$$y = c_0 - \frac{c_0}{2^2} x^2 + \frac{c_0}{2^2 4^2} x^4 - \frac{c_0}{2^2 4^2 6^2} x^6 + \cdots$$

$$= c_0 \sum_{n=0}^{\infty} \frac{1}{(n!)^2} \left(-\frac{x^2}{4}\right)^n. \tag{5.46}$$

It is unlikely that the reader is familiar with the series in (5.46). This series is often used in applied mathematics and is known as the Bessel function of index zero, $J_0(x)$. (We will study the properties of Bessel functions in Section 5.4.) Note also that the power series method has produced only *one* of the solutions of Eq. (5.43). To find the other solution, we can again proceed as in Example 5.2.4 (see also Exercise 3.3.7). Thus

$$y_2(x) = J_0(x) \int \frac{dx}{x J_0^2(x)}.$$

Finally, the general solution of (5.43) is given by

$$y(x) = AJ_0(x) + BJ_0(x) \int \frac{dx}{xJ_0^2(x)}.$$

In our next example we will meet a situation where the power series method fails to yield any solution.

Example 5.2.6 Consider the differential equation

$$x^2y'' + xy' + y = 0. \tag{5.47}$$

Making use of the series (5.26), (5.27), and (5.28), and multiplying by the appropriate powers of x, we have

$$\sum_{n=2}^{\infty} n(n - 1)c_nx^n + \sum_{n=1}^{\infty} nc_nx^n + \sum_{n=0}^{\infty} c_nx^n = 0$$

or

$$\sum_{n=0}^{\infty} (n^2 + 1)c_nx^n = 0. \tag{5.48}$$

Clearly, if we equate each of the coefficients of (5.48) to zero, all the coefficients c_n will vanish and $y \equiv 0$. Thus in this case the power series method fails completely in helping us find the general solution

$$y = A \cos (\ln |x|) + B \sin (\ln |x|) \tag{5.49}$$

of Eq. (5.47) (check!). However, the power series method will yield a particular solution of the nonhomogeneous equation

$$x^2y'' + xy' + y = \frac{1}{1 - x}, \qquad |x| < 1, \tag{5.50}$$

since (5.48) and the geometric series (5.10) will yield

$$\sum_{n=0}^{\infty} (n^2 + 1)c_nx^n = \sum_{n=0}^{\infty} x^n$$

or

$$\sum_{n=0}^{\infty} [(n^2 + 1)c_n - 1]x^n = 0.$$

Therefore,

$$c_n = \frac{1}{n^2 + 1}.$$

for all n, so that a particular solution of (5.50) is

$$y_p = 1 + \frac{x}{2} + \frac{x^2}{5} + \frac{x^3}{10} + \cdots + \frac{x^n}{n^2 + 1} + \cdots. \tag{5.51}$$

Using the ratio test (5.13), we can find the radius of convergence of the series (5.51):

$$\frac{1}{R} = \lim_{n \to \infty} \frac{n^2 + 1}{(n + 1)^2 + 1} = \lim_{n \to \infty} \frac{1 + 1/n^2}{(1 + 1/n)^2 + 1/n^2} = 1.$$

Hence the general solution of Eq. (5.50), valid for $|x| < 1$, is

$$y = A \cos (\ln |x|) + B \sin (\ln |x|) + \left(1 + \frac{x}{2} + \frac{x^2}{5} + \cdots \right).$$

EXERCISES 5.2

In Exercises 1 through 20 find the general solution of each equation. When initial conditions are specified, give the solution that satisfies them.

1. $y' = y$, $y(0) = 4$
2. $y' = 5y$, $y(0) = 1$
3. $y' = y - x$, $y(0) = 2$
4. $y' = -2y + x$
5. $y' = xy$
6. $y' = x^3 - 2xy$, $y(0) = 1$
7. $y' = y + e^x$
8. $xy' = y$
9. $xy' = ny$, n an integer
10. $y' = 1 + y^2$, $y(0) = 0$
11. $y'' + 4y = 0$
12. $y'' - 4y = 0$, $y(0) = 2$, $y'(0) = 0$
13. $y'' - y = 0$
14. $y'' + y = 0$, $y(0) = 1$, $y'(0) = 0$
15. $(1 + x^2)y'' + 2xy' - 2y = 0$
16. $y'' + 2y' + y = \sin x$
17. $y'' + y = e^{2x}$
18. $y'' - \frac{3}{x} y' = 0$
19. $x^2 y'' + xy' - 4y = 0$
20. $y'' - 2xy' + 2y = 0$
21. Does the power series method yield a solution to the equation
    a) $x^2 y' = y$?
    b) $x^3 y' = y$?
22. Solve $y' = y\sqrt{y^2 - 1}$.
23. Show that the power series method fails for
    $$x^2 y'' + x^2 y' + y = 0.$$

5.3 ORDINARY AND SINGULAR POINTS

The reader must have noticed in the last section that the power series method sometimes fails to yield a solution for one equation while working very well for an apparently similar equation. In this section we shall analyze this anomaly and discover its cause.

The main clue to the puzzle can be obtained by making the coefficient of the highest order derivative equal to 1. Thus, if we write each of the second order homogeneous equations in the examples of Section 5.2 in the form

$$y'' + a(x)y' + b(x)y = 0, \tag{5.51}$$

then the equations of Examples 5.2.5 and 5.2.6 become, respectively,

$$y'' + \frac{1}{x}y' + y = 0, \tag{5.52}$$

and

$$y'' + \frac{1}{x}y' + \frac{1}{x^2}y = 0. \tag{5.53}$$

Note that in each of these equations one or both of the terms $a(x)$ and $b(x)$ are not defined at $x = 0$. The important fact about Eq. (5.51) is that the behavior of its solution near the point $x = x_0$ depends on the behavior of the functions $a(x)$ and $b(x)$ near x_0. When both of the functions $a(x)$ and $b(x)$ are analytic at $x = x_0$ [that is, both $a(x)$ and $b(x)$ can be expressed as a power series in $(x - x_0)$ with positive radius of convergence], we call x_0 an *ordinary point* of Eq. (5.51). Any point that is not an ordinary point is called a *singular point* of Eq. (5.51).

Ordinary points Suppose that x_0 is an ordinary point of Eq. (5.51). Then both $a(x)$ and $b(x)$ can be expressed as power series in $(x - x_0)$ in some interval $|x - x_0| < R$. The next theorem asserts that under these circumstances the power series method will yield a power series expansion of the solution $y(x)$.

Theorem 5.2 Let x_0 be an ordinary point of the differential equation

$$y'' + a(x)y' + b(x)y = 0, \tag{5.54}$$

and let α and β be arbitrary constants. Then there exists a unique function $y(x)$, analytic at x_0, which satisfies Eq. (5.54) and the initial conditions $y(x_0) = \alpha$ and $y'(x_0) = \beta$. Furthermore, if the power series representations of $a(x)$ and $b(x)$ are valid for all x such that $|x - x_0| < R$, then so is the power series expansion of the solution $y(x)$.

The proof of the theorem is complicated, and its principal details are sketched in Exercise 27 at the end of this section.

Theorem 5.2 provides the justification for the use of the power series method in solving differential equations at an ordinary point. Examples 5.2.1 through 5.2.4 satisfy the hypotheses of Theorem 5.2 and in each case have $x = 0$ as an ordinary point.

Regular singular points   Consider the differential equation

$$y'' + \frac{2}{x} y' - y = 0,$$

which has a singular point at $x = 0$. It can easily be shown that $y = e^x/x$ is a solution of this equation. Although it is impossible to expand e^x/x as a power series in x, we can write it as a power of x times a power series in x:

$$\frac{e^x}{x} = \frac{1}{x}\left(1 + x + \frac{x^2}{2!} + \frac{x^3}{3!} + \cdots\right).$$

This suggests that we should try to find solutions of the form

$$y = x^r(c_0 + c_1 x + c_2 x^2 + c_3 x^3 + \cdots),$$

where r is any real or complex number, whenever $x = 0$ is a singular point of the differential equation. For one class of singular points, this modification of the power series method does yield solutions.

A *regular singular point* is a point x_0 such that the functions $(x - x_0)a(x)$ and $(x - x_0)^2 b(x)$ are analytic. For example, $x = 0$ is a regular singular point of both Eqs. (5.52) and (5.53), since in Eq. (5.52)

$$xa(x) = 1, \qquad x^2 b(x) = x^2, \tag{5.55}$$

and in Eq. (5.53)

$$xa(x) = 1, \qquad x^2 b(x) = 1. \tag{5.56}$$

All the functions in (5.55) and (5.56) are power series in x with all but one of the coefficients equal to zero, and are therefore analytic functions at $x = 0$.

A singular point that is not regular is called *irregular*.

Example 5.3.1   The point $x = 0$ is an irregular singular point of the two equations

$$y'' + \frac{1}{x^2} y' + y = 0,$$

$$y'' + \frac{1}{x} y' + \frac{1}{x^3} y = 0. \tag{5.57}$$

To simplify the explanation of the modified power series method, called the *method of Frobenius*, we shall assume that $x = 0$ is a regular singular point of the equation

$$y'' + a(x)y' + b(x)y = 0. \tag{5.58}$$

[If $x_0 \neq 0$ is a regular singular point, then the substitution $X = x - x_0$ will transform power series in $(x - x_0)$ into power series in X. The point $X = 0$ is then a regular singular point of the transformed differential equation.] We now assume that Eq. (5.58) has a solution of the form

$$y = x^r(c_0 + c_1x + c_2x^2 + \cdots) = \sum_{n=0}^{\infty} c_nx^{r+n}, \qquad c_0 = 1, \quad x > 0, \qquad (5.59)$$

where r is any real or complex number. We can assume that $c_0 = 1$, since the smallest power of x can always be factored out of an expression of the form (5.59) and any constant multiple of a solution is again a solution of the differential equation. In addition, the choice $c_0 = 1$ simplifies much of the following discussion. The restriction $x > 0$ is necessary to prevent difficulties for certain values of r, such as $r = \frac{1}{2}$ and $-\frac{1}{4}$, since we are not interested in imaginary solutions. [If we need to find a solution valid for $x < 0$, we can change variables by substituting $X = -x$ into Eq. (5.58) and solve the resulting equation for $X > 0$.] Since

$$y' = \sum_{n=0}^{\infty} c_n(r + n)x^{r+n-1} \qquad (5.60)$$

and

$$y'' = \sum_{n=0}^{\infty} c_n(r + n)(r + n - 1)x^{r+n-2}, \qquad (5.61)$$

Eq. (5.54) can be rewritten as

$$\sum_{n=0}^{\infty} c_n(r + n)(r + n - 1)x^{r+n-2}$$

$$+ a(x) \sum_{n=0}^{\infty} c_n(r + n)x^{r+n-1} + b(x) \sum_{n=0}^{\infty} c_nx^{r+n} = 0$$

or

$$\sum_{n=0}^{\infty} c_n[(r + n)(r + n - 1) + (r + n)xa(x) + x^2b(x)]x^{r+n-2} = 0. \qquad (5.62)$$

Since $x = 0$ is a regular singular point, both $xa(x)$ and $x^2b(x)$ can be expressed as power series in x with positive radii of convergence:

$$xa(x) = a_0 + a_1x + a_2x^2 + \cdots, \qquad (5.63)$$

$$x^2b(x) = b_0 + b_1x + b_2x^2 + \cdots. \qquad (5.64)$$

But $n \geqslant 0$, so that x^{r-2} is the smallest power of x in (5.62). Since the coefficients of a power series whose sum is zero must vanish, we have, for $n = 0$,

$$c_0[r(r - 1) + a_0 r + b_0] = 0. \tag{5.65}$$

By hypothesis $c_0 = 1$, so that we obtain the *indicial equation*

$$r(r - 1) + a_0 r + b_0 = 0, \tag{5.66}$$

whose roots, r_1 and r_2, are called the *exponents* of the differential equation (5.54). In what follows we shall see that one of the solutions of Eq. (5.54) will always be of the form (5.59) and that there are three possible forms for the second linearly independent solution corresponding to the following cases:

CASE 1. r_1 and r_2 do not differ by an integer.

CASE 2. $r_1 = r_2$.

CASE 3. r_1 and r_2 differ by a nonzero integer

We shall consider the three cases separately.

CASE 1. *r_1 and r_2 do not differ by an integer.* This is the easiest case, since Eq. (5.54) will have two solutions, for $x > 0$, of the forms

$$y_1(x) = x^{r_1}(c_0 + c_1 x + c_2 x^2 + \cdots), \qquad c_0 = 1,$$
$$y_2(x) = x^{r_2}(c_0^* + c_1^* x + c_2^* x^2 + \cdots), \qquad c_0^* = 1.$$

That y_1 and y_2 are linearly independent follows easily from the fact that y_1/y_2 cannot be constant, since if it were, the roots r_1 and r_2 would coincide. The coefficients $c_1, c_2, \ldots$ are obtained, as in Section 5.2, by replacing r by r_1 and setting the coefficients of each power of x equal to zero in Eq. (5.62). Similarly, to find $c_1^*, c_2^*, \ldots$, we repeat the process above for $r = r_2$. The procedure is demonstrated in the following example.

Example 5.3.2 Consider Eq. (5.53)

$$y'' + \frac{1}{x} y' + \frac{1}{x^2} y = 0.$$

By (5.56), the Maclaurin series for $xa(x)$ and $x^2 b(x)$ consist only of the term 1, so $a_0 = b_0 = 1$. Thus the indicial equation of (5.53) is

$$r(r - 1) + r + 1 = r^2 + 1 = 0,$$

with the roots $r_1 = i, r_2 = -i$, which do not differ by an integer. Setting $r = i$ in (5.62), we have

$$\sum_{n=0}^{\infty} c_n[(i + n)(i + n - 1) + (i + n) + 1]x^{i+n-2} = 0.$$

Equating all coefficients of this series to zero, we have, after combining terms,

$$0 = c_n[(i + n)^2 + 1] = c_n(n^2 + 2in), \qquad (5.67)$$

which holds only if $c_n = 0$ for $n > 0$. Thus

$$y_1(x) = c_0 x^i = e^{i(\ln x)} = [\cos(\ln x) + i\sin(\ln x)]. \qquad (5.68)$$

Similarly, substituting $r = -i$ into Eq. (5.62) yields the series

$$\sum_{n=0}^{\infty} c_n^*[(-i + n)(-i + n - 1) + (-i + n) + 1]x^{-i+n-2} = 0,$$

whose coefficients satisfy the condition

$$c_n^*[n^2 - 2in] = 0.$$

Thus $c_n^* = 0$ for $n > 0$. Hence

$$y_2(x) = c_0^* x^{-i} = [\cos(\ln x) - i\sin(\ln x)].$$

Finally, since linear combinations of solutions are solutions, the real and imaginary parts of y_1 and y_2,

$$y_1^*(x) = \frac{1}{2}(y_1 + y_2) = \cos(\ln x), \qquad y_2^*(x) = \frac{1}{2i}(y_1 - y_2) = \sin(\ln x),$$

are solutions of the equation (5.53). That y_1^* and y_2^* are linearly independent follows obviously by reason of the fact that $y_2^*/y_1^* = \tan(\ln x)$, which is non-constant. Hence Eq. (5.53) has the general solution

$$y = A\cos(\ln x) + B\sin(\ln x), \qquad x > 0.$$

(Compare this proof with that of Exercise 3.8.10.)

CASE 2. $r_1 = r_2$. Here we set $r = r_1$ and determine the coefficients $c_1, c_2, \ldots$ as in Case 1. We can then use formula (3.16) to find the second linearly independent solution, since one solution is known. Consider the following example.

Example 5.3.3 Recall Eq. (5.52)

$$y'' + \frac{1}{x}y' + y = 0.$$

By (5.55), the Maclaurin series are $xa(x) = 1$ and $x^2b(x) = x^2$; thus $a_0 = 1$ and $b_0 = 0$. Hence the indicial equation is

$$r(r - 1) + r = r^2 = 0$$

with the double root $r = 0$. Thus Eq. (5.52) has a power series solution, which was found in Example 5.2.5:

$$y_1(x) = J_0(x) = \sum_{n=0}^{\infty} \frac{1}{(n!)^2} \left(\frac{-x^2}{4}\right)^n.$$

Finally, as in Example 5.2.5,

$$y_2(x) = J_0(x) \int \frac{dx}{xJ_0^2(x)}. \tag{5.69}$$

Example 5.3.4 Consider the equation

$$y'' + y' + \frac{1}{4x^2} y = 0. \tag{5.70}$$

Here $xa(x) = x$ and $x^2 b(x) = \frac{1}{4}$, so that $a_0 = 0$ and $b_0 = \frac{1}{4}$. Hence the indicial equation is

$$r(r - 1) + \tfrac{1}{4} = (r - \tfrac{1}{2})^2 = 0$$

with $r = \frac{1}{2}$ as a double root. Setting $r = \frac{1}{2}$ in Eq. (5.62), we obtain

$$\sum_{n=0}^{\infty} c_n[(n + \tfrac{1}{2})(n - \tfrac{1}{2}) + (n + \tfrac{1}{2})x + \tfrac{1}{4}]x^{n-3/2} = 0$$

or

$$\sum_{n=0}^{\infty} c_n n^2 x^{n-3/2} + \sum_{n=0}^{\infty} c_n(n + \tfrac{1}{2})x^{n-1/2} = 0. \tag{5.71}$$

Since the first term of the first series vanishes, we can reindex this series by setting $n = k + 1$ and combine like powers of x to obtain

$$\sum_{k=0}^{\infty} [c_{k+1}(k + 1)^2 + c_k(k + \tfrac{1}{2})]x^{k-1/2} = 0. \tag{5.72}$$

Equating the coefficients of (5.72) to zero, we have the recurrence formula

$$c_{k+1}(k + 1)^2 + c_k(k + \tfrac{1}{2}) = 0.$$

Hence

$$c_1 = -\frac{c_0}{2}, \qquad c_2 = -\frac{c_1(\frac{3}{2})}{2^2} = \frac{3c_0}{2^2 \cdot 2^2}, \qquad c_3 = -\frac{5c_2}{2 \cdot 3^2} = -\frac{3 \cdot 5c_0}{2^3 \cdot 2^2 \cdot 3^2},$$

$$c_4 = -\frac{7c_3}{2 \cdot 4^2} = \frac{3 \cdot 5 \cdot 7c_0}{2^4 \cdot 2^2 \cdot 3^2 \cdot 4^2}, \dots,$$

so that

$$y_1(x) = x^{1/2}\left(c_0 - \frac{c_0}{2}x + \frac{3c_0}{2^2 \cdot 2^2}x^2 - \frac{3 \cdot 5c_0}{2^3 \cdot 2^2 \cdot 3^2}x^3 + \frac{3 \cdot 5 \cdot 7c_0}{2^4 \cdot 2^2 \cdot 3^2 \cdot 4^2}x^4 - \cdots\right)$$

$$= c_0 x^{1/2}\left[1 - \left(\frac{x}{2}\right) + \frac{3}{2^2}\left(\frac{x}{2}\right)^2 - \frac{3 \cdot 5}{2^2 \cdot 3^2}\left(\frac{x}{2}\right)^3 + \frac{3 \cdot 5 \cdot 7}{2^2 \cdot 3^2 \cdot 4^2}\left(\frac{x}{2}\right)^4 - \cdots\right]$$

$$= x^{1/2}\sum_{n=0}^{\infty}\frac{(2n)!}{(n!)^3}\left(-\frac{x}{4}\right)^n, \qquad x > 0. \tag{5.73}$$

To find the radius of convergence of the series in Eq. (5.73), we use the ratio test:

$$\frac{1}{R} = \lim_{n \to \infty}\left|\frac{(n!)^3 4^n(2n + 2)!}{(2n)!(n + 1)!^3 4^{n+1}}\right| = \lim_{n \to \infty}\left|\frac{(2n + 2)(2n + 1)}{4(n + 1)^3}\right|$$

$$= \lim_{n \to \infty}\left|\frac{2n + 1}{2(n + 1)^2}\right| = 0.$$

So the radius of convergence $R = \infty$. We can now use formula (3.16) to produce the second linearly independent solution or use the alternate procedure sketched below. We recall that in deriving formula (3.16) we let $y_2 = vy_1$, where v satisfies the differential equation (3.14) or, in our case,

$$\frac{v''}{v'} = -2\frac{y_1'}{y_1} - 1 = \frac{-2\left(\dfrac{1}{2\sqrt{x}}\right)\left[1 - 3\left(\dfrac{x}{2}\right) + \dfrac{3 \cdot 5}{2^2}\left(\dfrac{x}{2}\right)^2 - \cdots\right]}{\sqrt{x}\left[1 - \left(\dfrac{x}{2}\right) + \dfrac{3}{2^2}\left(\dfrac{x}{2}\right)^2 - \cdots\right]} - 1.$$

After finding the first few terms by long division (carried out exactly as in the division of one polynomial by another), we have

$$\frac{v''}{v'} = \frac{-1}{x}\left(1 - x + \frac{x^2}{4} - \cdots\right) - 1 = \frac{-1}{x} - \frac{x}{4} + \cdots. \tag{5.74}$$

Integrating both sides of (5.74), we obtain

$$\ln v' = -\ln x - \frac{x^2}{4} + \cdots$$

or

$$v' = \frac{1}{x}\exp\left(-\frac{x^2}{4} + \cdots\right)$$

$$= \frac{1}{x}\left[1 + \left(\frac{-x^2}{4} + \cdots\right) + \frac{1}{2!}\left(\frac{-x^2}{4} + \cdots\right)^2 + \cdots\right]. \tag{5.75}$$

After expanding the exponential as a power series in x, we integrate once more and find that v has the form

$$v = \ln x - \frac{x^2}{8} + \cdots .$$

Then

$$y_2 = vy_1 = \left(\ln x - \frac{x^2}{8} + \cdots \right) \cdot \sqrt{x} \left[1 - \left(\frac{x}{2} \right) + \frac{3}{2^2} \left(\frac{x}{2} \right)^2 - \cdots \right]$$

$$= (\ln x)y_1 + \sqrt{x} \left(-\frac{x^2}{8} + \cdots \right),$$

and the general solution of Eq. (5.70) has the form

$$y(x) = \sqrt{x} \left\{ (A + B \ln x) \left[1 - \left(\frac{x}{2} \right) + \cdots \right] + B \left(-\frac{x^2}{8} + \cdots \right) \right\}, \qquad x > 0.$$

Indeed, it is always true in such a case that the general solution has the form

$$y(x) = x^r \left(\sum_{n=0}^{\infty} c_n x^n + \ln x \sum_{n=0}^{\infty} c_n^* x^n \right), \qquad x > 0. \tag{5.76}$$

To verify this fact, we note that by long division

$$\frac{y_1'}{y_1} = \frac{(rc_0 x^{r-1} + (r + 1)c_1 x^r + \cdots)}{c_0 x^r + c_1 x^{r+1} + \cdots} = \left(\frac{r}{x} + \frac{c_1}{c_0} + \cdots \right), \qquad c_0 = 1, \tag{5.77}$$

and by (5.63),

$$a(x) = \frac{a_0}{x} + a_1 + a_2 x + \cdots . \tag{5.78}$$

Substituting (5.77) and (5.78) into Eq. (3.14), we have

$$-\frac{v''}{v'} = \frac{a_0 + 2r}{x} + (a_1 + 2c_1) + \cdots ,$$

and integrating both sides of this equation, we obtain

$$-\ln v' = (a_0 + 2r) \ln x + (a_1 + 2c_1)x + \cdots . \tag{5.79}$$

Since the indicial equation (5.66) has a double root

$$r = \frac{1 - a_0}{2},$$

it follows that $a_0 + 2r = 1$, so that we can rewrite (5.79) as

$$\frac{1}{v'} = x \exp\left[(a_1 + 2c_1)x + \cdots\right]$$

or

$$v' = \frac{1}{x} \exp\left[-(a_1 + 2c_1)x + \cdots\right]. \tag{5.80}$$

Expanding the exponential as a power series in x and integrating once more, we finally obtain

$$v(x) = \ln x - (a_1 + 2c_1)x + \cdots.$$

Since $y_2 = vy_1$ and y_1 has the form (5.59), the form of the general solution (5.76) is immediately obvious.

CASE 3. r_1 *and* r_2 *differ by a nonzero integer.* Suppose that $r_1 > r_2$. Then one solution of (5.54) will have the form

$$y_1 = x^{r_1}(c_0 + c_1 x + c_2 x^2 + \cdots), \qquad c_0 = 1, \qquad x > 0,$$

as in Case 1. In some instances it is not possible to determine y_2 as was done in Case 1, because the procedure regenerates the same series expansion we obtained for y_1 (in this case the first $r_1 - r_2$ coefficients c_i^* vanish). When this occurs, we proceed as in Case 2. These two possibilities are illustrated in the following two examples.

Example 5.3.5 *Bessel's equation of order $\frac{1}{2}$ is*

$$x^2 y'' + xy' + \left[x^2 - (\tfrac{1}{2})^2\right]y = 0$$

or, after division by x^2,

$$y'' + \frac{1}{x}y' + \left(1 - \frac{1}{4x^2}\right)y = 0. \tag{5.81}$$

Clearly $x = 0$ is a regular singular point, since the functions $xa(x) = 1$ and $x^2 b(x) = x^2 - \frac{1}{4}$ are both analytic at $x = 0$. The indicial equation for (5.81) is

$$r(r - 1) + r - \tfrac{1}{4} = r^2 - \tfrac{1}{4} = 0$$

with the roots $r_1 = \frac{1}{2}$ and $r_2 = -\frac{1}{2}$. Here the roots differ by the nonzero integer 1. Substituting $r = r_1$ in (5.62), we obtain

$$\sum_{n=0}^{\infty} c_n\left[(n + \tfrac{1}{2})(n - \tfrac{1}{2}) + (n + \tfrac{1}{2}) + (x^2 - \tfrac{1}{4})\right]x^{n-3/2} = 0$$

or

$$\sum_{n=0}^{\infty} c_n(n^2 + n)x^{n-3/2} + \sum_{n=0}^{\infty} c_n x^{n+1/2} = 0. \tag{5.82}$$

Noting that the first term of the first series vanishes, we can reindex the series (5.82) to obtain

$$\sum_{k=-1}^{\infty} c_{k+2}(k + 2)(k + 3)x^{k+1/2} + \sum_{k=0}^{\infty} c_k x^{k+1/2}$$

$$= 2c_1 x^{-1/2} + \sum_{k=0}^{\infty} [(k + 2)(k + 3)c_{k+2} + c_k]x^{k+1/2} = 0.$$

Setting the coefficients of this series equal to zero, we obtain $c_1 = 0$ and the recurrence relation

$$c_{k+2} = -\frac{c_k}{(k + 2)(k + 3)}. \tag{5.83}$$

Clearly all the coefficients with odd-numbered subscripts will vanish, and

$$c_2 = -\frac{c_0}{3!}, \quad c_4 = -\frac{c_2}{4 \cdot 5} = \frac{c_0}{5!}, \quad c_6 = -\frac{c_4}{6 \cdot 7} = -\frac{c_0}{7!}, \dots,$$

and we have the solution

$$y_1(x) = \sqrt{x}\left(c_0 - \frac{c_0}{3!}x^2 + \frac{c_0}{5!}x^4 - \frac{c_0}{7!}x^6 + \cdots\right)$$

$$= c_0\sqrt{x}\left(1 - \frac{x^2}{3!} + \frac{x^4}{5!} - \frac{x^6}{7!} + \cdots\right)$$

or, since $c_0 = 1$,

$$y_1(x) = \frac{1}{\sqrt{x}}\left(x - \frac{x^3}{3!} + \frac{x^5}{5!} - \frac{x^7}{7!} + \cdots\right).$$

Applying Eq. (5.17), we find that

$$y_1(x) = \frac{\sin x}{\sqrt{x}}, \quad x > 0.$$

Now we set $r = r_2$ in (5.62) to obtain

$$\sum_{n=0}^{\infty} c_n[(n - \tfrac{1}{2})(n - \tfrac{3}{2}) + (n - \tfrac{1}{2}) + (x^2 - \tfrac{1}{4})]x^{n-5/2} = 0$$

or

$$\sum_{n=0}^{\infty} c_n n(n-1)x^{n-5/2} + \sum_{n=0}^{\infty} c_n x^{n-1/2} = 0.$$

Since the first two terms of the first series vanish, reindexing and gathering like powers of x, we have

$$\sum_{k=0}^{\infty} [(k+2)(k+1)c_{k+2} + c_k]x^{k-1/2},$$

from which can be obtained the recurrence relation

$$c_{k+2} = -\frac{c_k}{(k+1)(k+2)} \tag{5.84}$$

Now by (5.84), we have

$$c_2 = -\frac{c_0}{2!}, \quad c_4 = -\frac{c_2}{3 \cdot 4} = \frac{c_0}{4!}, \quad c_6 = -\frac{c_4}{5 \cdot 6} = -\frac{c_0}{6!}, \ldots,$$

$$c_3 = -\frac{c_1}{3!}, \quad c_5 = -\frac{c_3}{4 \cdot 5} = \frac{c_1}{5!}, \quad c_7 = -\frac{c_5}{6 \cdot 7} = -\frac{c_1}{7!}, \ldots,$$

and

$$y_2(x) = x^{-1/2}\left[c_0 + c_1 x - c_0 \frac{x^2}{2!} - c_1 \frac{x^3}{3!} + c_0 \frac{x^4}{4!} + c_1 \frac{x^5}{5!} - \cdots \right]$$

$$= \frac{1}{\sqrt{x}}(c_0 \cos x + c_1 \sin x), \quad c_0 = 1. \tag{5.85}$$

Since the last term of y_2 is a multiple of y_1 and we are looking for linearly independent solutions, we may set $c_1 = 0$ to obtain

$$y_2 = \frac{\cos x}{\sqrt{x}}, \quad x > 0.$$

That y_1 and y_2 are linearly independent follows from the fact that $y_1/y_2 = \tan x$, which is nonconstant. Thus the general solution of (5.81) is

$$y = \frac{A}{\sqrt{x}}\cos x + \frac{B}{\sqrt{x}}\sin x, \quad x > 0.$$

Example 5.3.6 Consider *Bessel's equation of order 1*:

$$x^2 y'' + xy' + (x^2 - 1)y = 0$$

or

$$y'' + \frac{1}{x}y' + \left(1 - \frac{1}{x^2}\right)y = 0. \tag{5.86}$$

Again $x = 0$ is a regular singular point of (5.86), and the indicial equation for (5.86) is

$$r(r - 1) + r - 1 = r^2 - 1 = 0,$$

with roots $r_1 = 1, r_2 = -1$. Setting $r = 1$ in (5.62) yields

$$\sum_{n=0}^{\infty} c_n[n(n + 1) + (n + 1) + (x^2 - 1)]x^{n-1} = 0$$

or

$$3c_1 + \sum_{k=0}^{\infty} [c_{k+2}(k + 2)(k + 4) + c_k]x^{k+1} = 0. \tag{5.87}$$

By setting the coefficients of Eq. (5.87) to zero, we obtain $c_1 = 0$ and the recurrence relation

$$c_{k+2} = -\frac{c_k}{(k + 2)(k + 4)}.$$

Thus all the coefficients with odd-numbered subscripts vanish, and

$$c_2 = -\frac{c_0}{2 \cdot 4}, \quad c_4 = -\frac{c_2}{4 \cdot 6} = \frac{c_0}{2 \cdot 4^2 \cdot 6}, \quad c_6 = -\frac{c_4}{6 \cdot 8} = -\frac{c_0}{2 \cdot 4^2 \cdot 6^2 \cdot 8}, \ldots$$

Thus

$$y_1(x) = x\left(c_0 - \frac{c_0}{2 \cdot 4}x^2 + \frac{c_0}{2 \cdot 4^2 \cdot 6}x^4 - \frac{c_0}{2 \cdot 4^2 \cdot 6^2 \cdot 8}x^6 + \cdots\right)$$

$$= x\left(1 - \frac{1}{1!2!}\left(\frac{x}{2}\right)^2 + \frac{1}{2!3!}\left(\frac{x}{2}\right)^4 - \frac{1}{3!4!}\left(\frac{x}{2}\right)^6 + \cdots\right). \tag{5.88}$$

The series

$$1 - \frac{t}{1!2!} + \frac{t^2}{2!3!} - \frac{t^3}{3!4!} + \cdots + \frac{t^n}{n!(n + 1)!} + \cdots$$

has a radius of convergence $R = \infty$ since by the ratio test

$$\frac{1}{R} = \lim_{n \to \infty} \left|\frac{1/(n + 1)!(n + 2)!}{1/n!(n + 1)!}\right|$$

$$= \lim_{n \to \infty} \frac{n!(n + 1)!}{(n + 1)!(n + 2)!} = \lim_{n \to \infty} \frac{1}{(n + 1)(n + 2)} = 0.$$

Thus (5.88) is valid for all values $t = (x/2)^2$ and therefore holds for all values of x.

Now we shall see that the method above will not work for the second root $r_2 = -1$ of the indicial equation. Setting $r = -1$ in (5.62), we have

$$\sum_{n=0}^{\infty} c_n[(n-1)(n-2) + (n-1) + x^2 - 1]x^{n-3} = 0$$

or

$$-c_1 x^{-2} + c_0 x^{-1} + \sum_{k=0}^{\infty} [c_{k+3}(k+1)(k+3) + c_{k+1}]x^k = 0.$$

If we set all the coefficients in this series to zero, we will find that $c_0 = 0$, which contradicts the assumption that $c_0 = 1$ made in Eq. (5.59). Thus the method of Case 1 fails for this root. However, applying the method of Case 2, we can obtain from Eq. (5.79)

$$-\ln v' = 3 \ln x + \frac{2c_1}{c_0} x + \cdots, \qquad c_0 = 1,$$

since $xa(x) = 1 = a_0$ and $r = r_1 = 1$. Thus

$$v' = x^{-3} \exp(-2c_1 x + \cdots),$$

and expanding the exponential in powers of x, we obtain

$$v' = \frac{1}{x^3} - 2c_1 \frac{1}{x^2} + \frac{k_{-1}}{x} + k_0 + k_1 x + \cdots.$$

Integrating this series term by term, we find that

$$v = -\frac{1}{2x^2} + 2c_1 \frac{1}{x} + k_{-1} \ln x + k_0 x + \cdots$$

and

$$\begin{aligned}
y_2 = vy_1 &= \left(\frac{-1}{2x^2} + 2c_1 \frac{1}{x} + k_{-1} \ln x + k_0 x + \cdots\right) \\
&\quad \times \left\{x\left[1 - \frac{(x/2)^2}{1!2!} + \frac{(x/2)^4}{2!3!} - \cdots\right]\right\} \\
&= k_{-1}(\ln x)y_1 + \left(\frac{-1}{2x^2} + 2c_1 \frac{1}{x} + k_0 x + \cdots\right) \\
&\quad \times \left\{x\left[1 - \frac{(x/2)^2}{1!2!} + \frac{(x/2)^4}{2!3!} - \cdots\right]\right\} \\
&= k_{-1}(\ln x)y_1 + x^{-1}(-\tfrac{1}{2} + 2c_1 x + \cdots), \qquad x > 0. \quad (5.89)
\end{aligned}$$

In general, if the method of Case 1 fails for r_2, the procedure above will yield the solution

$$y_2 = k_{-1}(\ln x)y_1 + x^{r_2}(k_0 + k_1 x + \cdots), \qquad x > 0,$$

where $k_{-1} \neq 0$.

We gather all the facts we have proved in this section in one theorem:

Theorem 5.3 Let $x = 0$ be a regular singular point of the differential equation

$$y'' + a(x)y' + b(x)y = 0, \quad x \neq 0, \tag{5.90}$$

and let r_1 and r_2 be the roots of the indicial equation

$$r(r - 1) + a_0 r + b_0 = 0,$$

where a_0 and b_0 are given by the power series expansions

$$xa(x) = a_0 + a_1 x + a_2 x^2 + \cdots,$$
$$x^2 b(x) = b_0 + b_1 x + b_2 x^2 + \cdots.$$

Then Eq. (5.90) has two linearly independent solutions y_1 and y_2 whose form depends on r_1 and r_2 as follows:

CASE 1. If r_1 and r_2 do not differ by an integer, then

$$y_1(x) = |x|^{r_1} \left(\sum_{n=0}^{\infty} c_n x^n \right), \qquad c_0 = 1,$$

$$y_2(x) = |x|^{r_2} \left(\sum_{n=0}^{\infty} c_n^* x^n \right), \qquad c_0^* = 1.$$

(The absolute-value signs are needed to avoid the assumption that $x > 0$.)

CASE 2. If $r_1 = r_2 = r$, then

$$y_1(x) = |x|^r \left(\sum_{n=0}^{\infty} c_n x^n \right), \qquad c_0 = 1,$$

$$y_2(x) = |x|^r \left(\sum_{n=1}^{\infty} c_n^* x^n \right) + y_1(x) \ln |x|.$$

CASE 3. If $r_1 - r_2$ is a positive integer, then

$$y_1(x) = |x|^{r_1} \left(\sum_{n=0}^{\infty} c_n x^n \right), \qquad c_0 = 1,$$

$$y_2(x) = |x|^{r_2} \left(\sum_{n=0}^{\infty} c_n^* x^n \right) + c_{-1}^* y_1(x) \ln |x|, \qquad c_0^* = 1,$$

and c_{-1}^* may equal zero.

Furthermore, if the power series expansions for $xa(x)$ and $x^2b(x)$ are valid for x such that $|x| < R$, then the solutions y_1 and y_2 are valid for $0 < |x| < R$. The proof of this fact is left as an exercise (see Exercise 28).

EXERCISES 5.3

Find the general solutions of the differential equations in Exercises 1 through 22 by the method of Frobenius.

1. $y'' + \dfrac{1}{2x} y' + \dfrac{1}{4x} y = 0$

2. $y'' + \dfrac{2(1 - 2x)}{x(1 - x)} y' - \dfrac{2}{x(1 - x)} y = 0$

3. $y'' + \dfrac{6}{x} y' + \left(\dfrac{6}{x^2} - 1\right) y = 0$

4. $y'' + \dfrac{4}{x} y' + \left(1 + \dfrac{2}{x^2}\right) y = 0$

5. $y'' + \dfrac{3}{x} y' + 4x^2 y = 0$

6. $(x - 1)y'' - \left(\dfrac{4x^2 - 3x + 1}{2x}\right) y' + \left(\dfrac{2x^2 - x + 2}{2x}\right) y = 0$

7. $y'' + \dfrac{2}{x} y' - \dfrac{2}{x^2} y = 0$

8. $4xy'' + 2y' + y = 0$

9. $xy'' + 2y' + xy = 0$

10. $xy'' - y' + 4x^3 y = 0$

11. $xy'' + (1 - 2x)y' - (1 - x)y = 0$

12. $x(x + 1)^2 y'' + (1 - x^2)y' - (1 - x)y = 0$

13. $y'' - 2y' + \left(1 + \dfrac{1}{4x^2}\right) y = 0$

14. (*Euler's equation*) $x^2 y'' + Axy' + By = 0$, A and B constant

15. $x(x - 1)y'' - (1 - 3x)y' + y = 0$

16. $x^2(x^2 - 1)y'' - x(x^2 + 1)y' + (x^2 + 1)y = 0$

17. $y'' + \dfrac{y'}{x} - y = 0$

18. $2xy'' - (x - 3)y' - y = 0$

19. $y'' + \dfrac{x + 1}{2x} y' + \dfrac{3}{2x} y = 0$

20. $y'' + \dfrac{1}{2x} y' - \dfrac{x + 1}{2x^2} y = 0$

21. $x^2 y'' + x(x - 1)y' - (x - 1)y = 0$

22. $xy'' - (3 + x)y' + 2y = 0$

23. Prove by means of Eq. (3.14) that Eq. (5.69) has the form

$$y_2(x) = J_0(x) \ln x + \sum_{n=1}^{\infty} \dfrac{(-1)^{n+1}}{2^{2n}(n!)^2}\left(1 + \dfrac{1}{2} + \cdots + \dfrac{1}{n}\right) x^{2n}, \quad x > 0.$$

24. Can the method of Frobenius be used to solve Exercise 5.2.21?

25. Consider the differential equation

$$y'' - \frac{1}{x^2} y' + \frac{1}{x^3} y = 0.$$

a) Show that $x = 0$ is an irregular singular point of this equation.
b) Use the fact that $y_1 = x$ is a solution to find a second independent solution.
c) Show that the solution y_2 cannot be expressed as a series of the form (5.59). Thus this solution cannot be found by the method of Frobenius.

26. The differential equation

$$x^2 y'' + (4x - 1)y' + 2y = 0$$

has $x = 0$ as an irregular singular point.

a) Suppose that (5.59) is inserted into this equation. Show that $r = 0$ and the corresponding Frobenius method "solution" is

$$y = \sum_{n=0}^{\infty} (n + 1)! x^n.$$

b) Prove that the series above has radius of convergence $R = 0$. Hence even though a Frobenius series may formally satisfy a differential equation, it may not be a valid solution at an irregular singular point.

*27. Let $x_0 = 0$ and suppose that the Maclaurin series

$$a(x) = \sum_{n=0}^{\infty} a_n x^n, \qquad b(x) = \sum_{n=0}^{\infty} b_n x^n$$

are valid for $|x| < R$.

a) Show that if (5.54) has a power series solution

$$y(x) = \sum_{n=0}^{\infty} c_n x^n, \qquad c_0 = 1,$$

then the coefficients c_n satisfy the recursion formula

$$(n + 1)(n + 2)c_{n+2} = -\sum_{k=0}^{n} [(k + 1)a_{n-k}c_{k+1} + b_{n-k}c_k].$$

b) Use the root test formula (5.7) for finding the radius of convergence to show that for any r such that $0 < r < R$ there is a constant $M > 0$ such that

$$(n + 1)(n + 2)|c_{n+2}| \leq \frac{M}{r^n} \sum_{k=0}^{\infty} [(k + 1)|c_{k+1}| + |c_k|]r^k + M|c_{n+1}|r.$$

c) Use the ratio test to prove that the series

$$\sum_{n=0}^{\infty} |c_n| x^n$$

converges for $|x| < r$. Then by the comparison test for series, it follows that the series $y(x)$ converges for $|x| < r$. Since r is arbitrary, it follows that the series representation of the solution is valid for $|x| < R$.

28. Modify the argument in Exercise 27 to justify the last statement of Theorem 5.3.

5.4 BESSEL FUNCTIONS

The differential equation

$$x^2 y'' + xy' + (x^2 - p^2)y = 0, \tag{5.91}$$

which is known as *Bessel's equation of order* p (≥ 0), is one of the most important differential equations in applied mathematics. The equation (5.91) was first investigated in 1703 by Jakob Bernoulli (1654–1705) in connection with the oscillatory behavior of a hanging chain, and later by Friedrich Wilhelm Bessel (1784–1846) in his studies of planetary motion. Since then, the Bessel functions have been used in the studies of elasticity, fluid motion, potential theory, diffusion, and the propagation of waves. We shall present a few applications of the Bessel functions in Chapter 10 when we study the solutions of partial differential equations. (See Section 10.4.)

The reader may recall that in Section 5.3 we found the solution of Bessel's equation for $p = 0, \frac{1}{2}$, and 1 (see Examples 5.3.3, 5.3.5, and 5.3.6). In each of these examples, the method of Frobenius was an important tool, so we anticipate again the successful application of this procedure. We assume that a solution of the form

$$y(x) = |x|^r \sum_{n=0}^{\infty} c_n x^n, \qquad x \neq 0, \qquad c_0 \neq 0, \tag{5.92}$$

exists for Bessel's equation of order p. Then substituting (5.92) into Eq. (5.91), we have

$$|x|^r \left[\sum_{n=0}^{\infty} c_n(r+n)(r+n-1)x^n + \sum_{n=0}^{\infty} c_n(r+n)x^n + (x^2 - p^2) \sum_{n=0}^{\infty} c_n x^n \right] = 0$$

or

$$|x|^r \left[\sum_{n=0}^{\infty} c_n[(r+n)^2 - p^2]x^n + \sum_{n=0}^{\infty} c_n x^{n+2} \right] = 0. \tag{5.93}$$

The indicial equation is $r^2 - p^2 = 0$ with the roots $r_1 = p(\geqslant 0)$ and $r_2 = -p$. For any $p \geqslant 0$ we can obtain a first solution $y_1(x)$ for Eq. (5.91) by the method of Frobenius. Letting $r = p$ in (5.93), we have

$$|x|^p \left[(1 + 2p)c_1 x + \sum_{n=2}^{\infty} c_n n(n + 2p)x^n + \sum_{n=0}^{\infty} c_n x^{n+2} \right] = 0$$

or

$$|x|^p \left[(1 + 2p)c_1 x + \sum_{k=0}^{\infty} [c_{k+2}(k + 2)(k + 2 + 2p) + c_k]x^{k+2} \right] = 0. \quad (5.94)$$

Since the sum of the series in (5.94) is zero, the coefficients must all vanish, yielding $c_1 = 0$ and the recurrence relation

$$c_{k+2} = \frac{-c_k}{(k + 2)(k + 2 + 2p)}. \quad (5.95)$$

Hence all the coefficients with odd-numbered subscripts $c_{2j+1} = 0$, since by (5.95) they can all be expressed as a multiple of c_1. Letting $k = 2j$, we see that the coefficients with even-numbered subscripts satisfy the equation

$$c_{2(j+1)} = \frac{-c_{2j}}{2^2(j + 1)(p + j + 1)},$$

which yields

$$c_2 = \frac{-c_0}{2^2(p + 1)}, \qquad c_4 = \frac{-c_2}{2^2 \cdot 2(p + 2)} = \frac{c_0}{2^4 2!(p + 1)(p + 2)},$$

$$c_6 = \frac{-c_4}{2^2 \cdot 3(p + 3)} = \frac{-c_0}{2^6 3!(p + 1)(p + 2)(p + 3)}, \ldots$$

Hence the series (5.92) becomes

$$y_1(x) = |x|^p \left[c_0 - \frac{c_0}{2^2(p + 1)} x^2 + \frac{c_0}{2^4 2(p + 1)(p + 2)} x^4 - \cdots \right]$$

$$= c_0 |x|^p \sum_{n=0}^{\infty} (-1)^n \frac{x^{2n}}{2^{2n} n!(p + 1)(p + 2) \cdots (p + n)}. \quad (5.96)$$

To write (5.96) in a more compact form, we recall the definition of the *gamma function* (see Exercise 1.3.14), which we now extend to all values $p > -1$:

$$\Gamma(p + 1) = \int_0^{\infty} e^{-t} t^p \, dt. \quad (5.97)$$

Integrating $\Gamma(p + 1)$ by parts, we have

$$\Gamma(p + 1) = \int_0^\infty e^{-t}t^p \, dt = -e^{-t}t^p \Big|_0^\infty + p \int_0^\infty e^{-t}t^{p-1} \, dt.$$

The first expression on the right is zero, and the integral on the right-hand side is $\Gamma(p)$. We thus have the basic property of gamma functions:

$$\Gamma(p + 1) = p\Gamma(p). \tag{5.98}$$

Since

$$\Gamma(1) = \int_0^\infty e^{-t} \, dt = -e^{-t} \Big|_0^\infty = 1,$$

it follows that $\Gamma(2) = \Gamma(1) = 1!, \Gamma(3) = 2\Gamma(2) = 2!, \ldots$, and in general, $\Gamma(n + 1) = n!$. Thus, the gamma function is the extension to real numbers $p > -1$, of the factorial function.

It is customary in (5.96) to let $c_0 = [2^p\Gamma(p + 1)]^{-1}$. Then (5.96) becomes

$$J_p(x) = \left|\frac{x}{2}\right|^p \sum_{n=0}^\infty (-1)^n \frac{(x/2)^{2n}}{n!\Gamma(p + n + 1)}, \qquad x \neq 0, \tag{5.99}$$

which is known as the *Bessel function of the first kind of order p*. Thus $J_p(x)$ is the first solution of (5.91) guaranteed by Theorem 5.3.

To find the second solution promised by Theorem 5.3, we must consider the difference $r_1 - r_2 = 2p$. By Theorem 5.3 (Case 1), if p is *not* a multiple of $\frac{1}{2}$, we will again be able to apply the method of Frobenius with $r = -p$ to find the second solution. And the results of Examples 5.3.5 and 5.3.6 make it appear likely that we will obtain $\ln|x|$ terms only when p is an integer. Therefore, we set $r = -p$ in Eq. (5.93) and assume p is not an integer. Then we have

$$|x|^{-p}\left[(1 - 2p)c_1 x + \sum_{k=0}^\infty [c_{k+2}(k + 2)(k + 2 - 2p) + c_k]x^{k+2}\right] = 0. \tag{5.100}$$

Setting the coefficients of this series to zero, we have $(1 - 2p)c_1 = 0$, which implies that $c_1 = 0$ unless $p = \frac{1}{2}$ (a case that was already dealt with in Example 5.3.5), and the recurrence relation

$$c_{k+2} = \frac{-c_k}{(k + 2)(k + 2 - 2p)}. \tag{5.101}$$

We observe that the term $(k + 2 - 2p)$ will vanish for some odd integer k if p is a multiple of $\frac{1}{2}$. Since $c_1 = 0$, (5.101) implies that all the coefficients c_k with odd-numbered subscripts are equal to zero. Hence we use (5.101) only to calculate the coefficients with even-numbered subscripts. Then

$$c_2 = \frac{-c_0}{2^2(1 - p)}, \qquad c_4 = \frac{-c_2}{2^2 \cdot 2(2 - p)} = \frac{c_0}{2^4 2!(1 - p)(2 - p)}, \ldots,$$

and the second solution is

$$J_{-p}(x) = \left|\frac{x}{2}\right|^{-p} \sum_{n=0}^{\infty} (-1)^n \frac{(x/2)^{2n}}{n!\Gamma(n - p + 1)}. \tag{5.102}$$

To see that (5.99) and (5.102) are linearly independent, we obtain by long division

$$\frac{J_p(x)}{J_{-p}(x)} = \frac{|x/2|^p/\Gamma(p + 1) - |x/2|^{p+2}/2!\Gamma(p + 3) + \cdots}{|x/2|^{-p}/\Gamma(1 - p) - |x/2|^{2-p}/2!\Gamma(3 - p) + \cdots}$$

$$= \frac{|x/2|^{2p}}{\Gamma(1 + p)/\Gamma(1 - p)} + \frac{3p|x/2|^{2p+2}}{\Gamma(3 + p)/\Gamma(1 - p)} + \cdots,$$

which clearly is not a constant function.

The radius of convergence of the series

$$\sum_{n=0}^{\infty} (-1)^n \frac{t^n}{n!\Gamma(n \pm p + 1)}$$

is easily found by the ratio test:

$$\frac{1}{R} = \lim_{n \to \infty} \frac{n!\Gamma(n \pm p + 1)}{(n + 1)!\Gamma(n \pm p + 2)} = \lim_{n \to \infty} \frac{1}{(n + 1)(n \pm p + 1)} = 0,$$

implying that $R = \infty$. Hence (5.99) and (5.102) converge for all values. We have proved the following result.

Theorem 5.4 If p is not an integer, then

$$y(x) = AJ_p(x) + BJ_{-p}(x),$$

is the general solution of Bessel's equation for all values $x \neq 0$.

If p is an integer, then the term $(k + 2 - 2p)$ in the recurrence relation (5.101) will vanish for some even integer $k = 2m$. Since

$$(k + 2)(k + 2 - p)c_{k+2} = -c_k, \tag{5.103}$$

c_{2m} will vanish, and iterating Eq. (5.103) repeatedly, we see that $c_{2m-2} = c_{2m-4} = \cdots = c_2 = c_0 = 0$. But this contradicts the assumed form (5.92) of the solution. Thus the method of Frobenius cannot be used when p is a positive integer. By Theorem 5.3, the second linearly independent solution of (5.91) must have the form

$$y_2(x) = J_p(x) \ln |x| + |x|^{-p} \sum_{n=0}^{\infty} c_n^* x^n, \qquad c_0^* \neq 0. \tag{5.104}$$

Our task now is to calculate the coefficients c_n^*, which we will do by substituting

(5.104) with $x > 0$ for y in Eq. (5.91). Since

$$y_2' = J_p' \ln x + \frac{1}{x} J_p + \sum_{n=0}^{\infty} c_n^*(n - p)x^{n-p-1}$$

and

$$y_2'' = J_p'' \ln x + \frac{2}{x} J_p' - \frac{1}{x^2} J_p + \sum_{n=0}^{\infty} c_n^*(n - p)(n - p - 1)x^{n-p-2},$$

we obtain the equation

$$2xJ_p' + \sum_{n=0}^{\infty} c_n^* n(n - 2p)x^{n-p} + \sum_{n=0}^{\infty} c_n^* x^{n-p+2} = 0. \tag{5.105}$$

[The logarithmic term disappears because J_p is a solution of Eq. (5.91) and the two terms containing J_p cancel out.] Differentiating (5.99), we have

$$J_p'(x) = \sum_{n=0}^{\infty} (-1)^n \frac{(n + p/2)(x/2)^{2n+p-1}}{n!(p + n)!},$$

since $\Gamma(p + n + 1) = (n + p)!$. Replacing J_p' in (5.105) by this expansion and factoring out an x^{-p}, we obtain, after some algebra,

$$x^{-p} \left[\sum_{n=0}^{\infty} (-1)^n \frac{(2n + p)x^{2n+2p}}{2^{2n+p-1} n!(p + n)!} \right.$$

$$\left. + c_1^*(1 - 2p)x + \sum_{n=2}^{\infty} [c_n^* n(n - 2p) + c_{n-2}^*]x^n \right] = 0. \tag{5.106}$$

First, we show that all the coefficients c_n^* with odd-numbered subscripts n are zero. Clearly $c_1^* = 0$ since p is an integer and the first series does not contain odd-numbered powers of x. Using only the last term in (5.106) for $n = 2k + 1$ (i.e. n odd), we obtain the recurrence relation

$$(2k + 1)(2k + 1 - 2p)c_{2k+1}^* = -c_{2k-1}^*.$$

Thus all the terms $c_{2k+1}^* = 0$. For the even-numbered terms we obtain successively from (5.106) the identities

$$2(2 - 2p)c_2^* + c_0^* = 0,$$
$$4(4 - 2p)c_4^* + c_2^* = 0,$$
$$\vdots \tag{5.107}$$
$$(2p - 2)2c_{2p-2}^* + c_{2p-4}^* = 0.$$

And for even-numbered powers of x at least as large as $2p$, we have the recurrence relation

$$\frac{(-1)^n(2n + p)}{2^{2n+p-1}n!(p + n)!} + c^*_{2n+2p}(2n)(2n + 2p) + c^*_{2n+2p-2} = 0,$$

$$n = 0, 1, 2, \ldots. \qquad (5.108)$$

Thus

$$c^*_2 = \frac{c^*_0}{2^2(p - 1)}, \quad c^*_4 = \frac{c^*_0}{2^4 2!(p - 2)(p - 1)}, \ldots, c^*_{2p-2} = \frac{c^*_0}{2^{2p-2}(p - 1)!^2},$$

and by (5.108),

$$\frac{p}{2^{p-1}p!} + c^*_{2p-2} = 0,$$

$$\frac{-(p + 2)}{2^{p+1}(p + 1)!} + c^*_{2p+2}(4p + 4) + c^*_{2p} = 0, \quad \text{etc.}$$

Hence

$$\frac{c^*_0}{2^{2p-2}(p - 1)^2} = c^*_{2p-2} = \frac{-p}{2^{p-1}p} = \frac{-1}{2^{p-1}(p - 1)},$$

so that

$$c^*_0 = -2^{p-1}(p - 1), \qquad (5.109)$$

and all the remaining coefficients can now be calculated. After an extremely long (but straightforward) calculation we obtain

$$y_2(x) = J_p(x) \ln |x| - \frac{1}{2}\left[\sum_{k=0}^{p-1} \frac{(p - k - 1)!}{k!}\left(\frac{x}{2}\right)^{2k-p} + \frac{h_p}{p!}\left(\frac{x}{2}\right)^p \right.$$

$$\left. + \sum_{k=1}^{\infty} \frac{(-1)^k[h_k + h_{p+k}]}{k!(p + k)!}\left(\frac{x}{2}\right)^{2k+p}\right], \qquad (5.110)$$

where

$$h_p = 1 + \frac{1}{2} + \frac{1}{3} + \cdots + \frac{1}{p}$$

and p is a positive integer. A similar procedure will yield the result in Exercise 5.3.23 for $p = 0$.

It is customary to replace (5.110) by the combination

$$Y_p(x) = \frac{2}{\pi}[y_2(x) + (\gamma - \ln 2)J_p(x)], \qquad p = 0, 1, 2, \ldots,$$

where

$$\gamma = \lim_{p \to \infty} (h_p - \ln p) = 0.57721\ 56649 \ldots$$

is the so-called *Euler constant*. This particular solution is obviously independent of $J_p(x)$ and is called the *Bessel function of the second kind of order p* or *Neumann's function of order p*. It is defined by the formula

$$Y_p(x) = \frac{2}{\pi} J_p(x) \left(\ln \frac{x}{2} + \gamma \right)$$
$$- \frac{1}{\pi} \left[\sum_{k=0}^{p-1} \frac{(p-k-1)!}{k!} \left(\frac{x}{2}\right)^{2k-p} + \frac{h_p}{p!} \left(\frac{x}{2}\right)^p + \sum_{k=1}^{\infty} (-1)^k \frac{[h_k + h_{p+k}]}{k!(p+k)!} \left(\frac{x}{2}\right)^{2k+p} \right],$$

for all integers $p = 0, 1, 2, \ldots$.

The function Y_p may be extended to all *real* numbers $p \geqslant 0$ (see Exercise 24) by the formula

$$Y_p(x) = \frac{1}{\sin p\pi} [J_p(x) \cos p\pi - J_{-p}(x)], \qquad p \neq 0, 1, 2, \ldots. \tag{5.111}$$

Using this definition of Y_p, we have the following result:

Theorem 5.5 A general solution of Bessel's equation of order p is

$$y(x) = A J_p(x) + B Y_p(x), \qquad x \neq 0.$$

A graph of the functions Y_0, Y_1, and Y_2 is shown in Fig. 5.1.

Properties of Bessel's functions Now that we have the expansions for $J_p(x)$ and $Y_p(x)$, we can derive a number of important formulas involving Bessel functions and their derivatives. For simplicity we shall assume $x > 0$. The first two identities are immediate consequences of Eq. (5.99):

$$\frac{d}{dx} [x^p J_p(x)] = x^p J_{p-1}(x), \tag{5.112}$$

$$\frac{d}{dx} [x^{-p} J_p(x)] = -x^{-p} J_{p+1}(x). \tag{5.113}$$

To prove Eq. (5.112), we differentiate the product $x^p J_p$ term by term:

$$\frac{d}{dx} \sum_{n=0}^{\infty} (-1)^n \frac{2^p(x/2)^{2n+2p}}{n!\Gamma(p+n+1)} = \sum_{n=0}^{\infty} (-1)^n \frac{2^{p-1}(x/2)^{2n+2p-1}2(n+p)}{n!\Gamma(p+n+1)}$$

$$= x^p \sum_{n=0}^{\infty} (-1)^n \frac{(x/2)^{2n+p-1}}{n!\Gamma(p+n)} = x^p J_{p-1}(x),$$

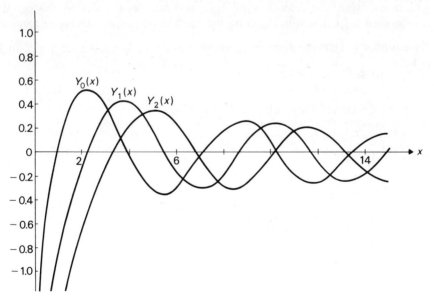

Figure 5.1

since $\Gamma(p + n + 1) = (p + n)\Gamma(p + n)$. The proof of Eq. (5.113) is similar (see Exercise 4). Expanding the left-hand sides of Eqs. (5.112) and (5.113), we have

$$x^p J_p' + p x^{p-1} J_p = x^p J_{p-1}$$

and

$$x^{-p} J_p' - p x^{p-1} J_p = -x^{-p} J_{p+1},$$

which may be simplified to yield the identities

$$x J_p' = x J_{p-1} - p J_p, \tag{5.114}$$

$$x J_p' = p J_p - x J_{p+1}. \tag{5.115}$$

Subtracting (5.115) from (5.114), we obtain the recursion relation

$$x J_{p+1} - 2p J_p + x J_{p-1} = 0. \tag{5.116}$$

Adding the two together yields

$$2 J_p' = J_{p-1} - J_{p+1}. \tag{5.117}$$

Formulas (5.112) to (5.117) are extremely important in solving problems involving the Bessel functions, since they allow us to express Bessel functions of higher order in terms of lower order functions. Note that for fixed x, $J_p(x)$ is a number, so that (5.116) is a second order homogeneous linear difference equation with nonconstant coefficients. Since $Y_p(x)$ also satisfies Eq. (5.116), this equation can be used for calculating $Y_p(x)$ by the method of Section 4.3 (see Exercise 7).

Example 5.4.1 Express $J_3(x)$ in terms of $J_0(x)$ and $J_1(x)$. We let $p = 2$ in Eq. (5.116):

$$x J_3 = 4 J_2 - x J_1.$$

Applying formula (5.116) with $p = 1$ to J_2 yields

$$x J_2 = 2 J_1 - x J_0.$$

Thus

$$J_3(x) = \frac{4}{x} J_2 - J_1 = \frac{4}{x^2}(2J_1 - xJ_0) - J_1$$

$$= \left(\frac{8}{x^2} - 1\right) J_1(x) - \frac{4}{x} J_0(x).$$

Example 5.4.2 Evaluate the integral

$$\int x^4 J_1(x) \, dx. \tag{5.118}$$

Integrating (5.118) by parts, we have by (5.112)

$$\int x^2 [x^2 J_1(x)] \, dx = x^2 (x^2 J_2) - \int x^2 J_2 \cdot 2x \, dx = x^4 J_2(x) - 2 \int x^3 J_2 \, dx.$$

Again applying (5.112) to the last integral, we obtain

$$\int x^4 J_1(x)\, dx = x^4 J_2(x) - 2x^3 J_3(x) + c.$$

In general, an integral of the form

$$\int x^m J_n(x)\, dx,$$

where m and n are integers such that $m + n \geqslant 0$, can be completely integrated if $m + n$ is odd. But if $m + n$ is even, then the result depends on the residual integral $\int J_0(x)\, dx$. It is not possible to reduce $\int J_0(x)\, dx$ [or $\int Y_0(x)\, dx$] any further and for this reason the functions

$$\int_0^x J_0(x)\, dx \qquad \text{and} \qquad \int_0^x Y_0(x)\, dx$$

have been tabulated.*

Example 5.4.3 Express $J_{3/2}(x)$ in terms of sin x and cos x. We recall from Example 5.3.5 that the general solution of Bessel's equation of order $\frac{1}{2}$ can be written in terms of sines and cosines. Multiplying $y_1(x)$ and $y_2(x)$ by $c_0 = [\sqrt{2}\Gamma(\frac{3}{2})]^{-1}$, we have

$$J_{1/2}(x) = \frac{1}{\sqrt{2}\Gamma(\frac{3}{2})} \frac{\sin x}{\sqrt{x}} = \sqrt{\frac{2}{\pi x}} \sin x \tag{5.119}$$

and

$$J_{-1/2}(x) = \frac{1}{\sqrt{2}\Gamma(\frac{3}{2})} \frac{\cos x}{\sqrt{x}} = \sqrt{\frac{2}{\pi x}} \cos x, \tag{5.120}$$

since $\Gamma(\frac{3}{2}) = (\frac{1}{2})\Gamma(\frac{1}{2}) = \sqrt{\pi}/2$ (see Appendix 1, formulas 55 and 56). By (5.116),

$$J_{3/2}(x) = \frac{1}{x} J_{1/2} - J_{-1/2} = \sqrt{\frac{2}{\pi x}} \left(\frac{\sin x}{x} - \cos x \right).$$

Similar results hold for $Y_p(x)$ (see Exercises 5, 6, 7).

Many differential equations with variable coefficients can be reduced to Bessel equations.

Example 5.4.4 Consider the equation

$$y'' + k^2 xy = 0. \tag{5.121}$$

The following substitution will reduce Eq. (5.121) to a Bessel equation. Let $u = y/\sqrt{x}$ and $z = 2kx^{3/2}/3$. Then

$$\frac{du}{dz} = \frac{du/dx}{dz/dx} = \frac{y'}{kx} - \frac{y}{2kx^2}$$

---

* A. N. Lowan and M. Abramwitz, "Tables of integrals of $\int_0^x J_0(t)\, dt$ and $\int_0^x Y_0(t)\, dt$," *J. Math. Phys.*, **22** (1943), 3–12.

and

$$\frac{d^2u}{dz^2} = \frac{\dfrac{d}{dx}\left(\dfrac{du}{dz}\right)}{dz/dx} = \frac{y''}{k^2x^{3/2}} - \frac{3y'}{2k^2x^{5/2}} + \frac{y}{k^2x^{7/2}}$$

Hence

$$z^2\frac{d^2u}{dz^2} + z\frac{du}{dz} = \frac{4}{9}x^{3/2}y'' + \frac{1}{9}\frac{y}{\sqrt{x}},$$

and using (5.121) for y'', we obtain

$$z^2\frac{d^2u}{dz^2} + z\frac{du}{dz} = -\frac{4}{9}k^2x^3\left(\frac{y}{\sqrt{x}}\right) + \frac{1}{9}\frac{y}{\sqrt{x}} = -\left(z^2 - \frac{1}{9}\right)u$$

or

$$z^2\frac{d^2u}{dz^2} + z\frac{du}{dz} + \left(z^2 - \frac{1}{9}\right)u = 0, \qquad (5.122)$$

the Bessel equation of order $\frac{1}{3}$. Since Eq. (5.122) has the general solution

$$u(z) = AJ_{1/3}(z) + BJ_{-1/3}(z),$$

Eq. (5.121) has the solution

$$y(x) = \sqrt{x}[AJ_{1/3}(\tfrac{2}{3}kx^{3/2}) + BJ_{-1/3}(\tfrac{2}{3}kx^{3/2})].$$

Additional facts concerning the orthogonality and zeros of Bessel functions will be developed in Chapter 9.

EXERCISES 5.4

1. Express $J_5(x)$ in terms of $J_0(x)$ and $J_1(x)$.
2. Express $J_{5/2}(x)$ in terms of $\sin x$ and $\cos x$.
3. Show that:
   a) $4J_p''(x) = J_{p+2}(x) - 2J_p(x) + J_{p-2}(x)$,
   b) $-8J_p'''(x) = J_{p+3}(x) - 3J_{p+1}(x) + 3J_{p-1}(x) - J_{p-3}(x)$.
4. Prove that $[x^{-p}J_p(x)]' = x^{-p}J_{p+1}(x)$.
5. Show that $[x^pY_p(x)]' = x^pY_{p-1}(x)$.
6. Prove that $[x^{-p}Y_p(x)]' = -x^{-p}Y_{p+1}(x)$.
7. Using the equations of Exercises 5 and 6, show that:
   a) $x(Y_{p+1} + Y_{p-1}) = 2pY_p$,
   b) $2Y_p' = Y_{p-1} - Y_{p+1}$.

8. Prove the following identities:

   a) $\int J_1(x) dx = -J_0(x) + c$
   b) $\int x^2 J_1(x) dx = 2xJ_1(x) - x^2 J_0(x) + c$
   c) $\int xJ_0(x) dx = xJ_1(x) + c$
   d) $\int x^3 J_0(x) dx = (x^3 - 4x)J_1(x) + 2x^2 J_0(x) + c$
   e) $\int J_0(x) \cos x \, dx = xJ_0(x) \cos x + xJ_1(x) \sin x + c$
   f) $\int J_0(x) \sin x \, dx = xJ_0(x) \sin x - xJ_1(x) \cos x + c$
   g) $\int J_1(x) \cos x \, dx = xJ_1(x) \cos x - (x \sin x + \cos x)J_0(x) + c$
   h) $\int J_1(x) \sin x \, dx = xJ_1(x) \sin x + (x \cos x - \sin x)J_0(x) + c$

9. Verify the identities:

   a) $\int x^2 J_0(x) dx = x^2 J_1(x) + xJ_0(x) - \int J_0(x) dx + c$
   b) $\int x^{-1} J_1(x) dx = -J_1(x) + \int J_0(x) dx + c$
   c) $\int xJ_1(x) dx = -xJ_0(x) + \int J_0(x) dx + c$
   d) $\int x^3 J_1(x) dx = 3x^2 J_1(x) - (x^3 - 3x)J_0(x) - 3\int J_0(x) dx + c$

10. Show that $\int J_0(\sqrt{x}) dx = 2xJ_1(\sqrt{x}) + c$.

11. Show that

$$\int xJ_0(x) \sin x \, dx = \tfrac{1}{3}\{x^2[J_0(x) \sin x - J_1(x) \cos x] + xJ_1(x) \sin x\} + c.$$

12. Show that

$$\int xJ_1(x) \cos x \, dx = \tfrac{1}{3}\{x^2[J_1(x) \cos x - J_0(x) \sin x] + 2xJ_1(x) \sin x\} + c.$$

13. Show that $\int J_0(x) dx = 2[J_1(x) + J_3(x) + J_5(x) + \cdots] + c$.

In Exercises 14 through 21 reduce each given equation to a Bessel equation and solve it.

14. $x^2 y'' + xy' + (a^2 x^2 - p^2)y = 0$ 15. $4x^2 y'' + 4xy' + (x^2 - p^2)y = 0$
16. $x^2 y'' + xy' + 4(x^4 - p^2)y = 0$ 17. $xy'' - y' + xy = 0$
18. $x^2 y'' + (x^2 + \tfrac{1}{4})y = 0$ 19. $xy'' + (1 + 2k)y' + xy = 0$
20. $y'' + k^2 x^2 y = 0$ 21. $y'' + k^2 x^4 y = 0$

22. Obtain the result of Exercise 5.3.23 by the methods of this section.

23. Obtain formula (5.110) for $p = 1$ by the methods of this section.

* 24. Prove that if $p \geqslant 0$ is an integer,

$$\lim_{q \to p} \frac{J_q(x) \cos q\pi - J_{-q}(x)}{\sin q\pi} = Y_p(x).$$

This is the extension (5.111) of Y_q to all real values $q \geqslant 0$.

25. a) Expand $e^{(x/2)[t-(1/t)]}$ as a power series in t by multiplying the series for $e^{xt/2}$ and $e^{-x/2t}$.
    b) Show that the coefficient of t^n in the expansion obtained in part (a) is $J_n(x)$.
    c) Conclude that

$$e^{(x/2)[t-(1/t)]} = J_0(x) + \sum_{n=1}^{\infty} J_n(x)[t^n + (-t)^{-n}].$$

This function is called the *generating function* of the Bessel functions.

d) Set $t = e^{i\theta}$ in the expression in part (c) and obtain the identities

$$\cos (x \sin \theta) = J_0(x) + 2 \sum_{n=1}^{\infty} J_{2n}(x) \cos 2n\theta$$

and

$$\sin (x \sin \theta) = 2 \sum_{n=1}^{\infty} J_{2n-1}(x) \sin (2n - 1)\theta.$$

26. Show that $J_0(x) = (1/\pi) \int_0^\pi \cos (x \cos t) \, dt$.

27. Show that $J_n(x) = (1/\pi) \int_0^\pi \cos (nt - x \sin t) \, dt$.

28. (*Modified Bessel functions*) The function $I_p(x) = i^{-p} J_p(ix)$, $i^2 = -1$, is called the *modified Bessel function of the first kind of order p*. Show that $I_p(x)$ is a solution of the differential equation

$$x^2 y'' + xy' - (x^2 + p^2)y = 0,$$

and obtain its series representation by the method of Frobenius.

29. Show that another solution of the differential equation in Exercise 28 is the *modified Bessel function of the second kind of order p*:

$$K_p(x) = \frac{\pi}{2 \sin p\pi} [I_{-p}(x) - I_p(x)].$$

30. Prove (see Exercise 25) that

$$e^{(x/2)[t + (1/t)]} = I_0(x) + \sum_{n=1}^{\infty} I_n(x)[t^n + t^{-n}].$$

31. Prove the identities:

a) $\dfrac{d}{dx} [x^p I_p(x)] = x^p I_{p-1}(x)$

b) $\dfrac{d}{dx} [x^{-p} I_p(x)] = x^{-p} I_{p+1}(x)$

c) $x(I_{p-1} - I_{p+1}) = 2p I_p$

32. a) Prove for all integers $n > 0$ that

$$\int_0^\infty J_{n+1}(x) \, dx = \int_0^\infty J_{n-1}(x) \, dx$$

by means of Eq. (5.117), given the approximation

$$J_n(x) \approx \sqrt{\frac{2}{\pi x}} \cos \left(x - \frac{\pi}{4} - \frac{n\pi}{2} \right).$$

b) Prove a similar fact for Y_n given the approximation

$$Y_n(x) \approx \sqrt{\frac{2}{\pi x}} \sin\left(x - \frac{\pi}{4} - \frac{n\pi}{2}\right).$$

c) Show that $\int_0^\infty J_n(x)\, dx = 1$.

d) Prove that $\int_0^\infty (J_n(x)/x)\, dx = 1/n$.

5.5 LEGENDRE POLYNOMIALS

Another very important differential equation that arises in many applications (see, for example, Sections 9.2 and 10.4) is *Legendre's differential equation*

$$(1 - x^2)y'' - 2xy' + p(p + 1)y = 0, \tag{5.123}$$

where p is a given real number. Any solution of Eq. (5.123) is called a *Legendre function*.*

Dividing Eq. (5.123) by $(1 - x^2)$, we obtain the equation

$$y'' - \frac{2x}{1 - x^2} y' + \frac{p(p + 1)}{1 - x^2} y = 0,$$

and we observe, using the geometric series, that the coefficient functions $a(x)$ and $b(x)$ of the form (5.51) are both analytic at $x = 0$ with radius of convergence $R = 1$:

$$a(x) = \frac{-2x}{1 - x^2} = -2x(1 + x^2 + x^4 + x^6 + \cdots)$$

and

$$b(x) = p(p + 1)(1 + x^2 + x^4 + x^6 + \cdots).$$

By Theorem 5.2 we can apply the power series method to obtain the general solution of (5.123). In addition, the general solution will have a power series representation valid in the interval $|x| < 1$. Substituting

$$y = \sum_{n=0}^\infty c_n x^n,$$

and its derivatives into (5.123), we have

$$(1 - x^2) \sum_{n=2}^\infty c_n n(n - 1)x^{n-2} - 2x \sum_{n=1}^\infty c_n n x^{n-1} + p(p + 1) \sum_{n=0}^\infty c_n x^n = 0$$

---

* Named after a famous French mathematician Adrien Marie Legendre (1752–1833).

or

$$\sum_{n=0}^{\infty} \{(n + 2)(n + 1)c_{n+2} - c_n[n(n + 1) - p(p + 1)]\}x^n = 0. \tag{5.124}$$

Setting the coefficients of the sum (5.124) to zero, we obtain the recurrence relation

$$(n + 2)(n + 1)c_{n+2} = c_n(n^2 + n - p^2 - p) = c_n(n - p)(n + p + 1).$$

Thus we have

$$c_{n+2} = -\frac{(p - n)(p + n + 1)}{(n + 2)(n + 1)} c_n. \tag{5.125}$$

Therefore

$$c_2 = -\frac{p(p + 1)}{2!} c_0, \qquad c_3 = -\frac{(p - 1)(p + 2)}{3!} c_1,$$

$$c_4 = -\frac{(p - 2)(p + 3)}{4 \cdot 3} c_2 = \frac{(p - 2)p(p + 1)(p + 3)}{4!} c_0,$$

$$c_5 = -\frac{(p - 3)(p + 4)}{5 \cdot 4} c_3 = \frac{(p - 3)(p - 1)(p + 2)(p + 4)}{5!} c_1, \quad \text{etc.}$$

Inserting these values for the coefficients into the power series expansion for $y(x)$ yields

$$y(x) = c_0 y_1(x) + c_1 y_2(x), \tag{5.126}$$

where

$$y_1(x) = 1 - p(p + 1)\frac{x^2}{2!} + (p - 2)p(p + 1)(p + 3)\frac{x^4}{4!} - \cdots, \tag{5.127}$$

$$y_2(x) = x - (p - 1)(p + 2)\frac{x^3}{3!} + (p - 3)(p - 1)(p + 2)(p + 4)\frac{x^5}{5!} - \cdots. \tag{5.128}$$

Dividing (5.128) by (5.127), we have

$$\frac{y_2(x)}{y_1(x)} = x + \frac{(p^2 + p + 1)}{3}x^3 + \cdots,$$

which obviously is nonconstant, implying that y_1 and y_2 are linearly independent. Thus (5.126) is the general solution of Legendre's equation (5.123) for $|x| < 1$.

In many applications the parameter p in Legendre's equation is a nonnegative integer. When this occurs, the right-hand side of (5.125) will vanish for $n = p$, implying that $c_{p+2} = c_{p+4} = c_{p+6} = \cdots = 0$. Thus one of the equations (5.127) or (5.128) reduces to a polynomial of degree p. (For even p it is y_1, for odd p it is

y_2.) These polynomials, multiplied by an appropriate constant, are called the *Legendre polynomials*. Because of their importance, we shall study these polynomials in greater detail. It is customary to set

$$c_p = \frac{(2p)!}{2^p(p!)^2}, \qquad p = 0, 1, 2, \ldots, \tag{5.129}$$

so that

$$c_{p-2} = -\frac{p(p-1)}{2(2p-1)} c_p = -\frac{(2p-2)!}{2^p(p-1)!(p-2)!},$$

$$c_{p-4} = -\frac{(p-2)(p-3)}{4(2p-3)} c_{p-2} = \frac{(2p-4)!}{2^p 2!(p-2)!(p-4)!}, \ldots$$

and in general

$$c_{p-2k} = \frac{(-1)^k(2p-2k)!}{2^p k!(p-k)!(p-2k)!}. \tag{5.130}$$

Then the *Legendre polynomials of degree p* is given by

$$P_p(x) = \sum_{k=0}^{M} \frac{(-1)^k(2p-2k)!}{2^p k!(p-k)!(p-2k)!} x^{p-2k}, \qquad p = 0, 1, 2, \ldots, \tag{5.131}$$

where M is the largest integer not greater than $p/2$. In particular, we have $P_0(x) = 1$, $P_1(x) = x$, $P_2(x) = \frac{1}{2}(3x^2 - 1)$, $P_3(x) = \frac{1}{2}(5x^3 - 3x)$, $P_4(x) = \frac{1}{8}(35x^4 - 30x^2 + 3)$, etc. As these particular results illustrate, as a consequence of the choice (5.129) of the value of c_p, we have $P_p(1) = 1$ and $P_p(-1) = (-1)^p$ for all integers $p \geqslant 0$. This fact will be proved in Example 5.5.3.

To obtain an even more concise form than (5.131) for the Legendre polynomials, we observe that we can write

$$P_p(x) = \sum_{k=0}^{M} \frac{(-1)^k}{2^p k!(p-k)!} \frac{d}{dx^p} (x^{2p-2k}),$$

since

$$\frac{d}{dx^p} (x^{2p-2k}) = (2p-2k) \frac{d}{dx^{p-1}} (x^{2p-2k-1}) = \cdots$$

$$= (2p-2k) \cdots (p-2k+1)x^{p-2k} = \frac{(2p-2k)!}{(p-2k)!} x^{p-2k}.$$

Hence

$$P_p(x) = \frac{1}{2^p p!} \frac{d^p}{dx^p} \sum_{k=0}^{M} (-1)^k \frac{p!}{k!(p-k)!} (x^2)^{p-k}.$$

We may now extend the range of this sum by letting k range from 0 to n. This extension will not affect the result, since the added terms are a polynomial of degree $< p$ so that the pth derivative will vanish. Thus

$$P_p(x) = \frac{1}{2^p p!} \frac{d^p}{dx^p} \sum_{k=0}^{n} \frac{p!}{k!(p-k)!} (x^2)^{p-k} (-1)^k,$$

and by the binomial formula we have

$$P_p(x) = \frac{1}{2^p p!} \frac{d^p}{dx^p} (x^2 - 1)^p, \qquad p = 0, 1, 2, \dots. \tag{5.132}$$

This formula is called *Rodrigues' formula** and provides an easy way of computing successive Legendre polynomials.

Example 5.5.1 Show that $P_2(x) = \frac{1}{2}(3x^2 - 1)$. By Rodrigues' formula

$$P_2(x) = \frac{1}{2^2 2!} \frac{d^2}{dx^2} (x^4 - 2x^2 + 1) = \frac{1}{8}(12x^2 - 4) = \frac{1}{2}(3x^2 - 1).$$

We can use Rodrigues' formula to obtain several useful recurrence relations. Observe that

$$P'_{p+1} = \frac{d}{dx} \left[\frac{1}{2^{p+1}(p+1)!} \frac{d^{p+1}}{dx^{p+1}} (x^2 - 1)^{p+1} \right]$$

$$= \frac{d}{dx} \left\{ \frac{1}{2^p p!} \frac{d^p}{dx^p} [x(x^2 - 1)^p] \right\} = \frac{1}{2^p p!} \frac{d^{p+1}}{dx^{p+1}} [x(x^2 - 1)^p]. \tag{5.133}$$

Hence taking the derivative of the term in brackets, we have

$$P'_{p+1} = \frac{1}{2^p p!} \frac{d^p}{dx^p} [(x^2 - 1)^p + 2px^2(x^2 - 1)^{p-1}]$$

$$= \frac{1}{2^p p!} \frac{d^p}{dx^p} [(2p + 1)(x^2 - 1)^p + 2p(x^2 - 1)^{p-1}]$$

$$= (2p + 1)P_p + P'_{p-1}, \qquad p = 1, 2, 3, \dots. \tag{5.134}$$

We can get another recurrence relation from (5.133) if we consider the effect of repeated differentiations on a product of the form $xf(x)$. Note that

$$\frac{d}{dx} [xf(x)] = x \frac{d}{dx} f(x) + f(x),$$

$$\frac{d^2}{dx^2} [xf(x)] = x \frac{d^2}{dx^2} f(x) + 2 \frac{d}{dx} f(x),$$

---

* Named after the French mathematician and banker Olinde Rodrigues (1794–1851).

and in general

$$\frac{d^{p+1}}{dx^{p+1}} [xf(x)] = x \frac{d^{p+1}}{dx^{p+1}} f(x) + (p + 1) \frac{d^p}{dx^p} f(x). \tag{5.135}$$

Applying (5.135) to the expression in brackets in (5.133), we obtain

$$P'_{p+1} = \frac{1}{2^p p!} \left[x \frac{d^{p+1}}{dx^{p+1}} (x^2 - 1)^p + (p + 1) \frac{d^p}{dx^p} (x^2 - 1)^p \right]$$

$$= xP'_p + (p + 1)P_p, \qquad p = 0, 1, 2, \ldots. \tag{5.136}$$

Thus we have proved the identities

$$(p + 1)P_p = P'_{p+1} - xP'_p,$$

$$(2p + 1)P_p = P'_{p+1} - P'_{p-1}. \tag{5.137}$$

Subtracting the top identity in (5.137) from the bottom one yields

$$pP_p = xP'_p - P'_{p-1}, \qquad p = 1, 2, \ldots. \tag{5.138}$$

Finally, we note that from (5.137) and (5.138) we can get

$$(p + 1)P_{p+1} - (2p + 1)xP_p + pP_{p-1}$$

$$= (xP'_{p+1} - P'_p) - x(P'_{p+1} - P'_{p-1}) + (P'_p - xP'_{p-1}),$$

so that we can eliminate all derivatives and obtain the relation

$$(p + 1)P_{p+1} + pP_{p-1} = (2p + 1)xP_p, \qquad p = 1, 2, \ldots. \tag{5.139}$$

The second order homogeneous linear difference Eq. (5.139) can be used to generate all the Legendre polynomials if P_0 and P_1 are given (see Chapter 4). We shall illustrate this iterative technique in the next example.

Example 5.5.2 Starting with $P_0 = 1$ and $P_1 = x$, calculate the polynomials P_2, P_3, and P_4. By (5.139)

$$P_{p+1} = \frac{(2p + 1)xP_p - pP_{p-1}}{p + 1},$$

so that

$$P_2 = \frac{3xP_1 - P_0}{2} = \frac{3x^2 - 1}{2},$$

$$P_3 = \frac{5xP_2 - 2P_1}{3} = \frac{15x^3 - 5x - 4x}{6} = \frac{5x^3 - 3x}{2},$$

$$P_4 = \frac{7xP_3 - 3P_2}{4} = \frac{35x^4 - 21x^2 - 9x^2 + 3}{8} = \frac{35x^4 - 30x^2 + 3}{8}.$$

Another extremely important identity, which is called the *generating function* for Legendre polynomials, is given in the following theorem:

Theorem 5.6

$$\frac{1}{\sqrt{1 - 2xz + z^2}} = P_0(x) + P_1(x)z + P_2(x)z^2 + \cdots + P_n(x)z^n + \cdots. \quad (5.140)$$

Proof. Using the binomial theorem (see Exercise 5.1.25), we expand the expression

$$[1 - (2xz - z^2)]^{-1/2} = 1 + \frac{1}{2}(2xz - z^2) + \frac{(\frac{1}{2})(\frac{3}{2})}{2!}(2xz - z^2)^2 + \cdots$$

$$+ \frac{(\frac{1}{2})(\frac{3}{2})(\frac{5}{2}) \cdots [(2p - 1)/2]}{p!}(2xz - z^2)^p + \cdots.$$

The power z^p can only occur in the terms going from the pth term $(2xz - z^2)^p$ $[= z^p(2x - z)^p]$ down. Thus by expanding the various powers of $(2x - z)$, we find that the coefficient of z^p is

$$\frac{(\frac{1}{2})(\frac{3}{2}) \cdots [(2p - 1)/2]}{p!}(2x)^p - \frac{(\frac{1}{2})(\frac{3}{2}) \cdots [(2p - 3)/2]}{(p - 1)!}(p - 1)(2x)^{p-2}$$

$$+ \frac{(\frac{1}{2})(\frac{3}{2}) \cdots [(2p - 5)/2]}{(p - 2)!}\frac{(p - 2)(p - 3)}{2!}(2x)^{p-4} - \cdots$$

or

$$\frac{(2p)!}{2^p(p!)^2}x^p - \frac{(2p - 2)!x^{p-2}}{2^p(p - 1)!(p - 2)!} + \frac{(2p - 4)!x^{p-4}}{2^p 2!(p - 2)!(p - 4)!} - \cdots. \quad (5.141)$$

Let M be the largest integer not greater than $p/2$. Note now that the highest power of z in the term $(2xz - z^2)^{M-1}$ is z^{2M-2}, and $2M - 2 \leqslant p - 2$. Thus the sum (5.141) includes only the terms

$$\sum_{k=0}^{M} \frac{(-1)^k(2p - 2k)!x^{p-2k}}{2^p k!(p - k)!(p - 2k)!} = P_p(x).$$

The proof is now complete.

Example 5.5.3 For all integers $p \geqslant 0$ we have the identities[*]

$$P_p(1) = 1 \quad \text{and} \quad P_p(-1) = (-1)^p.$$

To verify this fact, we set $x = \pm 1$ in Eq. (5.140), thus obtaining

$$\frac{1}{\sqrt{1 \mp 2z + z^2}} = P_0(\pm 1) + P_1(\pm 1)z + P_2(\pm 1)z^2 + \cdots + P_n(\pm 1)z^n + \cdots.$$

But $1 \mp 2z + z^2 = (1 \mp z)^2$, and the geometric series

$$\frac{1}{1 \mp z} = 1 \pm z + z^2 \pm \cdots + (\pm z)^n + \cdots$$

by Eq. (5.10). Thus the coefficients of the two series must agree, yielding the desired result.

Some additional properties of Legendre polynomials will be developed in Chapter 9 when we discuss the zeros and orthogonality of these functions. In that chapter we will show in particular that the Legendre polynomials give the best "least-squares" polynomial approximation to a given continuous function.

EXERCISES 5.5

1. Calculate P_5, P_6, P_7, and P_8 by means of Eq. (5.139).

2. Prove that the series (5.127) has a radius of convergence $R = 1$.

3. Prove that the series (5.128) has a radius of convergence $R = 1$.

4. Calculate P_4 by means of Rodrigues' formula.

5. Prove that $P_{2p+1}(0) = 0$ for all integers $p \geq 0$.

6. Prove that

$$P_{2p}(0) = \frac{(-1)^p(2p)!}{2^{2p}(p!)^2},$$

   for all $p \geq 0$.

7. Prove that for all integers $p \geq 0$

   a) $P'_{2p}(0) = 0$,

   b) $P'_{2p+1}(0) = \dfrac{(-1)^p(2p + 1)!}{2^{2p}(p!)^2}.$

8. Show that for all integers $p > 0$:

   a) $\displaystyle\int_0^1 P_p(x)\, dx = \frac{1}{p + 1} P_{p-1}(0)$

   b) $\displaystyle\int_0^1 P_{2p}(x)\, dx = 0$

   c) $\displaystyle\int_0^1 P_{2p+1}(x)\, dx = (-1)^p \frac{(2p)!(4p + 3)}{2^{2p+1}p!(p + 1)!}$

   d) Compute these integrals for $p = 0$.

9. Consider the differential equation

$$y'' - 2xy' + 2py = 0, \tag{5.142}$$

   which is known as *Hermite's equation*.

   a) Use the method of Frobenius to show that all solutions of (5.142) are of the form

$$y = c_0\left[1 + \sum_{n=1}^{\infty} \frac{2^n(-p)(-p + 2)\cdots(-p + 2n - 2)x^{2n}}{(2n)!}\right]$$

$$+ c_1\left[x + \sum_{n=1}^{\infty} \frac{2^n(1 - p)(1 - p + 2)\cdots(1 - p + 2n - 2)x^{2n+1}}{(2n + 1)!}\right],$$

   where c_0 and c_1 are arbitrary functions.

b) Show that (5.142) has a polynomial solution of degree p for a nonnegative integer p. These polynomials, denoted by $H_p(x)$, are called the *Hermite polynomials of degree p*.

c) Show that

$$H_p(x) = \sum_{n=0}^{M} \frac{(-1)^n p! (2x)^{p-2n}}{n!(p-2n)!},$$

where M is the greatest integer $\leqslant p/2$.

d) Calculate H_0, H_1, H_2, H_3, and H_4.

10. Consider *Laguerre's equation*

$$xy'' + (1-x)y' + py = 0. \tag{5.143}$$

a) Show that if p is a nonnegative integer, then there is a polynomial solution to (5.143) of the form

$$L_p(x) = \sum_{n=0}^{p} \frac{(-1)^n p! x^n}{(p-n)!(n!)^2}.$$

The functions $L_p(x)$ are known as the *Laguerre polynomials*.

b) Calculate $L_0(x)$, $L_1(x)$, $L_2(x)$, $L_3(x)$, and $L_4(x)$.

LAPLACE
TRANSFORMS

6

6.1 INTRODUCTION:
DEFINITION AND BASIC PROPERTIES OF
THE LAPLACE TRANSFORM

One of the most efficient methods of solving certain ordinary and partial differential equations is to use Laplace transforms. The effectiveness of the Laplace transform is due to its ability to convert a differential equation into an algebraic equation, whose solution yields the solution of the differential equation when the transformation is reversed. In many ways, this procedure is similar to the use of logarithms in performing multiplications and divisions.

As we shall see in a moment, the Laplace transform is defined as a certain integral over the range from zero to infinity. We recall that such an integral is called an *improper integral*. Formally, if t_0 is a given real number, then for any function $f(t)$, we define

$$\int_{t_0}^{\infty} f(t)\, dt = \lim_{A \to \infty} \int_{t_0}^{A} f(t)\, dt.$$

If this limit exists and is finite, we say that the improper integral *converges*. Otherwise, it *diverges*.

Example 6.1.1 Let $f(t) = e^{at}$, where $a \neq 0$. Then

$$\int_{0}^{\infty} e^{at}\, dt = \lim_{A \to \infty} \int_{0}^{A} e^{at}\, dt = \lim_{A \to \infty} \frac{1}{a} e^{at} \Big|_{0}^{A} = \lim_{A \to \infty} \frac{1}{a}(e^{aA} - 1).$$

Clearly this limit is finite if $a < 0$ and diverges to $+\infty$ if $a > 0$.

Example 6.1.2 Let $f(t) = \cos t$. Then

$$\int_{0}^{\infty} \cos t\, dt = \lim_{A \to \infty} \int_{0}^{A} \cos t\, dt = \lim_{A \to \infty} \sin t \Big|_{0}^{A} = \lim_{A \to \infty} \sin A.$$

But $\sin A$ has no limit as $A \to \infty$. Therefore, $\int_{0}^{\infty} \cos t\, dt$ diverges, even though $-1 \leqslant \int_{0}^{A} \cos t\, dt \leqslant 1$ for every $A \geqslant 0$.

Example 6.1.3 Let $f(t) = 1/t^p$, $p \neq 1$. Then

$$\int_{1}^{\infty} t^{-p}\, dt = \lim_{A \to \infty} \int_{1}^{A} t^{-p}\, dt$$

$$= \lim_{A \to \infty} \frac{1}{1 - p} t^{1-p} \Big|_{1}^{A} = \lim_{A \to \infty} \frac{1}{1 - p}(A^{1-p} - 1),$$

which converges to $1/(p - 1)$ if $p > 1$, and diverges if $p < 1$.

There are other kinds of improper integrals.

Example 6.1.4 Let $f(t) = t^{-1/2}$ and consider $\int_0^1 t^{-1/2} \, dt$. Since $f(t)$ is not defined at $t = 0$, we define

$$\int_0^1 t^{-1/2} \, dt = \lim_{A \to 0} \int_A^1 t^{-1/2} \, dt$$

$$= \lim_{A \to 0} 2t^{1/2} \bigg|_A^1 = \lim_{A \to 0} 2(1 - \sqrt{A}) = 2.$$

Therefore, $\int_0^1 t^{-1/2} \, dt$ converges.

Example 6.1.5 Let $f(t) = t^{-1}$. Then $\int_0^1 t^{-1} \, dt$ is an improper integral and

$$\int_0^1 t^{-1} \, dt = \lim_{A \to 0} \int_A^1 t^{-1} \, dt = \lim_{A \to 0} \ln A \bigg|_A^1$$

$$= \lim_{A \to 0} (\ln 1 - \ln A) = \lim_{A \to 0} (-\ln A) = +\infty.$$

Therefore, this integral diverges.

In the rest of this chapter we shall not formally calculate improper integrals as we did in these five examples, but the reader should always keep in mind the definition of an improper integral as a limit.

Let $f(t)$ be a real-valued function which is defined for $t \geqslant 0$. Suppose that $f(t)$ is multiplied by e^{-st} and the result is integrated with respect to t from zero to infinity. If the integral converges, it is a function of s:

$$F(s) = \int_0^\infty e^{-st} f(t) \, dt, \tag{6.1}$$

and it is called the *Laplace transform* of $f(t)$. *We shall denote the original function by a lower case letter and the transform by the same letter in capitals or by* $\mathscr{L}\{f(t)\}$:

$$F(s) = \mathscr{L}\{f(t)\}(s) = \int_0^\infty e^{-st} f(t) \, dt. \tag{6.2}$$

It should now be apparent why the word *transform* is associated with this operation. The operation "transforms" the original function $f(t)$ into a new function $\mathscr{L}\{f(t)\}(s)$. Since we shall be interested in reversing the procedure, we call the original function $f(t)$ the *inverse transform* of $F = \mathscr{L}\{f(t)\}$ and denote it by $f(t) = \mathscr{L}^{-1}\{F\}$. The term "inverse transform" is similar in meaning to the term "inverse of a function." For example, if $f(x) = \sqrt{x}$, then the inverse of f, denoted by $f^{-1}(x)$, is the function $f^{-1}(x) = x^2$. The meaning here is clear. If $x > 0$ and we take the positive square root of x and then square it, we will end up with the original number. In mathematical symbols, we have $f^{-1}(f(x)) = x$. Similarly for the Laplace transform, we may say that the inverse transform of F, $\mathscr{L}^{-1}\{F\}$, is the function whose transform is F. Note that the last statement translates into the mathematical symbols $\mathscr{L}^{-1}\{F\} = \mathscr{L}^{-1}\{\mathscr{L}\{f\}\} = f$.

Example 6.1.6 Let $f(t) = e^{at}$, where a is constant. Then

$$\mathscr{L}\{e^{at}\} = \int_0^\infty e^{-st}e^{at}\,dt = \left.\frac{e^{-(s-a)t}}{a-s}\right|_0^\infty,$$

which converges, if $s - a > 0$, to

$$\mathscr{L}\{e^{at}\} = \frac{1}{s-a}. \tag{6.3}$$

Thus the Laplace transform of the function e^{at} is the function $F(s) = 1/(s-a)$ for $s > a$. Note that we can also conclude that $\mathscr{L}^{-1}\{F(s)\} = e^{at}$.

The last example shows that $\mathscr{L}\{f(t)\}$ may not be defined for all values of s, but if it is defined, then it will exist for suitably large values of s. Indeed, (6.3) holds only for values $s > a$.

Example 6.1.7 Let $f(t) = 1/t$. Then

$$\int_0^\infty \frac{e^{-st}}{t}\,dt = \int_0^1 \frac{e^{-st}}{t}\,dt + \int_1^\infty \frac{e^{-st}}{t}\,dt.$$

But for t in the interval $0 \leqslant t \leqslant 1$, $e^{-st} \geqslant e^{-s}$ if $s > 0$. Thus,

$$\int_0^\infty \frac{e^{-st}}{t}\,dt \geqslant e^{-s}\int_0^1 \frac{dt}{t} + \int_1^\infty \frac{e^{-st}}{t}\,dt.$$

However, as we saw earlier, $\int_0^1 t^{-1}\,dt$ diverges. Thus $f(t) = 1/t$ has no Laplace transform. Here the problem is that $1/t$ is discontinuous at zero.

This example and the preceding discussion pinpoint the need for answers to the following questions:

1. Which functions $f(t)$ have Laplace transforms?
2. Can two functions $f(t)$ and $g(t)$ have the same Laplace transform?

In answering these questions, we shall be satisfied if a class of functions, large enough to contain virtually all functions that arise in practice, can be found for which the Laplace transforms exist and whose inverses are unique. The need for a unique inverse arises from the fact that we need to reverse the transformation to find the solution of a given problem and this step is not possible if the inverses within the given class of functions are not unique. We shall see in Section 6.3 that certain differential equations can be solved by first taking the Laplace transform of both sides of the equation, then finding the Laplace transform of the solution, and finally taking the unique inverse transform.

In order to give simple conditions that guarantee the existence of a Laplace transform, we shall require the following definitions.

A function $f(t)$ has a *jump discontinuity* at a point t^* if the function has different (finite) limits as it approaches t^* from the left and from the right. A function $f(t)$

defined on $[0,\infty)$ is *piecewise continuous* if it is continuous on every finite interval $0 \leqslant x \leqslant b$, except possibly at finitely many points where it has jump discontinuities. Such a function is illustrated in Fig. 6.1. The class of piecewise continuous functions includes every continuous function, as well as many important discontinuous functions, such as the unit step function, square waves, and the staircase function, which we shall encounter later in the chapter.

Figure 6.1

Theorem 6.1. Existence Theorem Let $f(t)$ be piecewise continuous on $t \geqslant 0$ and satisfy the condition

$$|f(t)| \leqslant Me^{at} \tag{6.4}$$

for $t \geqslant T$ and for fixed nonnegative constants a, M, and T. Then $\mathcal{L}\{f(t)\}$ exists for all $s > a$.

Proof. Since $f(t)$ is piecewise continuous, $e^{-st}f(t)$ is integrable over any finite interval on $t \geqslant 0$, and

$$|\mathcal{L}\{f(t)\}| = \left|\int_0^\infty e^{-st}f(t)\,dt\right| \leqslant \left|\int_0^T e^{-st}f(t)\,dt\right| + \int_T^\infty e^{-st}|f(t)|\,dt. \tag{6.5}$$

Thus the first integral on the right-hand side of Eq. (6.5) exists, and by (6.4)

$$\int_T^\infty e^{-st}|f(t)|\,dt \leqslant M\int_T^\infty e^{-(s-a)t}\,dt = \frac{Me^{-(s-a)t}}{a-s}\bigg|_T^\infty,$$

which converges to $Me^{-(s-a)T}/(s-a)$ if $s > a$. This completes the proof.

Here we used the *comparison test* for the convergence of integrals: Let $a(t)$ and $b(t)$ be piecewise continuous functions and suppose that $0 \leqslant a(t) \leqslant b(t)$ at all points t where a and b are continuous. If

$$\int_0^\infty b(t) \, dt < \infty,$$

then

$$\int_0^\infty a(t) \, dt < \infty.$$

The conditions in Theorem 6.1 are easy to test. For example:

$$\sinh t \leqslant e^t, \quad t^n \leqslant n!e^t, \quad \text{for } t > 0,$$

but

$$e^{t^2} > Me^{at}$$

for sufficiently large t, regardless of the choice of M and a.

These facts are easily shown as follows:

1. $\sinh t = \dfrac{e^t - e^{-t}}{2} < \dfrac{e^t}{2} < e^t$, since $-\dfrac{e^{-t}}{2} < 0.$

2. $e^t = 1 + t + \dfrac{t^2}{2!} + \cdots + \dfrac{t^n}{n!} + \cdots$, so that for $t > 0$, $\dfrac{t^n}{n!} \leqslant e^t$, or $t^n \leqslant n!e^t.$

3. Finally, let M and a be any fixed constants. Clearly, for sufficiently large t,

   $$t^2 > \ln M + at.$$

   (Simply choose t such that $t > a + \ln M/t$.) Exponentiation now yields

   $$e^{t^2} > e^{\ln M + at} = e^{\ln M}e^{at} = Me^{at}.$$

It should also be noted that there are functions having Laplace transforms that do not satisfy the hypotheses of Theorem 6.1. For example, $f(t) = 1/\sqrt{t}$ is infinite at $t = 0$; but setting $x^2 = st$, we have

$$\mathscr{L}\{t^{-1/2}\} = \int_0^\infty e^{-st}t^{-1/2} \, dt = \frac{2}{\sqrt{s}} \int_0^\infty e^{-x^2} \, dx = \sqrt{\frac{\pi}{s}},$$

according to formula 58 in Appendix 1.

The proof of uniqueness would take us too far afield at this point. However, it can be shown* that *two functions having the same Laplace transform cannot*

---

* See, for example, I. S. Sokolnikoff and R. M. Redheffer, *Mathematics of Physics and Modern Engineering*, McGraw-Hill Book Co., New York, 1966, p. 217.

differ over an interval of positive length, although they may differ at several isolated points. Thus two different piecewise continuous functions having the same Laplace transform can differ only at isolated points. Such differences are generally of no importance in applications. Hence the Laplace transform has an essentially unique inverse. In particular, different continuous functions have different Laplace transforms.

One of the most important properties of the Laplace transform is stated in the following theorem.

Theorem 6.2. Linearity $\mathscr{L}\{af(t) + bg(t)\} = a\mathscr{L}\{f(t)\} + b\mathscr{L}\{g(t)\}.$

Proof. By definition

$$\mathscr{L}\{af(t) + bg(t)\} = \int_0^\infty e^{-st}[af(t) + bg(t)]\,dt$$

$$= a\int_0^\infty e^{-st}f(t)\,dt + b\int_0^\infty e^{-st}g(t)\,dt$$

$$= a\mathscr{L}\{f(t)\} + b\mathscr{L}\{g(t)\}.$$

The linearity property allows us to deal with an equation term by term in order to obtain its Laplace transform.

Example 6.1.8 Let $f(t) = \cosh at = (e^{at} + e^{-at})/2$. Then using the result of Example 6.1.6, we find that

$$\mathscr{L}\{\cosh at\} = \frac{1}{2}\left[\frac{1}{s-a} + \frac{1}{s+a}\right] = \frac{s}{s^2 - a^2}, \tag{6.6}$$

for all $s > |a|$. Similarly,

$$\mathscr{L}\{\sinh at\} = \frac{a}{s^2 - a^2}, \qquad s > |a|. \tag{6.7}$$

Example 6.1.9 Recall from Eq. (3.28) that

$$e^{iat} = \cos at + i\sin at.$$

Repeating the same calculations as in Example 6.1.6, we see that the integral converges for $s > 0$ to

$$\mathscr{L}\{e^{iat}\} = \frac{1}{s - ia}. \tag{6.8}$$

Thus,

$$\mathscr{L}\{\cos at\} + i\mathscr{L}\{\sin at\} = \frac{1}{s - ia} = \frac{s + ia}{(s - ia)(s + ia)} = \frac{s + ia}{s^2 + a^2}.$$

Equating real and imaginary parts, we find that for $s > 0$,

$$\mathcal{L}\{\cos at\} = \frac{s}{s^2 + a^2}, \qquad \mathcal{L}\{\sin at\} = \frac{a}{s^2 + a^2}. \tag{6.9}$$

Example 6.1.10 Let $f(t) = t^n$, with $n = 0, 1, 2, 3, \ldots$. Then integrating by parts, we have

$$\mathcal{L}\{t^n\} = \int_0^\infty e^{-st} t^n \, dt = \frac{-e^{-st} t^n}{s} \bigg|_0^\infty + \frac{n}{s} \int_0^\infty e^{-st} t^{n-1} \, dt$$

$$= \frac{n}{s} \mathcal{L}\{t^{n-1}\} \qquad \text{for } s > 0. \tag{6.10}$$

Thus

$$\mathcal{L}\{t^n\} = \frac{n}{s} \mathcal{L}\{t^{n-1}\} = \frac{n(n-1)}{s^2} \mathcal{L}\{t^{n-2}\} = \cdots = \frac{n!}{s^n} \mathcal{L}\{t^0\}.$$

But $t^0 = 1 = e^{0t}$, so by Example 6.1.6, $\mathcal{L}\{t^0\} = 1/s$ and

$$\mathcal{L}\{t^n\} = \frac{n!}{s^{n+1}}. \tag{6.11}$$

These elementary results are summarized in Table 6.1. A more extensive list of Laplace transforms can be found in Appendix 2.

Table 6.1

$f(t)$	$\mathcal{L}\{f(t)\}$	Range of definition
t^n	$\dfrac{n!}{s^{n+1}}$	$s > 0$
e^{at}	$\dfrac{1}{s - a}$	$s > a$ (real)
$\sin at$	$\dfrac{a}{s^2 + a^2}$	$s > 0$
$\cos at$	$\dfrac{s}{s^2 + a^2}$	$s > 0$
$\sinh at$	$\dfrac{a}{s^2 - a^2}$	$s > |a|$
$\cosh at$	$\dfrac{s}{s^2 - a^2}$	$s > |a|$

EXERCISES 6.1

Find the Laplace transforms of the following functions where a and b are constants. For what values of s are these transforms defined?

1. $5t + 2$
2. $at + b$
3. $t^2 + at + b$
4. $t''(at + b)$
5. e^{5t+2}
6. $\sin(at + b)$
7. $\cos(at + b)$
8. $\sinh(at + b)$
9. $\cosh(at + b)$
10. te^t
11. $e^{-t}\sin t$
12. $e^{at}\cos bt$

13. Derive the equation

$$\mathscr{L}\{\sinh at\} = \frac{a}{s^2 - a^2}.$$

14. Derive the formulas in Eq. (6.9) directly by integrating by parts.

In Exercises 15 through 22 find $f(t)$ if $\mathscr{L}\{f(t)\}$ equals:

15. $\dfrac{7}{s^2}$
16. $\dfrac{18}{s^3} + \dfrac{7}{s}$
17. $\dfrac{a_1}{s} + \dfrac{a_2}{s^2} + \cdots + \dfrac{a_{n+1}}{s^{n+1}}.$

18. $\dfrac{s+1}{s^2+1}$
19. $\dfrac{7}{s-3}$
20. $\dfrac{s-2}{s^2-2}$

21. $\dfrac{s-2}{s^2+3}$
22. $\dfrac{1}{(s-1)^2}$

23. Suppose that $f(t) = \mathscr{L}^{-1}\{F(s)\}$ and that $g(t) = \mathscr{L}^{-1}\{G(s)\}$. Prove that $\alpha f(t) + \beta g(t) = \mathscr{L}^{-1}\{\alpha F(s) + \beta G(s)\}$, where α and β are any constants. This property is called the *linearity property* of the inverse Laplace transform.

*24. Let $\sum_{n=1}^{\infty} f_n(t)$ be a uniformly convergent series of functions, each of which has a Laplace transform defined for $s \geqslant a$ (that is, $\mathscr{L}\{f_n(t)\}$ exists for $s \geqslant a$). Prove that

$$f(t) = \sum_{n=1}^{\infty} f_n(t)$$

has a Laplace transform for $s \geqslant a$ defined by

$$\mathscr{L}\{f(t)\} = \sum_{n=1}^{\infty} \mathscr{L}\{f_n(t)\}.$$

6.2 SHIFTING THEOREMS AND LAPLACE TRANSFORMS OF DERIVATIVES AND INTEGRALS

In this section we will study a number of techniques for simplifying the task of computing the Laplace transform of more complicated functions.

The following function, which is extremely important for practical applications (for example, as a unit impulse in a mechanical control system), is known as the *unit step function* or *Heaviside function*:

$$H(t - a) = \begin{cases} 0, & t < a, \\ 1, & t > a, \end{cases} \tag{6.12}$$

where a is any fixed constant. In particular, if $a = 0$, we have

$$H(t) = \begin{cases} 0, & t < 0, \\ 1, & t > 0. \end{cases}$$

For $a \geqslant 0$ and $s > 0$, we obtain

$$\mathcal{L}\{H(t - a)\} = \int_0^\infty e^{-st} H(t - a)\, dt = \int_a^\infty e^{-st}\, dt = \frac{e^{-as}}{s}. \tag{6.13}$$

The unit step function can be used as a building block in the construction of other functions. For example,

$$f_1(t) = H(t - a) - H(t - b), \qquad a < b, \tag{6.14}$$

is a square wave between a and b, while

$$f_2(t) = H(t - a) + H(t - 2a) + H(t - 3a), \qquad a > 0,$$

yields a three step staircase. (See Fig. 6.2.) By the linearity property, we obtain

$$\mathcal{L}\{f_1(t)\} = \frac{1}{s}(e^{-as} - e^{-bs}), \qquad \mathcal{L}\{f_2(t)\} = \frac{1}{s}(e^{-as} + e^{-2as} + e^{-3as}).$$

We may generalize these ideas even further.

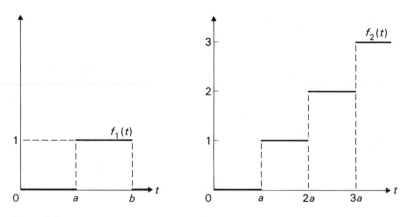

Figure 6.2

Example 6.2.1 The infinite staircase

$$f(t) = H(t) + H(t - a) + H(t - 2a) + H(t - 3a) + \cdots, \qquad a > 0, \quad (6.15)$$

has the Laplace transform

$$\mathscr{L}\{f(t)\} = \frac{1}{s}(1 + e^{-as} + e^{-2as} + e^{-3as} + \cdots) = \frac{1}{s(1 - e^{-as})}. \qquad (6.16)$$

In obtaining (6.16) we used the well-known formula for the sum of a geometric series

$$\sum_{n=0}^{\infty} x^n = 1 + x + x^2 + \cdots = \frac{1}{1-x}, \qquad |x| < 1, \qquad (6.17)$$

with $x = e^{-as}$, as well as the extension of Theorem 6.2 to an infinite series of functions (see Exercise 24 of Section 6.1).

Example 6.2.2 Let $f(t)$ be the periodic square wave shown in Fig. 6.3. Then we can write $f(t)$ in the form

$$f(t) = H(t) - 2H(t - a) + 2H(t - 2a) - 2H(t - 3a) + \cdots, \qquad (6.18)$$

from which it follows that

$$\mathscr{L}\{f(t)\} = \frac{1}{s}(1 - 2e^{-as} + 2e^{-2as} - 2e^{-3as} + \cdots)$$

$$= \frac{1}{s}\left(\frac{2}{1 + e^{-as}} - 1\right) = \frac{1 - e^{-as}}{s(1 + e^{-as})}. \qquad (6.19)$$

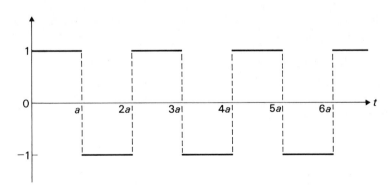

Figure 6.3

The next theorem presents a quick way of computing the Laplace transform $\mathscr{L}\{e^{at}f(t)\}$ when $\mathscr{L}\{f(t)\}$ is known.

Theorem 6.3. The First Shifting Theorem Let $F(s) = \mathscr{L}\{f(t)\}$. Then $\mathscr{L}\{e^{at}f(t)\}(s) = F(s - a)$.

Proof. By definition

$$\mathscr{L}\{e^{at}f(t)\}(s) = \int_0^\infty e^{-st}e^{at}f(t)\, dt = \int_0^\infty e^{-(s-a)t}f(t)\, dt$$
$$= \mathscr{L}\{f(t)\}(s - a) = F(s - a).$$

As the formula suggests, to find $\mathscr{L}\{e^{at}f(t)\}$, we simply replace each s in $\mathscr{L}\{f(t)\}$ by $s - a$.

Example 6.2.3 Applying the first shifting theorem to the formulas in Table 6.1, we obtain the results shown in Table 6.2.

Table 6.2

$f(t)$	$\mathscr{L}\{f(t)\}$	Range of definition
$e^{at}t^n$	$\dfrac{n!}{(s - a)^{n+1}}$	$s > a$
$e^{at}\sin bt$	$\dfrac{b}{(s - a)^2 + b^2}$	$s > a$
$e^{at}\cos bt$	$\dfrac{s - a}{(s - a)^2 + b^2}$	$s > a$
$e^{at}\sinh bt$	$\dfrac{b}{(s - a)^2 - b^2}$	$s > a + |b|$
$e^{at}\cosh bt$	$\dfrac{s - a}{(s - a)^2 - b^2}$	$s > a + |b|$

Theorem 6.4. The Second Shifting Theorem Let $a > 0$. Then

$$\mathscr{L}\{f(t - a)H(t - a)\} = e^{-as}\mathscr{L}\{f(t)\}.$$

Proof. Using the definition and the substitution $x = t - a$, we find that

$$\mathscr{L}\{f(t - a)H(t - a)\} = \int_0^\infty e^{-st}f(t - a)H(t - a)\, dt = \int_a^\infty e^{-st}f(t - a)\, dt$$
$$= \int_0^\infty e^{-s(x+a)}f(x)\, dx = e^{-as}\mathscr{L}\{f(t)\}.$$

In this theorem, we are shifting the function a units to the right and truncating it at zero.

Example 6.2.4 $\mathscr{L}\{\sin a(t - b)H(t - b)\} = e^{-bs}\mathscr{L}\{\sin at\} = ae^{-bs}/(s^2 + a^2).$

The most important property of Laplace transforms for solving differential equations concerns the transform of the derivative of a function $f(t)$. We prove below that differentiation of $f(t)$ roughly corresponds to multiplication of the transform by s.

Theorem 6.5. Differentiation Property   Let $f(t)$ satisfy the condition

$$|f(t)| \leqslant Me^{at} \tag{6.20}$$

for $t \geqslant T$, for fixed nonnegative constants a, M, and T, and suppose that $f'(t)$ is piecewise continuous for $t \geqslant 0$. Then the Laplace transform of $f'(t)$ exists for all $s > a$, and

$$\mathscr{L}\{f'(t)\} = s\mathscr{L}\{f(t)\} - f(0). \tag{6.21}$$

Proof. Since f is differentiable, it is also continuous, therefore satisfies the conditions of the existence theorem (6.1), and has a Laplace transform. Suppose, first, that $f'(t)$ is continuous on $t \geqslant 0$. Then integrating the definition of $\mathscr{L}\{f'(t)\}$ by parts, we obtain

$$\mathscr{L}\{f'(t)\} = \int_0^\infty e^{-st}f'(t)\, dt = e^{-st}f(t)\Big|_0^\infty + s\int_0^\infty e^{-st}f(t)\, dt. \tag{6.22}$$

Since $f(t)$ satisfies (6.20), the first term on the right-hand side in (6.22) vanishes at the upper limit when $s > a$, and by definition we obtain $\mathscr{L}\{f'\} = s\mathscr{L}\{f\} - f(0)$. When $f'(t)$ is piecewise continuous, the proof is similar. We simply break up the range of integration into parts on each of which $f'(t)$ is continuous, and integrate by parts as in (6.22). All first terms will cancel out or vanish except $-f(0)$, and the second terms will combine to yield $s\mathscr{L}\{f\}$.

Theorem 6.5 may be extended to apply to piecewise continuous functions $f(t)$ (Exercise 43 at the end of this section).

Equation (6.21) may be applied repeatedly to obtain the Laplace transform of higher order derivatives:

$$\mathscr{L}\{f''\} = s\mathscr{L}\{f'\} - f'(0) = s[s\mathscr{L}\{f\} - f(0)] - f'(0)$$

or

$$\mathscr{L}\{f''\} = s^2\mathscr{L}\{f\} - sf(0) - f'(0). \tag{6.23}$$

Similarly,

$$\mathscr{L}\{f'''\} = s^3\mathscr{L}\{f\} - s^2f(0) - sf'(0) - f''(0),$$

leading by induction to the following extension of Theorem 6.5.

Theorem 6.6   Let $f^{(k)}(t)$ satisfy (6.20) for $k = 0, 1, 2, \ldots, n - 1$ and suppose that $f^{(n)}(t)$ is piecewise continuous on $t \geqslant 0$. Then $\mathscr{L}\{f^{(n)}(t)\}$ exists and is given by

$$\mathscr{L}\{f^{(n)}(t)\} = s^n\mathscr{L}\{f\} - s^{n-1}f(0) - s^{n-2}f'(0) - \cdots - f^{(n-1)}(0). \tag{6.24}$$

Theorems 6.5 and 6.6 are important, since they are used to reduce the Laplace transform of a differential equation into an equation involving only the transform of the solution. This application will be considered in Section 6.3 and the following sections. However, these theorems are also useful in determining the transforms of certain functions.

Example 6.2.5 Let $f(t) = \sin^2 at$. Then

$$f'(t) = 2a \sin at \cos at = a \sin 2at,$$

so

$$\frac{2a^2}{s^2 + 4a^2} = \mathscr{L}\{f'\} = s\mathscr{L}\{f\} - f(0).$$

Since $f(0) = 0$, it follows that

$$\mathscr{L}\{\sin^2 at\} = \frac{2a^2}{s(s^2 + 4a^2)}, \qquad s > 0.$$

Example 6.2.6 Suppose that $f(t) = t \sin at$. Then

$$f'(t) = \sin at + at \cos at, \qquad f''(t) = 2a \cos at - a^2 t \sin at.$$

Thus, since $f(0) = f'(0) = 0$,

$$2a\mathscr{L}\{\cos at\} - a^2\mathscr{L}\{f(t)\} = \mathscr{L}\{f''\} = s^2\mathscr{L}\{f\},$$

so that

$$(s^2 + a^2)\mathscr{L}\{f\} = 2a\mathscr{L}\{\cos at\} = \frac{2as}{s^2 + a^2}$$

or

$$\mathscr{L}\{f(t)\} = \frac{2as}{(s^2 + a^2)^2}, \qquad s > 0.$$

Since differentiation of $f(t)$ corresponds to multiplication by s, it is not too surprising that integration of $f(t)$ corresponds to division by s.

Theorem 6.7. Integration Property If $f(t)$ is piecewise continuous and satisfies condition (6.20), then

$$\mathscr{L}\left\{\int_0^t f(u)\,du\right\} = \frac{1}{s}\mathscr{L}\{f(t)\}. \tag{6.25}$$

Proof. Since $f(t)$ satisfies the hypotheses of the existence theorem (6.1), $\mathscr{L}\{f(t)\}$ exists. We can assume $a > 0$, since if (6.20) holds for $a = 0$, then it also holds for

$a > 0$. Let

$$g(t) = \int_0^t f(u)\, du.$$

Then g is continuous (see Exercise 41) and

$$|g(t)| \leqslant \int_0^t |f(u)|\, du \leqslant M \int_0^t e^{au}\, du \leqslant \left(\frac{M}{a}\right) e^{at}.$$

Thus g also satisfies the existence theorem. And since $g'(t) = f(t)$ except at the points where f is discontinuous (by the fundamental theorem of calculus), we can use the differentiation property to obtain

$$\mathscr{L}\{f(t)\} = \mathscr{L}\{g'(t)\} = s\mathscr{L}\{g(t)\} - g(0).$$

Since $g(0) = 0$, the proof is complete.

Example 6.2.7 We shall use the integration property to find the Laplace transform of the integral

$$\int_0^t u \sin u\, du. \tag{6.26}$$

By (6.25) and Example 6.2.6, we have

$$\mathscr{L}\left\{\int_0^t u \sin u\, du\right\} = \frac{1}{s}\, \mathscr{L}\{t \sin t\} = \frac{2}{(s^2 + 1)^2}.$$

Example 6.2.8 The *unit impulse function* (also called the *Dirac delta function*) $\delta(t - a)$ is loosely described as a "function" which is zero everywhere except at $t = a$ and having the property that

$$\int_{-\infty}^{\infty} \delta(t - a)\, dt = 1. \tag{6.27}$$

As an illustration, we describe the Dirac delta function for $a = 0$. For any $\varepsilon > 0$, consider the function (see Fig. 6.4)

$$\delta_\varepsilon(t) = \begin{cases} 1/2\varepsilon, & -\varepsilon < t < \varepsilon, \\ 0, & |t| > \varepsilon. \end{cases}$$

Clearly, $\delta_\varepsilon(t)$ is piecewise continuous and

$$\int_{-\infty}^{\infty} \delta_\varepsilon(t)\, dt = \int_{-\varepsilon}^{\varepsilon} \frac{1}{2\varepsilon}\, dt = 1.$$

Then $\delta(t)$ may be defined by

$$\delta(t) = \lim_{\varepsilon \to 0} \delta_\varepsilon(t).$$

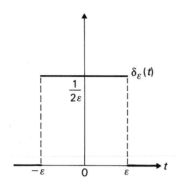

Figure 6.4

Of course, $\delta(t)$ is not a function. However, because it is the limit of piecewise continuous functions, we may treat $\delta(t)$ as if it were a legitimate function. We shall not prove this fact here but shall nevertheless make use of it in all further discussions of the delta function.

Since $\delta(t - a)$ concentrates all its "mass" at $t = a$, we see that

$$H(t - a) = \int_{-\infty}^{t} \delta(u - a)\, du, \tag{6.28}$$

since the integral in (6.28) vanishes if $t < a$ and equals 1 when $t > a$. By the integration property, if $a \geq 0$,

$$\frac{1}{s} \mathscr{L}\{\delta(t - a)\} = \mathscr{L}\left\{ \int_{0}^{t} \delta(u - a)\, du \right\} = \mathscr{L}\{H(t - a)\} = \frac{e^{-as}}{s}.$$

Thus $\mathscr{L}\{\delta(t - a)\} = e^{-as}$ for $a \geq 0$ and $s > 0$.

The differentiation and integration properties have other uses. For example, the function $f(t)$ shown in Fig. 6.5, has a square wave (6.14) as its derivative. Therefore,

$$\mathscr{L}\{H(t - a) - H(t - b)\} = s\mathscr{L}\{f(t)\}$$

or

$$\mathscr{L}\{f(t)\} = \frac{1}{s^2}(e^{-as} - e^{-bs}).$$

Example 6.2.9 Find the Laplace transform of the sawtooth function $g(t)$ shown in Fig. 6.6. Note that the derivative of $g(t)$ is the periodic square wave $f(t)$ of

Figure 6.5

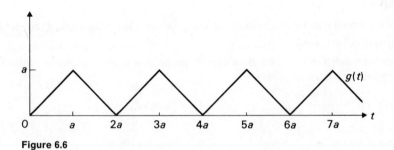

Figure 6.6

Example 6.2.2. Thus by Theorem 6.5 and Eq. (6.19),

$$s\mathcal{L}\{g(t)\} = \mathcal{L}\{f(t)\} = \frac{1 - e^{-as}}{s(1 + e^{-as})},$$

yielding

$$\mathcal{L}\{g(t)\} = \frac{1 - e^{-as}}{s^2(1 + e^{-as})}. \tag{6.29}$$

Example 6.2.10 Find

$$\mathcal{L}^{-1}\left\{\frac{2}{s^2(s^2 + 4)}\right\}.$$

From Example 6.2.5 we know that

$$\mathcal{L}\{\sin^2 t\} = \frac{2}{s(s^2 + 4)}.$$

Multiplying this equation on both sides by $1/s$ and applying the integration property, we have

$$\frac{2}{s^2(s^2 + 4)} = \frac{1}{s}\mathscr{L}\{\sin^2 t\} = \mathscr{L}\left\{\int_0^t \sin^2 u\, du\right\}$$

$$= \mathscr{L}\left\{\frac{t - \sin t \cos t}{2}\right\}.$$

Thus

$$\mathscr{L}^{-1}\left\{\frac{2}{s^2(s^2 + 4)}\right\} = \frac{t - \sin t \cos t}{2}.$$

EXERCISES 6.2

1. Verify the formulas in Table 6.2 directly by using the definition of the Laplace transform.

2. Evaluate $\mathscr{L}\{e^{at}H(t - b)\}$.

In Exercises 3 through 8 express each given hyperbolic function in terms of exponentials and apply the first shifting theorem to show:

3. $\mathscr{L}\{\cosh^2 at\} = \dfrac{s^2 - 2a^2}{s(s^2 - 4a^2)}$

4. $\mathscr{L}\{\sinh^2 at\} = \dfrac{2a^2}{s(s^2 - 4a^2)}$

5. $\mathscr{L}\{\cosh at \sin at\} = \dfrac{a(s^2 + 2a^2)}{s^4 + 4a^4}$

6. $\mathscr{L}\{\cosh at \cos at\} = \dfrac{s^3}{s^4 + 4a^4}$

7. $\mathscr{L}\{\sinh at \sin at\} = \dfrac{2a^2 s}{s^4 + 4a^4}$

8. $\mathscr{L}\{\sinh at \cos at\} = \dfrac{a(s^2 - 2a^2)}{s^4 + 4a^4}$

Using the method above, find the Laplace transforms in Exercises 9 through 14.

9. $\mathscr{L}\{\cosh at \cosh bt\}$

10. $\mathscr{L}\{\sinh at \sinh bt\}$

11. $\mathscr{L}\{\cosh at \sin bt\}$

12. $\mathscr{L}\{\cosh at \cos bt\}$

13. $\mathscr{L}\{\sinh at \sin bt\}$

14. $\mathscr{L}\{\sinh at \cos bt\}$

Using the second shifting theorem in Exercises 15 through 17, show that:

15. $\mathscr{L}\{\cos a(t - b)H(t - b)\} = \dfrac{e^{-bs}s}{s^2 + a^2}$

16. $\mathscr{L}\{\sinh a(t - b)H(t - b)\} = \dfrac{e^{-bs}a}{s^2 - a^2}$

17. $\mathscr{L}\{\cosh a(t - b)H(t - b)\} = \dfrac{e^{-bs}s}{s^2 - a^2}$

In Exercises 18 through 20, represent the graphed functions in terms of unit step functions and find their respective Laplace transform.

18.

19.

20.

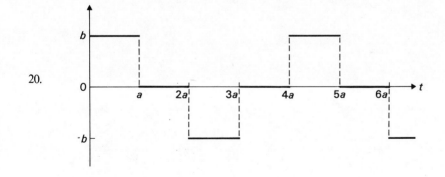

In Exercises 21 through 24 find the Laplace transform of each function by using the differentiation property.

21. $\cos^2 at$ 22. $t \cos at$ 23. $t^2 \sin at$

24. $t^2 \cos at$

In Exercises 25 through 27 use the method of Example 6.2.9 to find the Laplace transforms of the given functions.

25.

26.

27.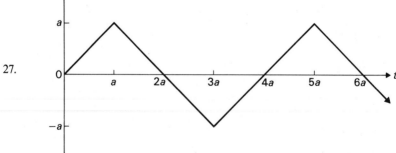

In Exercises 28 through 32 use the integration property to find:

28. $\mathscr{L}\left\{\int_0^t u^2 \sin au\, du\right\}$

29. $\mathscr{L}\left\{\int_0^t u^2 \cos au\, du\right\}$

30. $\mathscr{L}\left\{\int_0^t u \cos au\, du\right\}$

31. $\mathscr{L}\left\{\int_0^t \cos^2 au\, du\right\}$

32. $\mathscr{L}\left\{\int_0^t \sin^2 au\, du\right\}$

Find the inverse Laplace transform of each given function in Exercises 33 through 40 by the method of Example 6.2.10.

33. $\dfrac{1}{s(s-1)}$

34. $\dfrac{2}{s^3-4s}$

35. $\dfrac{2}{s^3+4s}$

36. $\dfrac{1}{s}\left(\dfrac{s-a}{s+a}\right)$

37. $\dfrac{1}{s^2}\left(\dfrac{s-a}{s+a}\right)$

38. $\dfrac{1}{s^3-s^2}$

39. $\dfrac{1}{s}\left(\dfrac{s+2}{s^2+4}\right)$

40. $\dfrac{1}{s}\left(\dfrac{s+1}{s^2-4}\right)$

41. Let $f(t)$ be a piecewise continuous function. Use the mean value theorem for integrals to show that

$$g(t) = \int_0^t f(u)\, du$$

is a continuous function of t. [*Hint*: Write the integral as a sum of integrals on each of which $f(t)$ is continuous.]

42. Show that the function in Exercise 41 has a derivative at all points of continuity of f and satisfies $g'(t) = f(t)$ except at points of discontinuity of f.

43. Let $f(t)$ be continuous, except for a jump discontinuity at $t = a\ (>0)$, and let it satisfy all other conditions of Theorem 6.5. Prove that

$$\mathscr{L}\{f'(t)\} = s\mathscr{L}\{f(t)\} - f(0) - e^{-as}[f(a+0) - f(a-0)],$$

where $f(a+0) = \lim_{h\to 0+} f(a+h)$ and $f(a-0) = \lim_{h\to 0-} f(a+h)$.

6.3 TRANSFORMING ORDINARY DIFFERENTIAL EQUATIONS

We will see in this section how linear ordinary differential equations with constant coefficients are converted to algebraic equations by means of Laplace transforms. From these algebraic equations we will then obtain the Laplace transform of the solution of the differential equation. Finally, reversing the process by taking the inverse Laplace transform, we will be able to obtain the solution of the differential equation.

Suppose that we wish to solve the nonhomogeneous differential equation with constant coefficients

$$y'' + ay' + by = f(t).$$

(In what follows, we will take Laplace transforms without worrying about their existence. Any solution so obtained must be checked by substitution into the original equation.) Taking the Laplace transforms of both sides, we obtain

$$\mathscr{L}\{y''\} + a\mathscr{L}\{y'\} + b\mathscr{L}\{y\} = \mathscr{L}\{f(t)\}.$$

Now by Theorems 6.5 and 6.6 (differentiation properties) we have

$$[s^2\mathscr{L}\{y\} - sy(0) - y'(0)] + a[s\mathscr{L}\{y\} - y(0)] + b\mathscr{L}\{y\} = \mathscr{L}\{f(t)\}.$$

Then

$$[s^2 + as + b]\mathscr{L}\{y\} - [sy(0) + ay(0) + y'(0)] = \mathscr{L}\{f(t)\},$$

so that

$$\mathscr{L}\{y\} = \frac{(s + a)y(0) + y'(0) + \mathscr{L}\{f(t)\}}{s^2 + as + b}. \tag{6.30}$$

Three facts are evident from (6.30):

1. Initial conditions must be given.
2. The function $f(t)$ must have a Laplace transform.
3. We have to be able to find $\mathscr{L}^{-1}$ of the right-hand side.

Thus Laplace transform methods are primarily intended for the solution of linear initial value problems with constant coefficients.

Example 6.3.1 Find the solution of the initial value problem

$$y'' - 4y = 0, \qquad y(0) = 1, \qquad y'(0) = 2. \tag{6.31}$$

Using the differentiation property, we transform (6.31) into the algebraic equation

$$[s^2\mathscr{L}\{y\} - sy(0) - y'(0)] - 4\mathscr{L}\{y\} = [s^2\mathscr{L}\{y\} - s - 2] - 4\mathscr{L}\{y\} = 0,$$

so that

$$\mathscr{L}\{y\} = \frac{s + 2}{s^2 - 4} = \frac{1}{s - 2}.$$

Thus, by Table 6.1, we have

$$y(t) = e^{2t}$$

which satisfies all the conditions in Eq. (6.31).

Example 6.3.2 Solve

$$y'' + \omega^2 y = 0, \qquad y(0) = A, \qquad y'(0) = B \tag{6.32}$$

Using the differentiation property, we obtain

$$s^2\mathcal{L}\{y\} - sy(0) - y'(0) + \omega^2\mathcal{L}\{y\} = s^2\mathcal{L}\{y\} - sA - B + \omega^2\mathcal{L}\{y\} = 0.$$

Solving for $\mathcal{L}\{y\}$, we have

$$\mathcal{L}\{y\} = \frac{As + B}{s^2 + \omega^2} = A\left(\frac{s}{s^2 + \omega^2}\right) + \frac{B}{\omega}\left(\frac{\omega}{s^2 + \omega^2}\right). \tag{6.33}$$

By reference to Table 6.1, we find that

$$y(t) = A\cos\omega t + \frac{B}{\omega}\sin\omega t,$$

which can readily be verified to be the solution of (6.32). In calculating the inverse transform, we used the result of Exercise 6.1.23.

This example indicates the necessity of writing $\mathcal{L}\{y\}$ as a linear combination of terms for which the inverse Laplace transforms are known. In the following examples, we shall make use of the method of *partial fractions* to obtain suitable linear combinations of terms with known inverses.

Example 6.3.3 Solve

$$y'' - 5y' + 4y = e^{2t}, \qquad y(0)'= 1, \qquad y'(0) = 0.$$

Making use of the differentiation property and Table 6.1, we have

$$[s^2\mathcal{L}\{y\} - sy(0) - y'(0)] - 5[s\mathcal{L}\{y\} - y(0)] + 4\mathcal{L}\{y\} = \mathcal{L}\{e^{2t}\}$$

or

$$[s^2\mathcal{L}\{y\} - s] - 5[s\mathcal{L}\{y\} - 1] + 4\mathcal{L}\{y\} = \frac{1}{s - 2},$$

so that

$$(s^2 - 5s + 4)\mathcal{L}\{y\} = s - 5 + \frac{1}{s - 2} = \frac{s^2 - 7s + 11}{s - 2}.$$

Then

$$\mathcal{L}\{y\} = \frac{s^2 - 7s + 11}{(s - 2)(s^2 - 5s + 4)} = \frac{s^2 - 7s + 11}{(s - 2)(s - 1)(s - 4)},$$

and we have obtained an expression where the denominator consists of unrepeated factors $(s - a)$. We wish to find constants A, B, and C such that

$$\frac{A}{s - 2} + \frac{B}{s - 1} + \frac{C}{s - 4} = \frac{s^2 - 7s + 11}{(s - 2)(s - 1)(s - 4)}. \tag{6.34}$$

There is an easy method for finding these constants:

$$A = \frac{s^2 - 7s + 11}{(s - 1)(s - 4)}\bigg|_{s=2} = -\frac{1}{2}, \quad B = \frac{s^2 - 7s + 11}{(s - 2)(s - 4)}\bigg|_{s=1} = \frac{5}{3},$$

$$C = \frac{s^2 - 7s + 11}{(s - 2)(s - 1)}\bigg|_{s=4} = -\frac{1}{6}.$$

Observe that we eliminate the denominator $(s - a)$ of each term on the left-hand side of (6.34) from the right-hand side of (6.34), and evaluate the resulting equation at $s = a$ to obtain the desired constant.

To understand why this procedure works, let us multiply both sides of (6.34) by $(s - 2)$. Then we have

$$A + (s - 2)\left[\frac{B}{s - 1} + \frac{C}{s - 4}\right] = \frac{s^2 - 7s + 11}{(s - 1)(s - 4)}. \tag{6.35}$$

Setting $s = 2$ on both sides eliminates all but the constant A on the left-hand side of (6.35), and therefore

$$A = \frac{s^2 - 7s + 11}{(s - 1)(s - 4)}\bigg|_{s=2}.$$

Returning to our problem, we see that

$$\mathcal{L}\{y\} = \frac{(-\frac{1}{2})}{s - 2} + \frac{(\frac{5}{3})}{s - 1} + \frac{(-\frac{1}{6})}{s - 4},$$

which implies, according to Table 6.1, that

$$y(t) = -\frac{e^{2t}}{2} + \frac{5e^t}{3} - \frac{e^{4t}}{6}.$$

Theorem 6.8. Unrepeated Factors $(s - a)$ If

$$\mathcal{L}\{y\} = \frac{G(s)}{(s - a_1)(s - a_2)\cdots(s - a_k)},$$

where $a_i \neq a_j$, for $i \neq j$, then

$$y(t) = A_1 e^{a_1 t} + A_2 e^{a_2 t} + \cdots + A_k e^{a_k t},$$

where

$$A_j = \frac{G(s)}{(s - a_1)\cdots(s - a_{j-1})(s - a_{j+1})\cdots(s - a_k)}\bigg|_{s=a_j}.$$

Example 6.3.4 Solve

$$y'' - 4y' + 4y = t^2, \qquad y(0) = 0, \qquad y'(0) = 1.$$

Using the differentiation property and Table 6.1, we find that

$$[s^2 \mathscr{L}\{y\} - 1] - 4s\mathscr{L}\{y\} + 4\mathscr{L}\{y\} = \frac{2}{s^3}.$$

Thus

$$(s^2 - 4s + 4)\mathscr{L}\{y\} = \frac{2}{s^3} + 1 = \frac{s^3 + 2}{s^3}$$

and

$$\mathscr{L}\{y\} = \frac{s^3 + 2}{s^3(s - 2)^2}. \tag{6.36}$$

The denominator now consists of repeated factors $(s - a)$. We now must seek constants $A_1, A_2, A_3, B_1,$ and B_2 such that

$$\frac{A_1}{s} + \frac{A_2}{s^2} + \frac{A_3}{s^3} + \frac{B_1}{s - 2} + \frac{B_2}{(s - 2)^2} = \frac{s^3 + 2}{s^3(s - 2)^2}. \tag{6.37}$$

Again there is an easy way to find these coefficients. Suppose that we wish to find A_2. In a manner similar to what we did before, we multiply both sides of (6.37) by s^3 to obtain

$$A_1 s^2 + A_2 s + A_3 + s^3 \left[\frac{B_1}{s - 2} + \frac{B_2}{(s - 2)^2} \right] = \frac{s^3 + 2}{(s - 2)^2}. \tag{6.38}$$

It is clear that we can obtain $A_3 = \frac{1}{2}$ by setting $s = 0$. To find A_2, we differentiate (6.38) with respect to s, obtaining

$$2A_1 s + A_2 + 3s^2 \left[\frac{B_1}{s - 2} + \frac{B_2}{(s - 2)^2} \right] + s^3 \left[\frac{B_1}{s - 2} + \frac{B_2}{(s - 2)^2} \right]'$$

$$= \frac{3s^2}{(s - 2)^2} - \frac{2(s^3 + 2)}{(s - 2)^3}. \tag{6.39}$$

Setting $s = 0$ yields $A_2 = \frac{1}{2}$. Differentiating (6.39) once more and letting $s = 0$, we have $2A_1 = \frac{3}{4}$ or $A_1 = \frac{3}{8}$. Similarly,

$$(s - 2)^2 \left[\frac{A_1}{s} + \frac{A_2}{s^2} + \frac{A_3}{s^3} \right] + B_1(s - 2) + B_2 = \frac{s^3 + 2}{s^3},$$

and $B_1 = -\frac{3}{8}$, $B_2 = \frac{5}{4}$. Thus, returning to (6.36), we have

$$\mathcal{L}\{y\} = \frac{(\frac{3}{8})}{s} + \frac{(\frac{1}{2})}{s^2} + \frac{(\frac{1}{2})}{s^3} + \frac{(-\frac{3}{8})}{s - 2} + \frac{(\frac{5}{4})}{(s - 2)^2}.$$

Hence, according to Tables 6.1 and 6.2,

$$y(t) = \frac{3}{8} + \frac{t}{2} + \frac{t^2}{4} - \frac{3}{8} e^{2t} + \frac{5}{4} te^{2t}.$$

Theorem 6.9. Repeated Factors $(s - a)^m$ If

$$\mathcal{L}\{y\} = \frac{G(s)}{(s - a)^m H(s)} = \frac{A_1}{s - a} + \frac{A_2}{(s - a)^2} + \cdots + \frac{A_m}{(s - a)^m} + F(s),$$

then

$$y(t) = e^{at}\left[A_1 + A_2 \frac{t}{(1!)} + \cdots + A_m \frac{t^{m-1}}{(m - 1)!} \right] + \mathcal{L}^{-1}\{F\},$$

where

$$A_j = \frac{1}{(m - j)!} \frac{d^{m-j}}{ds^{m-j}}\left[\frac{G(s)}{H(s)} \right]\bigg|_{s=a}, \qquad j = 1, \ldots, m.$$

Although Theorems 6.8 and 6.9 will also work, the next theorem is preferred for complex numbers a, since the solution is immediately given in terms of sines and cosines.

If the initial value problem with constant (real) coefficients yields an equation with complex factors in the denominator of the expression for $\mathcal{L}\{y\}$, then these factors must come in conjugate pairs. An example is Eq. (6.33),

$$\mathcal{L}\{y\} = \frac{As + B}{s^2 + \omega^2}.$$

Here it is convenient *not* to factor $s^2 + \omega^2$ into $(s + i\omega)(s - i\omega)$, but to express the transform as

$$\mathcal{L}\{y\} = A\left(\frac{s}{s^2 + \omega^2} \right) + \frac{B}{\omega}\left(\frac{\omega}{s^2 + \omega^2} \right),$$

which can be solved immediately by reference to the table in Appendix 2. A slight complication is encountered in the next problem.

Example 6.3.5 Solve

$$y'' + 2y' + 2y = f(t), \qquad y(0) = y'(0) = 1.$$

Using the differentiation property and solving for $\mathcal{L}\{y\}$, we obtain

$$\mathcal{L}\{y\} = \frac{s + 3 + \mathcal{L}\{f(t)\}}{s^2 + 2s + 2} = \frac{s + 3 + \mathcal{L}\{f(t)\}}{(s + 1)^2 + 1} \tag{6.40}$$

by completing the square in the denominator. If $f(t) = 0$, we may write (6.40) as

$$\mathcal{L}\{y\} = \frac{s + 1}{(s + 1)^2 + 1} + \frac{2}{(s + 1)^2 + 1},$$

and the solution follows from Table 6.2:

$$y(t) = e^{-t}(\cos t + 2 \sin t).$$

If $f(t) = t$, then (6.40) becomes

$$\mathcal{L}\{y\} = \frac{s^3 + 3s^2 + 1}{s^2[(s + 1)^2 + 1]} = \frac{A(s + 1) + B}{(s + 1)^2 + 1} + \frac{C_1}{s} + \frac{C_2}{s^2}. \tag{6.41}$$

We need to find the constants A, B, C_1, and C_2. To obtain A and B, we multiply both sides of (6.41) by $(s + 1)^2 + 1$:

$$\frac{s^3 + 3s^2 + 1}{s^2} = A(s + 1) + B + [(s + 1)^2 + 1]\left(\frac{C_1}{s} + \frac{C_2}{s^2}\right).$$

We then let s equal one of the roots of $(s + 1)^2 + 1 = 0$, say $s = -1 + i$. Then

$$\frac{s^3 + 3s^2 + 1}{s^2}\bigg|_{s = -1+i} = A(s + 1) + B\bigg|_{s = -1+i}$$

or $2 + \frac{3}{2}i = B + Ai$, so that $A = \frac{3}{2}$ and $B = 2$. Theorem 6.9 is used to obtain $C_1 = -\frac{1}{2}$ and $C_2 = \frac{1}{2}$. Finally,

$$y(t) = e^{-t}\left[\frac{3}{2} \cos t + 2 \sin t\right] - \frac{1}{2} + \frac{t}{2}.$$

Theorem 6.10. Unrepeated Complex Factor $(s - a)^2 + b^2$ If

$$\mathcal{L}\{y\} = \frac{G(s)}{[(s - a)^2 + b^2]H(s)} = \frac{A(s - a) + B}{(s - a)^2 + b^2} + F(s),$$

then

$$y(t) = \frac{e^{at}}{b}(bA \cos bt + B \sin bt) + \mathcal{L}^{-1}\{F\},$$

where B and bA are the real and imaginary parts of

$$B + ibA = \frac{G(s)}{H(s)}\bigg|_{s = a + ib}.$$

Returning to (6.40), we see that if $f(t) = e^{-t} \sin t$, then

$$\mathcal{L}\{y\} = \frac{s + 3 + [(s + 1)^2 + 1]^{-1}}{(s + 1)^2 + 1} = \frac{s^3 + 5s^2 + 8s + 7}{[(s + 1)^2 + 1]^2}$$

$$= \frac{A_1(s + 1) + B_1}{(s + 1)^2 + 1} + \frac{A_2(s + 1) + B_2}{[(s + 1)^2 + 1]^2}.$$

We again use the procedure of Theorem 6.9 and multiply both sides by $[(s + 1)^2 + 1]^2$, obtaining

$$s^3 + 5s^2 + 8s + 7 = [(s + 1)^2 + 1][A_1(s + 1) + B_1] + [A_2(s + 1) + B_2].$$

Evaluating this equation at $s = -1 + i$, we find $1 = B_2 + iA_2$, so that $B_2 = 1$ and $A_2 = 0$. Differentiating twice eliminates all terms involving A_2 and B_2:

$$6s + 10 = 6A_1(s + 1) + 2B_1.$$

Setting $s = -1 + i$ yields $A_1 = 1$, $B_1 = 2$, and

$$y(t) = e^{-t}\left(\cos t + \frac{5}{2}\sin t - \frac{t}{2}\cos t\right).$$

Thus in case we have repeated complex factors $[(s - a)^2 + b^2]^m$, we multiply $\mathcal{L}\{y\}$ by $[(s - a)^2 + b^2]^m$. Then we repeatedly evaluate the result at $s = a + ib$ and differentiate to obtain the necessary constants.

Two other formulas are often useful in problems that involve differentiating and integrating the Laplace transform of a function $f(t)$. By reversing the order in which the operations are performed, we see that

$$\frac{d}{ds}\mathcal{L}\{f(t)\} = \int_0^\infty \frac{d}{ds}(e^{-st})f(t)\, dt$$

$$= \int_0^\infty e^{-st}tf(t)\, dt = -\mathcal{L}\{tf(t)\} \tag{6.42}$$

and

$$\int_s^\infty \mathcal{L}\{f(t)\}(s)\, ds = \int_0^\infty \int_s^\infty e^{-st}\, ds\, f(t)\, dt$$

$$= \int_0^\infty e^{-st}\frac{f(t)}{t}\, dt = \mathcal{L}\left\{\frac{f(t)}{t}\right\}. \tag{6.43}$$

This reversal is valid whenever $\mathcal{L}\{f\}$ exists.* These two identities can sometimes (but not usually) be used to solve especially simple linear equations with variable coefficients by Laplace transform methods. We illustrate this by an example.

---

* See W. R. Derrick, *Introductory Complex Analysis and Applications*, Academic Press, New York, 1972, p. 177.

Example 6.3.6 Consider the equation with variable coefficients

$$ty'' - ty' - y = 0, \qquad y(0) = 0, \qquad y'(0) = 1.$$

If we let $Y(s) = \mathscr{L}\{y(t)\}$, then by the differentiation property and Eq. (6.42)

$$\mathscr{L}\{ty''\} = -\frac{d}{ds}\mathscr{L}\{y''\} = -\frac{d}{ds}\{s^2 Y(s) - sy(0) - y'(0)\}$$

$$= -s^2 Y' - 2sY$$

and

$$\mathscr{L}\{ty'\} = -\frac{d}{ds}\mathscr{L}\{y'\} = -\frac{d}{ds}\{sY - y(0)\} = -sY' - Y.$$

Substituting these expressions into the Laplace transform of the original equation yields

$$-s^2 Y' - 2sY + sY' + Y - Y = 0.$$

Rearranging and canceling terms, we have

$$(s^2 - s)Y' + 2sY = 0.$$

We now divide both sides by $s^2 - s = s(s - 1)$ to obtain

$$Y' + \frac{2}{s - 1} Y = 0.$$

Separating variables, we have

$$\frac{dY}{Y} = -\frac{2}{s - 1} ds,$$

and an integration yields

$$\ln Y = -2 \ln (s - 1) + c$$

or

$$Y(s) = \frac{c}{(s - 1)^2}.$$

By formula 2 of Appendix 2 (or Table 6.2),

$$y(t) = cte^t.$$

To find c, we use the initial condition $y'(0) = 1$. We note that

$$y'(t) = ce^t(t + 1),$$

so that $1 = y'(0) = c$. Thus the solution to the initial value problem is

$$y(t) = te^t.$$

We again caution the reader not to expect to be able to solve variable coefficient equations by this method. It will work only when

a) the coefficients $a_i(t)$ are polynomials in t,
b) the differential equation involving $Y(s)$ can be solved, and
c) the inverse transform of $Y(s)$ can be found.

It is rare that all these conditions can be met (see Exercise 42).

Further applications of Laplace transforms will be found in Sections 7.8
and 10.5, where techniques are developed to solve partial differential equations
and systems of ordinary differential equations.

EXERCISES 6.3

In Exercises 1 through 20 solve the given initial value problems.

1. $y'' + y = 0$, $y(0) = 1$, $y'(0) = 0$

2. $y'' + y' = 0$, $y(0) = 0$, $y'(0) = 1$

3. $y'' - a^2 y = 0$, $y(0) = A$, $y'(0) = B$

4. $y'' - ay' = 0$, $y(0) = 1$, $y'(0) = a$

5. $y'' + 2y' + 5y = 0$, $y(0) = y'(0) = 1$

6. $y'' - y' + y = 0$, $y(0) = y'(0) = 1$

7. $y'' - 4y' + 3y = 1$, $y(0) = 1$, $y'(0) = 4$

8. $y'' - 2y' - 3y = 5$, $y(0) = 0$, $y'(0) = 1$

9. $y'' - 9y = t$, $y(0) = 1$, $y'(0) = 2$

10. $y'' - 3y' - 4y = t^2$, $y(0) = 2$, $y'(0) = 1$

11. $y''' + y = 0$, $y(0) = y''(0) = 1$, $y'(0) = -1$

12. $y^{iv} - y = 0$, $y(0) = y''(0) = 1$, $y'(0) = y'''(0) = 0$

13. $y^{iv} - y = 0$, $y(0) = y''(0) = 0$, $y'(0) = y'''(0) = 1$

14. $y''' - 3y' - 2y = e^{2t}$, $y(0) = y'(0) = 0$, $y''(0) = 1$

15. $y'' + k^2 y = \cos kt$, $y(0) = 0$, $y'(0) = k$

16. $y'' + 4y = \cos t$, $y(0) = y'(0) = 0$

17. $y'' + a^2 y = \sin at$, $y(0) = a$, $y'(0) = a^2$

18. $y'' - y = te^t$, $y(0) = y'(0) = 1$

19. $y^{iv} - y = \cos t$, $y(0) = y''(0) = 1$, $y'(0) = y'''(0) = 0$

20. $y^{iv} - y = \sinh t$, $y(0) = y''(0) = 0$, $y'(0) = y'''(0) = 1$

Use formulas (6.42) and (6.43) to obtain the Laplace transforms of the functions in Exercises
21 through 36.

21. $t \cos kt$ 22. $t^2 \cos kt$ 23. $t \sin kt$

24. $t^2 \sin kt$ 25. $t \cosh t$ 26. $t^2 \cosh t$

27. $t \sinh t$ 28. $t^2 \sinh t$ 29. $\dfrac{\sin kt}{t}$

30. $\dfrac{1 - \cos at}{t}$ 31. $\dfrac{1 - \cosh at}{t}$ 32. $\displaystyle\int_0^t \dfrac{\sin ku}{u}\, du$

33. $\displaystyle\int_0^t \dfrac{1 - \cos au}{u}\, du$ 34. $\displaystyle\int_0^t \dfrac{1 - \cosh au}{u}\, du$

35. $\operatorname{erf}(t) = \dfrac{2}{\sqrt{\pi}} \displaystyle\int_0^t e^{-u^2}\, du$

36. $\dfrac{e^{-k^2/4t}}{\sqrt{\pi t}}$

Find the inverse Laplace transform of the functions in Exercises 37 through 40. Use derivatives and integrals.

37. $\ln\!\left(1 + \dfrac{a^2}{s^2}\right)$ 38. $\ln \dfrac{s - a}{s - b}$

39. $\arctan \dfrac{1}{s}$ 40. $\dfrac{1}{s} \arctan \dfrac{1}{s}$

* 41. Show that

$$\mathscr{L}\{t^n f(t)\} = (-1)^n \frac{d^n}{ds^n}\, \mathscr{L}\{f(t)\}.$$

[*Hint*: Generalize Eq. (6.42) by mathematical induction.]

* 42. Consider the equation

$$y'' + t^2 y = 0, \qquad y(0) = 0, \qquad y'(0) = 1.$$

a) Use the result of Exercise 41 to obtain a differential equation for $Y(s) = \mathscr{L}\{y(t)\}$.
b) Solve the differential equation and find $Y(s)$. (Note that it is not possible to invert this transform by the methods we have discussed.)

6.4 THE TRANSFORM OF CONVOLUTION INTEGRALS

It often occurs that in the process of solving a linear differential equation by transforms, we end up with a transform which is the product of two other transforms. Although we proved in Exercise 23 of Section 6.1 that

$$\mathscr{L}^{-1}\{F + G\} = \mathscr{L}^{-1}\{F\} + \mathscr{L}^{-1}\{G\},$$

it is not true that $\mathscr{L}^{-1}\{FG\} = \mathscr{L}^{-1}\{F\}\mathscr{L}^{-1}\{G\}$ (see Example 10 of Section 6.2). In this section we shall define the convolution of two functions f and g and show that $\mathscr{L}^{-1}\{FG\}$ is equal to the convolution of $\mathscr{L}^{-1}\{F\}$ and $\mathscr{L}^{-1}\{G\}$. We shall then apply this fact in a variety of ways.

Definition If $f(t)$ and $g(t)$ are piecewise continuous functions, then the *convolution* of f and g, written $f * g(t)$, is defined by

$$f * g(t) = \int_0^t f(t - u)g(u) \, du. \tag{6.44}$$

The notation $f * g(t)$ indicates that the convolution $f * g$ is a function of the independent variable t.

Using the change of variables $v = t - u$, we see that

$$f * g(t) = -\int_t^0 f(v)g(t - v) \, dv = \int_0^t g(t - v)f(v) \, dv$$

$$= g * f(t). \tag{6.45}$$

Hence $f * g(t) = g * f(t)$, and we can take the convolution in either order without altering the result. We may now state the main result of this section.

Theorem 6.11. Convolution Theorem for Laplace Transforms Let $F(s) = \mathcal{L}\{f(t)\}$ and $G(s) = \mathcal{L}\{g(t)\}$. Then $\mathcal{L}\{f * g(t)\} = F(s)G(s)$.

Proof. By definition

$$F(s)G(s) = \left(\int_0^\infty e^{-su}f(u) \, du \right) \left(\int_0^\infty e^{-sv}g(v) \, dv \right)$$

$$= \int_0^\infty \int_0^\infty e^{-s(u+v)}f(u)g(v) \, dv \, du. \tag{6.46}$$

If we make the change of variables $t = u + v$, then $dt = dv$ and the integral (6.46) is equal to

$$F(s)G(s) = \int_0^\infty \int_u^\infty e^{-st}f(u)g(t - u) \, dt \, du. \tag{6.47}$$

Changing the order of integration and noting that

$$\int_0^\infty \int_u^\infty dt \, du = \int_0^\infty \int_0^t du \, dt,$$

(check!), we have integral (6.47) equal to

$$F(s)G(s) = \int_0^\infty \int_0^t e^{-st}f(u)g(t - u) \, du \, dt$$

$$= \int_0^\infty e^{-st} \left[\int_0^t g(t - u)f(u) \, du \right] dt$$

$$= \int_0^\infty e^{-st}g * f(t) \, dt = \int_0^\infty e^{-st}f * g(t) \, dt$$

$$= \mathcal{L}\{f * g\},$$

and the proof is complete

Our first applications of this theorem are in the computation of inverse transforms.

Example 6.4.1 Compute $\mathcal{L}^{-1}\{s/(s^2 + 1)^2\}$. Since

$$\mathcal{L}\{\cos at\} = \frac{s}{s^2 + 1} \qquad \text{and} \qquad \mathcal{L}\{\sin t\} = \frac{1}{s^2 + 1},$$

we have

$$\mathcal{L}^{-1}\left\{\frac{s}{(s^2 + 1)^2}\right\} = \mathcal{L}^{-1}\left\{\frac{s}{s^2 + 1} \cdot \frac{1}{s^2 + 1}\right\}.$$

By Theorem 6.11, this is the convolution of $\cos t$ and $\sin t$. But

$$\sin t * \cos t = \int_0^t \sin(t - u)\cos u\,du$$

$$= \int_0^t (\sin t \cos u - \cos t \sin u)\cos u\,du$$

$$= \sin t \int_0^t \cos^2 u\,du - \cos t \int_0^t \sin u \cos u\,du,$$

which, according to the table of integrals (Appendix 1), is equal to

$$\sin t \left(\frac{\sin u \cos u + u}{2}\right) - \cos t \left.\frac{\sin^2 u}{2}\right|_{u=0}^{u=t} = \frac{t \sin t}{2}.$$

Therefore,

$$\mathcal{L}^{-1}\left\{\frac{s}{(s^2 + 1)^2}\right\} = \frac{t \sin t}{2}. \qquad (6.48)$$

Example 6.4.2 Compute $\mathcal{L}^{-1}\{1/s^2(s + 1)^2\}$. We have

$$\frac{1}{s^2} = \mathcal{L}\{t\} \qquad \text{and} \qquad \frac{1}{(s + 1)^2} = \mathcal{L}\{te^{-t}\}.$$

Therefore,

$$\mathcal{L}^{-1}\left\{\frac{1}{s^2(s + 1)^2}\right\} = \mathcal{L}^{-1}\left\{\frac{1}{s^2} \cdot \frac{1}{(s + 1)^2}\right\} = t * te^{-t}.$$

But

$$t * te^{-t} = \int_0^t (t - u)ue^{-u}\,du = t \int_0^t ue^{-u}\,du - \int_0^t u^2 e^{-u}\,du.$$

Integrating by parts, we obtain

$$t * te^{-t} = -te^{-u}(1 + u) + e^{-u}[u^2 + 2u + 2]\,\Big|_0^t = t - 2 + (t + 2)e^{-t}$$

or

$$\mathcal{L}^{-1}\left\{\frac{1}{s^2(s + 1)^2}\right\} = t - 2 + (t + 2)e^{-t}.$$

Although the convolution theorem is obviously very useful in calculating inverse transforms, it also has important applications in a very different area. In 1931 the Italian mathematician Vito Volterra* published a book that contained a fairly sophisticated model of population growth. It would be beyond the scope of this book to go into a derivation of Volterra's model. However, a central equation in this model is of the form

$$x(t) = f(t) + \int_0^t a(t - u)x(u)\, du. \tag{6.49}$$

An equation of this type, where $f(t)$ and $a(t)$ can be assumed to be continuous, is called a *Volterra integral equation*. Since the publication of Volterra's papers, many diverse phenomena in thermodynamics, electrical systems theory, nuclear reactor theory, and chemotherapy have been modeled with Volterra integral equations.

It is quite easy to see how Laplace transforms can be used to solve an equation in the form (6.49). Taking transforms on both sides of (6.49), using the convolution theorem, and denoting transforms by the appropriate capital letters, we obtain

$$X(s) = F(s) + A(s)X(s)$$

or

$$X(s) = \frac{F(s)}{1 - A(s)}. \tag{6.50}$$

Looking at Eq. (6.50), we immediately see that if $F(s)$ and $A(s)$ are defined for $s \geq s_0$, then $X(s)$ is similarly defined so long as $A(s) \neq 1$. Once $X(s)$ is known, we may (if possible) calculate the solution $x(t) = \mathscr{L}^{-1}\{X(s)\}$.

Example 6.4.3 Consider the integral equation

$$x(t) = t^2 + \int_0^t \sin(t - u)x(u)\, du. \tag{6.51}$$

Taking transforms, we have

$$X(s) = \frac{2}{s^3} + \frac{1}{s^2 + 1} \cdot X(s)$$

or

$$X(s) = \frac{2/s^3}{1 - 1/(s^2 + 1)} = \frac{2(s^2 + 1)}{s^5} = \frac{2}{s^3} + \frac{2}{s^5}.$$

---

*   V. Volterra, *Leçons sur la Théorie Mathématique de la Lutte pour la Vie*, Gauthier-Villars, Paris, 1931.

Hence the solution to Eq. (6.51) is given by

$$x(t) = t^2 + \tfrac{1}{12}t^4.$$

There are other applications of the very useful convolution theorem given in the exercises. The student, however, should always keep in mind that the greatest difficulty in using any of these methods is that it is frequently difficult to calculate inverse transforms. Unfortunately, most problems which arise lead to inverting transforms that do not fit into familiar patterns. For this reason methods have been devised for estimating such inverses. The interested reader should consult a more advanced book on Laplace transforms, such as the excellent book by Widder.*

EXERCISES 6.4

In Exercises 1 through 5 find the Laplace transform of each given convolution integral.

1. $f(t) = \displaystyle\int_0^t (t - u)^3 \sin u \, du$

2. $f(t) = \displaystyle\int_0^t e^{-(t-u)} \cos 2u \, du$

3. $f(t) = \displaystyle\int_0^t (t - u)^3 u^5 \, du$

4. $f(t) = \displaystyle\int_0^t \sinh 4(t - u) \cosh 5u \, du$

5. $f(t) = \displaystyle\int_0^t e^{17(t-u)} u^{19} \, du$

In Exercises 6 through 10 use the convolution theorem to calculate the inverse Laplace transforms of the given functions.

6. $F(s) = \dfrac{3}{s^4(s^2 + 1)}$

7. $F(s) = \dfrac{a}{s^2(s^2 + a^2)}$

8. $F(s) = \dfrac{1}{(s^2 + 1)^2}$

9. $F(s) = \dfrac{1}{s(s^2 + a^2)}$

10. $F(s) = \dfrac{1}{(s^2 + 1)^3}$

11. Solve the Volterra integral equation

$$x(t) = e^{-t} - 2 \int_0^t \cos (t - u)x(u) \, du.$$

12. Solve the Volterra integral equation

$$x(t) = t + \tfrac{1}{6} \int_0^t (t - u)^3 x(u) \, du.$$

---

* D. V. Widder, *The Laplace Transformation*, Princeton University Press, Princeton, N.J., 1941.

13. Consider the convolution integral equation

$$x(t) = \tfrac{1}{2} \sin 2t + \int_0^t x(t - u)x(u)\, du.$$

a) Show that

$$\mathscr{L}\{x(t)\} = \frac{1}{2}\left(\frac{\sqrt{s^2 + 4} + s}{\sqrt{s^2 + 4}}\right).$$

[*Hint*: It is necessary to solve a quadratic equation where functions of s are treated as constants.]

b) Using Appendix 2, obtain the solution $x(t) = J_1(2t)$, where J_1 is the Bessel function of order one.

c) Show that $\mathscr{L}\{x(t)\}$ can also be written as

$$\mathscr{L}\{x(t)\} = 1 - \frac{1}{2}\left(\frac{\sqrt{s^2 + 4} - s}{\sqrt{s^2 + 4}}\right)$$

and thereby conclude that

$$x(t) = \delta(t) - J_1(2t)$$

is also a solution, where $\delta(t)$ is the Dirac delta function defined in Section 6.2.

14. Another type of equation that arises frequently in applications is the *Volterra integro-differential equation*

$$x'(t) = Ax(t) + f(t) + \int_0^t a(t - u)x(u)\, du, \qquad x(t_0) = x_0, \tag{6.52}$$

where A is a constant and $f(t)$ and $a(t)$ are known continuous functions. Assuming that $a(t)$ and $f(t)$ have Laplace transforms for $s \geqslant s_0$, find $\mathscr{L}\{x(t)\}$ in terms of these transforms.

15. Solve the Volterra integro-differential equation

$$x'(t) = x(t) + e^{2t} + \int_0^t e^{2(t-u)}x(u)\, du, \qquad x(0) = 1.$$

6.5 LAPLACE TRANSFORM METHODS FOR DIFFERENCE EQUATIONS

It is easy to adapt the Laplace transform to solve linear difference equations with constant coefficients. Consider the equation

$$y_{n+2} + ay_{n+1} + by_n = f_n, \qquad y_0 = c_0, \qquad y_1 = c_1, \tag{6.53}$$

and define the functions of a real variable

$$y(t) = y_n, \qquad f(t) = f_n, \qquad \text{for } n \leqslant t < n + 1.$$

It is clear from this definition that if y_n and f_n are known for all n, then $y(t)$ and $f(t)$ are step functions of height y_n and f_n, respectively, in the interval $n \leqslant t < n + 1$

(see Fig. 6.7). Equation (6.53) then becomes

$$y(t + 2) + ay(t + 1) + by(t) = f(t), \qquad y(0) = c_0, \qquad y(1) = c_1. \tag{6.54}$$

Now by making use of the unit step function defined in Section 6.2, we can write

$$f(t) = \sum_{n=0}^{\infty} f_n[H(t - n) - H(t - (n + 1))]. \tag{6.55}$$

Assuming that the Laplace transform of Eq. (6.55) can be obtained term by term, we have

$$\begin{aligned}
\mathscr{L}\{f(t)\} &= \int_0^{\infty} e^{-st} f(t) \, dt \\
&= \sum_{n=0}^{\infty} f_n \left[\int_0^{\infty} e^{-st} H(t - n) \, dt - \int_0^{\infty} e^{-st} H(t - (n + 1)) \, dt \right] \\
&= \sum_{n=0}^{\infty} f_n \left[\frac{e^{-ns}}{s} - \frac{e^{-(n+1)s}}{s} \right] \\
&= \frac{1 - e^{-s}}{s} \sum_{n=0}^{\infty} f_n e^{-ns}. \tag{6.56}
\end{aligned}$$

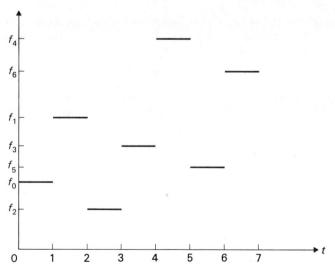

Figure 6.7

A similar result holds for $\mathcal{L}\{y(t)\}$, but a different computation is necessary for $\mathcal{L}\{y(t + 1)\}$ and $\mathcal{L}\{y(t + 2)\}$. Letting $u = t + 1$, we see that

$$
\begin{aligned}
\mathcal{L}\{y(t + 1)\} &= \int_0^\infty e^{-st} y(t + 1)\, dt = \int_1^\infty e^{-s(u-1)} y(u)\, du \\
&= e^s \left[\int_0^\infty e^{-su} y(u)\, du - \int_0^1 e^{-su} y(u)\, du \right] \\
&= e^s \left[\mathcal{L}\{y\} - \int_0^1 e^{-su} y(0)\, du \right] \\
&= e^s \mathcal{L}\{y(t)\} - \frac{y(0)}{s} (e^s - 1).
\end{aligned}
\tag{6.57}
$$

[Note the analogy with the differentiation property (Theorem 6.5).] We can find $\mathcal{L}\{y(t + 2)\}$ by applying Eq. (6.57) twice. Let $z(t + 1) = y(t + 2)$. Then

$$
\begin{aligned}
\mathcal{L}\{y(t + 2)\} &= \mathcal{L}\{z(t + 1)\} = e^s \mathcal{L}\{z(t)\} - \frac{z(0)}{s} (e^s - 1) \\
&= e^s \mathcal{L}\{y(t + 1)\} - \frac{y(1)}{s} (e^s - 1) \\
&= e^s \left[e^s \mathcal{L}\{y(t)\} - \frac{y(0)}{s} (e^s - 1) \right] - \frac{y(1)}{s} (e^s - 1) \\
&= e^{2s} \mathcal{L}\{y(t)\} - \frac{(y(0)e^s + y(1))}{s} (e^s - 1).
\end{aligned}
\tag{6.58}
$$

Applying these results to Eq. (6.54) and gathering like terms, we have

$$
\mathcal{L}\{y(t)\} = \frac{(c_0 e^s + c_1 + ac_0)(e^s - 1) + s\mathcal{L}\{f(t)\}}{s(e^{2s} + ae^s + b)},
$$

which, when inverted, yields $y(t)$. Finally, y_n is obtained by evaluating $y(n)$ for $n = 0, 1, 2, \ldots$.

Before solving difference equations by these methods, let us first compute several Laplace transforms useful in solving difference equations.

Example 6.5.1 Let $f_n = a^n$. Then by Eq. (6.56) and the formula for the sum of a geometric series (6.17), we have

$$
\begin{aligned}
\mathcal{L}\{f(t)\} &= \frac{1 - e^{-s}}{s} \sum_{n=0}^\infty a^n e^{-ns} = \frac{1 - e^{-s}}{s} \sum_{n=0}^\infty (ae^{-s})^n \\
&= \frac{1 - e^{-s}}{s} \cdot \frac{1}{1 - ae^{-s}} = \frac{1 - e^{-s}}{s(1 - ae^{-s})} = \frac{e^s - 1}{s(e^s - a)}.
\end{aligned}
\tag{6.59}
$$

(*Note.* This is valid when $ae^{-s} < 1$ or when $s > \ln(1/a) = -\ln a$.)

Example 6.5.2 If $f_n = na^n$, then

$$\mathcal{L}\{f(t)\} = \frac{1 - e^{-s}}{s} \sum_{n=0}^{\infty} n(ae^{-s})^n.$$

Let $P = \sum_{n=0}^{\infty} nx^n$. Then

$$(1 - x)P = P - xP = \sum_{n=1}^{\infty} x^n = \frac{1}{1 - x} - 1 = \frac{x}{1 - x},$$

so that $P = x/(1 - x)^2$. Therefore,

$$\mathcal{L}\{f(t)\} = \frac{ae^{-s}(1 - e^{-s})}{s(1 - ae^{-s})^2} = \frac{a(e^s - 1)}{s(e^s - a)^2}. \tag{6.60}$$

As in Example 6.5.1, this is valid when $s > -\ln a$.

Example 6.5.3 Let $f_n = \cos n\theta$ and $g_n = \sin n\theta$. Then we may define $h_n = f_n + ig_n = e^{in\theta}$ [according to Euler's formula (3.28)], which yields

$$\mathcal{L}\{h(t)\} = \mathcal{L}\{f(t) + ig(t)\} = \frac{1 - e^{-s}}{s} \sum_{n=0}^{\infty} e^{in\theta} e^{-ns}$$

$$= \frac{1 - e^{-s}}{s} \cdot \frac{1}{1 - e^{i\theta}e^{-s}}.$$

Multiplying the numerator and denominator by e^s, we obtain

$$\frac{1 - e^{-s}}{s} \cdot \frac{1}{1 - e^{i\theta}e^{-s}} = \frac{e^s - 1}{s(e^s - e^{i\theta})} = \frac{e^s - 1}{s[e^s - (\cos \theta + i \sin \theta)]}.$$

Then multiplying the numerator and denominator by $(e^s - \cos \theta + i \sin \theta)$ yields

$$\mathcal{L}\{f(t)\} + i\mathcal{L}\{g(t)\} = \frac{(e^s - 1)(e^s - \cos \theta + i \sin \theta)}{s[(e^s - \cos \theta)^2 + \sin^2 \theta]}.$$

Finally, equating real and imaginary terms, we have

$$\mathcal{L}\{f(t)\} = \frac{(e^s - 1)(e^s - \cos \theta)}{s[(e^s - \cos \theta)^2 + \sin^2 \theta]},$$

$$\mathcal{L}\{g(t)\} = \frac{\sin \theta \,(e^s - 1)}{s[(e^s - \cos \theta)^2 + \sin^2 \theta]}. \tag{6.61}$$

Example 6.5.4 Let

$$y_{n+2} - 3y_{n+1} + 2y_n = 0, \qquad y_0 = 1, \qquad y_1 = 2.$$

Using Eqs. (6.57) and (6.58), we obtain

$$\left[e^{2s} \mathscr{L}\{y\} - \frac{(e^s + 2)(e^s - 1)}{s} \right] - 3 \left[e^s \mathscr{L}\{y\} - \frac{e^s - 1}{s} \right] + 2\mathscr{L}\{y\} = 0.$$

Gathering all the terms involving $\mathscr{L}\{y\}$ yields

$$(e^{2s} - 3e^s + 2)\mathscr{L}\{y\} = \frac{e^s - 1}{s} (e^s + 2 - 3)$$

or

$$\mathscr{L}\{y\} = \frac{(e^s - 1)(e^s - 1)}{s(e^s - 1)(e^s - 2)} = \frac{e^s - 1}{2(e^s - 2)}.$$

Thus by (6.59), $y_n = 2^n$.

Example 6.5.5 $y_{n+2} + y_n = n$, $y_0 = 1$, $y_1 = 0$. Using (6.60) with $a = 1$, we have

$$\left[e^{2s} \mathscr{L}\{y\} - \frac{e^s(e^s - 1)}{s} \right] + \mathscr{L}\{y\} = \frac{1}{s(e^s - 1)}.$$

Then

$$(e^{2s} + 1)\mathscr{L}\{y\} = \frac{1}{s(e^s - 1)} + \frac{e^s(e^s - 1)}{s} = \frac{e^{3s} - 2e^{2s} + e^s + 1}{s(e^s - 1)}.$$

Using partial fractions, we obtain

$$\mathscr{L}\{y\} = \frac{e^{3s} - 2e^{2s} + e^s + 1}{s(e^s - 1)(e^{2s} + 1)} = \frac{3e^s(e^s - 1)}{2s(e^{2s} + 1)} + \frac{1}{2s(e^s - 1)} - \frac{1}{2s}.$$

Letting $\theta = \pi/2$ in Eq. (6.61) and $a = 1$ in (6.59) and (6.60), we have

$$y_n = \frac{3}{2} \cos \frac{n\pi}{2} + \frac{n}{2} - \frac{1}{2}.$$

EXERCISES 6.5

Using the methods of this section, solve the equations in Exercises 1 through 10.

1. $y_{n+2} + 4y_{n+1} + 6y_n = 0$, $y_0 = 0$, $y_1 = 1$
2. $y_{n+2} + 5y_{n+1} - 6y_n = 0$, $y_0 = 3$, $y_1 = 11$
3. $y_{n+2} - 6y_{n+1} + 9y_n = 0$, $y_0 = 5$, $y_1 = 12$
4. $6y_{n+2} + 5y_{n+1} + y_n = 0$, $y_0 = 1$, $y_1 = 0$
5. $y_{n+2} - 2y_{n+1} + 2y_n = 0$, $y_0 = 1$, $y_1 = 2$

6. $y_{n+2} - 3y_{n+1} + 2y_n = 1 + 5^{n+1}$, $y_0 = 1$, $y_1 = 0$

7. $y_{n+2} - 4y_{n+1} + 4y_n = n2^n$, $y_0 = y_1 = 0$

8. $y_{n+2} + 5y_n = \sin\dfrac{n\pi}{2}$, $y_0 = y_1 = 0$

9. $y_{n+2} - 5y_{n+1} + 6y_n = 4^n$, $y_0 = 0$, $y_1 = 1$

10. $y_{n+2} - 7y_{n+1} + 10y_n = 16n$, $y_0 = 1$, $y_1 = -3$

11. Obtain Eq. (6.58) directly by substituting $u = t + 2$ into the Laplace transform of $\mathscr{L}\{y(t + 2)\}$.

Another method that is very useful in solving difference equations is a discrete version of the Laplace transform. Define

$$L[y_n] = \sum_{n=0}^{\infty} \frac{y_n}{s^{n+1}} = \frac{y_0}{s} + \frac{y_1}{s^2} + \cdots. \tag{6.62}$$

Then the discrete transform of y_{n+1} is readily obtained, and yields a formula very similar to the differentiation property of Laplace transforms:

$$L[y_{n+1}] = \frac{y_1}{s} + \frac{y_2}{s^2} + \cdots = s\left[\frac{y_0}{s} + \frac{y_1}{s^2} + \frac{y_2}{s^3} + \cdots - \frac{y_0}{s}\right]$$

$$= sL[y_n] - y_0. \tag{6.63}$$

Applying (6.63) twice yields $L[y_{n+2}] = s^2L[y_n] - sy_0 - y_1$, and in a similar manner we can obtain $L[y_{n+3}]$, and so on.

To apply this technique, we need a list of discrete transforms of various expressions. For example,

$$L[a^n] = \frac{1}{s} + \frac{a}{s^2} + \frac{a^2}{s^3} + \cdots = \frac{1}{s}\left[1 + \frac{a}{s} + \left(\frac{a}{s}\right)^2 + \cdots\right]$$

$$= \frac{1}{s(1 - a/s)} = \frac{1}{s - a}. \tag{6.64}$$

Using procedures analogous to those above, prove that:

12. $L[na^n] = \dfrac{a}{(s - a)^2}$

13. $L[(n + 1)a^n] = \dfrac{s}{(s - a)^2}$

14. $L\left[\dfrac{1}{n!}\right] = \dfrac{e^{1/s}}{s}$

15. $L[\cos n\theta] = \dfrac{s - \cos\theta}{(s - \cos\theta)^2 + \sin^2\theta}$

16. $L[\sin n\theta] = \dfrac{\sin\theta}{(s - \cos\theta)^2 + \sin^2\theta}$

[*Hint*: Use DeMoivre's theorem, $e^{in\theta} = \cos n\theta + i \sin n\theta$, and Eq. (6.64) to derive the expressions in Exercises 15 and 16.]

Applying the formulas above, it is easy to solve the difference equation

$$y_{n+1} - y_n = 2^{-n}, \qquad y_0 = 2.$$

By (6.63) and (6.64), we have

$$sL[y_n] - 2 - L[y_n] = \frac{1}{s - \frac{1}{2}}.$$

Gathering like terms, we have

$$L[y_n] = \frac{2s}{(s-1)(s-\frac{1}{2})} = \frac{(\frac{4}{3})}{s-1} + \frac{(\frac{2}{3})}{s-\frac{1}{2}}.$$

Thus

$$y_n = \frac{4}{3} + \frac{2}{3}(2^{-n}).$$

Solve the following difference equations by the method above.

17. $y_{n+1} - 2y_n = 20, y_0 = 100$

18. $y_{n+2} + y_n = 0, y_0 = 0, y_1 = 1000$

19. $y_{n+2} - 3y_{n+1} + 2y_n = 1 + 5^{n+1}, y_0 = y_1 = 0$

20. $y_{n+2} + y_n = n + 1, y_0 = 1, y_1 = -\frac{1}{2}$

ELEMENTARY
SYSTEMS
OF
DIFFERENTIAL
EQUATIONS

7

7.1 INTRODUCTION

In the preceding chapters we discussed the problem of finding one unknown function, the solution of a single differential equation. In this chapter we will discuss systems of simultaneous first order differential equations. Such systems will arise in problems involving more than one unknown function, each of which is a function of a single independent variable, which often is time. For the sake of consistency throughout the chapter, we will usually denote the independent variable by t and the dependent variables (i.e., functions) by $x(t)$, $y(t)$, or by subscripted letters $x_1(t)$, $x_2(t)$, ..., $x_n(t)$.

This chapter is devoted to the examination of the derivation of some simple systems of equations and the calculation of their solutions. No use will be made of matrix methods in the first nine sections of this chapter, although a familiarity with some elementary properties of determinants (see Appendix 4) is desirable. Section 7.1 gives some examples of how simple systems can arise, together with an indication of how some of them can be solved. In later sections we will consider first order systems and illustrate why they are the only systems which need be discussed. Particular attention will be placed on obtaining solutions of linear first order systems. The discussion of systems from a matrix theory point of view is considered in Sections 7.10 through 7.13.

Example 7.1.1 Suppose that a chemical solution flows from one container, at a rate proportional to its volume, into a second container. It flows out from the second container at a constant rate. Let $x(t)$ and $y(t)$ denote the volumes of solution in the first and second containers, respectively, at time t. (The containers may be, for example, cells, in which case we are describing a diffusion process across a cell wall.) To establish the necessary equations, we note that the change in volume equals the difference between input and output in each container. The change in volume is the derivative of volume with respect to time. Since no chemical is flowing into the first container, the change in its volume equals the output:

$$\frac{dx}{dt} = -c_1 x,$$

where c_1 is a positive constant of proportionality. The amount of solution $c_1 x$ flowing out of the first container is the input of the second container. Let c_2 be the constant output of the second container. Then the change in volume in the second container equals the difference between its input and output:

$$\frac{dy}{dt} = c_1 x - c_2.$$

Thus we can describe the flow of solution by means of two differential equations. Since more than one differential equation is involved, we say that we have obtained a *system of differential equations*:

$$\frac{dx}{dt} = -c_1 x,$$

$$\frac{dy}{dt} = c_1 x - c_2,$$
(7.1)

where c_1 and c_2 are positive constants. By a solution of the system (7.1) we shall mean a pair of functions $x(t)$, $y(t)$ which simultaneously satisfy the two equations in (7.1). It is easy to solve this system by solving the two equations successively (this is not usually possible). If we denote the initial volumes in the two containers by $x(0)$ and $y(0)$, respectively, we see that the first equation has the solution

$$x(t) = x(0)e^{-c_1 t}.$$
(7.2)

Substituting (7.2) into the second equation of (7.1), we obtain the equation

$$\frac{dy}{dt} = c_1 x(0)e^{-c_1 t} - c_2,$$

which, upon integration, yields the solution

$$y(t) = y(0) + x(0)(1 - e^{-c_1 t}) - c_2 t.$$
(7.3)

Equations (7.2) and (7.3) together constitute the unique solution of system (7.1) which satisfies the given initial conditions.

Example 7.1.2 Let tank X contain 100 gallons of brine in which 100 pounds of salt is dissolved and tank Y contain 100 gallons of water. Suppose water flows into tank X at the rate of 2 gallons per minute, and the mixture flows from tank X into tank Y at 3 gallons per minute. From Y one gallon is pumped back to X (establishing *feedback*) while 2 gallons are flushed away. We wish to find the amount of salt in both tanks at all time t.

If we let $x(t)$ and $y(t)$ represent the number of pounds of salt in tanks X and Y at time t, and note that the change in weight equals the difference between input and output, we can again derive a system of linear first order equations. Tanks X and Y initially contain $x(0) = 100$ and $y(0) = 0$ pounds of salt, respectively, at time $t = 0$. The quantities $x/100$ and $y/100$ are, respectively, the amounts of salt contained in each gallon of water taken from tanks X and Y at time t. Three gallons are being removed from tank X and added to tank Y, while only one of the three gallons removed from tank Y is put in tank X. Thus we have the system

$$\frac{dx}{dt} = -3\frac{x}{100} + \frac{y}{100}, \qquad x(0) = 100$$

$$\frac{dy}{dt} = 3\frac{x}{100} - 3\frac{y}{100}, \qquad y(0) = 0.$$
(7.4)

We saw in Examples 1.1.5 and 3.11.1 how one of the dependent variables can be eliminated entirely: solving the second equation for x yields

$$x = y + \frac{100}{3} \frac{dy}{dt}. \tag{7.5}$$

Differentiating (7.5) and equating the left-hand side to the right-hand side of the first equation in (7.4), we have

$$\frac{-3x}{100} + \frac{y}{100} = \frac{dx}{dt} = \frac{dy}{dt} + \frac{100}{3} \frac{d^2y}{dt^2}. \tag{7.6}$$

Replacing the x term on the left-hand side of (7.6) by Eq. (7.5) produces the second order linear equation

$$\frac{100}{3} \frac{d^2y}{dt^2} + 2 \frac{dy}{dt} + \frac{2y}{100} = 0, \tag{7.7}$$

The initial conditions for (7.7) are obtained directly from the system (7.4), since $y(0) = 0$ and

$$y'(0) = 3 \frac{x(0)}{100} - 3 \frac{y(0)}{100} = 3. \tag{7.8}$$

Now using the methods of Chapter 3, it is easy to solve (7.7) and (7.8). We obtain

$$y(t) = 50\sqrt{3} \left[\exp\left(\frac{-3 + \sqrt{3}}{100} t \right) - \exp\left(\frac{-3 - \sqrt{3}}{100} t \right) \right]$$

and from (7.5),

$$x(t) = 50 \left[\exp\left(\frac{-3 + \sqrt{3}}{100} t \right) + \exp\left(\frac{-3 - \sqrt{3}}{100} t \right) \right].$$

In Sections 7.3 and 7.4 we will discuss in detail this method of eliminating one of the variables, and in Section 7.5 we will introduce a direct way of obtaining the solution without doing the elimination.

Example 7.1.3 As a third example, we consider the mass-spring system of Fig. 7.1, which is a direct generalization of the system described in Section 3.9. In this example we have two masses suspended by springs in series with spring constants λ_1 and λ_2 (see Section 3.9). If the vertical displacements from equilibrium of the two masses are denoted by $x_1(t)$ and $x_2(t)$, respectively, then using assumptions (a) and (b) (Hooke's law) of Section 3.9, we find that the net forces acting on the two masses are given by

$$F_1 = -\lambda_1 x_1 + \lambda_2(x_2 - x_1),$$
$$F_2 = -\lambda_2(x_2 - x_1).$$

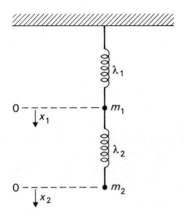

Figure 7.1

Here the positive direction is downward. Note that the first spring is compressed when $x_1 < 0$ and the second spring is compressed when $x_1 > x_2$. The equations of motion are

$$m_1 \frac{d^2 x_1}{dt^2} = -\lambda_1 x_1 + \lambda_2(x_2 - x_1),$$

$$m_2 \frac{d^2 x_2}{dt^2} = -\lambda_2(x_2 - x_1),$$
(7.9)

which comprise a system of two second order linear differential equations with constant coefficients. This system will be discussed further in the next section. From physical considerations we can expect that if the mass is set in motion with a given initial position and initial velocity, then the future motion of the mass on the spring can be predicted.

Example 7.1.4 In Example 1.1.5 we considered an ecosystem with a predator-prey interaction in which the predator species feeds exclusively on the prey while the prey population is assured of an ample food supply. A more general approach is to consider the generalization of the logistic equation (Example 1.1.4) to the case of two species competing for available resources in an ecological niche. More specifically, let $x(t)$ and $y(t)$ denote the populations of the two species at time t. Using Eq. (1.6) as a model, we obtain the system

$$x'(t) = x(\beta_1 - \delta_{11}x - \delta_{12}y),$$

$$y'(t) = y(\beta_2 - \delta_{21}x - \delta_{22}y).$$
(7.10)

The equations (7.10) are known as the *Lotka-Volterra* equations. They have been studied extensively in the field of mathematical ecology. This system is nonlinear, and in a manner analogous to that of Example 1.1.5, there is no way to calculate an explicit solution for (7.10). Nevertheless, a great deal of information about the solutions (which do exist although we don't know what they are) can be gleaned from knowledge of the constants β_1, β_2, δ_{11}, δ_{12}, δ_{21} and δ_{22}. For example, it is easy to find the *equilibrium populations* without solving the system. These are the populations for which there is no growth or decay. Obviously, one such pair of populations consists of the constant populations $x = y = 0$ (where both species are extinct). Other equilibrium populations can be obtained by setting $x' = y' = 0$ in (7.10). If population x or population y is extinct, then the equilibrium populations are $x = 0$, $y = \beta_2/\delta_{22}$ and $x = \beta_1/\delta_{11}$, $y = 0$, respectively. If neither population is extinct (that is, neither x nor y is zero), we divide the first equation of (7.10) by x and the second by y to obtain [using the fact that the right-hand sides of (7.10) are zero] the simultaneous equations

$$\delta_{11}x + \delta_{12}y = \beta_1,$$
$$\delta_{21}x + \delta_{22}y = \beta_2. \tag{7.11}$$

Now, if the determinant $D = \delta_{11}\delta_{22} - \delta_{12}\delta_{21}$ is nonzero, we obtain the unique solution to (7.11):

$$x = \frac{\delta_{22}\beta_1 - \delta_{12}\beta_2}{D}, \qquad y = \frac{\delta_{11}\beta_2 - \delta_{21}\beta_1}{D}. \tag{7.12}$$

The solution (7.12) represents an equilibrium population in which neither species is extinct if both terms in (7.12) are positive.

We will discuss further properties of the Lotka-Volterra equations in the discussion of stability in Chapter 12.

EXERCISES 7.1

1. In Example 7.1.1, suppose that the volume of solution flowing out of the second container is proportional to the volume present in the container. Modify Eqs. (7.1) to take this fact into account.

2. Using the method of Example 7.1.1, solve the system derived in Exercise 1 in terms of the initial volumes $x(0)$ and $y(0)$.

3. In Example 7.1.2, when does tank Y contain a maximal amount of salt? How much salt is in tank Y at that time?

4. Suppose, in Example 7.1.2, that the rate of flow from tank Y to tank X is 2 gallons per minute (instead of one) and all other facts are unchanged. Find the equations for the amount of salt in each tank at all times t.

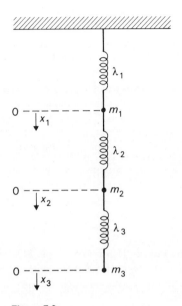

Figure 7.2

5. Consider the mass-spring system illustrated in Fig. 7.2. Here three masses are suspended in series by three springs with spring constants λ_1, λ_2, λ_3, respectively. Formulate a system of second order differential equations that describes this system.

6. Consider an ecosystem with three species competing for existing resources. (a) Derive the Lotka-Volterra equations for this system. (b) Find all equilibrium populations. (c) What conditions must hold in order that there be an equilibrium population with none of the three species extinct?

7. A community of n individuals is exposed to an infectious disease. At any given time t, the community is divided into three groups: group 1 with population $x_1(t)$ is the susceptible group; group 2 with a population of $x_2(t)$ is the group of infected individuals in circulation; and group 3, population $x_3(t)$, consists of those who are isolated, dead, or immune. It is reasonable to assume that initially $x_2(t)$ and $x_3(t)$ will be small compared to $x_1(t)$. Let α and β be positive constants denoting the rates at which susceptibles become infected and infected individuals join group 3, respectively. Then a reasonable model for the spread of the disease is given by the system

$$x_1'(t) = \alpha x_1(0)x_2,$$
$$x_2'(t) = \alpha x_1(0)x_2 - \beta x_2,$$
$$x_3'(t) = \beta x_2.$$

a) Solving the second of these equations first, find the solution in terms of $x_1(0)$, $x_2(0)$, and $x_3(0)$. Note that

$$x_1(0) + x_2(0) + x_3(0) = n.$$

b) Show that if $\alpha x(0) < \beta$, then the disease will not produce an epidemic.

c) What will happen if $\alpha x(0) > \beta$?

8. The presence of temperature inversions and low wind speeds will often trap air pollutants in a mountain valley for an extended period of time. Gaseous sulfur compounds are often a significant air pollution problem, and their study is complicated by their rapid oxidation. Hydrogen sulfide, H_2S, oxidizes into sulfur dioxide, SO_2, which in turn oxidizes into a sulphate. The following model has been proposed* for determining the concentrations $x(t)$ and $y(t)$ of H_2S and SO_2, respectively, in a fixed airshed. Let

$$\frac{dx}{dt} = -\alpha x + \gamma,$$

$$\frac{dy}{dt} = \alpha x - \beta y + \delta,$$

where the constants α and β are the conversion rates of H_2S into SO_2 and SO_2 into sulphate, respectively, and γ and δ are the production rates of H_2S and SO_2, respectively. Solve the equations sequentially and estimate the concentration levels that could be reached under a prolonged air pollution episode.

7.2 THE TRANSFORMATION OF HIGHER ORDER EQUATIONS TO FIRST ORDER SYSTEMS

In the next section we will consider some methods for solving first order linear systems; that is, systems of equations containing only first derivatives. Before doing that, however, it would be nice to know that first order systems are more than a very special case. In fact, we will show in this section that most differential equations (and systems) of *any* order can be reduced, by the introduction of new variables, to a system of first order equations. Once this has been done, it will be evident that the study of first order systems is of critical importance in the study of differential equations. Moreover, we will have good reason to consider only first order systems in the remainder of this chapter.

We begin with some simple examples.

Example 7.2.1 Consider the equation of harmonic motion (Example 3.5.1):

$$x'' + x = 0. \tag{7.13}$$

---

* R. L. Bohac, "A mathematical model for the conversion of sulphur compounds in the Missoula Valley Airshed," *Proceedings of the Montana Academy of Science* (1974).

If we let $y(t) = x'(t)$, then we have

$$y'(t) = x''(t) = -x(t).$$

Thus Eq. (7.13) has been transformed into the system

$$x' = y, \qquad y' = -x. \tag{7.14}$$

If a solution $(x(t), y(t)$ to system (7.14) is found, then the first component $x(t)$ will be a solution of the second order equation (7.13).

Example 7.2.2 Consider the Euler equation (see Section 3.8)

$$t^3x''' + 4t^2x'' - 8tx' + 8x = 0. \tag{7.15}$$

We define

$$x_1(t) = x(t), \qquad x_2(t) = x'(t), \qquad x_3(t) = x''(t).$$

Then $x_1'(t) = x_2(t)$, $x_2'(t) = x_3(t)$, and

$$x_3'(t) = x'''(t) = -\frac{1}{t^3}(4t^2x'' - 8tx' + 8x)$$

$$= -\frac{8}{t^3}x_1 + \frac{8}{t^2}x_2 - \frac{4}{t}x_3.$$

The third order equation (7.15) can now be expressed by the system of three first order equations

$$\begin{aligned}
x_1' &= x_2, \\
x_2' &= x_3, \\
x_3' &= -\frac{8}{t^3}x_1 + \frac{8}{t^2}x_2 - \frac{4}{t}x_3,
\end{aligned} \tag{7.16}$$

and the first component, x_1, of any solution to the system (7.16) is a solution of (7.15).

It is clear that the method of the above two examples can be easily generalized. Consider the most general nth order equation

$$f(t, x, x', x'', \ldots, x^{(n-1)}, x^{(n)}) = 0. \tag{7.17}$$

Suppose that Eq. (7.17) can be rewritten so that $x^{(n)}$ is a function of t and x and its derivatives up to the $(n-1)$st order. Then there exists a function g such that

$$x^{(n)} = g(t, x, x', x'', \ldots, x^{(n-1)}). \tag{7.18}$$

[It is not always possible to rewrite an equation of the form (7.17) into one of the form (7.18), but it is possible in most of the cases of interest. In particular, it is always possible when Eq. (7.17) is linear. A condition which guarantees the existence of such a function g is given by the *implicit function theorem*, discussed in most advanced calculus books.*] For our purposes, we will assume that the general nth order equation can be written in the form (7.18). We then have:

Theorem 7.1 The equation (7.18) can be written as a system of first order equations.

Proof. Define

$$x_1(t) = x(t),$$
$$x_2(t) = x'(t),$$
$$x_3(t) = x''(t),$$

$$\vdots$$

$$x_{n-1}(t) = x^{(n-2)}(t),$$
$$x_n(t) = x^{(n-1)}(t).$$

Then Eq. (7.18) becomes

$$x_1' = x_2,$$
$$x_2' = x_3,$$

$$\vdots \qquad\qquad\qquad\qquad\qquad\qquad\qquad (7.19)$$

$$x_{n-1}' = x_n,$$
$$x_n' = g(t, x_1, x_2, \ldots, x_n).$$

The situation is even easier when the equation is linear. Consider, for example, the equation

$$x^{(n)}(t) + a_1(t)x^{(n-1)}(t) + a_2(t)x^{(n-2)}(t) + \cdots$$
$$+ a_{n-1}(t)x'(t) + a_n(t)x(t) = f(t). \qquad (7.20)$$

Here the same introduction of variables leads to the system

$$x_1' = x_2,$$
$$x_2' = x_3,$$

$$\vdots \qquad\qquad\qquad\qquad\qquad\qquad\qquad (7.21)$$

$$x_{n-1}' = x_n,$$
$$x_n' = -a_n x_1 - a_{n-1} x_2 - \cdots - a_1 x_n + f.$$

---

* See, for example, R. Buck, *Advanced Calculus*, McGraw-Hill, New York, 1956, p. 222.

We have thus far discussed only the reduction of a single higher order equation to a first order system. The method can be applied equally well to systems of higher order equations.

Example 7.2.3 Consider the mass-spring system of Example 7.1.3,

$$x_1'' = -\left(\frac{\lambda_1 + \lambda_2}{m_1}\right)x_1 + \frac{\lambda_2}{m_1}x_2,$$

$$x_2'' = \frac{\lambda_2}{m_2}x_1 - \frac{\lambda_2}{m_2}x_2, \tag{7.22}$$

where we have gathered like terms and divided each equation by the mass.

Let us define the new variables $x_3 = x_1'$ and $x_4 = x_2'$. Then $x_3' = x_1''$, $x_4' = x_2''$ and (7.22) can be expressed as the system of four first order equations

$$x_1' = x_3,$$

$$x_2' = x_4,$$

$$x_3' = -\left(\frac{\lambda_1 + \lambda_2}{m_1}\right)x_1 + \frac{\lambda_2}{m_1}x_2, \tag{7.23}$$

$$x_4' = \frac{\lambda_2}{m_2}x_1 - \frac{\lambda_2}{m_2}x_2.$$

Before ending this section some discussion of initial conditions is needed. This is best done with examples.

Example 7.2.4 Consider the initial value problem

$$x'' + x = 0, \qquad x(0) = 1, \qquad x'(0) = 2.$$

Using the substitution $y = x'$ of Example 7.2.1, we obtain the system

$$x' = y, \qquad y' = -x,$$

with initial conditions $x(0) = 1$, and $y(0) = x'(0) = 2$.

Example 7.2.5 Consider the Euler equation of Example 7.2.2. Suppose we are given the initial conditions $x(2) = 3$, $x'(2) = -6$, $x''(2) = 14$. Then, since $x_1(t) = x(t)$, $x_2(t) = x'(t)$, and $x_3(t) = x''(t)$, Eq. (7.15) becomes the system

$$x_1' = x_2,$$

$$x_2' = x_3,$$

$$x_3' = -\frac{8}{t^3}x_1 + \frac{8}{t^2}x_2 - \frac{4}{t}x_3,$$

and

$$x_1(2) = 3, \qquad x_2(2) = -6, \qquad x_3(2) = 14.$$

It is clear that the method conveyed by the two preceding examples can be easily applied to the transformation of any initial value equation into a parallel initial value system.

EXERCISES 7.2

In Exercises 1 through 7 transform each given equation into a system of first order equations.

1. $x'' + 2x' + 3x = 0$

2. $x'' - 6tx' + 3t^3x = \cos t$

3. $x''' - x'' + (x')^2 - x^3 = t.$

4. $x^{iv} - \cos x(t) = t$

5. $x''' + xx'' - x'x^4 = \sin t$

6. $xx'x''x''' = t^5$

7. $x''' - 3x'' + 4x' - x = 0$

8. A mass m moves in xyz-space according to the equations of motion

$$mx'' = f(t,x,y,z),$$
$$my'' = g(t,x,y,z),$$
$$mz'' = h(t,x,y,z).$$

Transform these equations into a system of six first order equations.

9. Consider the system

$$x_1' = x_1, \qquad x_2' = x_2.$$

a) What is the general solution of this system?

b) Show that there is no second order equation equivalent to this system. [*Hint*: Show that any second order equation has solutions which are not solutions of this system.] This shows that first order systems are more general than higher order equations in the sense that any of the latter can be written as a first order system, but not vice versa.

7.3 THE METHOD OF ELIMINATION FOR LINEAR SYSTEMS WITH CONSTANT COEFFICIENTS

In this section we consider a simple method for finding a solution of the linear system

$$x_1' = a_{11}x_1 + a_{12}x_2 + \cdots + a_{1n}x_n + f_1(t),$$
$$x_2' = a_{21}x_1 + a_{22}x_2 + \cdots + a_{2n}x_n + f_2(t),$$
$$\vdots$$

$$(7.24)$$

$$x_n' = a_{n1}x_1 + a_{n2}x_2 + \cdots + a_{nn}x_n + f_n(t),$$

where the a_{ij} are constants ($i = 1, 2, \ldots, n$; $j = 1, 2, \ldots, n$). The technique involves the elimination of all but one of the dependent variables $x_1, \ldots, x_n$ by means of algebraic substitution and differentiation. Actually the reader should already have encountered this method, since it was used in reducing each of the first order systems in Examples 1.1.5, 3.11.1, and 7.1.2, to a second order differential equation. In essence, this procedure reverses the process described in Section 7.2.

For simplicity, we will only consider the linear system of two equations

$$x' = a_{11}x + a_{12}y + f_1,$$
$$y' = a_{21}x + a_{22}y + f_2. \tag{7.25}$$

Differentiating the first of these equations yields

$$x'' = a_{11}x' + a_{12}y' + f'_1.$$

We then substitute the expression for y' in (7.25) to obtain

$$x'' = a_{11}x' + a_{12}(a_{21}x + a_{22}y + f_2) + f'_1. \tag{7.26}$$

But according to the first equation (7.25),

$$y = \frac{1}{a_{12}}(x' - a_{11}x - f_1).$$

So upon simplification (7.26) becomes

$$x'' - (a_{11} + a_{22})x' + (a_{11}a_{22} - a_{12}a_{21})x = a_{12}f_2 - a_{22}f_1 + f'_1. \tag{7.27}$$

Equation (7.27) is a second order linear differential equation with constant coefficients, and can be solved by the methods of Chapter 3. Note that the computation is valid only when $a_{12} \neq 0$. However, the first equation in (7.25) can be solved immediately for x if $a_{12} = 0$, so that this is not a serious restriction. Finally, observe that once $x(t)$ is explicitly known, $y(t)$ can be easily computed from the first equation of (7.25).

Example 7.3.1 Consider the system

$$x' = x + y; \qquad x(0) = 1, \tag{7.28}$$

Differentiating the first equation and then substituting first into the second and then into the first equation of (7.28), we obtain

$$x'' = x' + y' = x' + x - y = x' + x - (x' - x)$$

or

$$x'' - 2x = 0.$$

Thus

$$x(t) = c_1 \cos \sqrt{2}t + c_2 \sin \sqrt{2}t.$$

Then according to the first equation of (7.28),

$$y = x' - x = -\sqrt{2}c_1 \sin \sqrt{2}t + \sqrt{2}c_2 \cos \sqrt{2}t - c_1 \cos \sqrt{2}t - c_2 \sin \sqrt{2}t$$
$$= (c_2\sqrt{2} - c_1) \cos \sqrt{2}t - (\sqrt{2}c_1 + c_2) \sin \sqrt{2}t.$$

Using the initial conditions, we find that

$$x(0) = c_1 = 1, \qquad y(0) = c_2\sqrt{2} - c_1 = 0, \qquad \text{or } c_2 = 1/\sqrt{2}.$$

Therefore, the unique solution of the system (7.28) is given by

$$x(t) = \cos \sqrt{2}t + \frac{1}{\sqrt{2}} \sin \sqrt{2}t,$$

$$y(t) = -\left(\sqrt{2} + \frac{1}{\sqrt{2}}\right) \sin \sqrt{2}t = -\frac{3}{\sqrt{2}} \sin \sqrt{2}t.$$

Example 7.3.2 Consider the system

$$x' = 2x + y + t,$$
$$y' = x + 2y + t^2. \tag{7.29}$$

Proceding as before, we obtain

$$x'' = 2x' + y' + 1 = 2x' + (x + 2y + t^2) + 1$$
$$= 2x' + x + (2x' - 4x - 2t) + t^2 + 1$$

or

$$x'' - 4x' + 3x = t^2 - 2t + 1 = (t - 1)^2. \tag{7.30}$$

The solution to the homogeneous part of Eq. (7.30) is $x(t) = c_1e^t + c_2e^{3t}$. A particular solution of (7.30) is easily found to be $\frac{1}{3}t^2 + \frac{2}{9}t + \frac{11}{27}$, so the general solution of Eq. (7.30) is

$$x(t) = c_1e^t + c_2e^{3t} + \tfrac{1}{3}t^2 + \tfrac{2}{9}t + \tfrac{11}{27}.$$

As before, since $y = x' - 2x - t$, we obtain

$$y(t) = c_1e^t + 3c_2e^{3t} + \tfrac{2}{3}t + \tfrac{2}{9} - 2c_1e^t - 2c_2e^{3t} - \tfrac{2}{3}t^2 - \tfrac{4}{9}t - \tfrac{22}{27} - t$$

or

$$y(t) = -c_1e^t + c_2e^{3t} - \tfrac{2}{3}t^2 - \tfrac{7}{9}t - \tfrac{16}{27}.$$

The reader should not be too surprised to discover that solutions of linear systems with constant coefficients seem to be very similar to the solutions of second and higher order equations with constant coefficients. The reason for this similarity will be brought out in the next two sections.

The method illustrated in the last two examples can easily be generalized to apply to linear systems with three or more equations. *A linear system of n first order equations reduces to an nth order linear differential equation,* because it generally requires one differentiation to eliminate each variable $x_2, \ldots, x_n$ from the system.

Finally, from Example 1.1.5 we recall that *some* nonlinear systems can also be reduced to a single (generally nonlinear) equation. The following example shows that this equation need not always be nonlinear.

Example 7.3.3 Consider the nonlinear system

$$x' = x + \sin x \cos x + 2y,$$
$$y' = \sin^2 x(x + \sin x \cos x + 2y) + x. \tag{7.31}$$

Proceeding as before, we obtain

$$\begin{aligned} x'' &= x'(1 + \cos^2 x - \sin^2 x) + 2y' \\ &= 2x' \cos^2 x + 2(x + \sin x \cos x + 2y) \sin^2 x + 2x \\ &= 2x' \cos^2 x + 2x' \sin^2 x + 2x = 2x' + 2x \end{aligned}$$

or

$$x'' - 2x' - 2x = 0.$$

EXERCISES 7.3

In Exercises 1 through 9 find the general solution of each given system of equations. When initial conditions are given, find the unique solution.

1. $x' = x + 2y$
   $y' = 3x + 2y$

2. $x' = x + 2y + t - 1$
   $y' = 3x + 2y - 5t - 2$
   $x(0) = 0, y(0) = 4$

3. $x' = -4x - y$
   $y' = x - 2y$

4. $x' = x + y$
   $y' = y$
   $x(0) = 1, y(0) = 0$

5. $x' = 8x - y$
   $y' = 4x + 12y$

6. $x' = 2x + y + 3e^{2t}$
   $y' = -4x + 2y + te^{2t}$

7. $x' = 3x + 3y + t$
   $y' = -x - y + 1$

8. $x' = 4x + y$
   $y' = -8x + 8y$
   $x(\pi/4) = 0, y(\pi/4) = 1$

9. $x' = 12x - 17y$
   $y' = 4x - 4y$

10. Describe a method of elimination to find the general solution of three first order differential equations:

$$x_1' = a_{11}x_1 + a_{12}x_2 + a_{13}x_3 + f_1,$$
$$x_2' = a_{21}x_1 + a_{22}x_2 + a_{23}x_3 + f_2,$$
$$x_3' = a_{31}x_1 + a_{32}x_2 + a_{33}x_3 + f_3.$$

In Exercises 11 and 12 use the method developed in Exercise 10 to find the general solution of each given system.

11. $x_1' = x_1$
    $x_2' = 2x_1 + x_2 - 2x_3$
    $x_3' = 3x_1 + 2x_2 + x_3$

12. $x_1' = x_1 + x_2 + x_3$
    $x_2' = 2x_1 + x_2 - x_3$
    $x_3' = -8x_1 - 5x_2 - 3x_3$

7.4 LINEAR SYSTEMS: THEORY

In this section we will consider the linear system of two first order equations

$$x' = a_{11}(t)x + a_{12}(t)y + f_1,$$
$$y' = a_{21}(t)x + a_{22}(t)y + f_2, \tag{7.32}$$

and the associated homogeneous system (i.e., $f_1 = f_2 = 0$)

$$x' = a_{11}(t)x + a_{12}(t)y,$$
$$y' = a_{21}(t)x + a_{22}(t)y. \tag{7.33}$$

The point of view here will emphasize the similarities between such systems and the linear second order equations discussed in Section 3.2. That there is a parallel between the two theories should not be surprising, since we have already shown in Sections 7.2 and 7.3 that any linear second order equation can always be transformed into a system of the form (7.32), and conversely (if the a_{ij} are constant).

By a *solution* of system (7.32) [or (7.33)] we will mean a *pair* of functions $\{x(t),y(t)\}$ which possess first derivatives and which satisfy the given equations. This is a formal statement of what we have been assuming all along. The following theorem, whose generalization to linear systems of n first order equations will be proved in Chapter 11, ensures the existence and uniqueness of solutions of the system (7.32).

Theorem 7.2 If the functions $a_{11}(t)$, $a_{12}(t)$, $a_{21}(t)$, $a_{22}(t)$, $f_1(t)$, and $f_2(t)$ are continuous, then given any numbers t_0, x_0, and y_0, there exists exactly one solution $\{x(t),y(t)\}$ of (7.32) that satisfies $x(t_0) = x_0$ and $y(t_0) = y_0$.

The pair of functions $\{x_3(t),y_3(t)\}$ is a *linear combination* of the pairs $\{x_1(t),y_1(t)\}$ and $\{x_2(t),y_2(t)\}$ if there exist constants c_1 and c_2 such that the following two equations hold:

$$x_3(t) = c_1 x_1(t) + c_2 x_2(t),$$
$$y_3(t) = c_1 y_1(t) + c_2 y_2(t). \tag{7.34}$$

The next theorem is the systems analogue of Theorem 3.2. Its easy proof is left as an exercise.

Theorem 7.3 If $\{x_1(t), y_1(t)\}$ and $\{x_2(t), y_2(t)\}$ are solutions of the homogeneous equation (7.33), then any linear combination of them is also a solution of the system (7.33).

Example 7.4.1 Consider the system

$$x' = -x + 6y,$$
$$y' = x - 2y. \tag{7.35}$$

It is easy to verify that $\{-2e^{-4t}, e^{-4t}\}$ and $\{3e^t, e^t\}$ are solutions of (7.35). Hence, by Theorem 7.3, the pair $\{-2c_1 e^{-4t} + 3c_2 e^t, c_1 e^{-4t} + c_2 e^t\}$ is a solution of (7.35) for any constants c_1 and c_2.

We define two pairs of functions $\{x_1(t), y_1(t)\}$ and $\{x_2(t), y_2(t)\}$ to be *linearly independent* if whenever the equations

$$c_1 x_1(t) + c_2 x_2(t) = 0,$$
$$c_1 y_1(t) + c_2 y_2(t) = 0, \tag{7.36}$$

hold for all values of t, then $c_1 = c_2 = 0$. In Example 7.4.1, the two given pairs of solutions are linearly independent since $c_1 e^{-4t} + c_2 e^t$ vanishes for all t only when $c_1 = c_2 = 0$.

Given two solutions $\{x_1(t), y_1(t)\}$ and $\{x_2(t), y_2(t)\}$, we define the *Wronskian* of the two solutions by the following determinant:

$$W(t) = \begin{vmatrix} x_1(t) & y_1(t) \\ x_2(t) & y_2(t) \end{vmatrix} = x_1(t)y_2(t) - x_2(t)y_1(t). \tag{7.37}$$

We can then prove the next theorem.

Theorem 7.4 If $W(t) \neq 0$ for every t, then (7.34) is the *general solution* of the homogeneous system (7.33) in the sense that given any solution $\{x^*, y^*\}$ of (7.33), there exist constants c_1 and c_2 such that

$$x^* = c_1 x_1 + c_2 x_2,$$
$$y^* = c_1 y_1 + c_2 y_2. \tag{7.38}$$

Proof. Let t_0 be given and consider the linear system of two equations in the unknown quantities c_1 and c_2:

$$c_1 x_1(t_0) + c_2 x_2(t_0) = x^*(t_0),$$
$$c_1 y_1(t_0) + c_2 y_2(t_0) = y^*(t_0).$$
(7.39)

The determinant of this system is $W(t_0)$, which is nonzero by assumption. Thus there is a unique pair of constants $\{c_1, c_2\}$ satisfying (7.39). By Theorem 7.3,

$$\{c_1 x_1(t) + c_2 x_2(t), \, c_1 y_1(t) + c_2 y_2(t)\}$$

is a solution of (7.33). But by (7.39), this solution satisfies the same initial conditions at t_0 as the solution $\{x^*(t), y^*(t)\}$. By the uniqueness part of Theorem 7.2, these solutions must be identical for all t, and therefore the theorem is proved.

Example 7.4.2 In Example 7.4.1 the Wronskian $W(t)$ is

$$W(t) = \begin{vmatrix} -2e^{-4t} & e^{-4t} \\ 3e^t & e^t \end{vmatrix} = -2e^{-3t} - 3e^{-3t} = -5e^{-3t} \neq 0.$$

Hence we need look no further for the general solution of system (7.35).

In view of the condition required in Theorem 7.4 that the Wronskian $W(t)$ never vanish, we shall consider the properties of the Wronskian more carefully. Let $\{x_1, y_1\}$ and $\{x_2, y_2\}$ be two solutions of the homogeneous system (7.33). Since $W(t) = x_1 y_2 - x_2 y_1$, we have

$$W'(t) = x_1 y_2' + x_1' y_2 - x_2 y_1' - x_2' y_1$$
$$= x_1(a_{21}x_2 + a_{22}y_2) + y_2(a_{11}x_1 + a_{12}y_1) - x_2(a_{21}x_1 + a_{22}y_1)$$
$$\quad - y_1(a_{11}x_2 + a_{12}y_1).$$

Multiplying these expressions through and canceling like terms, we obtain

$$W' = a_{11}x_1 y_2 + a_{22}x_1 y_2 - a_{11}x_2 y_1 - a_{22}x_2 y_1$$
$$= (a_{11} + a_{22})(x_1 y_2 - x_2 y_1) = (a_{11} + a_{22})W.$$

Thus

$$W(t) = W(t_0) \exp\left[\int_{t_0}^{t} [a_{11}(u) + a_{22}(u)] \, du\right].$$
(7.40)

We have shown the next theorem to be true.

Theorem 7.5 Let $\{x_1, y_1\}$ and $\{x_2, y_2\}$ be two solutions of the homogeneous system (7.33). Then the Wronskian $W(t)$ is either always zero or never zero in any interval (since $\exp x \neq 0$ for any x).

We are now ready to state the theorem that links linear independence with a nonvanishing Wronskian (see Lemma 3.3).

Theorem 7.6 Two solutions $\{x_1(t),y_1(t)\}$ and $\{x_2(t),y_2(t)\}$ are linearly independent if and only if $W(t) \neq 0$.

Proof. Let the solutions be linearly independent and suppose $W(t) = 0$. Then $x_1 y_2 = x_2 y_1$ or $x_1/x_2 = y_1/y_2 = c$ for some constant c. Then $x_1 = cx_2$ and $y_1 = cy_2$, so that the solutions are dependent, which is a contradiction. Hence $W(t) \neq 0$. Conversely, let $W(t) \neq 0$. If the solutions were dependent, then there would exist constants c_1 and c_2, not both zero, such that

$$c_1 x_1 + c_2 x_2 = 0,$$
$$c_1 y_1 + c_2 y_2 = 0.$$

Assuming that $c_1 \neq 0$, we then have $x_1 = cx_2$, $y_1 = cy_2$, where $c = -c_2/c_1$. But then

$$W(t) = x_1 y_2 - x_2 y_1 = cx_2 y_2 - cx_2 y_2 = 0,$$

which is again a contradiction. Therefore the solutions are linearly independent.

We may summarize the contents of the previous four theorems in the following statement. *Let $\{x_1,y_1\}$ and $\{x_2,y_2\}$ be solutions of the homogeneous linear system*

$$x' = a_{11}x + a_{12}y,$$
$$y' = a_{21}x + a_{22}y. \tag{7.41}$$

Then $\{c_1 x_1 + c_2 x_2, c_1 y_1 + c_2 y_2\}$ will be the general solution of the system (7.41) *provided that $W(t) \neq 0$*; that is, provided the solutions $\{x_1,y_1\}$ and $\{x_2,y_2\}$ are linearly independent.

Finally, let us consider the nonhomogeneous system (7.32). The following theorem is the direct analogue of Theorem 3.5. Its proof, easily patterned after that of Theorem 3.5, is left as an exercise.

Theorem 7.7 Let $\{x^*,y^*\}$ be the general solution of the system (7.32), and let $\{x_p,y_p\}$ be any solution of (7.32). Then $\{x^* - x_p, y^* - y_p\}$ is the general solution of the homogeneous equation (7.41). In other words, the general solution of (7.32) can be written as the sum of the general solution of the homogeneous system (7.41) and any particular solution of the nonhomogeneous system (7.32).

Example 7.4.3 Consider the system

$$x' = 3x + 3y + t,$$
$$y' = -x - y + 1. \tag{7.42}$$

We could solve this system by the methods of the previous section. Here we note first that $\{1, -1\}$ and $\{-3e^{2t}, e^{2t}\}$ are solutions to the homogeneous system

$$x' = 3x + 3y,$$
$$y' = -x - y.$$

A particular solution to (7.42) is $\{-\frac{1}{4}(t^2 + 9t + 3), \frac{1}{4}(t^2 + 7t)\}$. The general solution to (7.42) is, therefore,

$$\{x(t), y(t)\} = \{c_1 - 3c_2e^{2t} - \tfrac{1}{4}(t^2 + 9t + 3), -c_1 + c_2e^{2t} + \tfrac{1}{4}(t^2 + 7t)\}.$$

We close this section by noting that, as in Chapter 3, the theorems in this section can easily be generalized to apply to systems of three or more equations.

EXERCISES 7.4

1. a) Show that $\{e^{-3t}, -e^{-3t}\}$ and $\{(1 - t)e^{-3t}, te^{-3t}\}$ are solutions to

    $$x' = -4x - y,$$
    $$y' = x - 2y.$$

    b) Calculate the Wronskian and verify that the solutions are linearly independent.
    c) Write the general solution to the system.

2. a) Show that $\{e^{2t} \cos 2t, -2e^{2t} \sin 2t\}$ and $\{e^{2t} \sin 2t, 2e^{2t} \cos 2t\}$ are solutions of the system

    $$x' = 2x + y,$$
    $$y' = -4x + 2y.$$

    b) Calculate the Wronskian of these solutions and show that they are linearly independent.
    c) Show that $\{\frac{1}{4}te^{2t}, -\frac{11}{4}e^{2t}\}$ is a solution of the nonhomogeneous system

    $$x' = 2x + y + 3e^{2t},$$
    $$y' = -4x + 2y + te^{2t}.$$

    d) Combining (a) and (c), write the general solution of the nonhomogeneous equation in (c).

3. a) Show that $\{\sin t^2, 2t \cos t^2\}$ and $\{\cos t^2, -2t \sin t^2\}$ are solutions of the system

    $$x' = y$$
    $$y' = -4t^2x + \frac{1}{t}y.$$

    b) Show that the solutions are linearly independent.
    c) Show that $W(0) = 0$.
    d) Explain the apparent contradiction of Theorem 7.6.

4. a) Show that $\{\sin \ln t^2, (2/t)\cos \ln t^2\}$ and $\{\cos \ln t^2, -(2/t)\sin \ln t^2\}$ are linearly independent solutions of the system

$$x' = y,$$

$$y' = -\frac{4}{t^2}x - \frac{1}{t}y.$$

   b) Calculate the Wronskian $W(t)$.

5. Prove Theorem 7.3.

6. Prove Theorem 7.7.

7.5 THE SOLUTION OF HOMOGENEOUS LINEAR EQUATIONS WITH CONSTANT COEFFICIENTS: THE METHOD OF DETERMINANTS

As we saw in Section 7.3, the method of elimination can be used to solve systems of linear equations with constant coefficients. Since the algebraic manipulations required can get cumbersome, we shall develop in this section a more efficient method of obtaining the solution of homogeneous systems. Nonhomogeneous systems are discussed in Exercises 9 through 14. Consider the homogeneous system

$$x' = a_{11}x + a_{12}y,$$
$$y' = a_{21}x + a_{22}y, \tag{7.43}$$

where the a_{ij} are constants. Our main tool for solving second order linear homogeneous equations with constant coefficients involved obtaining an auxiliary equation by "guessing" that the solution had the form $y = e^{\lambda x}$.

Parallel to the method of Section 3.4, we guess that there is a solution to the system (7.43) of the form $\{\alpha e^{\lambda t}, \beta e^{\lambda t}\}$, where α, β, and λ are constants yet to be determined. Substituting $x(t) = \alpha e^{\lambda t}$ and $y(t) = \beta e^{\lambda t}$ into (7.43), we obtain

$$x' = \alpha \lambda e^{\lambda t} = a_{11}\alpha e^{\lambda t} + a_{12}\beta e^{\lambda t},$$
$$y' = \beta \lambda e^{\lambda t} = a_{21}\alpha e^{\lambda t} + a_{22}\beta e^{\lambda t}. \tag{7.44}$$

After dividing by $e^{\lambda t}$, we obtain the linear system

$$(a_{11} - \lambda)\alpha + a_{12}\beta = 0,$$
$$a_{21}\alpha + (a_{22} - \lambda)\beta = 0. \tag{7.45}$$

We would like to find values for λ such that the system of Eqs. (7.45) has a solution $\{\alpha, \beta\}$ where α and β are not both zero. According to the theory of determinants (see Appendix 4), such a solution will occur whenever the determinant of the system

$$D = \begin{vmatrix} a_{11} - \lambda & a_{12} \\ a_{21} & a_{22} - \lambda \end{vmatrix} = (a_{11} - \lambda)(a_{22} - \lambda) - a_{21}a_{12} \tag{7.46}$$

equals zero. Solving the equation $D = 0$, we obtain the quadratic equation

$$\lambda^2 - (a_{11} + a_{22})\lambda + (a_{11}a_{22} - a_{21}a_{12}) = 0. \tag{7.47}$$

We define this to be the *auxiliary equation* of the system (7.43). That we are using the same term again is no accident, as we shall now demonstrate. Suppose we differentiate the first equation in (7.43) and eliminate the function $y(t)$:

$$x'' = a_{11}x' + a_{12}(a_{21}x + a_{22}y). \tag{7.48}$$

Then

$$x'' - a_{11}x' - a_{12}a_{21}x = a_{22}a_{12}y = a_{22}(x' - a_{11}x),$$

and gathering like terms, we obtain the homogeneous equation

$$x'' - (a_{11} + a_{22})x' + (a_{11}a_{22} - a_{12}a_{21})x = 0. \tag{7.49}$$

The auxiliary equation for (7.49) is exactly the same as Eq. (7.47). Hence the algebraic steps needed to obtain Eq. (7.49) can be avoided by setting the determinant $D = 0$. The reader will observe that the nonhomogeneous system (7.25) leads to the nonhomogeneous second order equation (7.27), also with the auxiliary equation (7.47).

As in Chapter 3, there are three cases to consider depending on whether the two roots λ_1 and λ_2 of the auxiliary equation are real and distinct, real and equal, or complex conjugates. We will deal with the three cases separately.

CASE 1. *Distinct real roots.* If λ_1 and λ_2 are distinct real numbers, then corresponding to λ_1 and λ_2 we have the solution pairs to the system (7.43) $\{\alpha_1 e^{\lambda_1 t}, \beta_1 e^{\lambda_1 t}\}$ and $\{\alpha_2 e^{\lambda_2 t}, \beta_2 e^{\lambda_2 t}\}$, respectively. To find the constants α_1 and β_1 (not both zero), replace λ in the system of equations (7.45) by the value λ_1. The procedure is repeated for α_2, β_2, and λ_2. We note that these constants are not unique. In fact, for each number λ_1 or λ_2, there are an infinite number of pairs $\{\alpha, \beta\}$ which satisfy (7.45). To see this, we observe that if $\{\alpha, \beta\}$ is a solution pair, then so is $\{c\alpha, c\beta\}$ for any real number c. Finally, the solution pairs given above are linearly independent, since, if not, there exists a constant c such that $\alpha_2 e^{\lambda_2 t} = c\alpha_1 e^{\lambda_1 t}$ and $\beta_2 e^{\lambda_2 t} = c\beta_1 e^{\lambda_1 t}$, which is clearly impossible because $\lambda_1 \neq \lambda_2$. We therefore have proved:

Theorem 7.8 If λ_1 and λ_2 are distinct real roots of Eq. (7.47), then two linearly independent solutions of the system (7.43) are given by $\{\alpha_1 e^{\lambda_1 t}, \beta_1 e^{\lambda_1 t}\}$ and $\{\alpha_2 e^{\lambda_2 t}, \beta_2 e^{\lambda_2 t}\}$, where the pairs $\{\alpha_1, \beta_1\}$ and $\{\alpha_2, \beta_2\}$ are solutions of the system (7.45), with $\lambda = \lambda_1$ and $\lambda = \lambda_2$, respectively.

Example 7.5.1 Consider the system (see Examples 7.4.1 and 7.4.2)

$$x' = -x + 6y,$$
$$y' = x - 2y. \tag{7.50}$$

Here $a_{11} = -1$, $a_{12} = 6$, $a_{21} = 1$, $a_{22} = -2$, and Eq. (7.46) becomes

$$D = \begin{vmatrix} -1 - \lambda & 6 \\ 1 & -2 - \lambda \end{vmatrix} = (\lambda + 2)(\lambda + 1) - 6 = \lambda^2 + 3\lambda - 4 = 0,$$

which has the roots $\lambda_1 = -4$, $\lambda_2 = 1$. For $\lambda_1 = -4$ the system of equations (7.45) yields

$$3\alpha_1 + 6\beta_1 = 0,$$
$$\alpha_1 + 2\beta_1 = 0.$$

Noting that the first equation is a multiple of the second, we see that $\alpha_1 = -2\beta_1$, so that $\{-2,1\}$ is a solution. Hence the first solution pair is $\{-2e^{-4t}, e^{-4t}\}$. Similarly, with $\lambda_2 = 1$, we obtain the equations

$$-2\alpha_2 + 6\beta_2 = 0,$$
$$\alpha_2 - 3\beta_2 = 0,$$

which has a solution $\alpha_2 = 3$, $\beta_2 = 1$. Thus a second linearly independent solution is given by the pair $\{3e^t, e^t\}$. By Theorem 7.4, the general solution is given by the pair

$$\{x(t), y(t)\} = \{-2c_1e^{-4t} + 3c_2e^t, c_1e^{-4t} + c_2e^t\}.$$

CASE 2. *Two equal roots.* When $\lambda_1 = \lambda_2$, one obvious solution pair is $\{\alpha_1 e^{\lambda_1 t}, \beta_1 e^{\lambda_1 t}\}$. The other solution $\{\alpha_2 e^{\lambda_2 t}, \beta_2 e^{\lambda_2 t}\}$ given by Theorem 7.8 is not linearly independent unless $a_{11} = a_{22}$ and $a_{12} = a_{21} = 0$. In the latter case we have the uncoupled system of equations

$$x' = a_{11}x, \qquad y' = a_{22}y,$$

with the linearly independent solution pairs $\{\alpha_1 e^{\lambda_1 t}, 0\}$ and $\{0, \beta_2 e^{\lambda_1 t}\}$, $\lambda_1 = \lambda_2 = a_{11} = a_{22}$. (The equations are said to be uncoupled because each involves only one dependent variable.) On the basis of our results in Chapter 3, we would expect that a second linearly independent solution has the form $\{\alpha_2 te^{\lambda_1 t}, \beta_2 te^{\lambda_1 t}\}$. This, however, does not turn out to be the case. Rather, the second linearly independent solution has the form

$$\{x(t), y(t)\} = \{(\alpha_2 + \alpha_3 t)e^{\lambda_1 t}, (\beta_2 + \beta_3 t)e^{\lambda_1 t}\}. \tag{7.51}$$

To calculate the constants α_2, α_3, β_2, and β_3, it is necessary to substitute back into the original system (7.43). This is best shown by an example.

Example 7.5.2 Consider the system

$$x' = -4x - y,$$
$$y' = x - 2y. \tag{7.52}$$

Equation (7.46) is

$$D = \begin{vmatrix} -4 - \lambda & -1 \\ 1 & -2 - \lambda \end{vmatrix} = (\lambda + 4)(\lambda + 2) + 1 = \lambda^2 + 6\lambda + 9 = 0,$$

which has the double root $\lambda_1 = \lambda_2 = -3$. From the system (7.45), with $\lambda = -3$, we find that

$$-\alpha_1 - \beta_1 = 0,$$
$$\alpha_1 + \beta_1 = 0.$$

A nontrivial solution is $\alpha_1 = 1, \beta_1 = -1$, yielding the solution pair $\{e^{-3t}, -e^{-3t}\}$. We now try to find a solution of the form indicated by Eq. (7.51):

$$\{(\alpha_2 + \alpha_3 t)e^{-3t}, (\beta_2 + \beta_3 t)e^{-3t}\}. \tag{7.53}$$

Substituting the pair (7.53) into system (7.52), we obtain

$$e^{-3t}(\alpha_3 - 3\alpha_2 - 3\alpha_3 t) = -4(\alpha_2 + \alpha_3 t)e^{-3t} - (\beta_2 + \beta_3 t)e^{-3t}$$
$$e^{-3t}(\beta_3 - 3\beta_2 - 3\beta_3 t) = (\alpha_2 + \alpha_3 t)e^{-3t} - 2(\beta_2 + \beta_3 t)e^{-3t}.$$

Equating constant terms and multiples of t and dividing by e^{-3t}, we obtain the linearly dependent system of equations

$$\alpha_3 - 3\alpha_2 = -4\alpha_2 - \beta_2,$$
$$-3\alpha_3 = -4\alpha_3 - \beta_3,$$
$$\beta_3 - 3\beta_2 = \alpha_2 - 2\beta_2,$$
$$-3\beta_3 = \alpha_3 - 2\beta_3.$$

One solution is $\alpha_2 = 1, \beta_2 = -2, \alpha_3 = 1, \beta_3 = -1$. Thus a second solution of of (7.52) is $\{(1 + t)e^{-3t}, (-2 - t)e^{-3t}\}$. It is easy to verify that the solution pairs above are linearly independent, since $|W(t)| = 1$. We summarize these results by stating the following theorem.

Theorem 7.9 Let Eq. (7.47) have two equal real roots $\lambda_1 = \lambda_2$. Then there exist constants $\alpha_1, \alpha_2, \alpha_3, \beta_1, \beta_2,$ and β_3 such that two linearly independent solutions of (7.43) are given by

$$\{x_1(t), y_1(t)\} = \{\alpha_1 e^{\lambda_1 t}, \beta_1 e^{\lambda_1 t}\},$$
$$\{x_2(t), y_2(t)\} = \{(\alpha_2 + \alpha_3 t)e^{\lambda_1 t}, (\beta_2 + \beta_3 t)e^{\lambda_1 t}\}. \tag{7.54}$$

The constants α_1 and β_1 are found as a nontrivial solution of the homogeneous system of Eqs. (7.45), while the other constants are found by substituting the second equation of (7.54) back into the system (7.43).

Remark. In the substitution process, we always obtain a homogeneous system of four equations in the four unknowns α_2, β_2, α_3, and β_3. That this system has nontrivial solutions follows from the fact that the determinant of the system is zero. The proof is left as Exercise 8.

CASE 3. *Complex conjugate roots.* Let $\lambda_1 = a + ib$ and $\lambda_2 = a - ib$, where a and b are real and $b \neq 0$. Then the solution pairs $\{\alpha_1 e^{(a+ib)t}, \beta_1 e^{(a+ib)t}\}$ and $\{\alpha_2 e^{(a-ib)t}, \beta_2 e^{(a-ib)t}\}$ are linearly independent. However, the constants α_1, β_1, α_2, and β_2, obtained from the system (7.45), are complex numbers. To obtain real solution pairs we must proceed as follows. Let $\alpha_1 = A_1 + iA_2$, $\beta_1 = B_1 + iB_2$ and apply Euler's formula, $e^{i\theta} = \cos \theta + i \sin \theta$, to the first complex solution pair, obtaining

$$x(t) = (A_1 + iA_2)e^{at}(\cos bt + i \sin bt),$$
$$y(t) = (B_1 + iB_2)e^{at}(\cos bt + i \sin bt). \tag{7.55}$$

Multiplying and remembering that $i^2 = -1$, we obtain the equations

$$x(t) = e^{at}[(A_1 \cos bt - A_2 \sin bt) + i(A_1 \sin bt + A_2 \cos bt)],$$
$$y(t) = e^{at}[(B_1 \cos bt - B_2 \sin bt) + i(B_1 \sin bt + B_2 \cos bt)]. \tag{7.56}$$

Now, since the coefficients of the system (7.43) are *real*, the only way $\{x,y\}$ can be a solution pair is for all the real terms and similarly the imaginary terms to cancel out. Thus the real parts of x and y must form a solution pair of (7.43), as must the imaginary parts:

$$\{x_1(t), y_1(t)\} = \{e^{at}(A_1 \cos bt - A_2 \sin bt), e^{at}(B_1 \cos bt - B_2 \sin bt)\},$$
$$\{x_2(t), y_2(t)\} = \{e^{at}(A_1 \sin bt + A_2 \cos bt), e^{at}(B_1 \sin bt + B_2 \cos bt)\}. \tag{7.57}$$

The Wronskian of the solution pairs (7.57) is easily computed:

$$W(t) = e^{2at}(A_1 B_2 - A_2 B_1).$$

We want to show that the pairs (7.57) are linearly independent. Suppose otherwise. Then $W(t) = 0$, which means that $A_1 B_2 = A_2 B_1$. This implies that $B_2 \alpha_1 = A_2 \beta_1$ (according to the definition of α_1 and β_1). Now, neither α_1 nor β_1 vanishes, since if either were zero, so would be the other, and the solution pair (7.55) would be trivial. Also A_2 can't vanish, since if it did, so would B_2, and the first equation of (7.45) would prevent λ_1 from being complex. Multiplying the first equation of (7.45) by A_2, using the identity $B_2 \alpha_1 = A_2 \beta_1$, and dividing by α_1, we have

$$(a_{11} - \lambda_1)A_2 + a_{12}B_2 = 0.$$

But then λ_1 again is not complex. Therefore, it is impossible that $W(t)$ could vanish, and we have proved the following theorem.

Theorem 7.10 If Eq. (7.47) has the complex roots $\lambda_1 = a + ib$ and $\lambda_2 = a - ib$, then two linearly independent solution pairs of the system (7.43) are given by (7.57).

Example 7.5.3 Consider the system

$$x' = 4x + y,$$
$$y' = -8x + 8y,$$

(7.58)

with

$$D = \begin{vmatrix} 4 - \lambda & 1 \\ -8 & 8 - \lambda \end{vmatrix} = \lambda^2 - 12\lambda + 40 = 0.$$

The roots of the latter equation are $\lambda_1 = 6 + 2i$ and $\lambda_2 = 6 - 2i$, so that Theorem 7.10 yields the linearly independent solution pairs

$$\{x_1(t), y_1(t)\} = \{e^{6t}(A_1 \cos 2t - A_2 \sin 2t), e^{6t}(B_1 \cos 2t - B_2 \sin 2t)\},$$
$$\{x_2(t), y_2(t)\} = \{e^{6t}(A_1 \sin 2t + A_2 \cos 2t), e^{6t}(B_1 \sin 2t + B_2 \cos 2t)\}.$$

Substituting the first equation into (7.58) yields, after a great deal of algebra, the system of equations

$$(2A_1 - 2A_2 - B_1) \cos 2t - (2A_1 + 2A_2 - B_2) \sin 2t = 0,$$
$$(8A_1 - 2B_1 - 2B_2) \cos 2t - (8A_2 + 2B_1 - 2B_2) \sin 2t = 0.$$

Since t is arbitrary and the functions $\sin 2t$ and $\cos 2t$ are linearly independent, the terms in parentheses must all vanish. A choice of values that will accomplish this goal is $A_1 = 1$, $A_2 = \frac{1}{2}$, $B_1 = 1$, and $B_2 = 3$. Thus two linearly independent solution pairs to (7.58) are $\{e^{6t}(\cos 2t - \frac{1}{2} \sin 2t), e^{6t}(\cos 2t - 3 \sin 2t)\}$ and $\{e^{6t}(\sin 2t + \frac{1}{2} \cos 2t), e^{6t}(\sin 2t + 3 \cos 2t)\}$. The general solution of (7.58) is a linear combination of these two pairs.

Example 7.5.4 Most biological systems are controlled by the production of enzymes or hormones that stimulate or inhibit the secretion of some compound. For example, the pancreatic hormone glucagon stimulates the release of glucose from the liver to the plasma. A rise in blood glucose inhibits the secretion of glucagon but causes an increase in the production of the hormone insulin. Insulin, in turn, aids in the removal of glucose from the blood and in its conversion to glycogen in the muscle tissue. Let G and I be the deviations of plasma glucose and plasma insulin from the normal (fasting) level, respectively. We then have the system

$$\frac{dG}{dt} = -k_{11}G - k_{12}I,$$
$$\frac{dI}{dt} = k_{21}G - k_{22}I,$$

(7.59)

where the positive constants k_{ij} are model parameters, some of which may be determined experimentally. It is known that the system (7.59) exhibits a strongly damped oscillatory behavior, since direct injection of glucose into the blood will produce a fall of blood glucose to a level below fasting in about one and a half hour followed by a rise slightly above the fasting level in about three hours. Hence, the auxiliary equation of (7.59),

$$D = \begin{vmatrix} -k_{11} - \lambda & -k_{12} \\ k_{21} & -k_{22} - \lambda \end{vmatrix} = (k_{11} + \lambda)(k_{22} + \lambda) + k_{12}k_{21}$$

$$= \lambda^2 + (k_{11} + k_{22})\lambda + (k_{11}k_{22} + k_{12}k_{21}) = 0, \tag{7.60}$$

must have complex conjugate roots $-a \pm ib$, with $a = (k_{11} + k_{22})/2$ and $b = \sqrt{k_{12}k_{21} - (k_{11} - k_{22})^2/4}$, since only complex roots can lead to oscillatory behavior. By Theorem 7.10, we have the solution pairs

$$\{G_1,I_1\} = \{e^{-at}(A_1 \cos bt - A_2 \sin bt), e^{-at}(B_1 \cos bt - B_2 \sin bt)\},$$
$$\{G_2,I_2\} = \{e^{-at}(A_1 \sin bt + A_2 \cos bt), e^{-at}(B_1 \sin bt + B_2 \cos bt)\}. \tag{7.61}$$

Since the period of the oscillation is approximately three hours, we may set $b = 2\pi/3$ and measure time in hours. Substituting the first equation of (7.61) into (7.59) we obtain the equations

$$(-aA_1 - bA_2 + k_{11}A_1 + k_{12}B_1) \cos bt$$
$$+ (aA_2 - bA_1 - k_{11}A_2 - k_{12}B_2) \sin bt = 0,$$
$$(-aB_1 - bB_2 - k_{21}A_1 + k_{22}B_1) \cos bt$$
$$+ (aB_2 - bB_1 + k_{21}A_2 - k_{22}B_2) \sin bt = 0.$$

These equations must hold for all t. Thus all the terms in parentheses must vanish. A choice of values for which this occurs is $A_1 = 1$, $A_2 = 0$, $B_1 = (k_{22} - k_{11})/2k_{12}$, and $B_2 = -b/k_{12}$. Then the general solution of the system (7.59) is given by the pair $\{G(t),I(t)\}$ with

$$G(t) = e^{-at}[c_1 \cos bt + c_2 \sin bt],$$

$$\tag{7.62}$$

$$I(t) = e^{-at}\left[\frac{k_{22} - k_{11}}{2k_{12}}(c_1 \cos bt + c_2 \sin bt) + \frac{b}{k_{12}}(c_1 \sin bt - c_2 \cos bt)\right].$$

Assume now that the glucose injection was administered at a time when plasma insulin and glucose were at fasting levels and that the glucose was diffused completely in the blood before the insulin level began to increase ($t = 0$). Then $G(0) = G_0$ equals the ratio of the volume of glucose administered to blood volume, and $I(0) = 0$. Since $G(t)$ is at a maximum when $t = 0$, it follows that $c_1 = G_0$ and $c_2 = 0$. Hence

$$G(t) = G_0 e^{-at} \cos bt. \tag{7.63}$$

But

$$0 = I(0) = \frac{k_{22} - k_{11}}{2k_{12}} G_0,$$

so that $k_{11} = k_{22}$, $b = \sqrt{k_{12}k_{21}} = 2\pi/3$, and

$$I(t) = G_0 \frac{b}{k_{12}} e^{-at} \sin bt. \tag{7.64}$$

If the minimum level $G(\frac{3}{2})$ (<0) is known, then by (7.63),

$$e^{3a/2} = |G(\tfrac{3}{2})|/G_0,$$

so that

$$k_{11} = a = \frac{2}{3} \ln \frac{|G(\frac{3}{2})|}{G_0}.$$

If we determine the plasma insulin at any given time $t_0 > 0$, we can then evaluate the parameters k_{12} and k_{21}.

EXERCISES 7.5

In Exercises 1 through 7 use the method of determinants to find two linearly independent solutions for each given system.

1. $x' = 4x - 3y$
   $y' = 5x - 4y$
2. $x' = 7x + 6y$
   $y' = 2x + 6y$
3. $x' = -x + y$
   $y' = -5x + 3y$
4. $x' = x + y$
   $y' = -x + 3y$
5. $x' = -4x - y$
   $y' = x - 2y$
6. $x' = 4x - 2y$
   $y' = 5x + 2y$
7. $x' = 4x - 3y$
   $y' = 8x - 6y$

8. Substituting the second solution pair (7.54) into the system (7.43), obtain the homogeneous system of linear equations

$$\begin{aligned}
(\lambda - a_{11})\alpha_2 + \alpha_3 - a_{12}\beta_2 &= 0, \\
(\lambda - a_{11})\alpha_3 - a_{12}\beta_3 &= 0, \\
-a_{21}\alpha_2 + (\lambda - a_{22})\beta_2 + \beta_3 &= 0, \\
-a_{21}\alpha_3 + (\lambda - a_{22})\beta_3 &= 0.
\end{aligned} \tag{7.65}$$

a) Show that since $\lambda_1 = \lambda_2 = (a_{11} + a_{22})/2$, the second and fourth equations of (7.65) are identical.

b) Prove that if two equations of a linear system are multiples of each other, then the determinant of the system is zero.

c) Conclude from parts (a) and (b) that the determinant of the system (7.65) is zero, and from this that (7.65) has nontrivial solutions.

9. Consider the nonhomogeneous equations

$$x' = a_{11}x + a_{12}y + f_1,$$
$$y' = a_{21}x + a_{22}y + f_2. \tag{7.66}$$

Let $\{x_1,y_1\}$ and $\{x_2,y_2\}$ be two linearly independent solution pairs of the homogeneous system (7.43). Show that

$$x_p(t) = v_1(t)x_1(t) + v_2(t)x_2(t),$$
$$y_p(t) = v_1(t)y_1(t) + v_2(t)y_2(t),$$

is a particular solution of the system (7.66) if v_1 and v_2 satisfy the equations

$$v_1'x_1 + v_2'x_2 = f_1,$$
$$v_1'y_1 + v_2'y_2 = f_2.$$

This process for finding a particular solution of the nonhomogeneous system (7.65) is called the *variation of constants method for systems*. Note the close parallel between this method and the method of Section 3.7.

In Exercises 10 through 14 use the variation of constants method to find a particular solution for each given nonhomogeneous system.

10. $x' = 2x + y + 3e^{2t}$
    $y' = -4x + 2y + te^{2t}$

11. $x' = 3x + 3y + t$
    $y' = -x - y + 1$

12. $x' = -2x + y$
    $y' = -3x + 2y + 2\sin t$

13. $x' = -x + y + \cos t$
    $y' = -5x + 3y$

14. $x' = 3x - 2y + t$
    $y' = 2x - 2y + 3e^t$

15. In an experiment of cholesterol turnover in humans, radioactive cholesterol $-4 - C^{14}$ was injected intravenously and the total plasma cholesterol and radioactivity were measured. It was discovered that the turnover of cholesterol behaves like a two compartment system.* The compartment consisting of the organs and blood has a rapid turnover, while the turnover in the other compartment is much slower. Assume that the body intakes and excretes all cholesterol through the first compartment. Let $x(t)$ and $y(t)$ denote the deviations from normal cholesterol levels in each compartment. Suppose that the daily fractional transfer coefficient from compartment x is 0.134, of which 0.036 is the input to compartment y, and that the transfer coefficient from compartment y is 0.02.

---

* D. S. Goodman and R. P. Noble, "Turnover of plasma cholesterol in man," *J. Clin. Invest.*, **47** (1968), 231–241.

a) Describe the problem discussed above as a system of homogeneous linear differential equations.
b) Obtain the general solution of the system.

7.6 ELECTRIC CIRCUITS WITH SEVERAL LOOPS

We shall make use of the concepts developed in Sections 2.8 and 4.7 to study electrical networks with two or more coupled closed circuits. The two fundamental principles governing such networks are the two previously stated laws of Kirchhoff:

1. The algebraic sum of all voltage drops around any closed circuit is zero.
2. The algebraic sum of the currents flowing into any junction in the network is zero.

Consider the circuit in Fig. 7.3. There are two loops. By Kirchhoff's voltage law, we obtain

$$L\frac{dI_L}{dt} + RI_R = E, \tag{7.67}$$

$$\frac{Q_C}{C} - RI_R = 0. \tag{7.68}$$

Since $I = dQ/dt$, the second equation may be rewritten as

$$\frac{I_C}{C} - R\frac{dI_R}{dt} = 0. \tag{7.69}$$

Figure 7.3

By Kirchhoff's current law, we have

$$I_L = I_C + I_R, \tag{7.70}$$

which, if substituted into Eq. (7.69), yields, together with (7.67), the nonhomogeneous system of linear first order differential equations

$$\frac{dI_L}{dt} = -\frac{R}{L} I_R + \frac{E}{L},$$

$$\frac{dI_R}{dt} = \frac{I_L}{RC} - \frac{I_R}{RC}. \tag{7.71}$$

The auxiliary equation of this system is

$$D = \begin{vmatrix} -\lambda & -R/L \\ 1/RC & -\lambda - 1/RC \end{vmatrix} = \lambda(\lambda + 1/RC) + 1/LC$$

$$= \lambda^2 + \lambda/RC + 1/LC = 0. \tag{7.72}$$

The roots of (7.72) are $(-L \pm \sqrt{L^2 - 4R^2LC})/2RLC$.

If $L > 4R^2C$, the roots are different negative numbers λ_1, λ_2 (because $\sqrt{L^2 - 4R^2LC} < \sqrt{L^2} = L$) and we obtain the solution pairs

$$\{I_{L1}, I_{R1}\} = \{\alpha_1 e^{\lambda_1 t}, \beta_1 e^{\lambda_1 t}\},$$

$$\{I_{L2}, I_{R2}\} = \{\alpha_2 e^{\lambda_2 t}, \beta_2 e^{\lambda_2 t}\}, \tag{7.73}$$

of the homogeneous system

$$\frac{dI_L}{dt} = -\frac{R}{L} I_R,$$

$$\frac{dI_R}{dt} = \frac{I_L}{RC} - \frac{I_R}{RC}. \tag{7.74}$$

The coefficients α_j and β_j satisfy Eqs. (7.45) so that the first equation becomes

$$-\lambda_j \alpha_j - (R/L)\beta_j = 0, \qquad j = 1, 2, \tag{7.75}$$

which implies that $\alpha_j = -R\beta_j/L\lambda_j$. Hence the general solution of the homogeneous system (7.74) is given by

$$\{R(k_1 e^{\lambda_1 t} + k_2 e^{\lambda_2 t}), -L(k_1 \lambda_1 e^{\lambda_1 t} + k_2 \lambda_2 e^{\lambda_2 t})\}.$$

To find the general solution of (7.71), we must obtain a particular solution of (7.71). Suppose E is constant (i.e., a battery or transformer). Then the pair $\{E/R, E/R\}$ is easily seen to be a particular solution of (7.71). Thus the general solution of the nonhomogeneous system (7.71) is

$$\{I_L, I_R\} = \{R(k_1 e^{\lambda_1 t} + k_2 e^{\lambda_2 t}) + E/R, -L(k_1 \lambda_1 e^{\lambda_1 t} + k_2 \lambda_2 e^{\lambda_2 t}) + E/R\}.$$

From this equation it is clear that as time increases, both currents I_L and I_R tend to E/R since λ_1 and $\lambda_2 < 0$, and hence I_C tends to zero, by (7.70).

If $L = 4R^2 C$, we have a double root $\lambda_1 = -1/2RC$. By Theorem 7.9, the homogeneous equation has the solution pairs

$$\{I_{L1}, I_{R1}\} = \{\alpha_1 e^{\lambda_1 t}, \beta_1 e^{\lambda_1 t}\},$$
$$\{I_{L2}, I_{R2}\} = \{(\alpha_2 t + \alpha_3)e^{\lambda_1 t}, (\beta_2 t + \beta_3)e^{\lambda_1 t}\}.$$

The coefficients α_1, β_1 satisfy Eq. (7.75), so $\alpha_1 = \beta_1/2$. To obtain the other four coefficients α_2, α_3, β_2, and β_3, we must substitute $\{I_{L2}, I_{R2}\}$ into the system (7.74). After some algebra we obtain $\alpha_2 = \beta_2/2$, $\beta_2 + \lambda_1(2\alpha_3 - \beta_3) = 0$. The selection $\alpha_2 = \frac{1}{2}$, $\alpha_3 = 2RC$, $\beta_2 = 1$, $\beta_3 = 2RC$ satisfies these two equations, so that (7.71) has the general solution

$$\{I_L, I_R\} = \left\{\frac{1}{2} e^{\lambda_1 t}[k_1 + k_2(t + 4RC)] + \frac{E}{R}, e^{\lambda_1 t}[k_1 + k_2(t + 2RC)] + \frac{E}{R}\right\}.$$

As time increases, again I_L and I_R tend to E/R while I_C tends to 0.

Finally, if $L < 4R^2 C$, λ_1 and λ_2 are complex conjugates and the solution pairs of the homogeneous system (7.74) are

$$\{I_{L1}, I_{R1}\} = \{e^{at}(A_1 \cos bt - A_2 \sin bt), e^{at}(B_1 \cos bt - B_2 \sin bt)\},$$
$$\{I_{L2}, I_{R2}\} = \{e^{at}(A_1 \sin bt + A_2 \cos bt), e^{at}(B_1 \sin bt + B_2 \cos bt)\},$$

where $a = -1/2RC < 0$, and $b = \sqrt{4R^2 LC - L^2}/2RLC$. By (7.70) these equations yield damped oscillations with I_L and I_R tending to E/R and I_C tending to zero as t tends to infinity.

Example 7.6.1 Suppose $R = 100$ ohms, $L = 1$ henry, $C = 10^{-4}$ farads, and $E = 100$ volts. Clearly $L < 4R^2 C$, so the roots of the auxiliary equation, $a \pm ib$, are complex conjugates. Here

$$a = -\frac{1}{2RC} = -50, \qquad b = \frac{\sqrt{4R^2 LC - L^2}}{2RLC} = 50\sqrt{3},$$

and the two independent solution pairs of the homogeneous equation (7.74) are

$$\{I_{L1}, I_{R1}\}$$
$$= \{e^{-50t}(A_1 \cos 50\sqrt{3}t - A_2 \sin 50\sqrt{3}t), e^{-50t}(B_1 \cos 50\sqrt{3}t - B_2 \sin 50\sqrt{3}t)\},$$
$$\{I_{L2}, I_{R2}\}$$
$$= \{e^{-50t}(A_1 \sin 50\sqrt{3}t + A_2 \cos 50\sqrt{3}t), e^{-50t}(B_1 \sin 50\sqrt{3}t + B_2 \cos 50\sqrt{3}t)\}.$$

Substituting these equations into (7.74) yields the homogeneous system of equations

$$
\begin{aligned}
-50A_1 - 50\sqrt{3}A_2 + 100B_1 &&&= 0, \\
-50\sqrt{3}A_1 + 50A_2 && -100B_2 &= 0, \\
-100A_1 && + 50B_1 - 50\sqrt{3}B_2 &= 0, \\
&100A_2 - 50\sqrt{3}B_1 - 50B_2 &&= 0.
\end{aligned} \tag{7.76}
$$

A selection of values satisfying (7.76) is provided by setting $A_1 = 2$, $A_2 = 0$, $B_1 = 1$, and $B_2 = -\sqrt{3}$. Thus the general solution has the form

$$
\begin{aligned}
\{I_L, I_R\} = \{ &2e^{-50t}(k_1 \cos 50\sqrt{3}t + k_2 \sin 50\sqrt{3}t) + 1, \\
&e^{-50t}[(k_1 - \sqrt{3}k_2) \cos 50\sqrt{3}t + (k_2 + \sqrt{3}k_1) \sin 50\sqrt{3}t] + 1 \}.
\end{aligned}
$$

We have used the fact that in this example $E/R = 1$.

EXERCISES 7.6

1. Let $R = 100$ ohms, $L = 4$ henries, $C = 10^{-4}$ farads, and $E = 100$ volts in the network of Fig. 7.3. Suppose the currents I_R and I_L are both zero at time $t = 0$. Find the currents when $t = 0.001$ second.

2. Let $L = 1$ henry in Exercise 1 and suppose all the other facts are unchanged. Find the currents when $t = 0.001$ second and 0.1 second.

3. Let $L = 8$ henries in Exercise 1 and suppose all the other facts are unchanged. Find the currents when $t = 0.001$ second and 0.1 second.

4. Suppose $E = 100e^{-1000t}$ volts and all the other values are unchanged in Exercise 1. Do:
   a) Exercise 1
   b) Exercise 2
   c) Exercise 3

5. Repeat Exercise 4 for $E = 100 \sin 60\pi t$.

6. Find the current at time t in each loop of the network in Fig. 7.4 given that $E = 100$ volts, $R = 10$ ohms, and $L = 10$ henries.

Figure 7.4

7. Repeat Exercise 6 for $E = 10 \sin t$.

8. Consider the air-core transformer network shown in Fig. 7.5 with $E = 10 \cos t$, $R = 1$, $L = 2$, and mutual inductance $L_* = -1$ (which depends on the relative modes of winding of the two coils involved). Treating the mutual inductance as an inductance for each circuit, find the two circuit currents at all times t assuming they are zero at $t = 0$.

Figure 7.5

7.7 SYSTEMS OF NONLINEAR DIFFERENTIAL EQUATIONS

In Example 7.1.4 we encountered a nonlinear system of differential equations. As mentioned there, it is almost always impossible to derive an explicit solution for such systems. Nevertheless, as we saw in the analysis of the Lotka-Volterra equations, a good deal of information can be derived from an analysis of the equations, without calculating a solution. In this section we shall present another example of a nonlinear system, together with a discussion of some of the properties of its solution. The techniques described here are intended only as an indication of some of the ways in which one can attack a nonlinear problem. A much more detailed analysis of nonlinear systems will appear in Chapter 12.

A. Rapoport describes in his book*, a model for the arms race between two nations. Let x and y denote the expenditures (which could be measured in billions of dollars or rubles, for example) of two nations on arms. Rapoport postulates that the competing nations would each adopt some form of the following strategy: decrease spending at a rate proportional to the amount it is now spending but increase it at a rate proportional to what the opponent is spending, for small expenditures of the opponent, and proportional to the square of the opponent's

---

* A. Rapoport, *Flights, Games and Debates*, University of Michigan Press, Ann Arbor, Michigan, 1960.

expenditures when the latter are larger. A mathematical model for this strategy is given by the system

$$x' = -a_{11}x + a_{12}y + a_{13}y^2,$$
$$y' = a_{21}x - a_{22}y + a_{23}x^2,$$

(7.77)

where $a_{11}, a_{12}, a_{13}, a_{21}, a_{22}$, and a_{23} are positive constants.

The first thing to notice about the system (7.77) is that for "small" values of x and y, it is close to the linear system

$$x' = -a_{11}x + a_{12}y,$$
$$y' = a_{21}x - a_{22}y.$$

(7.78)

To see this, note that if x and y are much smaller than one, then x^2 and y^2 are, by comparison, practically negligible. Thus properties of the system (7.77) can be gleaned from an analysis of the homogeneous, linear system (7.78). In differential equations terminology we say that the system (7.77) is a *perturbation* of system (7.78). A discussion of perturbations of linear systems properly belongs to the theory of stability in Chapter 12 (see Section 12.3). Of course, for larger values of x and y, the square terms in Eqs. (7.77) tend to dominate and the solutions of these equations can be expected to be very different from the solutions of the system (7.78).

Another approach is to find the points of equilibrium. In the arms race problem we can interpret these points to be those at which the arms spending is fixed on both sides. Such equilibrium points, often called *critical points* of the system, occur whenever both derivatives x' and y' simultaneously vanish. One obvious point of equilibrium is $x = y = 0$; that is, neither nation spends anything on arms. It is much more interesting (but less pleasant) to calculate the equilibrium points where both x and y are positive. To find such points, we must solve the nonlinear system of equations

$$-a_{11}x + a_{12}y + a_{13}y^2 = 0,$$
$$a_{21}x + a_{23}x^2 - a_{22}y = 0.$$

(7.79)

The system (7.79) can be solved sequentially. The first equation yields

$$x = \frac{1}{a_{11}}(a_{12}y + a_{13}y^2).$$

(7.80)

Substituting (7.80) into the second equation of (7.79) gives us the equation

$$\frac{a_{21}}{a_{11}}(a_{12}y + a_{13}y^2) + \frac{a_{23}}{a_{11}^2}(a_{12}y + a_{13}y^2)^2 - a_{22}y = 0$$

or, after multiplying and rearranging,

$$\frac{a_{23}a_{13}^2}{a_{11}^2}\,y^4 + \frac{2a_{23}a_{12}a_{13}}{a_{11}^2}\,y^3 + \frac{a_{23}a_{12}^2 + a_{21}a_{13}a_{11}}{a_{11}^2}$$

$$+ \frac{a_{21}a_{12} - a_{22}a_{11}}{a_{11}}\,y = 0.$$

Dividing by y (which is assumed not to be zero), we obtain

$$\frac{a_{23}a_{13}^2}{a_{11}^2}\,y^3 + \frac{2a_{23}a_{12}a_{13}}{a_{11}^2}\,y^2 + \frac{a_{23}a_{12}^2 + a_{21}a_{13}a_{11}}{a_{11}^2}\,y$$

$$+ \frac{a_{21}a_{12} - a_{22}a_{11}}{a_{11}} = 0. \qquad (7.81)$$

Equation (7.81) has three solutions y_1, y_2, y_3, one or more of which may be a positive real number. We can substitute each such positive real solution y^* of (7.81) back into Eq. (7.80) to find x^*, and if x^* is also positive, then we have found an equilibrium point (x^*, y^*). (Unfortunately in mathematics, as in real life, such equilibrium points exist all too rarely.)

Finally, another way to analyze the system (7.77) is to calculate how the spending level of country y varies with the spending level of country x; that is, what is dy/dx? By the chain rule, since

$$\frac{dy}{dx} = \frac{dy}{dt} \cdot \frac{dt}{dx} = \frac{dy/dt}{dx/dt},$$

we have

$$\frac{dy}{dx} = \frac{a_{21}x - a_{22}y + a_{23}x^2}{-a_{11}x + a_{12}y + a_{13}y^2}. \qquad (7.82)$$

Equation (7.82) is a nonlinear differential equation which cannot be solved by any of the methods of Chapter 2. However, we can still use Eq. (7.82) to calculate the *relative* changes of y versus x at different levels of spending. That is, given y and x, we can calculate dy/dx (or dx/dy, if desired).

In Chapter 12 we will delve in more detail into other methods for squeezing information out of nonlinear systems of differential equations. Numerical techniques that can be used to approximate the solution of nonlinear differential equations and systems are discussed in Chapter 8. As we have seen, information is available, but it is often a tedious process to extract it. This is the major reason that so much more time is spent on linear systems, which often have nice, clean, findable solutions. (Unfortunately, this practice is sometimes analogous to the case of the man who looks for his lost coin under the lamppost where there is enough light, rather than in the dark alley where he dropped it,)

EXERCISES 7.7

1. In 1934 Gause* investigated a model for the interaction of two competing species of para-
   mecium in the presence of a limited food supply. He assumed the equations governing
   the population of each species to be a combination of the logistic equation (due to the
   limited food supply) and a death rate proportional to the population of the competing
   species:

   $$\frac{dx}{dt} = a_1 x - b_1 x^2 - c_1 xy,$$

   $$\frac{dy}{dt} = a_2 y - b_2 y^2 - c_2 xy,$$

                    (7.83)

   where the constants a_1, a_2, b_1, b_2, c_1, and c_2 are all nonnegative.

   a) Eliminate one of the variables and obtain a nonlinear second order equation.
   b) Find all the nonzero equilibrium points for Eq. (7.83).
   c) Find a solution to (7.83), given $a_1 = a_2 = b_1 = b_2 = c_1 = c_2 = 1$.
   d) Calculate dy/dx and find a solution, given $a_1 = a_2 = 0$. To what situation would
      this case correspond?

2. Suppose that a population is subject to a measles epidemic. Let $x(t)$ and $y(t)$ be the numbers
   of susceptible and infected individuals at time t, respectively. Bailey† postulated the
   following simple mathematical model:

   $$\frac{dx}{dt} = -\beta xy + \mu,$$

   $$\frac{dy}{dt} = \beta xy - \gamma y,$$

                    (7.84)

   where μ is the birth parameter of new susceptible individuals and β and γ are infection
   and removal (i.e., immunity or death) rates, respectively. Generally, once infected, an
   individual gains immunity from further infection.

   a) Find the positive equilibrium points (x_0, y_0) for Eq. (7.84).
   b) Assume a small deviation from such an equilibrium point: $x = x_0(1 + u)$, $y = y_0(1 + v)$, where u and v are relatively small. Substitute these values into (7.84)
      to obtain the system

   $$\frac{\gamma}{\beta\mu}\frac{du}{dt} = -(u + v + uv),$$

   $$\frac{1}{\gamma}\frac{dv}{dt} = u(1 + v).$$

                    (7.85)

   c) Ignoring the products uv (since they are very small), eliminate u from (7.85) and
      solve the resulting linear system.

---

* G. F. Gause, *The Struggle for Existence*, Williams and Wilkins, Baltimore, 1934.
† N. T. J. Bailey, *The Mathematical Theory of Epidemics*, Hafner, New York, 1957, p. 136.

d) Assuming that the incubation period $1/\gamma$ is approximately two weeks, find the period (as a function of β and μ) from peak to peak between outbreaks of measles.

3. Consider an ideal pendulum consisting of a weightless rod of length l supported at one end and attached to a particle of positive weight at the other end. Suppose that the pendulum is displaced by an angle θ_0 and released. (See Fig. 7.6.) It will then swing in a vertical plane. Since the length of the arc subtended by an angle θ is given by $s = l\theta$, the angular velocity and acceleration are $l\theta'$ and $l\theta''$, respectively. Thus the angular acceleration must equal the component of gravitational force in the tangential direction:

$$l\frac{d^2\theta}{dt^2} = -g \sin \theta, \qquad \theta(0) = \theta_0, \qquad \theta'(0) = 0. \tag{7.86}$$

This nonlinear second order differential equation may be reduced to the nonlinear system of first order equations

$$\frac{d\theta}{dt} = \psi,$$

$$\frac{d\psi}{dt} = -\frac{g}{l} \sin \theta, \tag{7.87}$$

where $\theta(0) = \theta_0$, $\psi(0) = \theta'(0) = 0$.

a) Find the equilibrium points of (7.87) and give their meaning.

b) Multiplying Eqs. (7.87) together, show that

$$\psi^2 = \frac{2g}{l}(\cos \theta - \cos \theta_0) \tag{7.88}$$

and plot this curve for $\theta_0 = 0$, $\pi/4$, $\pi/2$, π, and 2π in the $\theta\psi$-plane (usually called the *phase plane*). What is the meaning of these curves?

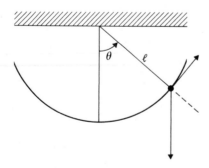

Figure 7.6

c) For small θ_0, replace $\sin \theta$ by θ in (7.87) and solve the resulting linear system. Find the (approximate) time for a complete oscillation of the pendulum.

d) Using the first equation of (7.87) together with (7.88), show that the exact time for each oscillation is

$$T = 4 \sqrt{\frac{l}{2g}} \int_0^{\theta_0} \frac{d\theta}{\sqrt{\cos \theta - \cos \theta_0}}.$$

4. The van der Pol equation

$$x'' + \mu(x^2 - 1)x' + x = 0, \qquad \mu \neq 0, \tag{7.89}$$

arises in the study of oscillatory vacuum tube circuits.

a) Show that (7.89) can be expressed as the system

$$\frac{dx}{dt} = y - \mu\left(\frac{x^3}{3} - x\right),$$

$$\frac{dy}{dt} = -x, \tag{7.90}$$

and obtain an equation in the xy-plane for dy/dx.

*b) Without solving the equation obtained for dy/dx in part (a), plot the graph of a solution of this equation in the phase plane by recalling that $-dx/dy$ is the slope of the normal to the curve.

7.8 LAPLACE TRANSFORM METHODS FOR SYSTEMS

Laplace transform techniques are very useful for solving systems of differential equations with given initial conditions. Consider the system

$$\frac{dx}{dt} = a_{11}x + a_{12}y + f(t),$$

$$\frac{dy}{dt} = a_{21}x + a_{22}y + g(t), \tag{7.91}$$

with initial conditions $x(0) = x_0$, $y(0) = y_0$. Taking the Laplace transform of both equations in (7.91) and letting the corresponding capital letter represent the Laplace transforms of the functions x, y, f, and t, we obtain

$$sX - x_0 = a_{11}X + a_{12}Y + F,$$

$$sY - y_0 = a_{21}X + a_{22}Y + G. \tag{7.92}$$

Gathering all the terms involving X and Y on the left-hand side, we obtain from

(7.92) the system of simultaneous equations

$$(s - a_{11})X - a_{12}Y = F + x_0,$$
$$-a_{21}X + (s - a_{22})Y = G + y_0.$$

(7.93)

If $F + x_0$ and $G + y_0$ are not both identically zero, we may solve (7.93) simultaneously, obtaining

$$X = \frac{(s - a_{22})(F + x_0) + a_{12}(G + y_0)}{s^2 - (a_{11} + a_{22})s + (a_{11}a_{22} - a_{12}a_{21})},$$

$$Y = \frac{(s - a_{11})(G + y_0) + a_{21}(F + x_0)}{s^2 - (a_{11} + a_{22})s + (a_{11}a_{22} - a_{12}a_{21})}.$$

(7.94)

Note that the denominators of the fractions in (7.94) are identical to the auxiliary equation for the system (7.91) which we obtained in Section 7.5. Thus we may rewrite (7.94) in the form

$$X = \frac{(s - a_{22})(F + x_0) + a_{12}(G + y_0)}{(s - \lambda_1)(s - \lambda_2)},$$

$$Y = \frac{(s - a_{11})(G + y_0) + a_{21}(F + x_0)}{(s - \lambda_1)(s - \lambda_2)},$$

(7.95)

where λ_1 and λ_2 are the solutions of the auxiliary equation

$$\lambda^2 - (a_{11} + a_{22})\lambda + (a_{11}a_{22} - a_{12}a_{21}) = 0.$$

We can now find the solution $\{x,y\}$ of the system (7.91) with given initial conditions by inverting (if possible) the equations (7.95).

It should be apparent from the discussion above that the Laplace transform allows us to convert a system of differential equations with given initial conditions to a system of simultaneous equations. This method clearly can be generalized to apply to systems of n linear first order differential equations with constant coefficients, thus yielding a corresponding system of n simultaneous linear equations. As in the case of a single differential equation (see Section 6.3), Laplace transform methods are primarily intended for the solution of systems of linear differential equations with constant coefficients and given initial conditions. Any deviation from these conditions can so complicate the problem as to make no solution obtainable.

Example 7.8.1 Consider the initial value problem

$$x' = x - y + e^t, \qquad x(0) = 1,$$
$$y' = 2x + 3y + e^{-t}, \qquad y(0) = 0.$$

(7.96)

Using the differentiation property of Laplace transforms, we have

$$sX - 1 = X - Y + 1/(s - 1),$$
$$sY = 2X + 3Y + 1/(s + 1). \tag{7.97}$$

The system (7.97) may be rewritten as

$$(s - 1)X + Y = s/(s - 1),$$
$$-2X + (s - 3)Y = 1/(s + 1), \tag{7.98}$$

from which we find that

$$X = \frac{s^2 - 3s - 1}{(s + 1)[(s - 2)^2 + 1]} = \frac{A_1}{s + 1} + \frac{A_2(s - 2) + A_3}{(s - 2)^2 + 1},$$
$$Y = \frac{3s - 1}{(s + 1)[(s - 2)^2 + 1]} = \frac{B_1}{s + 1} + \frac{B_2(s - 2) + B_3}{(s - 2)^2 + 1}. \tag{7.99}$$

Using the methods developed in Theorems 6.8 and 6.10, we obtain $A_1 = \frac{3}{10}$, $A_2 = \frac{7}{10}$, $A_3 = -\frac{11}{10}$, $B_1 = -\frac{2}{5}$, $B_2 = \frac{2}{5}$, and $B_3 = \frac{9}{5}$. Therefore, we have the solution

$$\left\{ \frac{3}{10} e^{-t} + \frac{e^{2t}}{10} (7 \cos t - 11 \sin t), -\frac{2}{5} e^{-t} + \frac{e^{2t}}{5} (2 \cos t + 9 \sin t) \right\}.$$

EXERCISES 7.8

Solve the following initial value problems involving systems of differential equations by the Laplace transform method.

1. $x' = y, x(0) = 1$
   $y' = x, y(0) = 0$
2. $x' = -3x + 4y, x(0) = 3$
   $y' = -2x + 3y, y(0) = 2$
3. $x' = 4x - 2y, x(0) = 2$
   $y' = 5x + 2y, y(0) = -2$
4. $x' = x - 2y, x(0) = 1$
   $y' = 4x + 5y, y(0) = -2$
5. $x' = -3x + 4y + \cos t, x(0) = 0$
   $y' = -2x + 3y + t, y(0) = 1$
6. $x' = 4x - 2y + e^t, x(0) = 1$
   $y' = 5x + 2y - t, y(0) = 0$
7. $x' = x - 2y + t^2, x(0) = 1$
   $y' = 4x + 5y - e^t, y(0) = -1$

7.9 SYSTEMS OF LINEAR DIFFERENCE EQUATIONS

In this section we give a brief description of systems of linear difference equations. It is clear from the parallel development of Chapters 3 and 4 that there should be a difference equations analogue for all the material in this chapter on systems of differential equations, as indeed there is. However, in the interest of brevity, we will present here only a discussion of systems of two linear first order difference equations with constant coefficients and the method of solution by elimination.

It is easy to show (see Exercise 1) that any second or higher order linear difference equation can be reduced to a system of first order equations, as was shown for differential equations in Section 7.2. Therefore, it is necessary to discuss only systems of first order equations. The most general system of two first order linear equations is given by

$$x_{n+1} = a_{11,n}x_n + a_{12,n}y_n + f_n,$$
$$y_{n+1} = a_{21,n}x_n + a_{22,n}y_n + g_n. \tag{7.100}$$

The notation x_n, f_n, $a_{ij,n}$ indicates that the given term is the nth value of the sequence $\{x_n\}$, $\{f_n\}$, or $\{a_{ij,n}\}$. If all the terms in each of the sequences $\{a_{ij,n}\}$ are identical, we may write

$$x_{n+1} = a_{11}x_n + a_{12}y_n + f_n,$$
$$y_{n+1} = a_{21}x_n + a_{22}y_n + g_n. \tag{7.101}$$

This is a linear system *with constant coefficients*. If all the terms f_n and g_n are zero, we say the system (7.100) [or (7.101)] is *homogeneous*; otherwise, it is *nonhomogeneous*.

To solve the constant coefficients equation, we begin by noting that

$$x_{n+2} = a_{11}x_{n+1} + a_{12}y_{n+1} + f_{n+1}. \tag{7.102}$$

Then using the second equation of (7.101) and substituting the expression for y_n into the first equation, we have

$$x_{n+2} = a_{11}x_{n+1} + a_{12}\left[a_{21}x_n + \frac{a_{22}}{a_{12}}(x_{n+1} - a_{11}x_n - f_n) + g_n\right] + f_{n+1}.$$

Rearranging and combining terms, we obtain

$$x_{n+2} - (a_{11} + a_{22})x_{n+1} + (a_{11}a_{22} - a_{12}a_{21})x_n = f_{n+1} + a_{12}g_n - a_{22}f_n. \tag{7.103}$$

This is a linear second order difference equation with constant coefficients that can be solved by the methods of Chapter 4. Once x_n is known, we can easily substitute it into (7.101) to find y_n. Note the analogy between this technique and the technique described in Section 7.3 for systems of differential equations. We shall illustrate this technique with a number of examples.

Example 7.9.1 Consider the homogeneous system

$$x_{n+1} = 2x_n + y_n,$$
$$y_{n+1} = x_n + 2y_n. \tag{7.104}$$

Following the procedure outlined above, we have

$$x_{n+2} = 2x_{n+1} + y_{n+1} = 2x_{n+1} + x_n + 2y_n$$
$$= 2x_{n+1} + x_n + 2(x_{n+1} - 2x_n)$$

or

$$x_{n+2} - 4x_{n+1} + 3x_n = 0. \tag{7.105}$$

The auxiliary equation corresponding to the second order equation (7.105) is

$$\lambda^2 - 4\lambda + 3 = 0,$$

with the roots $\lambda_1 = 3$, $\lambda_2 = 1$. Therefore, the general solution for x_n is

$$x_n = c_1 3^n + c_2. \tag{7.106}$$

Since $x_{n+1} = c_1 3^{n+1} + c_2$, we have, from the first equation of (7.104),

$$y_n = x_{n+1} - 2x_n = (c_1 3^{n+1} + c_2) - 2(c_1 3^n + c_2) = c_1 3^n - c_2.$$

The general solution to (7.104) is therefore

$$x_n = c_1 3^n + c_2, \qquad y_n = c_1 3^n - c_2.$$

Note that if initial values are given, it is easy to find the unique solution. For example, if $x_0 = 1$ and $y_0 = 2$, then

$$x_0 = c_1 + c_2 = 1, \qquad y_0 = c_1 - c_2 = 2,$$

so that $c_1 = \frac{3}{2}$, $c_2 = -\frac{1}{2}$, and the unique solution to the initial value problem is

$$x_n = \tfrac{3}{2}3^n - \tfrac{1}{2} = \tfrac{1}{2}(3^{n+1} - 1),$$
$$y_n = \tfrac{3}{2}3^n + \tfrac{1}{2} = \tfrac{1}{2}(3^{n+1} + 1).$$

Example 7.9.2 Consider the homogeneous system

$$x_{n+1} = x_n + y_n,$$
$$\tag{7.107}$$
$$y_{n+1} = -4x_n + 5y_n.$$

Proceeding as before, we have

$$x_{n+2} = x_{n+1} + y_{n+1} = x_{n+1} - 4x_n + 5y_n = x_{n+1} - 4x_n + 5x_{n+1} - 5x_n$$

or

$$x_{n+2} - 6x_{n+1} + 9x_n = 0.$$

The auxiliary equation is

$$\lambda^2 - 6\lambda + 9 = 0,$$

which has the double root $\lambda_1 = \lambda_2 = 3$. Thus

$$x_n = c_1 3^n + c_2 n 3^n.$$

But
$$y_n = x_{n+1} - x_n = c_1 3^{n+1} + c_2(n + 1)3^{n+1} - c_1 3^n - c_2 n 3^n$$
$$= (2c_1 + 3c_2 + 2c_2 n)3^n.$$

The general solution is therefore
$$x_n = 3^n(c_1 + c_2 n),$$
$$y_n = 3^n(2c_1 + 3c_2 + 2c_2 n).$$

Example 7.9.3 Consider the nonhomogeneous system
$$x_{n+1} = -x_n + 2y_n + 5^{n+1},$$
$$y_{n+1} = -3x_n + 4y_n + \tfrac{1}{2}. \tag{7.108}$$

Eliminating the y_n terms, we obtain the nonhomogeneous second order equation
$$x_{n+2} - 3x_{n+1} + 2x_n = 1 + 5^{n+1}.$$

In Example 4.6.1 the solution of this equation was found to be
$$x_n = c_1 + c_2 2^n - n + \frac{5^{n+1}}{12}. \tag{7.109}$$

Now we have $y_n = (x_{n+1} + x_n - 5^{n+1})/2$, so that
$$y_n = c_1 + 3c_2 2^{n-1} - n - \frac{1}{2} - \frac{5^{n+1}}{4}. \tag{7.110}$$

The general solution of (7.108) is given by Eqs. (7.109) and (7.110).

Example 7.9.4 Snow geese mate in pairs in late spring of each year. Each female lays an average of five eggs, of which approximately twenty percent are claimed by predators and foul weather. The goslings, of which sixty percent are female, mature rapidly and are fully developed by the time the annual migration begins. Hunters and disease claim about 300,000 geese and 200,000 ganders annually. Are snow geese in danger of extinction?

Let x_n be the number of adult geese and y_n the number of adult ganders at the end of the nth year. The snow geese population is determined by the following model
$$x_{n+1} = 2.4 \min (x_n, y_n) + x_n - 300,000,$$
$$y_{n+1} = 1.6 \min (x_n, y_n) + y_n - 200,000, \tag{7.111}$$

where the minimum indicates that single males or females do not reproduce. Since (7.111) is nonlinear, we assume that $y_n \leqslant x_n$, and obtain the linear system
$$x_{n+1} = x_n + 2.4y_n - 300,000,$$
$$y_{n+1} = 2.6y_n - 200,000, \tag{7.112}$$

which may be solved sequentially. We solve the second equation of (7.112) first. By (2.67),

$$y_{n+1} = (2.6)^{n+1}y_0 - 200,000 \sum_{k=0}^{n} (2.6)^{n-k}$$

$$= (2.6)^{n+1}y_0 - 200,000 \left[\frac{(2.6)^{n+1} - 1}{(2.6) - 1} \right],$$

so that

$$y_{n+1} = (2.6)^{n+1}(y_0 - 125,000) + 125,000. \tag{7.113}$$

Substituting (7.113) into the first equation of the system (7.112) yields the equation

$$x_{n+1} = x_n + (2.6)^n(2.4y_0 - 300,000), \tag{7.114}$$

with the solution

$$x_{n+1} = x_0 + (2.4y_0 - 300,000) \sum_{k=0}^{n} (2.6)^k$$

$$= \tfrac{3}{2}(y_0 - 125,000)(2.6)^{n+1} + [x_0 + \tfrac{3}{2}(y_0 - 125,000)]. \tag{7.115}$$

From (7.113) and (7.115) it is clear that extinction will occur if $y_0 < 125,000$, an equilibrium will hold for $y_0 = 125,000$, and a population explosion will result if $y_0 > 125,000$. Of course, since the data employed is based on averages, no single number (for example, $y_0 = 125,000$) is critical. However, should the number of ganders fall substantially below 125,000, there would be a definite need to protect the species.

Example 7.9.5 Box models are a very useful and simple tool for describing the spread of air pollution. To construct such a model, we divide the geographical area of interest into a number of possibly unequal rectangular boxes. We shall assume that within each box the concentration of the pollutant is uniform. (This assumption is not realistic unless the box size is very small, but as the reader will note, cruder assumptions will be made later on.) It is desirable to have the following information within each box: production rate and initial concentration of the pollutant, average wind speed and direction at regular (say, hourly) intervals of time, and the height of the atmosphere in which the pollutant mixes (*mixing height*). Generally, only a small part of the information required is available, so further crude assumptions are necessary. The wind speed and direction determines the transfer of pollutants from box to box. This is done by shifting a given box in the direction and distance traveled by the wind over a fixed time interval. The percentage of the moving box that is thus superimposed on the other box yields the fraction of the pollutant transferred from the former to the latter box. The mass of the pollutant in each box is found by multiplying the concentration by the volume (the product of the area and the mixing height). Consider the following situation consisting of three boxes each containing x_n, y_n, and z_n grams of sulfur dioxide, SO_2, at hour n. (See Fig. 7.7.) We then have the equations

$$x_{n+1} = a_{1,n}x_n + b_{1,n}y_n + c_{1,n}z_n + f_n,$$
$$y_{n+1} = a_{2,n}x_n + b_{2,n}y_n + c_{2,n}z_n + g_n, \qquad (7.116)$$
$$z_{n+1} = a_{3,n}x_n + b_{3,n}y_n + c_{3,n}z_n + h_n,$$

where $a_{j,n}$, $b_{j,n}$, and $c_{j,n}$ are the fractional transfer coefficients determined by the wind velocity, and f_n, g_n, and h_n are the amounts of SO_2 produced in each box during the nth hour. Note that $a_{1,n} + a_{2,n} + a_{3,n} \leq 1$, since shifting box x in any direction but East will allow part of the shifted box to lie outside the given three boxes. This is a realistic consideration, since air pollution is seldom a closed system. The system (7.116) generally does not have a closed form solution. However, given x_0, y_0, z_0, the fractional transfer coefficients, and the amounts produced in each box, we can easily calculate the amounts and the uniform concentration of SO_2 in each box for as many hours as desired.

Figure 7.7

EXERCISES 7.9

1. Consider the second order difference equation

$$x_{n+2} + a_n x_{n+1} + b_n x_n = f_n. \qquad (7.117)$$

Show that the substitution $y_n = x_{n+1}$ will transform the equation into a first order system of two difference equations.

In Exercises 2 through 6 transform each given equation or system into a system of first order equations.

2. $x_{n+2} - 3x_{n+1} + 4x_n = 0$

3. $x_{n+2} - nx_{n+1} + n^2 x_n = 3^n$

4. $x_{n+3} + x_{n+2} + x_{n+1} + x_n = \sin n$

5. $x_{n+2} + x_{n+1}x_n = 0$

6. $x_{n+2} = 3x_{n+1} - 4y_{n+1} - 2x_n + y_n$
   $y_{n+2} = -2x_{n+1} + 5y_{n+1} + x_n - 4y_n + 3^n$

In Exercises 7 through 12 solve each given system of difference equations. Find a particular solution when initial conditions are given.

7. $x_{n+1} = -2x_n + y_n$
   $y_{n+1} = 4x_n + y_n$

8. $x_{n+1} = -4x_n + y_n$
   $y_{n+1} = -10x_n + 2y_n$
   $x_0 = 1, y_0 = 4$

9. $x_{n+1} = -x_n - 3y_n$
   $y_{n+1} = \frac{1}{3}x_n + y_n$
   $x_0 = 0, y_0 = 1$

10. $x_{n+1} = x_n - y_n + n$
    $y_{n+1} = 5x_n + y_n + 1$

11. $x_{n+1} = 2x_n - y_n - 1$
    $y_{n+1} = -x_n + 2y_n + 2$
    $x_0 = 1, y_0 = 0$

12. $x_{n+1} = -y_n + 2^n$
    $y_{n+1} = -x_n + 2^n$

13. Let the initial concentrations in boxes x, y, and z of Example 7.9.5 be $2 \times 10^{-7}, 3 \times 10^{-7}$, and 10^{-7} grams per cubic meter (g/m^3), respectively. Suppose that each box is 10 kilometers on a side, the mixing height being 200 meters, and the production rates are $f_n = g_n = h_n = 100$ kilograms per hour for all n. Assume that the following wind data apply to all three boxes.

Hour	0	1	2	3
Average wind speed (kilometers/hr)	5	7	10	5
Mean direction (degrees)	90°	45°	30°	270°

Find the concentrations in each box at the end of one, two, and three hours.

---

*7.10 MATRICES AND SYSTEMS OF LINEAR FIRST ORDER EQUATIONS

In this and the next three sections we shift our emphasis radically and examine systems of differential equations by means of the theory of vectors and matrices. For this reason, we assume from here on that the reader is familiar with the elementary properties of vectors and matrices,* including vector and matrix addition and scalar multiplication, matrix multiplication, the notion of linear dependence and independence of vectors, and the calculation of the inverse of an invertible matrix. Before discussing the relationship between matrices and systems of equations, let us consider the notion of a vector and matrix function.

---

* In the remainder of this chapter, vectors will be represented by boldface lower case letters and matrices by capital letters.

An n-component vector function

$$\mathbf{v}(t) = \begin{bmatrix} v_1(t) \\ v_2(t) \\ \cdot \\ \cdot \\ \cdot \\ v_n(t) \end{bmatrix} \tag{7.118}$$

is an n-vector, each of whose components is a function (usually assumed to be continuous). An $n \times n$ matrix function $A(t)$ is an $n \times n$ matrix

$$A(t) = \begin{bmatrix} a_{11}(t) & a_{12}(t) & \cdots & a_{1n}(t) \\ a_{21}(t) & a_{22}(t) & \cdots & a_{2n}(t) \\ \cdot \\ \cdot \\ \cdot \\ a_{n1}(t) & a_{n2}(t) & \cdots & a_{nn}(t) \end{bmatrix}, \tag{7.119}$$

each of whose n^2 components is a function. We may add and multiply vector and matrix functions in the same way that we add and multiply constant vectors and matrices. Thus if, for example,

$$A(t) = \begin{pmatrix} a_{11}(t) & a_{12}(t) \\ a_{21}(t) & a_{22}(t) \end{pmatrix} \quad \text{and} \quad B(t) = \begin{pmatrix} b_{11}(t) & b_{12}(t) \\ b_{21}(t) & b_{22}(t) \end{pmatrix},$$

then

$$A(t)B(t) = \begin{pmatrix} a_{11}(t)b_{11}(t) + a_{12}(t)b_{21}(t) & a_{11}(t)b_{12}(t) + a_{12}(t)b_{22}(t) \\ a_{21}(t)b_{11}(t) + a_{22}(t)b_{21}(t) & a_{21}(t)b_{12}(t) + a_{22}(t)b_{22}(t) \end{pmatrix}.$$

We may also differentiate and integrate vector and matrix functions, componentwise. Thus if $\mathbf{v}(t)$ is given by (7.118), then

$$\mathbf{v}'(t) = \begin{bmatrix} v_1'(t) \\ v_2'(t) \\ \cdot \\ \cdot \\ \cdot \\ v_n'(t) \end{bmatrix} \quad \text{and} \quad \int_{t_0}^{t} \mathbf{v}(s)\, ds = \begin{bmatrix} \int_{t_0}^{t} v_1(s)\, ds \\ \int_{t_0}^{t} v_2(s)\, ds \\ \cdot \\ \cdot \\ \cdot \\ \int_{t_0}^{t} v_n(s)\, ds \end{bmatrix}.$$

Similarly, if $A(t)$ is given by (7.119), then

$$A'(t) = \begin{bmatrix} a'_{11}(t) & a'_{12}(t) & \cdots & a'_{1n}(t) \\ a'_{21}(t) & a'_{22}(t) & \cdots & a'_{2n}(t) \\ \cdot \\ \cdot \\ \cdot \\ a'_{n1}(t) & a'_{n2}(t) & \cdots & a'_{nn}(t) \end{bmatrix}$$

and

$$\int_{t_0}^{t} A(s)\,ds = \begin{bmatrix} \int_{t_0}^{t} a_{11}(s)\,ds & \int_{t_0}^{t} a_{12}(s)\,ds & \cdots & \int_{t_0}^{t} a_{1n}(s)\,ds \\ \int_{t_0}^{t} a_{21}(s)\,ds & \int_{t_0}^{t} a_{22}(s)\,ds & \cdots & \int_{t_0}^{t} a_{2n}(s)\,ds \\ \cdot \\ \cdot \\ \cdot \\ \int_{t_0}^{t} a_{n1}(s)\,ds & \int_{t_0}^{t} a_{n2}(s)\,ds & \cdots & \int_{t_0}^{t} a_{nn}(s)\,ds \end{bmatrix}.$$

Example 7.10.1   Let

$$\mathbf{v}(t) = \begin{bmatrix} t \\ t^2 \\ \sin t \\ e^t \end{bmatrix}.$$

Then

$$\mathbf{v}'(t) = \begin{bmatrix} 1 \\ 2t \\ \cos t \\ e^t \end{bmatrix} \quad \text{and} \quad \int_0^t \mathbf{v}(s)\,ds = \begin{bmatrix} t^2/2 \\ t^3/3 \\ 1 - \cos t \\ e^t - 1 \end{bmatrix}.$$

In earlier sections we discussed the general system of n first order equations

$$x'_1 = a_{11}(t)x_1 + a_{12}(t)x_2 + \cdots + a_{1n}(t)x_n + f_1(t),$$
$$x'_2 = a_{21}(t)x_1 + a_{22}(t)x_2 + \cdots + a_{2n}(t)x_n + f_2(t),$$

$$\cdot$$
$$\cdot \qquad\qquad\qquad\qquad\qquad\qquad\qquad\qquad (7.120)$$
$$\cdot$$

$$x'_n = a_{n1}(t)x_1 + a_{n2}(t)x_2 + \cdots + a_{nn}(t)x_n + f_n(t),$$

which is nonhomogeneous if at least one of the functions $f_i(t)$, $i = 1, 2, \ldots, n$, is not the zero function, and the associated homogeneous system

$$
\begin{aligned}
x_1' &= a_{11}(t)x_1 + a_{12}(t)x_2 + \cdots + a_{1n}(t)x_n, \\
x_2' &= a_{21}(t)x_1 + a_{22}(t)x_2 + \cdots + a_{2n}(t)x_n, \\
&\;\;\vdots \\
x_n' &= a_{n1}(t)x_1 + a_{n2}(t)x_2 + \cdots + a_{nn}(t)x_n.
\end{aligned}
\tag{7.121}
$$

As was shown in Section 7.2, systems containing higher order equations can always be reduced to systems of first order equations. Hence we shall restrict our discussion to the systems of first order equations (7.120) and (7.121).

We now define the vector function $\mathbf{x}(t)$, the matrix function $A(t)$, and the vector function $\mathbf{f}(t)$ as follows:

$$
\mathbf{x}(t) = \begin{bmatrix} x_1(t) \\ x_2(t) \\ \vdots \\ x_n(t) \end{bmatrix}, \quad
A(t) = \begin{bmatrix} a_{11}(t) & a_{12}(t) & \cdots & a_{1n}(t) \\ a_{21}(t) & a_{22}(t) & \cdots & a_{2n}(t) \\ \vdots & & & \\ a_{n1}(t) & a_{n2}(t) & \cdots & a_{nn}(t) \end{bmatrix}, \quad
\mathbf{f}(t) = \begin{bmatrix} f_1(t) \\ f_2(t) \\ \vdots \\ f_n(t) \end{bmatrix}.
\tag{7.122}
$$

Then, using (7.122), we can rewrite the system (7.120) as the *vector differential equation*

$$
\mathbf{x}'(t) = A(t)\mathbf{x}(t) + \mathbf{f}(t).
\tag{7.123}
$$

The system (7.121) becomes

$$
\mathbf{x}'(t) = A(t)\mathbf{x}(t).
\tag{7.124}
$$

It is clear from what has already been said that any linear differential equation or system can be written in the form (7.124), if it is homogeneous, and in the form (7.123) if it is not. The reason for writing such system in the form (7.123) or (7.124) is that, besides the obvious advantage of compactness of notation, (7.123) and (7.124) behave very much like first order linear differential equations, as we shall see. It will be very easy to work with systems of equations in this way once we get used to the notation.

Example 7.10.2 Consider the system

$$
\begin{aligned}
x_1' &= (2t)x_1 + (\sin t)x_2 - e^t x_3 + e^t, \\
x_2' &= -t^3 x_1 + e^{\sin t} x_2 - (\ln t)x_3 + \cos t, \\
x_3' &= 2x_1 - 5tx_2 + 2tx_3 - \tan t.
\end{aligned}
$$

It can be rewritten as

$$
\begin{pmatrix} x_1 \\ x_2 \\ x_3 \end{pmatrix} = \begin{pmatrix} 2t & \sin t & -e^t \\ -t^3 & e^{\sin t} & -\ln t \\ 2 & -5 & 2t \end{pmatrix} \begin{pmatrix} x_1 \\ x_2 \\ x_3 \end{pmatrix} + \begin{pmatrix} e^t \\ \cos t \\ \tan t \end{pmatrix}.
$$

Example 7.10.3 Consider the general second order linear differential equation

$$
x'' + a(t)x' + b(t)x = f(t).
$$

Defining $x_1 = x$ and $x_2 = x'$, we obtain the equivalent system

$$
x_1' = x_2,
$$
$$
x_2' = -b(t)x_1 - a(t)x_2 + f(t),
$$

which can be rewritten as

$$
\mathbf{x}' = A(t)\mathbf{x} + \mathbf{f}(t),
$$

where

$$
\mathbf{x} = \begin{pmatrix} x_1 \\ x_2 \end{pmatrix}, \qquad A(t) = \begin{pmatrix} 0 & 1 \\ -b(t) & -a(t) \end{pmatrix}, \qquad \mathbf{f}(t) = \begin{pmatrix} 0 \\ f(t) \end{pmatrix}.
$$

Example 7.10.4 Consider the third order equation with constant coefficients

$$
x''' - 6x'' + 11x' - 6x = 0. \tag{7.125}
$$

Defining $x_1 = x$, $x_2 = x'$, and $x_3 = x''$, we obtain the system

$$
x_1' = x_2,
$$
$$
x_2' = x_3,
$$
$$
x_3' = 6x_1 - 11x_2 + 6x_3,
$$

or

$$
\mathbf{x}' = A\mathbf{x}, \tag{7.126}
$$

where

$$
\mathbf{x} = \begin{pmatrix} x_1 \\ x_2 \\ x_3 \end{pmatrix} \qquad \text{and} \qquad A = \begin{pmatrix} 0 & 1 & 0 \\ 0 & 0 & 1 \\ 6 & -11 & 6 \end{pmatrix}.
$$

Let us now consider the initial value problem

$$
\mathbf{x}(t) = A(t)\mathbf{x}(t) + \mathbf{f}(t), \qquad \mathbf{x}(t_0) = \mathbf{x}_0, \tag{7.127}
$$

where

$$\mathbf{x}(t_0) = \begin{bmatrix} x_1(t_0) \\ x_2(t_0) \\ \cdot \\ \cdot \\ \cdot \\ x_n(t_0) \end{bmatrix} \quad \text{and} \quad \mathbf{x}_0 = \begin{bmatrix} x_{10} \\ x_{20} \\ \cdot \\ \cdot \\ \cdot \\ x_{n0} \end{bmatrix}.$$

We say that a vector function

$$\boldsymbol{\varphi}(t) = \begin{bmatrix} \varphi_1(t) \\ \cdot \\ \cdot \\ \cdot \\ \varphi_n(t) \end{bmatrix}$$

is a *solution* to (7.127) if $\boldsymbol{\varphi}$ is differentiable and satisfies the differential equation and the given initial condition. The following theorem will be proved in Chapter 11.

Theorem 7.11 Let $A(t)$ and $\mathbf{f}(t)$ be continuous matrix and vector functions, respectively, on some interval $[a,b]$. (That is, the component functions of both $A(t)$ and $\mathbf{f}(t)$ are continuous). Then there exists a unique vector function $\boldsymbol{\varphi}(t)$ which is a solution to the initial value problem (7.127) on the entire interval $[a,b]$.

Example 7.10.5 Referring to Example 7.10.4, we find by the methods of Chapter 3 that e^t, e^{2t}, and e^{3t} are solutions of Eq. (7.125). Since a solution vector for this problem is

$$\boldsymbol{\varphi}(t) = \begin{pmatrix} x(t) \\ x'(t) \\ x''(t) \end{pmatrix},$$

we see that three vector solutions of (7.126) are

$$\boldsymbol{\varphi}_1(t) = \begin{pmatrix} e^t \\ e^t \\ e^t \end{pmatrix}, \quad \boldsymbol{\varphi}_2(t) = \begin{pmatrix} e^{2t} \\ 2e^{2t} \\ 4e^{2t} \end{pmatrix}, \quad \boldsymbol{\varphi}_3(t) = \begin{pmatrix} e^{3t} \\ 3e^{3t} \\ 9e^{3t} \end{pmatrix}.$$

If we specify the initial condition

$$\mathbf{x}(0) = \begin{pmatrix} 2 \\ -3 \\ 5 \end{pmatrix},$$

then the unique solution vector is easily seen to be

$$\varphi(t) = \begin{pmatrix} -11e^t + 31e^{2t} - 18e^{3t} \\ -11e^t + 62e^{2t} - 54e^{3t} \\ -11e^t + 124e^t - 108e^{3t} \end{pmatrix}.$$

Example 7.10.6 The system

$$x_1' = -4x_1 - x_2,$$
$$x_2' = x_1 - 2x_2,$$
$$x_1(0) = 1, \qquad x_2(0) = 2,$$

can be written as

$$\begin{pmatrix} x_1 \\ x_2 \end{pmatrix} = \begin{pmatrix} -4 & -1 \\ 1 & -2 \end{pmatrix} \begin{pmatrix} x_1 \\ x_2 \end{pmatrix}; \qquad \begin{pmatrix} x_1(0) \\ x_2(0) \end{pmatrix} = \begin{pmatrix} 1 \\ 2 \end{pmatrix}.$$

It can easily be verified that

$$\varphi_1(t) = \begin{pmatrix} e^{-3t} \\ -e^{-3t} \end{pmatrix} \quad \text{and} \quad \varphi_2(t) = \begin{pmatrix} (1 - t)e^{-3t} \\ te^{-3t} \end{pmatrix}$$

are solution vectors. It can also be immediately verified that the unique solution vector that satisfies the given initial conditions is

$$\varphi(t) = \begin{pmatrix} (1 - 3t)e^{-3t} \\ (2 + 3t)e^{-3t} \end{pmatrix}.$$

The central problem of the next two sections is to derive properties of vector solutions and, where possible, to calculate them. In the next section we will show how all solutions to the homogeneous system (7.124) can be represented in a convenient form, and in Section 7.12 how information about the solutions to this homogeneous system can be used to find a particular solution to the nonhomogeneous system (7.123).

EXERCISES 7.10

In Exercises 1 through 6 write each given equation or system in the matrix-vector form (7.123), (7.124), or (7.127).

1. $x_1' = 2x_1 + 3x_2$
    $x_2' = 4x_1 - 6x_2$

2. $x_1' = (\cos t)x_1 - (\sin t)x_2 + e^{t^2}$
    $x_2' = e^t x_1 + 2tx_2 - \ln t$
    $x_1(2) = 3, x_2(2) = 7$

3. $x''' - 2x'' + 4tx' - x = \sin t$

4. $x^{(iv)} + 2x''' - 3x'' + 4x' - 7x = 0$
$x(0) = 1,\ x'(0) = 2,\ x''(0) = 3,\ x'''(0) = 4$

5. $x_1' = 2tx_1 - 3t^2 x_2 + (\sin t)x_3$
$x_2' = 2x_1 - 4x_3 - \sin t$
$x_3' = 17x_2 + 4tx_3 + e^t$

6. $x''' + a(t)x'' + b(t)x' + c(t)x = f(t)$
$x(t_0) = d_1,\ x'(t_0) = d_2,\ x''(t_0) = d_3$

In Exercises 7 through 14 verify that each given vector function is a solution to the given system.

7. $\mathbf{x}' = \begin{pmatrix} 1 & 1 \\ -3 & -1 \end{pmatrix}\mathbf{x}, \quad \varphi(t) = \begin{pmatrix} \cos\sqrt{2}t \\ \sqrt{2}\sin\sqrt{2}t - \cos\sqrt{2}t \end{pmatrix}$

8. $\mathbf{x}' = \begin{pmatrix} 2 & 1 \\ 1 & 2 \end{pmatrix}\mathbf{x} + \begin{pmatrix} t \\ t^2 \end{pmatrix}, \quad \varphi(t) = \begin{pmatrix} e^t + \frac{1}{3}t^2 + \frac{2}{9}t + \frac{11}{27} \\ -e^t - \frac{2}{3}t^2 - \frac{7}{9}t - \frac{16}{27} \end{pmatrix}$

9. $\mathbf{x}' = \begin{pmatrix} -1 & 6 \\ 1 & -2 \end{pmatrix}\mathbf{x}, \quad \varphi(t) = \begin{pmatrix} 3e^t \\ e^t \end{pmatrix}$

10. $\mathbf{x}' = \begin{pmatrix} -4 & -1 \\ 1 & -2 \end{pmatrix}\mathbf{x}, \quad \varphi(t) = \begin{pmatrix} (1 + t)e^{-3t} \\ (-2 - t)e^{-3t} \end{pmatrix}$

11. $\mathbf{x}' = \begin{pmatrix} 4 & 1 \\ -8 & 8 \end{pmatrix}\mathbf{x}, \quad \varphi(t) = \begin{pmatrix} e^{6t}(\cos 2t - \frac{1}{2}\sin 2t) \\ e^{6t}(\cos 2t - 3\sin 2t) \end{pmatrix}$

12. $\mathbf{x}' = \begin{pmatrix} 1 & 1 & 1 \\ -1 & -1 & 0 \\ -1 & 0 & 1 \end{pmatrix}\mathbf{x}, \quad \varphi(t) = \begin{pmatrix} \sin t - \cos t \\ \cos t \\ \cos t \end{pmatrix}$

13. $\mathbf{x}' = \begin{pmatrix} 1 & -1 & 1 & -1 \\ 0 & -1 & 2 & -2 \\ 0 & 0 & 2 & -3 \\ 0 & 0 & 0 & -2 \end{pmatrix}\mathbf{x}, \quad \varphi(t) = \begin{pmatrix} e^{-2t} \\ 2e^{-2t} \\ 3e^{-2t} \\ 4e^{-2t} \end{pmatrix}$

14. $\mathbf{x}' = \begin{pmatrix} 3 & 2 & 1 \\ -1 & 0 & -1 \\ 1 & 1 & 2 \end{pmatrix}\mathbf{x}, \quad \varphi(t) = \begin{pmatrix} e^{2t} + te^{2t} \\ -te^{2t} \\ te^{2t} \end{pmatrix}$

15. Let $\varphi_1(t)$ and $\varphi_2(t)$ be any two vector solutions of the homogeneous system $\mathbf{x}' = A(t)\mathbf{x}$. Show that $\varphi(t) = c_1\varphi_1(t) + c_2\varphi_2(t)$ is also a solution.

16. Let $\varphi_1(t)$ and $\varphi_2(t)$ be vector solutions of the nonhomogeneous system (7.123). Show that their difference,

$$\varphi(t) = \varphi_1(t) - \varphi_2(t),$$

is a solution to the homogeneous system (7.124).

17. Find the derivative and integral of each of the following vector and matrix functions:

a) $\mathbf{x}(t) = (t, \sin t)$

b) $\mathbf{y}(t) = \begin{pmatrix} e^t \\ \cos t \\ \tan t \end{pmatrix}$

c) $A(t) = \begin{pmatrix} \sqrt{t} & t^2 \\ e^{2t} & \sin 2t \end{pmatrix}$

d) $B(t) = \begin{pmatrix} \ln t & e^t \sin t & e^t \cos t \\ t^{5/2} & -\cos t & -\sin t \\ 1/t & te^{t^2} & t^2 e^{t^3} \end{pmatrix}$

---

*7.11 FUNDAMENTAL SETS AND FUNDAMENTAL MATRIX SOLUTIONS OF A HOMOGENEOUS SYSTEM OF DIFFERENTIAL EQUATIONS

In this section we will discuss properties of the homogeneous system

$$\mathbf{x}' = A(t)\mathbf{x}, \tag{7.128}$$

where $\mathbf{x}(t)$ is an n-vector and $A(t)$ is an $n \times n$ matrix.

Let $\boldsymbol{\varphi}_1(t), \boldsymbol{\varphi}_2(t), \ldots, \boldsymbol{\varphi}_m(t)$ be m vector solutions of (7.128). We say that they are *linearly independent* if the equation

$$c_1 \boldsymbol{\varphi}_1(t) + c_2 \boldsymbol{\varphi}_2(t) + \cdots + c_m \boldsymbol{\varphi}_m(t) = \mathbf{0}$$

holds only for $c_1 = c_2 = \cdots = c_m = 0$. Since the system (7.128) is equivalent to an nth order equation, it is natural for us to seek n linearly independent solutions to the system. *Any set of n linearly independent solutions of (7.128) is called a fundamental set of solutions.*

In Sections 7.3 and 7.5 we saw how a fundamental set of solutions (that is, two linearly independent solutions) could be found in the case where A was a 2×2 constant matrix. In particular, by Theorem 7.6, the vectors

$$\boldsymbol{\varphi}_1(t) = \begin{pmatrix} x_1(t) \\ y_1(t) \end{pmatrix} \quad \text{and} \quad \boldsymbol{\varphi}_2(t) = \begin{pmatrix} x_2(t) \\ y_2(t) \end{pmatrix}$$

are a fundamental set of solutions if and only if the Wronskian $W(t)$, defined by (7.37), is nonzero.

Example 7.11.1 Let $\mathbf{x}' = A\mathbf{x}$ where

$$A = \begin{pmatrix} -1 & 6 \\ 1 & -2 \end{pmatrix}.$$

In Example 7.5.1 we verified that

$$\varphi_1(t) = \begin{pmatrix} -2e^{-4t} \\ e^{-4t} \end{pmatrix} \quad \text{and} \quad \varphi_2(t) = \begin{pmatrix} 3e^t \\ e^t \end{pmatrix}$$

are solution vectors. To show that they are fundamental (as we did in Example 7.5.1), suppose that $c_1\varphi_1(t) + c_2\varphi_2(t) = 0$ for every t. Then, in particular, the second component yields $c_1e^{-4t} + c_2e^t = 0$ for all t. For $t = 0$, this implies that $c_1 + c_2 = 0$ or $c_1 = -c_2$. Then $c_1(e^{-4t} - e^t) = 0$ for all t. This is impossible unless $c_1 = -c_2 = 0$, which proves the independence of φ_1 and φ_2.

Example 7.11.2 Let $\mathbf{x}' = A\mathbf{x}$ where

$$A = \begin{pmatrix} -4 & -1 \\ 1 & -2 \end{pmatrix}.$$

In Example 7.5.2 we obtained the two solution vectors

$$\varphi_1(t) = \begin{pmatrix} e^{-3t} \\ -e^{-3t} \end{pmatrix} \quad \text{and} \quad \varphi_2(t) = \begin{pmatrix} (1 + t)e^{-3t} \\ (-2 - t)e^{-3t} \end{pmatrix}.$$

If $c_1\varphi_1(t) + c_2\varphi_2(t) = 0$, then the first component yields $e^{-3t}[c_1 + c_2(1 + t)] = 0$ for all t. Since $e^{-3t} \neq 0$ if $t = 0$, we have $c_1 + c_2 = 0$, while $t = 1$ yields the equation $c_1 + 2c_2 = 0$. These equations are satisfied only when $c_1 = c_2 = 0$, so that $\varphi_1(t)$ and $\varphi_2(t)$ form a fundamental set of solutions.

Example 7.11.3 Consider the system $\mathbf{x}' = A\mathbf{x}$ where

$$A = \begin{pmatrix} 4 & 1 \\ -8 & 8 \end{pmatrix}$$

(see Example 7.5.3). Two solution vectors are

$$\varphi_1(t) = \begin{pmatrix} e^{6t}(\cos 2t - \tfrac{1}{2}\sin 2t) \\ e^{6t}(\cos 2t - 3\sin 2t) \end{pmatrix}, \quad \varphi_2(t) = \begin{pmatrix} e^{6t}(\sin 2t + \tfrac{1}{2}\cos 2t) \\ e^{6t}(\sin 2t + 3\cos 2t) \end{pmatrix}.$$

To show that they are linearly independent, suppose that $c_1\varphi_1(t) + c_2\varphi_2(t) = 0$ for every t. Then for $t = 0$, we have

$$\begin{pmatrix} c_1 \\ c_1 \end{pmatrix} + \begin{pmatrix} \tfrac{1}{2}c_2 \\ 3c_2 \end{pmatrix} = 0$$

or

$$c_1 + \tfrac{1}{2}c_2 = 0,$$
$$c_1 + 3c_2 = 0.$$

Clearly the only solution of this system is $c_1 = c_2 = 0$, and so φ_1 and φ_2 are, indeed, linearly independent.

Example 7.11.4   Let $\mathbf{x}' = A\mathbf{x}$ where

$$A = \begin{pmatrix} 1 & 1 & -2 \\ -1 & 2 & 1 \\ 0 & 1 & -1 \end{pmatrix}.$$

Then it is easy to show that

$$\boldsymbol{\varphi}_1(t) = \begin{pmatrix} e^{-t} \\ 0 \\ e^{-t} \end{pmatrix}, \qquad \boldsymbol{\varphi}_2(t) = \begin{pmatrix} 3e^t \\ 2e^t \\ e^t \end{pmatrix}, \qquad \boldsymbol{\varphi}_3(t) = \begin{pmatrix} e^{2t} \\ 3e^{2t} \\ e^{2t} \end{pmatrix}$$

are a fundamental set of solutions.

Let $\boldsymbol{\varphi}_1, \boldsymbol{\varphi}_2, \ldots, \boldsymbol{\varphi}_n$ be n-vector solutions of (7.128). Let $\Phi(t)$ be the matrix whose columns are the vectors $\boldsymbol{\varphi}_1, \boldsymbol{\varphi}_2, \ldots, \boldsymbol{\varphi}_n$. That is,

$$\Phi(t) = (\boldsymbol{\varphi}_1(t), \ldots, \boldsymbol{\varphi}_n(t)) = \begin{bmatrix} \varphi_{11}(t) & \varphi_{12}(t) & \cdots & \varphi_{1n}(t) \\ \varphi_{21}(t) & \varphi_{22}(t) & \cdots & \varphi_{2n}(t) \\ \cdot & & & \\ \cdot & & & \\ \cdot & & & \\ \varphi_{n1}(t) & \varphi_{n2}(t) & \cdots & \varphi_{nn}(t) \end{bmatrix}. \tag{7.129}$$

Such a matrix is called a *matrix solution* of the system (7.128). Equivalently, *an $n \times n$ matrix function $\Phi(t)$ is a matrix solution of (7.128) if and only if each of its columns is a solution vector of (7.128).* If the vectors $\boldsymbol{\varphi}_1, \boldsymbol{\varphi}_2, \ldots, \boldsymbol{\varphi}_n$ form a fundamental set of solutions (that is, if they are linearly independent), then $\Phi(t)$ is called a *fundamental matrix solution.* In what follows, we shall show that fundamental matrix solutions play a central role in the theory of linear systems of differential equations.

Example 7.11.5   In the four previous examples of this section, the fundamental matrix solutions were found to be, respectively:

1. $\begin{pmatrix} -2e^{-4t} & 3e^t \\ e^{-4t} & e^t \end{pmatrix}$

3. $\begin{pmatrix} e^{6t}(\cos 2t - \frac{1}{2}\sin 2t) & e^{6t}(\sin 2t + \frac{1}{2}\cos 2t) \\ e^{6t}(\cos 2t - 3\sin 2t) & e^{6t}(\sin 2t + 3\cos 2t) \end{pmatrix}$

2. $\begin{pmatrix} e^{-3t} & (1+t)e^{-3t} \\ -e^{-3t} & (-2-t)e^{-3t} \end{pmatrix}$

4. $\begin{pmatrix} e^{-t} & 3e^t & e^{2t} \\ 0 & 2e^t & 3e^{2t} \\ e^{-t} & e^t & e^{2t} \end{pmatrix}$

Note that fundamental matrix solutions are not unique, due to the fact that a solution vector may be multiplied by any constant and still remain a solution. In addition, any linear combination of solutions is again a solution (see Exercise 6). However, we have uniqueness in the *principal matrix solution* $\Psi(t)$, which is defined as that fundamental matrix solution which satisfies the condition

$$\Psi(t_0) = I,\qquad(7.130)$$

where I is the $n \times n$ identity matrix.

We will show later in this section that if $A(t)$ is continuous, then (7.128) always has a unique principal matrix solution. But first, we will demonstrate an easy way to determine whether or not a given matrix solution is a fundamental matrix solution.

Let $\Phi(t)$ be a matrix solution of (7.128). We define the *Wronskian of* $\Phi(t)$, written $W(t)$, *as*

$$W(t) = \det \Phi(t).\qquad(7.131)$$

We should remind the reader of two facts from matrix theory. First, a matrix A is invertible if and only if $\det A \neq 0$. Second, $\det A \neq 0$ if and only if the columns of A are linearly independent. From these facts it follows that $\Phi(t)$ will be a fundamental matrix solution if and only if $W(t)$ is nonzero for some t. We will see in the next theorem that $W(t)$ is either always zero or never zero, so that we can calculate $W(t)$ for some especially simple value of t, say $t = 0$, to determine whether Φ is a fundamental matrix solution. Note that many of these theorems are similar to those proven in Chapter 3. In particular, the reader should compare the present definition (7.131) of the Wronskian with the definitions of the Wronskian in Sections 3.2 and 7.4.

Theorem 7.12. Abel's Formula Let $W(t)$ be the Wronskian of the matrix solution $\Phi(t)$ of the system (7.128). Then

$$W(t) = W(t_0) \exp \left(\int_{t_0}^{t} \operatorname{tr} A(s)\, ds \right),\qquad(7.132)$$

where the trace of A, written $\operatorname{tr} A(t)$, is the sum of the diagonal elements of the matrix $A(t)$:

$$\operatorname{tr} A(t) = a_{11}(t) + a_{22}(t) + \cdots + a_{nn}(t).\qquad(7.133)$$

Proof. We prove this theorem for the case of $A(t)$ being a 2×2 matrix (see Theorem 7.5). The proof for the $n \times n$ case is similar (but more complicated) and is left as an exercise. In the 2×2 case, the system (7.128) and the matrix solution may be written as

$$\begin{pmatrix} x_1 \\ x_2 \end{pmatrix}' = \begin{pmatrix} a_{11}(t) & a_{12}(t) \\ a_{21}(t) & a_{22}(t) \end{pmatrix} \begin{pmatrix} x_1 \\ x_2 \end{pmatrix}, \qquad \Phi(t) = (\boldsymbol{\varphi}_1, \boldsymbol{\varphi}_2) = \begin{pmatrix} \varphi_{11} & \varphi_{12} \\ \varphi_{21} & \varphi_{22} \end{pmatrix}.$$

Since $W(t) = \varphi_{11}\varphi_{22} - \varphi_{12}\varphi_{21}$, the derivative

$$W' = \varphi_{11}\varphi_{22}' + \varphi_{11}'\varphi_{22} - \varphi_{12}\varphi_{21}' - \varphi_{12}'\varphi_{21}$$

may be written in determinant form as

$$W' = \begin{vmatrix} \varphi_{11} & \varphi_{12} \\ \varphi'_{21} & \varphi'_{22} \end{vmatrix} + \begin{vmatrix} \varphi'_{11} & \varphi'_{12} \\ \varphi_{21} & \varphi_{22} \end{vmatrix}. \tag{7.134}$$

But $\varphi'_{11} = a_{11}\varphi_{11} + a_{12}\varphi_{21}$, since $\boldsymbol{\varphi}_1$ is a vector solution, and similarly for φ'_{12}, φ'_{21}, φ'_{22}. Replacing these derivatives in (7.134), we obtain

$$W' = \begin{vmatrix} \varphi_{11} & \varphi_{12} \\ a_{21}\varphi_{11} + a_{22}\varphi_{21} & a_{21}\varphi_{12} + a_{22}\varphi_{22} \end{vmatrix}$$

$$+ \begin{vmatrix} a_{11}\varphi_{11} + a_{12}\varphi_{21} & a_{21}\varphi_{11} + a_{22}\varphi_{21} \\ \varphi_{21} & \varphi_{22} \end{vmatrix}$$

$$= D_1 + D_2. \tag{7.135}$$

But according to the theory of determinants (see Appendix 4), a determinant is unchanged when a multiple of one row is added to another row. Additionally, multiplication of every element in one row by a given constant is equivalent to mutliplying the determinant by that constant. Hence we may multiply the first row of D_1 by $-a_{21}$ and add it to the second row. Then

$$D_1 = \begin{vmatrix} \varphi_{11} & \varphi_{12} \\ a_{22}\varphi_{21} & a_{22}\varphi_{22} \end{vmatrix} = a_{22} \begin{vmatrix} \varphi_{11} & \varphi_{12} \\ \varphi_{21} & \varphi_{22} \end{vmatrix} = a_{22}W.$$

Similarly, $D_2 = a_{11}W$. Thus

$$W'(t) = [a_{11}(t) + a_{22}(t)]W(t) = [\text{tr } A(t)]W(t). \tag{7.136}$$

Equation (7.136) is a first order (scalar) differential equation that has the solution

$$W(t) = W(t_0) \exp\left(\int_{t_0}^{t} \text{tr } A(s)\, ds\right).$$

This completes the proof for the 2×2 case. Although the proof is much longer than that of Theorem 7.5, it can be immediately extended to the $n \times n$ case (see Exercises 11 and 12).

Example 7.11.6 We consider the four matrix solutions of Example 7.11.5. Evaluating each at $t = 0$, we obtain:

1. $W(0) = \begin{vmatrix} -2 & 3 \\ 1 & 1 \end{vmatrix} = -5$

   2. $W(0) = \begin{vmatrix} 1 & 1 \\ -1 & -2 \end{vmatrix} = -1$

3. $W(0) = \begin{vmatrix} 1 & \frac{1}{2} \\ 1 & 3 \end{vmatrix} = \frac{5}{2}$

   4. $W(0) = \begin{vmatrix} 1 & 3 & 1 \\ 0 & 2 & 3 \\ 1 & 1 & 1 \end{vmatrix} = 6$

Therefore, without direct verification of linear independence, we nevertheless see that all four matrix solutions are fundamental matrix solutions.

We are now ready to prove the theorem mentioned earlier, namely that principal matrix solutions exist and are unique. Since principal matrix solutions are fundamental matrix solutions, this theorem also proves the existence of fundamental matrix solutions.

Theorem 7.13 Let $A(t)$ be continuous on some interval $[a,b]$. Then for any t_0, $a \leqslant t_0 \leqslant b$, there exists a unique fundamental matrix solution $\Psi(t)$ of the system (7.128) satisfying the condition $\Psi(t_0) = I$.

Proof. Let δ_i, $i = 1, 2, \ldots, n$, denote the n-column vector that has a one in the ith position (row) and a zero everywhere else:

$$\delta_1 = \begin{bmatrix} 1 \\ 0 \\ 0 \\ . \\ . \\ . \\ 0 \end{bmatrix}, \quad \delta_2 = \begin{bmatrix} 0 \\ 1 \\ 0 \\ . \\ . \\ . \\ 0 \end{bmatrix}, \quad \ldots, \quad \delta_n = \begin{bmatrix} 0 \\ 0 \\ . \\ . \\ 0 \\ 1 \end{bmatrix}.$$

By the basic existence-uniqueness Theorem 7.11 with $\mathbf{f}(t) \equiv \mathbf{0}$, there exists a unique vector solution $\varphi_i(t)$ of (7.128) that satisfies $\varphi_i(t_0) = \delta_i$, $i = 1, 2, \ldots, n$. Define the matrix function

$$\Psi(t) = (\varphi_1(t), \varphi_2(t), \ldots, \varphi_n(t)). \tag{7.137}$$

Then $\Psi(t)$ is the matrix whose columns are the vector solutions φ_i, $i = 1, 2, \ldots, n$. Clearly $\Psi(t)$ is a matrix solution and $\Psi(t_0) = (\varphi_1(t_0), \ldots, \varphi_n(t_0)) = (\delta_1, \delta_2, \ldots, \delta_n) = I$. It remains to be shown that $\Psi(t)$ is a fundamental matrix solution. This is easy to do. We simply note that $\det \Psi(t_0) = \det I = 1 \neq 0$, and the theorem is proved.

The calculation of a fundamental or principal matrix solution is often tedious when $A(t)$ is a constant matrix and generally impossible if $A(t)$ is nonconstant. The difficulty that occurs when $A(t)$ is not constant has already been encountered in Chapter 3, when we tried to solve second order linear equations with variable coefficients. When $A(t)$ is constant, we can use the methods of Sections 7.3 and 7.5 to find fundamental sets of solutions, which immediately yield fundamental matrix solutions. In the remainder of this section we will show that all solutions of the system (7.128) can be expressed in terms of a single fundamental matrix solution.

First, we define the *associated matrix equation* to the system (7.128):

$$X'(t) = A(t)X(t). \tag{7.138}$$

We now seek a matrix (instead of a vector) solution of Eq. (7.138). The following fact is easy to prove and is left as an exercise: *$X(t)$ is a solution of the associated matrix equation* (7.138) *if and only if every column of $X(t)$ is a solution of the system* (7.128).

Theorem 7.14 Let Φ be a matrix solution of (7.128) and let C be any constant matrix. Then $\Phi_1 = \Phi C$ is also a matrix solution of (7.128).

Proof. Since a matrix solution of (7.128) is also a solution of (7.138), and conversely, Φ is a solution of (7.138). We must show that Φ_1 is also a solution of (7.138). But $\Phi_1' = (\Phi C)' = \Phi'C + \Phi C' = \Phi'C$, since $C' = 0$, C being constant. Finally, since Φ is a solution, $\Phi_1' = \Phi'C = A\Phi C = A\Phi_1$, and the proof is complete.

Example 7.11.7 Consider Example 7.11.1, with the fundamental matrix solution

$$\Phi(t) = \begin{pmatrix} -2e^{-4t} & 3e^t \\ e^{-4t} & e^t \end{pmatrix}.$$

Let

$$C_1 = \begin{pmatrix} 1 & 2 \\ 3 & 4 \end{pmatrix} \quad \text{and} \quad C_2 = \begin{pmatrix} 1 & 2 \\ 2 & 4 \end{pmatrix}$$

Then

$$\Phi_1 = \Phi C_1 = \begin{pmatrix} -2e^{-4t} + 9e^t & -4e^{-4t} + 12e^t \\ e^{-4t} + 3e^t & 2e^{-4t} + 4e^t \end{pmatrix}$$

and

$$\Phi_2 = \Phi C_2 = \begin{pmatrix} -2e^{-4t} + 6e^t & -4e^{-4t} + 12e^t \\ e^{-4t} + 2e^t & 2e^{-4t} + 4e^t \end{pmatrix}.$$

It is easily verified that both Φ_1 and Φ_2 are matrix solutions. Note that while Φ_1 is another fundamental matrix solution, Φ_2 is not, since det $\Phi_2(0) = 0$. Can you explain this fact (see Exercise 10)?

Theorem 7.14 gives us an easy way of finding a principal matrix solution when a fundamental matrix solution is known. To see this, let $\Phi(t)$ be a fundamental matrix solution. Since det $\Phi(t_0) \neq 0$, $\Phi(t_0)$ is invertible, and we define $C = \Phi^{-1}(t_0)$ and $\Psi(t) = \Phi(t)C$. By Theorem 7.14, $\Psi(t)$ is a matrix solution and $\Psi(t_0) = \Phi(t_0)C = \Phi(t_0)\Phi^{-1}(t_0) = I$ so that $\Psi(t)$ is a principal matrix solution. Thus, if $\Phi(t)$ is a fundamental matrix solution, we can always obtain a principle matrix solution by multiplying $\Phi(t)$ on the right-hand side by $\Phi^{-1}(t_0)$.

Example 7.11.8 A fundamental matrix solution of the system of Example 7.11.1 is [see Example 7.11.5(1)]:

$$\Phi(t) = \begin{pmatrix} -2e^{-4t} & 3e^t \\ e^{-4t} & e^t \end{pmatrix}, \qquad \Phi(0) = \begin{pmatrix} -2 & 3 \\ 1 & 1 \end{pmatrix}.$$

Then

$$C = \Phi^{-1}(0) = \begin{pmatrix} -\frac{1}{5} & \frac{3}{5} \\ \frac{1}{5} & \frac{2}{5} \end{pmatrix} = \frac{1}{5}\begin{pmatrix} -1 & 3 \\ 1 & 2 \end{pmatrix},$$

so that

$$\Psi(t) = \Phi(t)C = \frac{1}{5}\begin{pmatrix} 2e^{-4t} + 3e^t & -6e^{-4t} + 6e^t \\ -e^{-4t} + e^t & 3e^{-4t} + 2e^t \end{pmatrix}$$

is a principal matrix solution.

Theorem 7.15 Let $\Phi(t)$ be a fundamental matrix solution and let $X(t)$ be any other matrix solution of the system (7.128). Then there exists a constant matrix C such that $X(t) = \Phi(t)C$. That is, *any solution vector of (7.128) can be written as a linear combination of vectors in a fundamental set.* (Compare this theorem with Theorem 3.4.)

Before giving the proof, we should warn the reader that it is important to state on which side we are multiplying the matrix Φ by C, since matrix multiplication is not in general commutative.

Proof. Since $\Phi(t)$ is a fundamental matrix solution, det $\Phi(t) \neq 0$ and $\Phi^{-1}(t)$ exists for every t. We will show that

$$\frac{d}{dt}[\Phi^{-1}(t)X(t)] = 0.$$

This will imply that $\Phi^{-1}(t)X(t)$ is a constant matrix C and the theorem will be proved. First, we calculate

$$\frac{d}{dt}[\Phi^{-1}(t)].$$

Using the product rule of differentiation (which can easily be shown to apply to matrix products), we have

$$0 = \frac{dI}{dt} = \frac{d}{dt}(\Phi\Phi^{-1}) = \frac{d\Phi}{dt}\Phi^{-1} + \Phi\frac{d\Phi^{-1}}{dt} \tag{7.139}$$

or, after multiplying both sides of (7.139) on the left by Φ^{-1} and solving for $d\Phi^{-1}/dt$,

$$\frac{d\Phi^{-1}}{dt} = -\Phi^{-1}\frac{d\Phi}{dt}\Phi^{-1}. \tag{7.140}$$

Note the analogy between (7.140) and the identity

$$\frac{d}{dt}\left(\frac{1}{f(t)}\right) = -\frac{f'(t)}{[f(t)]^2}.$$

Now, by the product formula of derivatives,

$$\frac{d}{dt}(\Phi^{-1}X) = \left(\frac{d}{dt}\Phi^{-1}\right)X + \Phi^{-1}\frac{dX}{dt}, \tag{7.141}$$

and since both Φ and X are solutions of (7.138), Eq. (7.141) becomes

$$\frac{d}{dt}(\Phi^{-1}X) = \left(-\Phi^{-1}\frac{d\Phi}{dt}\Phi^{-1}\right)X + \Phi^{-1}(AX)$$

$$= -\Phi^{-1}A\Phi\Phi^{-1}X + \Phi^{-1}AX = -\Phi^{-1}AX + \Phi^{-1}AX = 0.$$

Example 7.11.9   Consider the system $\mathbf{x}' = A\mathbf{x}$ where

$$A = \begin{pmatrix} 1 & -2 \\ 2 & -3 \end{pmatrix}.$$

It is easy to verify that

$$\Phi_1(t) = \begin{pmatrix} e^{-t} & (2t+2)e^{-t} \\ e^{-t} & (2t+1)e^{-t} \end{pmatrix} = e^{-t}\begin{pmatrix} 1 & 2t+2 \\ 1 & 2t+1 \end{pmatrix},$$

is a fundamental matrix solution. Another matrix solution is

$$\Phi_2(t) = e^{-t}\begin{pmatrix} 4t+7 & 8t+1 \\ 4t+5 & 8t-3 \end{pmatrix}.$$

There is a matrix C such that $\Phi_2 = \Phi_1 C$. But $\Phi_1^{-1}(t)\Phi_2(t) = C$ holds for every value of t, in particular for $t = 0$. Thus

$$C = \Phi_1^{-1}(0)\Phi_2(0) = \begin{pmatrix} -1 & 2 \\ 1 & -1 \end{pmatrix}\begin{pmatrix} 7 & 1 \\ 5 & -3 \end{pmatrix} = \begin{pmatrix} 3 & -7 \\ 2 & 4 \end{pmatrix}.$$

Example 7.11.10   Consider the system $\mathbf{x}' = A\mathbf{x}$ where

$$A = \begin{pmatrix} 3 & -1 & 1 \\ -1 & 5 & -1 \\ 1 & -1 & 3 \end{pmatrix}.$$

A fundamental matrix solution is

$$\Phi(t) = \begin{pmatrix} e^{2t} & e^{3t} & e^{6t} \\ 0 & e^{3t} & -2e^{6t} \\ -e^{2t} & e^{3t} & e^{6t} \end{pmatrix}.$$

Another matrix solution is

$$X(t) = \begin{pmatrix} e^{2t} + 2e^{3t} + 3e^{6t} & e^{2t} - 3e^{3t} - 2e^{6t} & 2e^{2t} + 5e^{3t} + 7e^{6t} \\ 2e^{3t} - 6e^{6t} & -3e^{3t} + 4e^{6t} & 5e^{3t} - 14e^{6t} \\ -e^{2t} + 2e^{2t} + 3e^{6t} & -e^{2t} - 3e^{3t} - 2e^{6t} & -2e^{2t} + 5e^{3t} + 7e^{6t} \end{pmatrix}.$$

As in the previous example, a matrix C such that $X(t) = \Phi(t)C$ is given by

$$C = \Phi^{-1}(0)X(0) = \frac{1}{6}\begin{pmatrix} 3 & 0 & -3 \\ 2 & 2 & 2 \\ 1 & -2 & 1 \end{pmatrix}\begin{pmatrix} 6 & -4 & 14 \\ -4 & 1 & -9 \\ 4 & -6 & 10 \end{pmatrix}$$

$$= \frac{1}{6}\begin{pmatrix} 6 & 6 & 12 \\ 12 & -18 & 30 \\ 18 & -12 & 42 \end{pmatrix} = \begin{pmatrix} 1 & 1 & 2 \\ 2 & -3 & 5 \\ 3 & -2 & 7 \end{pmatrix}.$$

Theorem 7.16 Let $\Phi(t)$ be a fundamental matrix solution of (7.128) and let $\mathbf{x}(t)$ be any solution of (7.128). Then there exists a constant vector $\mathbf{c}$ such that

$$\mathbf{x}(t) = \Phi(t)\mathbf{c}. \tag{7.142}$$

This theorem is an immediate consequence of Theorem 7.15, if we form the matrix solution $X(t) = (\mathbf{x}(t), \ldots, \mathbf{x}(t))$, whose n columns are each the vector solution $\mathbf{x}(t)$. Then a matrix C exists such that $X(t) = \Phi(t)C$. Every column of C is the vector $\mathbf{c}$.

Example 7.11.11 In Example 7.11.3 we found the fundamental matrix solution

$$\Phi(t) = e^{6t}\begin{pmatrix} \cos 2t - \frac{1}{2}\sin 2t & \sin 2t + \frac{1}{2}\cos 2t \\ \cos 2t - 3\sin 2t & \sin 2t + 3\cos 2t \end{pmatrix}.$$

The vector

$$\mathbf{x}(t) = e^{6t}\begin{pmatrix} -5\sin 2t \\ -10\cos 2t - 10\sin 2t \end{pmatrix}$$

is easily seen to be a solution. From (7.142) we obtain

$$\mathbf{c} = \Phi^{-1}(t)\mathbf{x}(t) = \Phi^{-1}(0)\mathbf{x}(0) = \frac{2}{5}\begin{pmatrix} 3 & -\frac{1}{2} \\ -1 & 1 \end{pmatrix}\begin{pmatrix} 0 \\ -10 \end{pmatrix}$$

$$= \frac{2}{5}\begin{pmatrix} 5 \\ -10 \end{pmatrix} = \begin{pmatrix} 2 \\ -4 \end{pmatrix}.$$

Example 7.11.12 Consider the system

$$x_1' = \frac{-t}{1 - t^2}x_1 + \frac{1}{1 - t^2}x_2,$$

$$x_2' = \frac{1}{1 - t^2}x_1 - \frac{t}{1 - t^2}x_2$$

or

$$\mathbf{x}' = \begin{bmatrix} \dfrac{-t}{1-t^2} & \dfrac{1}{1-t^2} \\ \dfrac{1}{1-t^2} & \dfrac{-t}{1-t^2} \end{bmatrix} \mathbf{x}. \tag{7.143}$$

It is easy to verify that

$$\Phi_1(t) = \begin{pmatrix} t \\ 1 \end{pmatrix} \quad \text{and} \quad \Phi_2(t) = \begin{pmatrix} 1 \\ t \end{pmatrix}$$

are linearly independent solutions and that

$$\Phi(t) = \begin{pmatrix} t & 1 \\ 1 & t \end{pmatrix}$$

is a fundamental matrix solution. First, note that while $W(0) = -1 \neq 0$, we also have $W(1) = 0$. A cursory inspection will help to explain this apparent contradiction of Theorem 7.12. The matrix $A(t)$ is undefined at $t = 1$; thus there cannot even be a solution to (7.143) at $t = 1$. The theorem about Wronskians is, of course, only valid in an interval over which the solution is defined. Continuing with the example, let us find a solution $\mathbf{x}(t)$ that satisfies the initial conditions

$$\mathbf{x}(0) = \begin{pmatrix} x_1(0) \\ x_2(0) \end{pmatrix} = \begin{pmatrix} 2 \\ -3 \end{pmatrix}.$$

Then

$$\mathbf{c} = \Phi^{-1}(0)\mathbf{x}(0) = \begin{pmatrix} 0 & 1 \\ 1 & 0 \end{pmatrix} \begin{pmatrix} 2 \\ -3 \end{pmatrix} = \begin{pmatrix} -3 \\ 2 \end{pmatrix}.$$

[Note that $\Phi(0) = \begin{pmatrix} 0 & 1 \\ 1 & 0 \end{pmatrix}$ is a matrix that is its own inverse.] Thus

$$\mathbf{x}(t) = \Phi(t)\mathbf{c} = \begin{pmatrix} t & 1 \\ 1 & t \end{pmatrix} \begin{pmatrix} -3 \\ 2 \end{pmatrix} = \begin{pmatrix} 2 - 3t \\ -3 + 2t \end{pmatrix}$$

is a solution vector of (7.143) that satisfies the given initial conditions.

EXERCISES 7.11

In Exercises 1 through 5 decide whether each given set of solution vectors constitutes a fundamental set of the given system by (a) determining whether the vectors are linearly independent, and (b) using the method of Wronskians to determine whether or not $W(t)$ is zero.

1. $\mathbf{x}' = \begin{pmatrix} 2 & 5 \\ 0 & 2 \end{pmatrix} \mathbf{x}, \quad \varphi_1(t) = \begin{pmatrix} e^{2t}(1 + 10t) \\ 2e^{2t} \end{pmatrix}, \quad \varphi_2(t) = \begin{pmatrix} e^{2t}(-3 + 20t) \\ 4e^{2t} \end{pmatrix}$

2. $\mathbf{x}' = \begin{pmatrix} 4 & -13 \\ 2 & -6 \end{pmatrix} \mathbf{x}, \quad \boldsymbol{\varphi}_1(t) = \begin{pmatrix} e^{-t}(13 \cos t - 26 \sin t) \\ e^{-t}(7 \cos t - 9 \sin t) \end{pmatrix},$

$\boldsymbol{\varphi}_2(t) = \begin{pmatrix} e^{-t}(26 \cos t - 52 \sin t) \\ e^{-t}(14 \cos t - 18 \sin t) \end{pmatrix}$

3. $\mathbf{x}' = \begin{pmatrix} 1 & 1 \\ 4 & 1 \end{pmatrix} \mathbf{x}, \quad \boldsymbol{\varphi}_1(t) = \begin{pmatrix} e^{3t} - e^{-t} \\ 2e^{3t} + 2e^{-t} \end{pmatrix}, \qquad \boldsymbol{\varphi}_2(t) = \begin{pmatrix} 2e^{3t} \\ 4e^{3t} \end{pmatrix}$

4. $\mathbf{x}' = \begin{pmatrix} 1 & -1 & 4 \\ 3 & 2 & -1 \\ 2 & 1 & -1 \end{pmatrix} \mathbf{x}, \quad \boldsymbol{\varphi}_1(t) = \begin{pmatrix} e^t + 2e^{-t} + 3e^{3t} \\ -4e^t - 2e^{-2t} + 6e^{3t} \\ -e^t - 2e^{-2t} + 3e^{3t} \end{pmatrix},$

$\boldsymbol{\varphi}_2(t) = \begin{pmatrix} -2e^t + 2e^{-2t} \\ 8e^t - 2e^{-2t} \\ 2e^t - 2e^{-2t} \end{pmatrix}, \quad \boldsymbol{\varphi}_3(t) = \begin{pmatrix} 3e^t - 6e^{-2t} + 3e^{3t} \\ -12e^t + 6e^{-2t} + 6e^{3t} \\ -3e^t + 6e^{-2t} + 3e^{3t} \end{pmatrix}$

5. $\mathbf{x}' = \begin{pmatrix} 3 & 2 & 1 \\ -1 & 0 & -1 \\ 1 & 1 & 2 \end{pmatrix} \mathbf{x}, \quad \boldsymbol{\varphi}_1(t) = \begin{pmatrix} -e^t + te^{2t} + e^{2t} \\ e^t - te^{2t} \\ te^{2t} \end{pmatrix},$

$\boldsymbol{\varphi}_2(t) = \begin{pmatrix} 2e^{2t} + te^{2t} \\ -e^{2t} - te^{2t} \\ e^{2t} + te^{2t} \end{pmatrix}, \quad \boldsymbol{\varphi}_3(t) = \begin{pmatrix} -e^t + 3e^{2t} + 2te^{2t} \\ e^t - e^{2t} - 2te^{2t} \\ e^{2t} + 2te^{2t} \end{pmatrix}$

6. Let $\boldsymbol{\varphi}_1(t), \boldsymbol{\varphi}_2(t), \ldots, \boldsymbol{\varphi}_m(t)$ be m solutions of the homogeneous system (7.128). Show that

$$\boldsymbol{\varphi}(t) = c_1 \boldsymbol{\varphi}_1(t) + c_2 \boldsymbol{\varphi}(t) + \cdots + c_m \boldsymbol{\varphi}_m(t)$$

is also a solution.

In each of Exercises 7 through 9 two matrix functions Φ_1 and Φ_2 are given. Find a matrix C such that $\Phi_2(t) = \Phi_1(t)C$.

7. $\Phi_1(t) = e^{6t} \begin{pmatrix} \cos 2t - \frac{1}{2} \sin t & \sin 2t + \frac{1}{2} \cos 2t \\ \cos 2t - 3 \sin 2t & \sin 2t + 3 \cos 2t \end{pmatrix}$,

$\Phi_2(t) = e^{6t} \begin{pmatrix} \frac{1}{2} \cos t - \frac{3}{2} \sin 2t & -\frac{3}{2} \cos 2t + 2 \sin 2t \\ -2 \cos 2t - 4 \sin 2t & \cos 2t + 7 \sin 2t \end{pmatrix}$

8. $\Phi_1(t) = \begin{pmatrix} \sin e^t & \cos e^t \\ e^t \cos e^t & -e^t \sin e^t \end{pmatrix}, \quad \Phi_2(t) = \begin{pmatrix} \cos e^t & 3 \sin e^t + 2 \cos e^t \\ -e^t \sin e^t & 3e^t \cos e^t - 2e^t \sin e^t \end{pmatrix}$

9. $\Phi_1(t) = \begin{pmatrix} e^{-t} & 3e^t & e^{2t} \\ 0 & 2e^t & 3e^{2t} \\ e^{-t} & e^t & e^{2t} \end{pmatrix}, \quad \Phi_2(t) = \begin{pmatrix} e^{-t} + e^{2t} & -e^{-t} + 3e^t & e^{-t} + 3e^t \\ 3e^{2t} & 2e^t & 2e^t \\ e^{-t} + e^{2t} & -e^{-t} + e^t & e^{-t} + e^t \end{pmatrix}$

10. Let $\Phi_1(t)$ be a fundamental matrix solution of $\mathbf{x}' = A\mathbf{x}$. Then $\Phi_2 = \Phi_1 C$ is a matrix solution for any constant matrix C. Show that Φ_2 is a fundamental matrix solution if and only if C is nonsingular.

11. Let $\Phi(t)$ be a matrix solution of the system $\mathbf{x}' = A(t)\mathbf{x}$ where $A(t)$ is a 3×3 matrix.

Prove that

$$W(t) = W(t_0) \exp\left[\int_{t_0}^{t} [a_{11}(s) + a_{22}(s) + a_{33}(s)]\, ds\right].$$

* 12. Using the result of Exercise 11, prove Theorem 7.12 for the case of $A(t)$ being an $n \times n$ matrix.

In Exercises 13 through 16 find the principal matrix solution $\Psi(t)$ for each given fundamental matrix solution $\Phi(t)$. Assume that $t_0 = 0$.

13. $\Phi(t) = e^{6t} \begin{pmatrix} \cos 2t - \frac{1}{2}\sin t & \sin 2t + \frac{1}{2}\cos 2t \\ \cos 2t - 3\sin 2t & \sin 2t + 3\cos 2t \end{pmatrix}$

14. $\Phi(t) = \begin{pmatrix} \sin e^t & \cos e^t \\ e^t \cos e^t & -e^t \sin e^t \end{pmatrix}$

15. $\Phi(t) = \begin{pmatrix} e^{-t} & 3e^t & e^{2t} \\ 0 & 2e^t & 3e^{2t} \\ e^{-t} & e^t & e^{2t} \end{pmatrix}$

16. $\Phi(t) = \begin{pmatrix} 2e^t + te^t & te^t + e^t & e^t \\ 0 & e^t & e^t \\ -3e^t - te^t & -te^t - 2e^t & -e^t \end{pmatrix}$

17. Consider the problem

$$\mathbf{x}' = \begin{pmatrix} 3 & -2 \\ 2 & -1 \end{pmatrix} \mathbf{x}.$$

It is easy to verify that

$$\Phi(t) = \begin{pmatrix} 2te^t + e^t & 2te^t \\ 2te^t & -e^t + 2te^t \end{pmatrix}$$

is a fundamental matrix solution. Find a solution that satisfies each of the following initial conditions:

a) $\mathbf{x}(0) = \begin{pmatrix} 1 \\ 2 \end{pmatrix}$ b) $\mathbf{x}(0) = \begin{pmatrix} -2 \\ 3 \end{pmatrix}$ c) $\mathbf{x}(1) = \begin{pmatrix} 0 \\ 1 \end{pmatrix}$

d) $\mathbf{x}(-1) = \begin{pmatrix} 2 \\ 1 \end{pmatrix}$ e) $\mathbf{x}(3) = \begin{pmatrix} 3 \\ 3 \end{pmatrix}$ f) $\mathbf{x}(a) = \begin{pmatrix} b \\ c \end{pmatrix}$

18. In Example 7.11.5 we saw that a fundamental matrix solution to the system

$$\mathbf{x}' = \begin{pmatrix} 1 & 1 & -2 \\ -1 & 2 & 1 \\ 0 & 1 & -1 \end{pmatrix} \mathbf{x},$$

of Example 7.11.4 was

$$\Phi(t) = \begin{pmatrix} e^{-t} & 3e^t & e^{2t} \\ 0 & 2e^t & 3e^{2t} \\ e^{-t} & e^t & e^{2t} \end{pmatrix}.$$

Find particular solutions that satisfy the following conditions:

a) $\mathbf{x}(0) = \begin{pmatrix} 1 \\ -1 \\ 2 \end{pmatrix}$ b) $\mathbf{x}(0) = \begin{pmatrix} 3 \\ 1 \\ 2 \end{pmatrix}$

c) $\mathbf{x}(1) = \begin{pmatrix} 1 \\ 0 \\ 1 \end{pmatrix}$ d) $\mathbf{x}(-1) = \begin{pmatrix} 2 \\ -3 \\ 5 \end{pmatrix}$

19. Consider the second order equation

$$x'' + a(t)x' + b(t)x = 0. \tag{7.144}$$

a) Write Eq. (7.144) in the form $\mathbf{x}' = A(t)\mathbf{x}$.
b) Given that

$$\Phi(t) = \begin{pmatrix} \varphi_1 & \varphi_2 \\ \varphi_1' & \varphi_2' \end{pmatrix}$$

is a fundamental matrix solution, show that

$$\det \Phi(t) = \det \Phi(t_0) \exp\left[-\int_{t_0}^{t} a(s)\, ds \right].$$

c) Show that the formula in part (b) can be rearranged as

$$\varphi_2' - \frac{\varphi_1'}{\varphi_1} \varphi_2 = \frac{\det \Phi(t_0)}{\varphi_1} \exp\left[-\int_{t_0}^{t} a(s)\, ds \right]. \tag{7.145}$$

Therefore, if one solution $\varphi_1(t)$ of (7.144) is known, then another solution can be calculated by solving Eq. (7.145).

20. Given that $\varphi_1(t) = \sin(\ln t)$, find a second linearly independent solution of

$$x'' + \frac{1}{t}x' + \frac{1}{t^2}x = 0.$$

21. Given that $\varphi_1(t) = e^{t^2}$ is a solution of

$$x'' - 2tx' - 2x = 0,$$

find a second linearly independent solution.

22. Given that $\varphi_1(t) = \sin t^2$ is a solution of

$$tx'' - x' + 4t^3x = 0,$$

find a second linearly independent solution.

In Exercises 23 through 29 use the method of Section 7.5 to find a fundamental matrix solution of the system $\mathbf{x}' = A\mathbf{x}$ for each given matrix A. Then use this fundamental matrix solution to obtain a principal matrix solution. Assume that $t_0 = 0$.

23. $A = \begin{pmatrix} 4 & -3 \\ 5 & -4 \end{pmatrix}$ 24. $A = \begin{pmatrix} 7 & 6 \\ 2 & 6 \end{pmatrix}$ 25. $A = \begin{pmatrix} -1 & 1 \\ -5 & 3 \end{pmatrix}$ 26. $A = \begin{pmatrix} 1 & 1 \\ -1 & 3 \end{pmatrix}$

27. $A = \begin{pmatrix} -4 & -1 \\ 1 & -2 \end{pmatrix}$ 28. $A = \begin{pmatrix} 4 & -2 \\ 5 & 2 \end{pmatrix}$ 29. $A = \begin{pmatrix} 4 & -3 \\ 8 & -6 \end{pmatrix}$

---

*7.12 NONHOMOGENEOUS SYSTEMS

We shall now present a method for solving the nonhomogeneous system

$$\mathbf{x}' = A(t)\mathbf{x} + \mathbf{f}(t), \tag{7.146}$$

given that a fundamental matrix solution $\Phi(t)$ for the homogeneous system

$$\mathbf{x}' = A(t)\mathbf{x} \tag{7.147}$$

is known. Such a solution can always be found if $A(t)$ is a constant matrix (by the methods of Section 7.3 or 7.5).

Theorem 7.17 Let $\boldsymbol{\varphi}_p(t)$ and $\boldsymbol{\varphi}_q(t)$ be two solutions of the system (7.146). Then their difference,

$$\boldsymbol{\varphi}(t) = \boldsymbol{\varphi}_p(t) - \boldsymbol{\varphi}_q(t),$$

is a solution of (7.147).

Proof. $\boldsymbol{\varphi}' = (\boldsymbol{\varphi}_p - \boldsymbol{\varphi}_q)' = (A\boldsymbol{\varphi}_p + \mathbf{f}) - (A\boldsymbol{\varphi}_q + \mathbf{f}) = A(\boldsymbol{\varphi}_p - \boldsymbol{\varphi}_q) = A\boldsymbol{\varphi}$. Thus, as in the case of linear equations (Chapter 3), it is necessary only to find one particular solution of (7.146). If $\boldsymbol{\varphi}_p(t)$ is such a solution, then *the general solution to the nonhomogeneous system* (7.146) *is of the form*

$$\boldsymbol{\varphi}(t) = \Phi(t)\mathbf{c} + \boldsymbol{\varphi}_p(t), \tag{7.148}$$

where $\mathbf{c}$ *is a vector of arbitrary constants and* $\Phi(t)$ *is a fundamental matrix solution of the homogeneous equation* (7.147). That Eq. (7.148) is a solution can be verified as follows:

$$\boldsymbol{\varphi}'(t) = \Phi'(t)\mathbf{c} + \boldsymbol{\varphi}_p'(t)$$
$$= [A(t)\Phi(t)\mathbf{c}] + [A(t)\boldsymbol{\varphi}_p(t) + \mathbf{f}(t)],$$

since Φ is a solution of the associated matrix equation and $\boldsymbol{\varphi}_p$ is a particular solution of (7.147). Combining terms, we have

$$\boldsymbol{\varphi}'(t) = A(t)[\Phi(t)\mathbf{c} + \boldsymbol{\varphi}_p(t)] + \mathbf{f}(t) = A(t)\boldsymbol{\varphi}(t) + \mathbf{f}(t),$$

and $\boldsymbol{\varphi}$ is a solution of (7.147).

Let us now derive a simple formula for finding the solution of the initial value problem

$$\mathbf{x}' = A(t)\mathbf{x} + \mathbf{f}(t), \qquad \mathbf{x}(t_0) = \mathbf{x}_0. \tag{7.149}$$

This solution is given in terms of a fundamental matrix solution of the homogeneous system (7.147). Let $\boldsymbol{\varphi}(t)$ be the solution of (7.149). Then $\boldsymbol{\varphi}(t_0) = \mathbf{x}_0$ and

$$\boldsymbol{\varphi}' = A\boldsymbol{\varphi} + \mathbf{f}. \tag{7.150}$$

Let $\Phi(t)$ be a fundamental matrix solution for (7.147). Then $\Phi' = A\Phi$, Φ^{-1} exists, and $\Phi\Phi^{-1} = I$ so that (7.150) can be written as

$$\boldsymbol{\varphi}' = A\Phi\Phi^{-1}\boldsymbol{\varphi} + \mathbf{f} = \Phi'\Phi^{-1}\boldsymbol{\varphi} + \mathbf{f}$$

or

$$-\Phi'\Phi^{-1}\boldsymbol{\varphi} + \boldsymbol{\varphi}' = \mathbf{f}. \tag{7.151}$$

We may multiply both sides of (7.151) by Φ^{-1} to obtain

$$-\Phi^{-1}\Phi'\Phi^{-1}\boldsymbol{\varphi} + \Phi^{-1}\boldsymbol{\varphi}' = \Phi^{-1}\mathbf{f}. \tag{7.152}$$

But according to (7.140),

$$\frac{d}{dt}\Phi^{-1} = -\Phi^{-1}\Phi'\Phi^{-1}.$$

Therefore, (7.152) becomes

$$\left(\frac{d}{dt}\Phi^{-1}\right)\boldsymbol{\varphi} + \Phi^{-1}\frac{d}{dt}\boldsymbol{\varphi} = \Phi^{-1}\mathbf{f}$$

or

$$\frac{d}{dt}(\Phi^{-1}\boldsymbol{\varphi}) = \Phi^{-1}\mathbf{f}. \tag{7.153}$$

We may now integrate both sides of (7.153) from t_0 to t to obtain

$$\Phi^{-1}(t)\boldsymbol{\varphi}(t)\Big|_{t_0}^{t} = \int_{t_0}^{t} \Phi^{-1}(s)\mathbf{f}(s)\,ds$$

or

$$\Phi^{-1}(t)\boldsymbol{\varphi}(t) - \Phi^{-1}(t_0)\boldsymbol{\varphi}(t_0) = \int_{t_0}^{t} \Phi^{-1}(s)\mathbf{f}(s)\,ds.$$

Multiplying both sides by $\Phi(t)$, rearranging, and taking note of the initial condition $\boldsymbol{\varphi}(t_0) = \mathbf{x}_0$, we have

$$\boldsymbol{\varphi}(t) = \Phi(t)\Phi^{-1}(t_0)\mathbf{x}_0 + \Phi(t)\int_{t_0}^{t} \Phi^{-1}(s)\mathbf{f}(s)\,ds. \tag{7.154}$$

Formula (7.154) is called the *variation of constants* formula for the initial value problem (7.149). If $\mathbf{x}_0$ is unspecified, then (7.154) indicates that all solutions to the nonhomogeneous Eq. (7.146) can be expressed in terms of a fundamental matrix solution for the associated homogeneous system (7.147) and the vector $\mathbf{f}(t)$. We note that since $\Phi^{-1}(t_0)\mathbf{x}_0$ is a constant vector, $\Phi(t)\Phi^{-1}(t_0)\mathbf{x}_0$ is a solution vector to the homogeneous system (7.147). Thus

$$\boldsymbol{\varphi}_p(t) = \Phi(t) \int_{t_0}^t \Phi^{-1}(s)\mathbf{f}(s) \, ds$$

must be a particular solution to the nonhomogeneous system (7.147). We also note that if $\Phi(t)$ is a principal matrix solution to (7.147), then $\Phi(t_0) = \Phi^{-1}(t_0) = I$ so that (7.154) takes the simpler form

$$\boldsymbol{\varphi}(t) = \Phi(t)\mathbf{x}_0 + \Phi(t) \int_{t_0}^t \Phi^{-1}(s)\mathbf{f}(s) \, ds$$
$$= \Phi(t)\mathbf{x}_0 + \boldsymbol{\varphi}_p(t).$$

We summarize these results in the theorem below.

Theorem 7.18 Let $\Phi(t)$ be a fundamental matrix solution for the homogeneous system

$$\mathbf{x}' = A(t)\mathbf{x}. \tag{7.155}$$

Then the solution to the initial value problem

$$\mathbf{x}' = A(t)\mathbf{x} + \mathbf{f}(t), \qquad \mathbf{x}(t_0) = \mathbf{x}_0, \tag{7.156}$$

is given by

$$\boldsymbol{\varphi}(t) = \Phi(t)\Phi^{-1}(t_0)\mathbf{x}_0 + \boldsymbol{\varphi}_p(t),$$

where

$$\boldsymbol{\varphi}_p(t) = \Phi(t) \int_{t_0}^t \Phi^{-1}(s)\mathbf{f}(s) \, ds \tag{7.157}$$

is a particular solution of the nonhomogeneous system

$$\mathbf{x}' = A(t)\mathbf{x} + \mathbf{f}(t).$$

If $\Psi(t)$ is the principal matrix solution of the homogeneous system (7.155), then the solution of the initial value problem (7.156) is

$$\boldsymbol{\varphi}(t) = \Psi(t)\mathbf{x}_0 + \Psi(t) \int_{t_0}^t \Psi^{-1}(s)\mathbf{f}(s) \, ds.$$

Example 7.12.1 Consider the system

$$\mathbf{x}' = \begin{pmatrix} x_1 \\ x_2 \end{pmatrix}' = \begin{pmatrix} 4 & 2 \\ 3 & 3 \end{pmatrix} \begin{pmatrix} x_1 \\ x_2 \end{pmatrix} + \begin{pmatrix} e^t \\ e^{2t} \end{pmatrix} = A\mathbf{x} + \mathbf{f}(t).$$

A fundamental matrix solution for the homogeneous system is

$$\Phi(t) = \begin{pmatrix} 2e^t & e^{6t} \\ -3e^t & e^{6t} \end{pmatrix}.$$

Then

$$\Phi^{-1}(t) = \frac{1}{5} \begin{pmatrix} e^{-t} & -e^{-t} \\ 3e^{-6t} & 2e^{-6t} \end{pmatrix},$$

and by (7.156), we have the particular solution

$$\boldsymbol{\varphi}_p = \frac{1}{5} \begin{pmatrix} 2e^t & e^{6t} \\ -3e^t & e^{6t} \end{pmatrix} \int_0^t \begin{pmatrix} e^{-s} & -e^{-s} \\ 3e^{-6s} & 2e^{-6s} \end{pmatrix} \begin{pmatrix} e^s \\ e^{2s} \end{pmatrix} ds$$

$$= \frac{1}{5} \begin{pmatrix} 2e^t & e^{6t} \\ -3e^t & e^{6t} \end{pmatrix} \int_0^t \begin{pmatrix} 1 - e^s \\ 3e^{-5s} + 2e^{-4s} \end{pmatrix} ds.$$

Since the integral of a vector function is the vector of integrals,

$$\boldsymbol{\varphi}_p(t) = \frac{1}{5} \begin{pmatrix} 2e^t & e^{6t} \\ -3e^t & e^{6t} \end{pmatrix} \begin{pmatrix} t - e^t + 1 \\ -e^{-5t} - \dfrac{e^{-4t}}{2} + \dfrac{8}{15} \\ \dfrac{15}{} \end{pmatrix}$$

$$= \begin{pmatrix} \frac{2}{5}te^t - \frac{1}{2}e^{2t} + \frac{29}{75}e^t + \frac{8}{75}e^{6t} \\ -\frac{3}{5}te^t + \frac{7}{10}e^{2t} - \frac{46}{75}e^t + \frac{8}{75}e^{6t} \end{pmatrix}.$$

Example 7.12.2 Consider the initial value problem (see Example 3.7.2)

$$x'' + x = 2\cos t, \qquad x(0) = 5, \qquad x'(0) = 2.$$

Using the substitution $x_1 = x$, $x_2 = x'$, we can write this in matrix form

$$\begin{pmatrix} x_1 \\ x_2 \end{pmatrix}' = \begin{pmatrix} 0 & 1 \\ -1 & 0 \end{pmatrix} \begin{pmatrix} x_1 \\ x_2 \end{pmatrix} + \begin{pmatrix} 0 \\ 2\cos t \end{pmatrix},$$

where the homogeneous system has the principal matrix solution

$$\Psi(t) = \begin{pmatrix} \cos t & \sin t \\ -\sin t & \cos t \end{pmatrix}.$$

Then

$$\Psi^{-1}(t) = \begin{pmatrix} \cos t & -\sin t \\ \sin t & \cos t \end{pmatrix},$$

and

$$\int_0^t \Psi^{-1}(s)\mathbf{f}(s)\, ds = \int_0^t \begin{pmatrix} \cos s & -\sin s \\ \sin s & \cos s \end{pmatrix} \begin{pmatrix} 0 \\ 2\cos s \end{pmatrix} ds$$

$$= \int_0^t \begin{pmatrix} -2\sin s \cos s \\ 2\cos^2 s \end{pmatrix} ds = \begin{pmatrix} \cos^2 t - 1 \\ t + \sin t \cos t \end{pmatrix}.$$

Thus, since the solution $\varphi(t)$ satisfies the initial conditions

$$\varphi(0) = \begin{pmatrix} 5 \\ 2 \end{pmatrix},$$

we obtain by Theorem 7.18

$$\varphi(t) = \begin{pmatrix} \cos t & \sin t \\ -\sin t & \cos t \end{pmatrix} \begin{pmatrix} 5 \\ 2 \end{pmatrix} + \begin{pmatrix} \cos t & \sin t \\ -\sin t & \cos t \end{pmatrix} \begin{pmatrix} \cos^2 t - 1 \\ t + \sin t \cos t \end{pmatrix}$$

$$= \begin{pmatrix} 5 \cos t + 2 \sin t + t \sin t \\ -4 \sin t + 2 \cos t + t \cos t \end{pmatrix},$$

where we have used the fact that $\cos^3 t + \sin^2 t \cos t = \cos t (\cos^2 t + \sin^2 t) = \cos t$.

Example 7.12.3 Consider the system

$$\begin{pmatrix} x_1 \\ x_2 \\ x_3 \end{pmatrix}' = \begin{pmatrix} 0 & 1 & 0 \\ 0 & 0 & 1 \\ -2/t^3 & 2/t^3 & 1/t \end{pmatrix} \begin{pmatrix} x_1 \\ x_2 \\ x_3 \end{pmatrix} + \begin{pmatrix} 2t^2 \\ -t^3 \\ t^5 \end{pmatrix} \qquad (7.158)$$

with initial conditions $x_1(1) = 2$, $x_2(1) = 0$, $x_3(1) = -1$. A fundamental matrix solution (see Example 3.8.3) is

$$\Phi(t) = \begin{pmatrix} t & 1/t & t^2 \\ 1 & -1/t^2 & 2t \\ 0 & 2/t^3 & 2 \end{pmatrix}.$$

Note that this solution is valid only for $t > 0$ since $\Phi(t)$ is not defined at $t = 0$. Then

$$\Phi^{-1}(t) = \frac{1}{6} \begin{pmatrix} 6/t & 0 & -3t \\ 2t & -2t^2 & t^3 \\ -2/t^2 & 2/t & 2 \end{pmatrix}.$$

Hence, by (7.154), the solution to the initial value problem (7.158) is

$$\varphi(t) = \Phi(t)\Phi^{-1}(1) \begin{pmatrix} 2 \\ 0 \\ -1 \end{pmatrix} + \Phi(t) \int_1^t \Phi^{-1}(s)\mathbf{f}(s)\, ds = \Phi(t)\mathbf{c} + \varphi_p(t).$$

Setting $t = 1$, we have

$$\Phi^{-1}(1) = \frac{1}{6} \begin{pmatrix} 6 & 0 & -3 \\ 2 & -2 & 1 \\ -2 & 2 & 2 \end{pmatrix}$$

so that

$$\Phi(t)\Phi^{-1}(1)\begin{pmatrix} 2 \\ 0 \\ -1 \end{pmatrix} = \frac{1}{6}\begin{pmatrix} 9t - 5/t - 2t^2 \\ 9 - 5/t^2 - 4t \\ 10/t^3 - 4 \end{pmatrix}.$$

Therefore, after a great deal of arithmetic, we have

$$\varphi(t) = \frac{1}{6}\begin{pmatrix} 9t - 5/t - 2t^2 \\ 9 - 5/t^2 - 4t \\ 10/t^3 - 4 \end{pmatrix} + \frac{1}{6}\begin{pmatrix} t & 1/t & t^2 \\ 1 & -1/t^2 & 2t \\ 0 & 2/t^3 & 2 \end{pmatrix}\int_1^t \begin{pmatrix} 12s - 3s^6 \\ 4s^3 + 2s^5 + s^8 \\ -4 - 2s^2 + 2s^5 \end{pmatrix} ds$$

$$= \begin{bmatrix} \dfrac{7}{27t} + \dfrac{33}{21}t - \dfrac{5}{18}t^2 + \dfrac{1}{2}t^3 - \dfrac{t^5}{18} + \dfrac{t^8}{378} \\[2mm] \dfrac{-7}{27t^2} + \dfrac{33}{21} - \dfrac{5}{9}t + \dfrac{3}{2}t^2 - \dfrac{5}{18}t^4 + \dfrac{4}{189}t^7 \\[2mm] \dfrac{14}{27t^3} - \dfrac{5}{9} + 3t - \dfrac{10}{9}t^3 + \dfrac{4}{27}t^6 \end{bmatrix}.$$

EXERCISES 7.12

In each of Exercises 1 through 9 calculate a fundamental matrix solution for the associated homogeneous system and then use (7.157) and the variation of constants formula (7.154) to obtain a particular solution to the given nonhomogeneous system. Where initial conditions are given, find the unique solution that satisfies them.

1. $\mathbf{x}' = \begin{pmatrix} -2 & -2 \\ -5 & 1 \end{pmatrix}\mathbf{x} + \begin{pmatrix} e^t \\ e^{2t} \end{pmatrix}$

2. $\mathbf{x}' = \begin{pmatrix} -12 & 7 \\ -7 & 2 \end{pmatrix}\mathbf{x} + \begin{pmatrix} t \\ 2 \end{pmatrix}$, $x_1(0) = 1, x_2(0) = 0$

3. $\mathbf{x}' = \begin{pmatrix} 2 & -1 \\ 5 & -2 \end{pmatrix}\mathbf{x} + \begin{pmatrix} \sin t \\ \cos t \end{pmatrix}$, $x_1(0) = 0, x_2(0) = 1$

4. $\mathbf{x}' = \begin{pmatrix} 3 & 2 \\ -5 & 1 \end{pmatrix}\mathbf{x} + \begin{pmatrix} 2\sin 3t \\ \cos 3t \end{pmatrix}$

5. $\mathbf{x}' = \begin{pmatrix} 1 & 1 & -2 \\ -1 & 2 & 1 \\ 0 & 1 & -1 \end{pmatrix}\mathbf{x} + \begin{pmatrix} e^t \\ e^{2t} \\ e^{3t} \end{pmatrix}$, $x_1(0) = 0, x_2(0) = 1, x_3(0) = -1$

6. $\mathbf{x}' = \begin{pmatrix} 0 & 1 & 0 \\ 0 & 0 & 1 \\ 1 & -3 & 3 \end{pmatrix}\mathbf{x} + \begin{pmatrix} t^2 \\ 0 \\ 1 \end{pmatrix}$

7. $\mathbf{x}' = \begin{pmatrix} 4 & 6 & 6 \\ 1 & 3 & 2 \\ -1 & -5 & -2 \end{pmatrix} \mathbf{x} + \begin{pmatrix} 1 - 4t - 6t^2 - 6t^4 \\ t - 3t^2 - 2t^4 \\ t + 5t^2 + 4t^3 + 2t^4 \end{pmatrix}$

8. $\mathbf{x}' = \begin{pmatrix} 1 & -1 & -1 \\ 1 & -1 & 0 \\ 1 & 0 & -1 \end{pmatrix} \mathbf{x} + \begin{pmatrix} 2 \\ e^{-t} \\ e^{-t} \end{pmatrix}$, $x_1(0) = 1, x_2(0) = -1, x_3(0) = 0$

9. $\mathbf{x}' = \begin{pmatrix} 2 & -5 \\ 1 & -2 \end{pmatrix} \mathbf{x} + \begin{pmatrix} 0 \\ \cot t \end{pmatrix}, 0 < t < \pi$

In each of Exercises 10 through 12 one homogeneous solution to a given system is given. Use the method of Exercise 7.11.19 to obtain a fundamental matrix solution. Then use this solution to find the general solution of the nonhomogeneous system.

10. $\mathbf{x}' = \begin{pmatrix} 0 & 1 \\ -1/4t^2 & 0 \end{pmatrix} \mathbf{x} + \begin{pmatrix} \sqrt{t} \\ 2 \end{pmatrix}, \boldsymbol{\varphi}_1(t) = \begin{pmatrix} \sqrt{t} \\ -1/2\sqrt{t} \end{pmatrix}$

11. $\mathbf{x}' = \begin{pmatrix} 0 & 1 \\ -1/t^2 & -3/t \end{pmatrix} \mathbf{x} + \begin{pmatrix} t \\ e^t \end{pmatrix}, \boldsymbol{\varphi}_1(t) = \begin{pmatrix} 1/t \\ -1/t^2 \end{pmatrix}$

12. $\mathbf{x}' = \begin{pmatrix} \dfrac{-t}{1 - t^2} & \dfrac{1}{1 - t^2} \\ \dfrac{1}{1 - t^2} & \dfrac{-t}{1 - t^2} \end{pmatrix} \mathbf{x} + \begin{pmatrix} t^2 \\ t^3 \end{pmatrix}, -1 < t < 1, \boldsymbol{\varphi}_1(t) = \begin{pmatrix} 1 \\ t \end{pmatrix}$

13. Let $\boldsymbol{\varphi}_1(t)$ be a solution to $\mathbf{x}'(t) = A\mathbf{x}(t) + \mathbf{b}_1(t), \boldsymbol{\varphi}_2(t)$ a solution to $\mathbf{x}'(t) = A\mathbf{x}(t) + \mathbf{b}_2(t), \ldots$, and $\boldsymbol{\varphi}_n(t)$ a solution to $\mathbf{x}'(t) = A\mathbf{x}(t) + \mathbf{b}_n(t)$. Prove that $\boldsymbol{\varphi}_1(t) + \boldsymbol{\varphi}_2(t) + \cdots + \boldsymbol{\varphi}_n(t)$ is a solution to

$$\mathbf{x}'(t) = A\mathbf{x}(t) + \mathbf{b}_1(t) + \mathbf{b}_2(t) + \cdots + \mathbf{b}_n(t).$$

This again is called the *principle of superposition*.

*7.13 AN APPLICATION OF NONHOMOGENEOUS SYSTEMS: FORCED OSCILLATIONS

Consider the nonhomogeneous system with constant coefficients

$$\mathbf{x}'(t) = A\mathbf{x}(t) + \mathbf{b}(t). \tag{7.159}$$

The system (7.159) can be considered in the following way. If A is fixed, then given an *input vector* $\mathbf{b}(t)$, we can obtain an *output vector* $\mathbf{x}(t)$. That is, with fixed A, the output of the system (the solution) is determined by the input to the system. In this context, vector $\mathbf{b}(t)$ is called the *forcing vector* of the system. With this terminology, the system (7.159) is called an *input-output system*.

It often occurs in practice that the forcing term is periodic in nature (for example, electrical circuits forced by an alternating current, such as in Examples

2.8.2, 3.10.1, and 3.10.2). An important question may now be asked: if the forcing term is periodic, does there exist a periodic solution to (7.159) with the same period? An affirmative answer to this question would tell us that an oscillatory input function (an alternating current, for example) will lead to oscillatory behavior of the system governed by the differential equation.

Accordingly, we suppose that

$$\mathbf{b}(t) = e^{i\beta t}\mathbf{b} = (\cos \beta t + i \sin \beta t)\mathbf{b}, \tag{7.160}$$

where $\mathbf{b}$ is a constant vector and $e^{i\beta t}$ is periodic with period $2\pi/\beta$. Suppose that there is a solution $\mathbf{x}(t)$ to (7.159) which is periodic with period $2\pi/\beta$. Then we may write

$$\mathbf{x}(t) = e^{i\beta t}\mathbf{x}, \tag{7.161}$$

where $\mathbf{x}$ is a constant vector. Substituting (7.160) and (7.161) into (7.159), we obtain

$$i\beta e^{i\beta t}\mathbf{x} = (e^{i\beta t}\mathbf{x})' = e^{i\beta t}A\mathbf{x} + e^{i\beta t}\mathbf{b}. \tag{7.162}$$

Dividing both sides of (7.162) by $e^{i\beta t}$ and rearranging terms, we obtain,

$$(A - i\beta I)\mathbf{x} = -\mathbf{b}. \tag{7.163}$$

Equation (7.163) is a nonhomogeneous system of n equations in n unknowns which has a unique solution if and only if

$$\det (A - i\beta I) \neq 0.$$

But if A is a 2×2 matrix, we may write this as

$$D = \begin{vmatrix} a_{11} - i\beta & a_{12} \\ a_{21} & a_{22} - i\beta \end{vmatrix} = (a_{11} - i\beta)(a_{22} - i\beta) - a_{12}a_{21} \neq 0$$

[see Eq. (7.46)]. In other words, there is a unique solution if and only if $i\beta$ is *not* a root of the auxiliary equation (7.47). We therefore have the following theorem.

Theorem 7.19 In the system (7.159), let the forcing term $\mathbf{b}(t)$ be periodic with the form $\mathbf{b}(t) = e^{i\beta t}\mathbf{b}$. Then, if $i\beta$ is not a root of the auxiliary equation (7.47), there exists a unique periodic solution $\mathbf{x}(t) = e^{i\beta t}\mathbf{x}$ of (7.159) such that $\mathbf{x} = -(A - i\beta I)^{-1}\mathbf{b}$. If $i\beta$ is a root of Eq. (7.47), then there are either no such solutions or an infinite number of them. (Note that the differential equation will always have a solution. This theorem tells us something about the nature of the solutions.)

Remark. If $i\beta$ is not a root of (7.47), then a periodic (or oscillatory) input will give rise to a periodic output. This phenomenon is called *forced oscillations* (see Section 3.8).

Example 7.13.1 Consider the system

$$\mathbf{x}'(t) = \begin{pmatrix} x_1 \\ x_2 \end{pmatrix}' = \begin{pmatrix} 2 & 1 \\ 1 & 0 \end{pmatrix}\begin{pmatrix} x_1 \\ x_2 \end{pmatrix} + e^{2it}\begin{pmatrix} 2 \\ 1 \end{pmatrix} = A\mathbf{x}(t) + \mathbf{b}(t).$$

Here the forcing term is periodic with period π. Since $2i$ is not a root of (7.47), we may use (7.163) to obtain a periodic solution of period π:

$$\mathbf{x}(t) = e^{2it}\mathbf{x},$$

where

$$\mathbf{x} = -(A - 2iI)^{-1}\begin{pmatrix} 2 \\ 1 \end{pmatrix} = -\begin{pmatrix} 2 - 2i & 1 \\ 1 & -2i \end{pmatrix}^{-1}\begin{pmatrix} 2 \\ 1 \end{pmatrix}$$

$$= \frac{-1}{5 + 4i}\begin{pmatrix} 2i & 1 \\ 1 & -2 + 2i \end{pmatrix}\begin{pmatrix} 2 \\ 1 \end{pmatrix}.$$

Multiplying the numerator and denominator by $5 - 4i$, we have

$$\mathbf{x} = \frac{-5 + 4i}{41}\begin{pmatrix} 1 + 4i \\ 2i \end{pmatrix} = \frac{-1}{41}\begin{pmatrix} 21 + 16i \\ 8 + 10i \end{pmatrix}.$$

We would like to write this solution in a more illuminating form. We recall that (see Section 3.5)

$$A + Bi = \sqrt{A^2 + B^2}\left(\frac{A}{\sqrt{A^2 + B^2}} + i\frac{B}{\sqrt{A^2 + B^2}}\right) = \sqrt{A^2 + B^2}e^{i\theta},$$

where $\tan\theta = B/A$ (and $\cos\theta = A/\sqrt{A^2 + B^2}$, $\sin\theta = B/\sqrt{A^2 + B^2}$). Therefore, we have

$$\mathbf{x} = \frac{-1}{41}\begin{pmatrix} \sqrt{697}e^{i\theta_1} \\ \sqrt{164}e^{i\theta_2} \end{pmatrix},$$

where $\tan\theta_1 = \frac{16}{21}$ and $\tan\theta_2 = \frac{5}{4}$. Hence

$$\mathbf{x}(t) = \frac{-1}{41}\begin{pmatrix} \sqrt{697}e^{i(\theta_1 + 2t)} \\ \sqrt{164}e^{i(\theta_2 + 2t)} \end{pmatrix}.$$

Although $\mathbf{x}(t)$ has the same period, π, as the forcing term, the coordinate functions have been shifted by θ_1 and θ_2, respectively. Such a phenomenon is called a *phase shift*.

Now, we answer the question, What happens if $i\beta$ is a root of the auxiliary equation of A? According to the discussion in Section 7.5 (Case 3), a fundamental matrix solution $\Phi(t)$ of the homogeneous system $\mathbf{x}' = A\mathbf{x}$ contains terms of the form $e^{i\beta t}$ so that $\Phi^{-1}(t)$ contains terms of the form $e^{-i\beta t}$ (since $\Phi\Phi^{-1} = I$). Using Eq. (7.157), we find a particular solution of (7.159):

$$\varphi_p(t) = \Phi(t)\int_{t_0}^t \Phi^{-1}(s)\mathbf{b}e^{i\beta s}\,ds.$$

But the product of $\Phi^{-1}(s)$ and $e^{i\beta s}$ will contain constant terms (since $e^{-i\beta s}e^{i\beta s} = 1$) and the integral of these constant terms will be of the form ct, which becomes unbounded as t tends to $\pm\infty$. Such a phenomenon is called *resonance* (see Section 3.9). If $i\beta$ is a root of the auxiliary equation of A, then β is called a *natural frequency* of the system. We can summarize the above discussion by stating that *in general, resonance will occur when the frequency of the input vector is a natural frequency of the system.*

Example 7.13.2 Consider the circuit shown in Fig. 7.8. Applying Kirchhoff's voltage law (see Section 2.8), we obtain the second order nonhomogeneous equation

$$L\frac{d^2I}{dt^2} + \frac{I}{C} = E'(t). \tag{7.164}$$

Letting $x_1 = I$, $x_2 = I'$, we have

$$x_2' = I'' = \frac{E'}{L} - \frac{I}{LC} = \frac{E'}{L} - \frac{x_1}{LC},$$

so we may rewrite (7.164) in the form

$$\begin{pmatrix} x_1 \\ x_2 \end{pmatrix}' = \begin{pmatrix} 0 & 1 \\ -1/LC & 0 \end{pmatrix}\begin{pmatrix} x_1 \\ x_2 \end{pmatrix} + \begin{pmatrix} 0 \\ E'/L \end{pmatrix}. \tag{7.165}$$

Suppose that $E = E_0e^{i\omega t}$ (see Exercise 3.10.6). Then (7.165) may be treated by the method of this section, and the solution $\mathbf{x}(t)$ is of the form $e^{i\omega t}\mathbf{x}$, where

$$\mathbf{x} = -(A - i\omega I)^{-1}\mathbf{b} \quad\text{and}\quad \mathbf{b} = \begin{pmatrix} 0 \\ i\omega E_0/L \end{pmatrix}.$$

But

$$\det(A - i\omega I) = \det\begin{pmatrix} -i\omega & 1 \\ -1/LC & -i\omega \end{pmatrix} = \frac{1}{LC} - \omega^2,$$

Figure 7.8

which is nonzero if $\omega \neq 1/\sqrt{LC}$. Therefore,

$$\mathbf{x} = \frac{-1}{(1/LC) - \omega^2} \begin{pmatrix} -i\omega & -1 \\ 1/LC & -i\omega \end{pmatrix} \begin{pmatrix} 0 \\ i\omega E_0/L \end{pmatrix} = \frac{\omega C E_0}{LC\omega^2 - 1} \begin{pmatrix} -i \\ \omega \end{pmatrix}$$

and

$$\mathbf{x}(t) = \frac{\omega C E_0}{LC\omega^2 - 1} \begin{pmatrix} -i \\ \omega \end{pmatrix} e^{i\omega t}. \tag{7.166}$$

Thus the current is

$$x_1(t) = I(t) = -i\omega C E_0 e^{i\omega t}$$

or

$$I(t) = \omega C E_0 (\sin \omega t - i \cos \omega t).$$

Of course, currents are not imaginary. The complex term here simply tells us that the real and imaginary parts of the solution are the responses to the real and imaginary parts of the forcing function.

If $\omega = 1/\sqrt{LC}$ (the natural frequency), then the circuit is in resonance. Since $A^2 = (-1/LC)I$, it follows after a short computation that the principal matrix solution is

$$\Psi(t) = \begin{pmatrix} \cos(t/\sqrt{LC}) & \sqrt{LC} \sin(t/\sqrt{LC}) \\ (-1/\sqrt{LC}) \sin(t/\sqrt{LC}) & \cos(t/\sqrt{LC}) \end{pmatrix},$$

and by Eq. (7.157)

$$\mathbf{x}(t) = \Psi(t)\mathbf{x}(0) + \Psi(t) \int_0^t \Psi^{-1}(s)\mathbf{b}(s)\, ds. \tag{7.167}$$

To calculate the integral, we observe that since $\cos(-s) = \cos s$ and $\sin(-s) = -\sin s$, we have

$$\Psi^{-1}(s)\mathbf{b}(s) = \begin{bmatrix} \cos\dfrac{s}{\sqrt{LC}} & -\sqrt{LC}\sin\dfrac{s}{\sqrt{LC}} \\ \dfrac{1}{\sqrt{LC}}\sin\dfrac{s}{\sqrt{LC}} & \cos\dfrac{s}{\sqrt{LC}} \end{bmatrix} \begin{bmatrix} 0 \\ \dfrac{iE_0 e^{is/\sqrt{LC}}}{L\sqrt{LC}} \end{bmatrix}$$

$$= \begin{bmatrix} \dfrac{E_0}{L}\sin\dfrac{s}{\sqrt{LC}}\left(\sin\dfrac{s}{\sqrt{LC}} - i\cos\dfrac{s}{\sqrt{LC}}\right) \\ \dfrac{E_0}{L\sqrt{LC}}\cos\dfrac{s}{\sqrt{LC}}\left(-\sin\dfrac{s}{\sqrt{LC}} + i\cos\dfrac{s}{\sqrt{LC}}\right) \end{bmatrix}.$$

Using formulas 19, 24, and 27 of Appendix 1, we see that the integral in (7.167) becomes

$$\int_0^t \Psi^{-1}(s)\mathbf{b}(s)\,ds = \frac{E_0}{2L} \begin{bmatrix} \left[t - \sqrt{LC}\sin\dfrac{t}{\sqrt{LC}} \right] e^{it/\sqrt{LC}} \\ \left[\dfrac{t}{\sqrt{LC}} - 1 + \cos\dfrac{t}{\sqrt{LC}} \right] e^{it/\sqrt{LC}} \end{bmatrix}. \tag{7.168}$$

Adding this vector to $\mathbf{x}(0)$ and multiplying $\Psi(t)$ by the result will give the solution of the system (7.165). Note that the vector (7.168) has components that become arbitrarily large as t approaches $+\infty$. This is an effect of resonance on the system.

EXERCISES 7.13

In Exercises 1 through 4 find a solution $\mathbf{x}(t)$, if possible, of each given system that has the same period of oscillation as the forcing term. If the natural and forcing frequencies coincide, show that a particular solution is unbounded as t approaches $\pm\infty$ (that is, resonance occurs).

1. $\mathbf{x}' = \begin{pmatrix} 2 & -5 \\ 0 & 3 \end{pmatrix} \mathbf{x} + \begin{pmatrix} 1 \\ 4 \end{pmatrix} e^{it}$

2. $\mathbf{x}' = \begin{pmatrix} 1 & -2 \\ \frac{5}{2} & -1 \end{pmatrix} \mathbf{x} + \begin{pmatrix} 2 \\ 3 \end{pmatrix} e^{2it}$

3. $\mathbf{x}' = \begin{pmatrix} 3 & 4 \\ -1 & -2 \end{pmatrix} \mathbf{x} + \begin{pmatrix} -1 \\ 4 \end{pmatrix} e^{-5it}$

4. $\mathbf{x}' = \begin{pmatrix} -1 & -7 \\ 5 & 1 \end{pmatrix} \mathbf{x} + \begin{pmatrix} -2 \\ 7 \end{pmatrix} e^{4it}$

5. Suppose that $\mathbf{b}(t) = \sum_{k=1}^m \mathbf{b}_k e^{i\beta_k t}$ (a finite sum of periodic inputs). Assume that $i\beta_k$ is not a root of the auxiliary equation of A for $k = 1, 2, \ldots, m$. Use the principle of super-position (see Exercise 7.12.13) to find a solution $\mathbf{x}(t)$ for (7.159) that can be written in the form

$$\mathbf{x}(t) = \sum_{k=1}^m \mathbf{x}_k e^{i\beta_k t}.$$

6. Use Euler's formula (3.28) to show that

$$2\cos\beta t = e^{i\beta t} + e^{-i\beta t} \quad \text{and} \quad 2i\sin\beta t = e^{i\beta t} - e^{-i\beta t}.$$

Then use the results of Exercise 5 to obtain periodic solutions to the following equations (if possible):

a) $x'' - x' - 2x = \sin t$
b) $x'' + 4x = \cos t$
c) $x'' + 4x = \cos 2t$
d) $x'' + 2x' - 15x = \sin 4t$

7. Verify the form of the principal matrix solution $\Psi(t)$ if ω equals the natural frequency in Example 7.13.2.

8. Verify Eq. (7.168).

9. a) Express the forced vibration equation (3.61) for a spring-weight system with $\mu = 0$ as a 2×2 system of differential equations.

   b) Obtain a solution by the method of this section, given that $\omega \neq \sqrt{\lambda/m}$.

   c) What happens if $\omega = \sqrt{\lambda/m}$? Justify your answer.

NUMERICAL
METHODS

8

In the preceding chapters we developed several techniques for solving differential and difference equations. Unfortunately, these methods are not always applicable to the equations that arise in practical problems. Additionally, even when they do apply, the computational labor of solving, for example, a system of many simultaneous first order equations may be formidable. It is for these reasons that techniques have been developed for computing to any degree of accuracy the numerical solution of almost any such problem. In this chapter we shall discuss some of these methods.

Before discussing particular numerical methods, we stress that care must always be exercised in the utilization of any scheme for the numerical solution of an equation. The accuracy of any solution depends not only on the "correctness" of the numerical method being employed, but also on the precision of the hand calculator or computer used for the tedious arithmetic computations. A detailed discussion of the kinds of error one can expect to encounter will appear in Sections 8.1, 8.2, and 8.5.

8.1 FIRST ORDER EQUATIONS

In this section we shall assume that the initial value problem

$$\frac{dy}{dx} = f(x,y), \qquad y(x_0) = y_0 \tag{8.1}$$

has a unique solution $y(x)$. The three techniques we will describe below approximate this solution $y(x)$ only at a finite number of points

$$x_0, \quad x_1 = x_0 + h, \quad x_2 = x_0 + 2h, \ldots, \quad x_n = x_0 + nh,$$

where h is some (nonzero) real number. The methods provide a value y_k that is an approximation to the exact value $y(x_k)$ for $k = 0, 1, \ldots, n$. As the notation suggests, what we are doing is approximating the equation (8.1) by a suitable difference equation.

Euler's Method

This procedure is crude but very simple. The idea is to obtain y_1 by assuming that $f(x,y)$ varies so little on the interval $x_0 \leqslant x \leqslant x_1$ that only a very small error is made by replacing it by the constant value $f(x_0,y_0)$. Integrating

$$\frac{dy}{dx} = f(x,y)$$

from x_0 to x_1, we obtain

$$y(x_1) - y_0 = y(x_1) - y(x_0) = \int_{x_0}^{x_1} f(x,y) \, dx \approx f(x_0,y_0)(x_1 - x_0) \tag{8.2}$$

or, since $h = x_1 - x_0$,

$$y_1 = y_0 + hf(x_0, y_0). \tag{8.3}$$

Repeating the process with (x_1, y_1) to obtain y_2, etc., we obtain the difference equation

$$y_{n+1} = y_n + hf(x_n, y_n). \tag{8.4}$$

It should be pointed out that we do not intend to solve this difference equation by the methods of Chapter 2; instead, we shall solve it iteratively, that is, by first finding y_1, then using (8.4) to find y_2, and so on.

The geometric meaning of Eq. (8.4) is shown in Fig. 8.1, where the smooth curve is the unknown exact solution of (8.1), which is being approximated by the broken line. Remember that $f(x_0, y_0)$ is the slope of the tangent to $y = y(x)$ at the point (x_0, y_0). The differences Δ_k are errors at the kth stage in the process.

Example 8.1.1 Solve

$$\frac{dy}{dx} = y + x^2, \qquad y(0) = 1. \tag{8.5}$$

We wish to find $y(1)$ by approximating the solution at $x = 0.0, 0.2, 0.4, 0.6, 0.8,$ and 1.0. Here $h = 0.2$, $f(x_n, y_n) = y_n + x_n^2$, and Euler's method (8.4) yields

$$y_{n+1} = y_n + h \cdot f(x_n, y_n).$$

Figure 8.1

Since $y_0 = y(0) = 1$, we obtain

$$y_1 = y_0 + h \cdot f(x_0, y_0) = 1 + 0.2(1 + 0^2) = 1.2,$$
$$y_2 = y_1 + h \cdot f(x_1, y_1) = 1.2 + 0.2[1.2 + (0.2)^2] = 1.448 \approx 1.45,$$

and so on. We arrange our work as shown in Table 8.1. The value $y_5 = 2.77$, corresponding to $x_5 = 1.0$, is our approximate value for $y(1)$.

Table 8.1

x_n	y_n	$f(x_n, y_n) = y_n + x_n^2$	$y_{n+1} = y_n + h \cdot f(x_n, y_n)$
0.0	1.00	1.00	1.20
0.2	1.20	1.24	1.45
0.4	1.45	1.61	1.77
0.6	1.77	2.13	2.20
0.8	2.20	2.84	2.77
1.0	2.77		

If we solve the equation, we will find that $y = 3e^x - x^2 - 2x - 2$, so that $y(1) = 3e - 5 \approx 3.154$. Thus the Euler's method estimate was off by about twelve percent. This is not surprising because we treated the derivative as a constant over intervals of length 0.2 units. The error that arises in this way is called *discretization error*, because the "discrete" function $f(x_n, y_n)$ was substituted for the "continuously valued" function $f(x, y)$. It should be intuitively obvious that if we reduce the step size h, then we can improve the accuracy of our answer, since, then, the "discretized" function $f(x_n, y_n)$ will be closer to the true value of $f(x, y)$ over the interval $[0, 1]$. This is illustrated in Fig. 8.2 with $h \doteq 0.2$ and $h = 0.1$. Indeed, carrying out similar calculations with $h = 0.1$ yields an approximation of $y(1)$ of 3.07, which is a good deal more accurate (an error of about three percent).

Thus, in general, reducing step size will improve accuracy. However, a warning must be attached to this. Reducing the step size will obviously increase the amount of work that has to be done. Moreover, at every stage of the computation *round-off errors* are introduced. For example, in our calculations with $h = 0.2$, we rounded off the exact value 1.448 to the value 1.45 (correct to two decimal places). The rounded off value was then used to calculate further values of y_n. It is not unusual for a computer solution of a more complicated differential equation to take several thousand individual computations, thus having several thousand round-off errors. In some problems (see Section 8.5) the accumulated round-off error can be so large that the resulting computed solution will be sufficiently inaccurate to invalidate the result. Fortunately, this usually does not occur since round-off errors can be positive or negative and tend to cancel one another out. This statement is made under the assumption (usually true) that the average of the round-off errors is

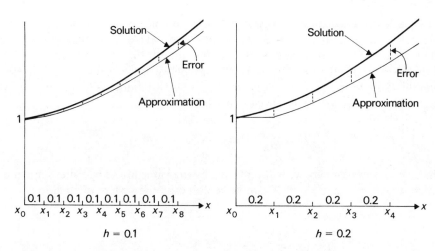

Figure 8.2

zero. In any event, it should be clear that reducing the step size, thereby increasing the number of computations, is a procedure that should be carried out carefully. In general, each problem has an optimal step size, and a smaller than optimal step size will yield a greater error due to accumulated round-off errors.

Returning to our example, it should be apparent that a program using the capabilities of a computer will be needed to get an answer good to three or four decimal places. The following is a typical BASIC program that will perform the needed calculations:

```
10  DEF FNF(X,Y) = X*X + Y
20  INPUT X, Y, H, N
30  FOR K = 0 TO N
40  Y = Y + H*FNF(X,Y)
50  X = X + H
60  NEXT K
70  PRINT Y
80  END
```

This program can be adapted to any equation of the form (8.1) by changing the definition of the function $f(x,y)$ in line 10. In line 20, where the user's teletype produces a "?", one must supply the values x_0, y_0, h, and the number of steps N, where $x_N = x_0 + Nh$.

Improved Euler Method

This method has better accuracy than Euler's method and so is more valuable for hand computation. It is based on the intuitively evident fact that an improvement will result if we average the values of the slope at the left and right endpoints of the interval, thereby reducing the difference between $f(x,y)$ and $f(x_n, y_n)$ in each interval of the form $x_n \leqslant x < x_{n+1}$ (see Fig. 8.3). This amounts to approximating the integral in Eq. (8.2) by the *trapezoidal rule*:

$$\int_{x_0}^{x_1} f(x,y)\, dx \approx \frac{h}{2}\{f(x_0, y_0) + f(x_1, y(x_1))\}.$$

Since $y(x_1)$ is not known, we will replace it by the value found by Euler's method, which we call z_1; then Eq. (8.2) can be replaced by the system of equations

$$z_1 = y_0 + hf(x_0, y_0),$$

$$y_1 = y_0 + \frac{h}{2}[f(x_0, y_0) + f(x_1, z_1)].$$

(8.6)

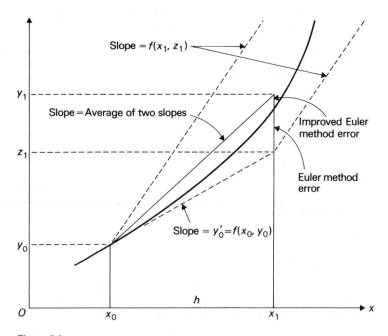

Figure 8.3

This gives us the general procedure

$$z_{n+1} = y_n + hf(x_n, y_n),$$

$$y_{n+1} = y_n + \frac{h}{2}[f(x_n, y_n) + f(x_{n+1}, z_{n+1})].$$

(8.7)

Since $x_0 = 0$ and $y_0 = 1$ in Example 8.1.1, we obtain, with $h = 0.2$,

$$z_1 = y_0 + hf(x_0, y_0) = 1 + 0.2(1 + 0^2) = 1.2,$$

$$y_1 = y_0 + \frac{h}{2}[f(x_0, y_0) + f(x_1, z_1)] = 1 + 0.1[(1 + 0^2) + 1.2 + 0.2^2]$$

$$= 1 + 0.1[2.24] = 1.224 \approx 1.22,$$

$$z_2 = y_1 + hf(x_1, y_1) = 1.22 + 0.2[1.22 + (0.02)^2] = 1.472 \approx 1.47,$$

$$y_2 = y_1 + \frac{h}{2}[f(x_1, y_1) + f(x_2, z_2)]$$

$$= 1.22 + 0.1[1.22 + (0.2)^2 + 1.47 + (0.4)^2] = 1.509 \approx 1.51,$$

and so on. Table 8.2 shows the approximate values of the solution of the Eq. (8.5)

Table 8.2

x_n	y_n	$f(x_n,y_n) = y_n + x_n^2$	z_{n+1}	$f(x_{n+1},z_{n+1}) = z_{n+1} + x_{n+1}^2$	y_{n+1}
0.0	1.0	1.0	1.20	1.24	1.22
0.2	1.22	1.26	1.47	1.63	1.51
0.4	1.51	1.67	1.84	2.20	1.90
0.6	1.90	2.26	2.35	2.99	2.43
0.8	2.43	3.07	3.04	4.04	3.14
1.0	3.14				

used in Example 8.1.1. The error this time is less than one percent. A BASIC program that performs these calculations is given below:

```
10  DEF FNF(X,Y) = X * X + Y
20  INPUT X, Y, H, N
30  FOR K = 0 TO N
35  Z = Y + H * FNF(X,Y)
40  Y = Y + H * (FNF(X,Y) + FNF(X + H,Z))/2
50  X = X + H
60  NEXT K
70  PRINT Y
80  END
```

This program can be obtained from the last one by adding a new line (35) and changing line 40. Changing line 10 will allow us to adapt this program to any equation of the form (8.1).

Runge-Kutta Method

This powerful method gives accurate results without a large number of steps (that is, without the need to make the step size too small). The efficiency is obtained by using a version of Simpson's rule (a frequently used method of numerical integration) in evaluating the integral in Eq. (8.2). We will derive the formula in Section 8.3, but for now we merely describe its use. Let

$$y_{n+1} = y_n + \tfrac{1}{6}(m_1 + 2m_2 + 2m_3 + m_4), \tag{8.8}$$

where

$$m_1 = hf(x_n, y_n),$$

$$m_2 = hf\left(x_n + \frac{h}{2}, y_n + \frac{m_1}{2}\right),$$

$$m_3 = hf\left(x_n + \frac{h}{2}, y_n + \frac{m_2}{2}\right), \tag{8.9}$$

$$m_4 = hf(x_n + h, y_n + m_3).$$

Note that the values m_1/h, m_2/h, m_3/h, and m_4/h are four slopes between $x_n \leqslant x \leqslant x_{n+1}$, so that (8.8) is a weighted average of these slopes—a procedure similar to the one we used in the improved Euler method. Observe that when $f(x,y)$ is a function of x alone, $m_2 = m_3$ and Eq. (8.8) becomes *Simpson's rule*.

We apply the Runge-Kutta method to Eq. (8.5) of Example 8.1.1 with $h = 1$ and $n = 1$:

$$m_1 = f(0,1) = 1,$$
$$m_2 = f(\tfrac{1}{2}, \tfrac{3}{2}) = \tfrac{7}{4},$$
$$m_3 = f(\tfrac{1}{2}, \tfrac{15}{8}) = \tfrac{17}{8},$$
$$m_4 = f(1, \tfrac{25}{8}) = \tfrac{33}{8}.$$

Thus

$$y_1 = 1 + \tfrac{1}{6}(1 + \tfrac{7}{2} + \tfrac{17}{4} + \tfrac{33}{8}) = \tfrac{151}{48} \approx 3.146.$$

In one step this method got us even closer to the correct value than the improved Euler method. A BASIC program to perform the Runge-Kutta method is not hard to write. A FORTRAN program for Eq. (8.5) is given below.

```
      DIMENSION ZM(4)
1     FORMAT (3F,I)
2     FORMAT (2F12.4)
      ACCEPT 1, X, Y, H, N
      DO 7 I = 1,N
      K = 1
      XK = X
      YK = Y
      GO TO 5
3     XK = X + H/2.0
      YK = Y + ZM(K-1)/2.0
      GO TO 5
4     XK = X + H
      YK = Y + ZM(3)
5     ZM(K) = H*(XK*XK+YK)
      K = K + 1
      IF (K-4) 3, 4, 6
6     Y = Y + (ZM(1) + 2.*ZM(2) + 2.*ZM(3) + ZM(4))/6.0
      X = X + H
7     CONTINUE
8     TYPE 2, X, Y
      STOP
      END
```

This program can be adapted to other equations of type (8.1) by introducing the appropriate function $f(x,y)$ in the line identified by statement number 5. To get output at each interval in the procedure, we simply interchange the order of the lines numbered 7 and 8.

The three methods described in this section illustrate the many different numerical schemes available for finding approximate solutions to differential equations. The Euler method is the easiest to apply but requires many steps (a small step size) to achieve any reasonable degree of accuracy. The accuracy of the Euler method as a function of step size will be discussed in detail in the next section. The improved Euler method is more accurate (with the same step size) but involves more arithmetic calculations at each iteration. Finally, the Runge-Kutta method (8.8) usually yields greater accuracy at the cost of a great deal more work at each step. The choice of method often is determined by the accuracy needed. If only an approximate (one or two decimal place) answer is needed, then the Euler method or the improved Euler method will be sufficient. However, for greater accuracy, a more sophisticated numerical scheme is required.

EXERCISES 8.1

1. Apply the Euler method to the following initial value problems.

   a) $\dfrac{dy}{dx} = x + y$, $y(0) = 1$. Find $y(1)$ with $h = 0.2$.

   b) $\dfrac{dy}{dx} = x - y$, $y(1) = 2$. Find $y(3)$ with $h = 0.4$.

   c) $\dfrac{dy}{dx} = \dfrac{x - y}{x + y}$, $y(2) = 1$. Find $y(1)$ with $h = -0.2$.

   Check the accuracy of your answers by finding the exact values.

2. Apply the improved Euler method and the Runge-Kutta method (8.8) to the equations in Exercise 1. Check the accuracy of both schemes by finding the exact value.

3. Let $dy/dx = (x - y)/(x + y)$, $y(2) = 0$. Use Euler's method to find $y(1)$, with $h = -0.2$, and compare your answer with the exact value. Explain why Euler's method fails in this case. Will the Runge-Kutta method work?

4. Let $y' = e^{xy}$, $y(0) = 1$. Use either the BASIC program for Euler's method, with $H = 0.1$ and $N = 40$, or hand calculation with $h = 1$, $N = 4$ to obtain a value for $y(4)$. What difficulties are encountered? How much confidence do you have in your answer?

5. Write a FORTRAN program for the improved Euler method.

6. Write a BASIC program for the Runge-Kutta method.

*8.2 AN ERROR ANALYSIS FOR EULER'S METHOD

In this section we discuss only the discretization errors encountered in the use of Euler's method. Round-off errors depend not only on the method and the number of steps in the calculation, but also on the type of instrument (hand calculator, slide rule, computer, pencil and paper, etc.) used for computing the answer. Round-off error will not be discussed in this section (see Section 8.5), although it never should be ignored.

Let us again consider the first order initial value problem

$$y' = f(x,y), \qquad y(x_0) = y_0, \tag{8.10}$$

and use the iteration scheme

$$y_{n+1} = y_n + hf(x_n, y_n), \tag{8.11}$$

where h is a fixed step size.

We shall assume for the remainder of this section that $f(x,y)$ possesses continuous first partial derivatives. Then on any finite interval $\partial f(x,y)/\partial y$ is bounded by some constant which we denote by L (a continuous function is always bounded on a closed, bounded interval). Since $y'(x) = f(x,y)$, we obtain, by the chain rule,

$$y''(x) = \frac{\partial f}{\partial x}(x,y) + \frac{\partial f}{\partial y}(x,y)y'(x),$$

which must be continuous since it is the sum of continuous functions. Hence $y''(x)$ must be bounded on the interval $x_0 \leqslant x \leqslant a$. So we assume that $|y''(x)| < M$ for some positive constant M.

We now wish to estimate the error e_n at the nth step of the iteration defined by (8.11). Since $y(x_n)$ is the exact value of the solution $y(x)$ at the point $x_n = x_0 + nh$, and y_n is the approximate value at that point, the error at the nth step is given by

$$e_n = y_n - y(x_n). \tag{8.12}$$

Note that $y_0 = y(x_0)$, so that $e_0 = 0$.

Now $y(x_{n+1}) = y(x_n + h)$ and $y''(x)$ is continuous. So we may use Taylor's theorem with remainder to obtain

$$y(x_{n+1}) = y(x_n + h) = y(x_n) + hy'(x_n) + \frac{h^2}{2}y''(\xi_n), \tag{8.13}$$

where $x_n \leqslant \xi_n \leqslant x_{n+1}$. We may now state the main result of this section.

Theorem 8.1 Let $f(x,y)$ have continuous first partial derivatives and let y_n be the approximate solution of (8.10) generated by Euler's method (8.11). Suppose that $y(x)$ is defined and the inequalities

$$\left| \frac{\partial f}{\partial y}(x,y) \right| < L \quad \text{and} \quad |y''(x)| < M$$

hold on the bounded interval $x_0 \leqslant x \leqslant a$. Then the error $e_n = y_n - y(x_n)$ satisfies the inequality

$$|e_n| \leqslant \frac{hM}{2L}(e^{(x_n - x_0)L} - 1) = \frac{hM}{2L}(e^{nhL} - 1). \tag{8.14}$$

In particular, since $x_n - x_0 \leqslant a - x_0$ (which is finite), every $|e_n|$ tends to zero as h tends to zero.

Proof. Subtracting (8.13) from (8.11) yields

$$y_{n+1} - y(x_{n+1}) = y_n - y(x_n) + h[f(x_n,y_n) - y'(x_n)] - \frac{h^2}{2}y''(\xi_n)$$

or

$$e_{n+1} = e_n + h\{f(x_n,y_n) - f(x_n,y(x_n))\} - \frac{h^2}{2}y''(\xi_n). \tag{8.15}$$

By the mean value theorem of differential calculus,

$$f(x_n,y_n) - f(x_n,y(x_n)) = \frac{\partial f}{\partial y}(x_n,\hat{y}_n)[y_n - y(x_n)] = \frac{\partial f}{\partial y}(x_n,\hat{y}_n)e_n, \tag{8.16}$$

where $\hat{y}_n$ is between y_n and $y(x_n)$. We substitute (8.16) into (8.15) to obtain

$$e_{n+1} = e_n + h\frac{\partial f}{\partial y}(x_n,\hat{y}_n)e_n - \frac{h^2}{2}y''(\xi_n). \tag{8.17}$$

But $|\partial f/\partial y| \leqslant L$ and $|y''| \leqslant M$, so that taking the absolute value of both sides of (8.17) and using the triangle inequality, we obtain

$$|e_{n+1}| \leqslant |e_n| + hL|e_n| + \frac{h^2}{2}M = (1 + hL)|e_n| + \frac{h^2}{2}M. \tag{8.18}$$

We now consider the first order difference equation

$$r_{n+1} = (1 + hL)r_n + \frac{h^2}{2}M, \qquad r_0 = 0, \tag{8.19}$$

and claim that if r_n is the solution to (8.19), then $|e_n| \leqslant r_n$. We show this by induction. It is true for $n = 0$, since $e_0 = r_0 = 0$. We assume it is true for $n = k$ and prove it for $n = k + 1$. That is, we assume that $|e_m| \leqslant r_m$, for $m = 0, 1, \ldots, k$. Then

$$r_{k+1} = (1 + hL)r_k + \frac{h^2}{2}M \geqslant (1 + hL)|e_k| + \frac{h^2}{2}M \geqslant |e_{k+1}|,$$

and the claim is proved. [The last step follows from (8.18).] We can solve the constant coefficient first order nonhomogeneous difference equation (8.19) by the methods of Section 2.6. Since $a_j = 1 + hL$ and $f_j = h^2M/2$ for all j, and $r_0 = 0$, using Eq. (2.67) and the sum of the first n terms of a geometric progression, we obtain

$$r_n = \frac{hM}{2L}(1 + hL)^n - \frac{hM}{2L}. \tag{8.20}$$

Now, $e^{hL} = 1 + hL + h^2L^2/2! + \cdots$, so that

$$1 + hL \leqslant e^{hL} \qquad \text{and} \qquad (1 + hL)^n \leqslant (e^{hL})^n = e^{nhL}.$$

Thus

$$|e_n| \leqslant r_n \leqslant \frac{hM}{2L}e^{nhL} - \frac{hM}{2L} = \frac{hM}{2L}(e^{nhL} - 1). \tag{8.21}$$

But $x_n = x_0 + nh$, so that $x_n - x_0 = nh$ and (8.21) becomes

$$|e_n| \leqslant \frac{hM}{2L}(e^{(x_n - x_0)L} - 1),$$

and the theorem is proved.

Theorem 8.1 not only shows that the errors get small as h tends to zero, but also tells us *how fast* the errors decrease. If we define the constant k by

$$k = \frac{M}{2L} \left| e^{(a-x_0)L} - 1 \right|, \tag{8.22}$$

then we have

$$|e_n| \leqslant kh. \tag{8.23}$$

Thus the error is bounded by a *linear* function of h. (Note that $|e_n|$ is bounded by a term that depends only on h, not on n.) Roughly speaking, this implies that the error decreases at a rate proportional to the decrease in the step size. If, for example, we halve the step size, then we can expect at least to halve the error. Actually, since the estimates used in arriving at (8.23) were very crude, we can often do better, as in Example 8.1.1, where we halved the step size and decreased the error by a factor of four. Nevertheless, it is useful to have an upper bound for the error. It should be noted, however, that this bound may be difficult to obtain, since it is frequently difficult to find a bound for $y''(x)$.

Example 8.2.1 Consider the equation $y' = y$, $y(0) = 1$. We have

$$f(x,y) = y \quad \text{and} \quad \left| \frac{\partial f}{\partial y}(x,y) \right| = 1 = L.$$

Since the solution of the problem is $y(x) = e^x$, we have $|y''| \leqslant e^1 = M$ on the interval $0 \leqslant x \leqslant 1$. Then Eq. (8.22) becomes

$$k = \frac{e}{2} |e - 1| = \frac{e^2 - e}{2} \approx 2.33,$$

so that

$$|e_n| \leqslant 2.33h.$$

Therefore, using a step size of $h = 0.1$, say, we can expect to have an error at each step of less than 0.233 (see Table 8.3). We note that the greatest actual error is about half of the maximum possible error according to (8.23).

It turns out that it is possible to derive error estimates like (8.21) or (8.23) for every method we shall discuss in this chapter for solving differential equations numerically. To actually derive these estimates would take us beyond the scope of this book*, but we should mention that for the Runge-Kutta method discussed

---

* For a more detailed analysis, see, for example, C. W. Gear, *Numerical Initial Value Problems in Ordinary Differential Equations*, Prentice-Hall, Englewood Cliffs, N.J., 1971.

Table 8.3

x_n	$y_n' = f(x_n,y_n)$	$y_{n+1} = y_n + hy_n'$	$y(x_n) = e^{x_n}$	$e_n = y_n - y(x_n)$
0.0	1	1.1	1	0
0.1	1.1	1.21	1.11	−0.01
0.2	1.21	1.33	1.22	− .01
0.3	1.33	1.46	1.35	− .02
0.4	1.46	1.61	1.49	− .03
0.5	1.61	1.77	1.65	− .04
0.6	1.77	1.95	1.82	− .05
0.7	1.95	2.15	2.01	− .06
0.8	2.15	2.37	2.23	− .08
0.9	2.37	2.61	2.46	− .09
1.0	2.61		2.72	− .11

in Section 8.1, the discretization error e_n is of the form

$$|e_n| \leqslant kh^4,$$

for some appropriate constant k. Thus halving the step size for example, has the effect of decreasing the bound on the error by a factor of $2^4 = 16$. However, the price for this greater accuracy is to have to calculate $f(x,y)$ at four points [see (8.9)] for each step in the iteration.

EXERCISES 8.2

1. Consider the differential equation $y' = -y$, $y(0) = 1$. We wish to find $y(1)$.
   a) Calculate an upper bound on the error of Euler's method as a function of h.
   b) Calculate this bound for $h = 0.1$ and $h = 0.2$.
   c) Perform the iterations for $h = 0.2$ and $h = 0.1$ and compare the actual error with the maximum error.

2. Consider the equation of Exercise 1. Ignoring round-off error, how many iterations would have to be performed in order to guarantee that the calculation of $y(1)$ obtained by Euler's method be correct to:
   a) five decimal places?
   b) six decimal places?

3. Answer the questions in Exercise 2 for the equation

   $$y' = 3y - x^2, \qquad y(1) = 2$$

   if we wish to find $y(1.5)$.

8.3 RUNGE-KUTTA FORMULAS

In this section we will derive some techniques, similar to the Runge-Kutta method (8.8), (8.9), for the numerical calculation of a solution to an initial value problem. In Euler's method we used one value for the derivative $y' = f(x,y)$ for each iteration. In the improved Euler's method we used two values. In the Runge-Kutta formulas we shall discuss, we shall have to make use of four values of the derivative for each iteration. We will show how to find the four "best" values in a sense to be made precise later.

To begin, we need to recall the Taylor series expansions of functions of one or two variables:

$$y(x_0 + h) = y(x_0) + hy'(x_0) + \frac{h^2}{2!} y''(x_0) + \frac{h^3}{3!} y'''(x_0) + \cdots, \tag{8.24}$$

$$\begin{aligned} f(x_0 + mh, y_0 + nh) &= f(x_0, y_0) + h(mf_x + nf_y) \\ &+ \frac{h^2}{2!}(m^2 f_{xx} + 2mn f_{xy} + n^2 f_{yy}) \\ &+ \frac{h^3}{3!}(m^3 f_{xxx} + 3m^2 n f_{xxy} + 3mn^2 f_{xyy} + n^3 f_{yyy}) \\ &+ \cdots, \end{aligned} \tag{8.25}$$

where the partials are all evaluated at the point (x_0, y_0). Since

$$y' = f(x,y), \tag{8.26}$$

we find that

$$y'' = f_x + (f_y)y' = f_x + ff_y, \tag{8.27}$$

$$y''' = f_{xx} + 2ff_{xy} + f^2 f_{yy} + f_y(f_x + ff_y), \tag{8.28}$$

and so on. (Note that $f_{xy} = f_{yx}$ when these derivatives are continuous.) Thus (8.24) can be written in the form

$$\begin{aligned} y_1 - y_0 &= hf + \frac{h^2}{2}(f_x + ff_y) \\ &+ \frac{h^3}{6}[f_{xx} + 2ff_{xy} + f^2 f_{yy} + f_y(f_x + ff_y)] + \cdots. \end{aligned} \tag{8.29}$$

The main idea is somehow to select several values (x,y) so that the Taylor series expansions (8.25) of the corresponding $f(x,y)$ terms coincide with the terms on the right-hand side of Eq. (8.29). Suppose we let

$$m_1 = hf(x_0,y_0),$$
$$m_2 = hf(x_0 + nh,y_0 + nm_1),$$
$$m_3 = hf(x_0 + ph,y_0 + pm_2),$$
$$m_4 = hf(x_0 + qh,y_0 + qm_3).$$

(The reason for doing this will be made clear shortly.) Using (8.25), we may write these values as

$$m_1 = hf,$$
$$m_2 = h\left[f + nh(f_x + ff_y) + \frac{(nh)^2}{2}(f_{xx} + 2ff_{xy} + f^2f_{yy}) + \cdots \right],$$
$$m_3 = h\left\{ f + ph(f_x + ff_y) + \frac{h^2}{2}[p^2(f_{xx} + 2ff_{xy} + f^2f_{yy}) + 2npf_y(f_x + ff_y)] + \cdots \right\},$$
$$m_4 = h\left\{ f + qh(f_x + ff_y) + \frac{h^2}{2}[q^2(f_{xx} + 2ff_{xy} + f^2f_{yy}) + 2pqf_y(f_x + ff_y)] + \cdots \right\},$$

where all functions are evaluated at the point (x_0,y_0).

We now consider an expression of the form

$$am_1 + bm_2 + cm_3 + dm_4$$

and try to equate it to the right-hand side of Eq. (8.29). This will have the effect of giving us a numerical scheme that agrees with the solution to (8.26) up to and including third order terms. Then the error will be no greater than terms like kh^4, and so on. Matching like expressions, we find

$$a + b + c + d = 1,$$
$$bn + cp + dq = \tfrac{1}{2},$$
$$bn^2 + cp^2 + dq^2 = \tfrac{1}{3}, \qquad (8.30)$$
$$cnp + dpq = \tfrac{1}{6}.$$

Now any solution of these equations will produce a method where there is no error up to the third order terms. Suppose we take $n = p = \tfrac{1}{2}$ and $q = 1$. Then (8.30) reduces to the system of equations

$$a + b + c + d = 1,$$
$$b + c + 2d = 1,$$
$$3b + 3c + 12d = 4,$$
$$3c + 6d = 2,$$

which has the solution $a = d = \tfrac{1}{6}$, $b = c = \tfrac{1}{3}$. Thus the Runge-Kutta formula

(8.8) described in the last section agrees with $y_1 - y_0$ for all terms up to and including the terms in h^3. Actually, with quite a bit more work, one can show that they agree in the h^4 terms, too. Thus the error (if any) involves only terms in h^5 and higher. Hence, for small h, we should expect to get very good results.

Other formulas are also readily derivable. Suppose we choose $n = \frac{1}{3}, p = \frac{2}{3}, q = 1$. Then (8.30) yields

$$a + b + c + d = 1,$$
$$2b + 4c + 6d = 3,$$
$$b + 4c + 9d = 3,$$
$$4c + 12d = 3,$$

which has the solution $a = d = \frac{1}{8}, b = c = \frac{3}{8}$. The formula, known as the *Kutta-Simpson $\frac{3}{8}$-rule*, may be written as

$$y_1 = y_0 + \frac{1}{8}(m_1 + 3m_2 + 3m_3 + m_4),$$

where

$$
\begin{aligned}
m_1 &= hf(x_0,y_0)\\
m_2 &= hf(x_0 + h/3, y_0 + m_1/3),\\
m_3 &= hf(x_0 + 2h/3, y_0 + 2m_2/3),\\
m_4 &= hf(x_0 + h, y_0 + m_3).
\end{aligned}
\tag{8.31}
$$

Similarly, the choice $n = \frac{1}{3}, p = \frac{2}{3}, q = 1$ also yields the solution $a = c = 0$, $b = \frac{3}{4}, d = \frac{1}{4}$, and hence

$$y_1 = y_0 + \frac{1}{4}(3m_2 + m_4),$$

where m_2 and m_4 are defined as in (8.31). Since the number of possible choices of n, p, and q is infinite, the reader may amuse himself by solving (8.30) for a, b, c, d with whatever values of n, p, and q that he selects.

EXERCISES 8.3

1. Set $d = 0$ in Eq. (8.30) and let $n = \frac{1}{3}, p = \frac{2}{3}$. Then find the coefficients a, b, c for these choices. The resulting formula is called *Heun's formula*.

2. Set $a = d = 0, b = \frac{3}{4}, c = \frac{1}{4}$ and find n and p.

3. Set $a = \frac{2}{8}, b = c = \frac{3}{8}, d = 0$ and find n and p.

4. Prove that the Runge-Kutta formula agrees with (8.29) up to and including the h^4 terms.

5. Explain why the choice $n = \frac{1}{3}, p = \frac{2}{3}, q = 1$ yielded two sets of solutions a, b, c, d. Are other solutions possible? Does the choice $n = p = \frac{1}{2}, q = 1$ allow for multiple solutions?

8.4 PREDICTOR-CORRECTOR FORMULAS

The simplest type of predictor-corrector formula is one that we have already used, namely the improved Euler method. Recall that in this method, we first obtain a value for y_n by Euler's method (the *predicting* part of the process) and then improve on the accuracy of the method by applying a trapezoidal rule (the *correcting* phase of the process). In this section we will indicate how such methods are derived, and provide several procedures that will perform their tasks with a high degree of accuracy.

We begin by developing a procedure for obtaining *quadrature formulas*. These formulas are designed to obtain the approximate value of a definite integral by using equally spaced values of the integrand. Two well-known quadrature formulas are the trapezoidal rule and Simpson's rule,* which can be written in the forms

$$y_1 - y_0 = \frac{h}{2}[y_0' + y_1'], \tag{8.32}$$

$$y_2 - y_0 = \frac{h}{3}[y_0' + 4y_1' + y_2'], \tag{8.33}$$

respectively, where y' is the integrand and h is the interval between values of the integrand. Note that, unlike in previously discussed methods, here we have slightly shifted the point of view by representing the right-hand side as a sum of derivatives instead of values of the function f.

In (8.32) we are only using the points y_0' and y_1' while in (8.33) we also use y_2'. For this reason, (8.32) is called a 2-point quadrature formula, while (8.33) is a 3-point quadrature formula. In general, the more points involved, the higher the accuracy. In the present case, (8.32) provides the exact integral only for straight lines, while (8.33) is also exact for quadratic polynomials (see Exercises 5 and 6).

We shall now develop a method for obtaining 3-point quadrature formulas. The method is easily adapted to n-point quadrature formulas.

3-Point Methods

Assume that the function $y(x)$ is a quadratic polynomial

$$y(x) = a_0 + a_1x + a_2x^2. \tag{8.34}$$

Let h and x_0 be fixed real numbers and define $x_k = x_0 + kh$, for all integers k. Let $y_k = y(x_k)$. We wish to find coefficients A_1, A_2, A_3 such that

$$y_j - y_i = h[A_0y_0' + A_1y_1' + A_2y_2'], \tag{8.35}$$

---

* These rules can be found in most calculus books.

where $i, j = 0, 1, 2$ and $i \neq j$. This will lead to an integration scheme that is exact for quadratic polynomials. For example, if $i = 0$, then the left-hand side of (8.35) becomes

$$y_j - y_0 = a_1(x_j - x_0) + a_2(x_j^2 - x_0^2)$$
$$= a_1 jh + a_2[2x_0 jh + (jh)^2] = j(a_1 h + 2a_2 x_0 h) + j^2(a_2 h^2). \quad (8.36)$$

On the other hand, using the facts that $y' = a_1 + 2a_2 x$ and $x_k = x_0 + kh$, we find that the right-hand side of (8.35) becomes

$$h[A_0 y_0' + A_1 y_1' + A_2 y_2']$$
$$= h[A_0(a_1 + 2a_2 x_0) + A_1(a_1 + 2a_2 x_1) + A_2(a_1 + 2a_2 x_2)]$$
$$= (A_0 + A_1 + A_2)(a_1 h + 2a_2 x_0 h) + (2A_1 + 4A_2)(a_2 h^2). \quad (8.37)$$

Equating (8.36) and (8.37), we obtain the simultaneous equations

$$A_0 + A_1 + A_2 = j,$$
$$2A_1 + 4A_2 = j^2. \quad (8.38)$$

Since this system is underdetermined, it has an infinite number of solutions A_0, A_1, A_2, each leading to a different quadrature formula. For example, if $j = 1$, we can pick $A_0 = \frac{5}{6}, A_1 = -\frac{1}{6}, A_2 = \frac{1}{3}$ and arrive at the formula

$$y_1 - y_0 = \frac{h}{6}(5y_0' - y_1' + 2y_2'). \quad (8.39)$$

If $j = 2$, then selecting $A_0 = \frac{1}{3}, A_1 = \frac{4}{3}, A_2 = \frac{1}{3}$ will yield Simpson's rule (8.33). In addition, quadrature formulas may be added or subtracted to yield new formulas. For example, subtracting (8.39) from (8.33) yields

$$y_2 - y_1 = \frac{h}{2}[-y_0' + 3y_1']. \quad (8.40)$$

Let us now illustrate how predictor-corrector formulas are employed. Suppose we are given the initial value problem

$$\frac{dy}{dx} = f(x,y), \qquad y(x_0) = y_0. \quad (8.41)$$

The value y_0 is given, and y_0' can be obtained by evaluating (8.41) at $x = x_0, y = y_0$. We next perform an improved Euler method computation to obtain y_1 [and y_1' by means of (8.41)]. At this point it is often desirable to repeatedly apply the trapezoidal rule of the process until the value of y_1 stabilizes (that is, remains unchanged to a given number of decimal places; see Example 8.4.1). Then since y_0, y_0', y_1, y_1' are all known, we may use Eq. (8.40) to *predict* the value of y_2. This value is then used in (8.41) to obtain y_2', and Simpson's rule (8.33) is used to *correct* the y_2 value previously determined. The process is now repeated to predict and correct y_3 and y_3' in terms of the known values y_1, y_1', y_2, y_2'. We use

$$y_{n+2} - y_{n+1} = \frac{h}{2}\left[-y_n' + 3y_{n+1}'\right] \tag{8.42}$$

to *predict* the value of y_{n+2}, (8.41) to obtain y_{n+2}',

$$y_{n+2} - y_n = \frac{h}{3}\left[y_n' + 4y_{n+1}' + y_{n+2}'\right] \tag{8.43}$$

to *correct* the value of y_{n+2}, and (8.41) again to obtain y_{n+2}'.

Example 8.4.1 Let $dy/dx = y + x^2$, $y(0) = 1$, and suppose we wish to find $y(1)$, with $h = 0.2$, by the predictor-corrector formulas (8.42) and (8.43). Then $y_0 = 1$, and to find y_1, we use the improved Euler method repeatedly:

x_0	y_0	y_0'	$\bar{y}_1$	$\bar{y}_1'$	y_1
0.0	1.0	1.0	1.20	1.24	1.22
			1.22	1.26	1.23
			1.23	1.27	1.23

The first entry in the $\bar{y}_1$ column is obtained by Euler's formula: $y_1 = y_0 + hy_0'$. The $\bar{y}_1'$ column is obtained from the differential equation $\bar{y}_1' = f(x_1, \bar{y}_1) = \bar{y}_1 + x_1^2$. The last column is found by the improved Euler formula

$$y_1 = y_0 + \frac{h}{2}(y_0' + \bar{y}_1'),$$

and the corrected value for y_1 is transferred to the $\bar{y}_1$ column. The process is then repeated. We see that the process stabilizes with $y_1 = 1.23$, $y_1' = 1.27$ (provided that we are using two decimal place accuracy). We now apply (8.42) and (8.43):

P: $\quad y_2 = y_1 + \frac{h}{2}(-y_0' + 3y_1') = 1.23 + (0.1)(-1.0 + 3.81) = 1.51,$

$\quad y_2' = f(x_2, y_2) = y_2 + x_2^2 = 1.67$

C: $\quad y_2 = y_0 + \frac{h}{3}(y_0' + 4y_1' + y_2') = 1.0 + \left(\frac{0.2}{3}\right)(1.0 + 5.08 + 1.67)$

$\quad = 1.52,$

$\quad y_2' = y_2 + x_2^2 = 1.68$

P: $y_3 = y_2 + \dfrac{h}{2}(-y_1' + 3y_2') = 1.52 + (0.1)(-1.27 + 5.04) = 1.90,$

   $y_3' = y_3 + x_3^2 = 2.26$

C: $y_3 = y_1 + \dfrac{h}{3}(y_1' + 4y_2' + y_3') = 1.23 + \left(\dfrac{0.2}{3}\right)(1.27 + 6.72 + 2.26)$

   $= 1.91,$

   $y_3' = y_3 + x_3^2 = 2.27$

P: $y_4 = y_3 + \dfrac{h}{2}(-y_2' + 3y_3') = 2.42,$

   $y_4' = 3.06$

C: $y_4 = y_2 + \dfrac{h}{3}(y_2' + 4y_3' + y_4') = 2.44,$

   $y_4' = 3.08$

P: $y_5 = y_4 + \dfrac{h}{2}(-y_3' + 3y_4') = 3.14,$

   $y_5' = 4.14$

C: $y_5 = y_3 + \dfrac{h}{3}(y_3' + 4y_4' + y_5') = 3.16$

The result is slightly better than the answer obtained by the improved Euler method (with a lot more work).

A very accurate predictor-corrector method due to Milne uses a 4-point quadrature formula as a predictor and Simpson's rule as a corrector:

P: $y_{n+4} - y_n = \dfrac{4h}{3}\left[2y_{n+1}' - y_{n+2}' + 2y_{n+3}'\right]$ (8.44)

C: $y_{n+4} - y_{n+2} = \dfrac{h}{3}\left[y_{n+2}' + 4y_{n+3}' + y_{n+4}'\right]$

To apply this method, it is necessary to have good values for $y_0, y_0', y_1, y_1', y_2, y_2', y_3, y_3'$.

Predictor-corrector methods have advantages and disadvantages from the point of view of accuracy and the amount of work involved. Clearly the "correction" makes each step, and therefore the final answer, more accurate—with the same step size. However, the greater number of calculations can be expensive (in terms of increased computer costs) and lead to greater round-off error.

EXERCISES 8.4

1. Obtain the trapezoidal rule (8.32) by using the 3-point quadrature formulas (8.35) and (8.38).

2. Obtain the 3-point quadrature formulas:

   a) $y_1 - y_0 = \dfrac{h}{12}(5y_0' + 8y_1' - y_2')$

   b) $y_2 - y_0 = \dfrac{h}{8}(y_0' + 14y_1' + y_2')$

   c) $y_2 - y_0 = \dfrac{h}{4}(y_0' + 6y_1' + y_2')$

3. Obtain the 3-point quadrature formula

   $$y_3 - y_0 = \frac{h}{2}[-y_0' + 5y_1' + 2y_2'].$$

4. Use the formula obtained in Exercise 3 and Simpson's rule as a predictor-corrector to solve the initial value problem

   $$y' = x + y, \qquad y(0) = 1,$$

   for $y(1)$ with $h = 0.2$. Use the improved Euler method to find y_1 and y_2.

5. Let $y(x) = ax + b$. Show that the trapezoidal rule (8.32) provides the exact value for $\int_0^1 y(x)\, dx$ for step sizes $h = \frac{1}{2}$ and $\frac{1}{3}$.

6. Let $y(x) = ax^2 + bx + c$. Show that Simpson's rule (8.33) provides the exact value for $\int_0^1 y(x)\, dx$ for step sizes $h = \frac{1}{4}$ and $\frac{1}{8}$.

7. Show that the equations for 4-point quadrature formulas analogous to (8.38) are

   $$A_0 + A_1 + A_2 + A_3 = j,$$
   $$2A_1 + 4A_2 + 6A_3 = j^2,$$
   $$3A_1 + 12A_2 + 27A_3 = j^3.$$

   Use these equations to derive Milne's equation (8.44).

8. Obtain the underdetermined system of equations for 5-point quadrature formulas analogous to those in Exercise 7.

9. Use Milne's equation (8.44) to solve the equation in Exercise 4 with $h = 0.2$.

8.5 THE PROPAGATION OF ROUND-OFF ERROR: AN EXAMPLE OF NUMERICAL INSTABILITY

In this section we will show how a theoretically very accurate method can produce results that are useless. A *multistep method* is a method (such as predictor-corrector formulas) that involves information about the solution at more than one point.

Consider the multistep method given by the equation

$$y_{n+1} = y_{n-1} + 2hf(x_n,y_n). \tag{8.45}$$

Here it is necessary to use both the nth and the $(n - 1)$st iterate to obtain the $(n + 1)$st iterate. It can be shown* that this method has the following error estimate:

$$|e_n| = |y_n - y(x_n)| \leqslant kh^2.$$

In particular, since for Euler's method $|e_n| \leqslant kh$, we would theoretically expect more accuracy in solving our initial value problem by using Eq. (8.45) than by Euler's method. However, this does not always turn out to be the case.

Example 8.5.1 Consider the initial value problem

$$y' = -y + 2, \qquad y(0) = 1.$$

The solution to this equation is easily obtained: $y(x) = 2 - e^{-x}$. Let us obtain $y(5)$ by Euler's method and the method of (8.45). To use the latter, we need two initial values y_0 and y_1. Since we know the solution, we use the exact value $y_1 = y(x_1) = 2 - e^{-x_1}$. Table 8.4 illustrates the computation with a step size $h = 0.25$. The second column is the correct value of $y(x_n)$ to four decimal places. Column three gives the Euler iterates, while column four gives the iterates obtained by the two-step method (8.45). Column five is the Euler error, $e_n^{(\text{Euler})} = y_n^{(\text{Euler})} - y(x_n)$, and column six is the error of the two-step method, $e_n^{(2s)} = y_n^{(2s)} - y(x_n)$.

It is evident that the two-step method (8.45) produces a smaller error for small values of x_n than Euler's method. However, as x_n increases, the error in Euler's method decreases, while the error in the two-step method not only increases but does so with oscillating sign. This phenomenon is called *numerical instability*. As we shall see, it is due to a propagation of round-off errors.

Let us now explain what will lead to this instability. In the example, $f(x_n,y_n) = -y_n + 2$, so that Eq. (8.45) is

$$y_{n+1} = y_{n-1} + 2h(2 - y_n)$$

or

$$y_{n+1} + 2hy_n - y_{n-1} = 4h, \qquad y_0 = 1. \tag{8.46}$$

Equation (8.46) is a linear nonhomogeneous second order difference equation with constant coefficients that can be solved by the methods of Section 4.6. The general solution is given by

$$y_n = c_1\lambda_1^n + c_2\lambda_2^n + 2, \tag{8.47}$$

---

* See S. D. Conte, *Elements of Numerical Analysis*, McGraw-Hill, New York, 1965, Section 6.6.

Table 8.4

x_n	$y(x_n) =$ $2 - e^{-x_n}$	$y_n^{(E)} =$ $y_{n-1}^{(E)} + h(2 - y_{n-1}^{(E)})$	$y_n^{(2s)} =$ $y_{n-2}^{(2s)} + 2h(2 - y_{n-1}^{(2s)})$	$e_n^{(E)}$	$e_n^{(2s)}$
0.00	1.0000	1.0000	1.0000	0.0000	0.0000
0.25	1.2212	1.2500	1.2212	.0288	0.0000
0.50	1.3935	1.4375	1.3894	.0440	− .0041
0.75	1.5276	1.5781	1.5265	.0505	− .0011
1.00	1.6321	1.6836	1.6262	.0515	− .0059
1.25	1.7135	1.7627	1.7134	.0492	− .0001
1.50	1.7769	1.8220	1.7695	.0453	− .0074
1.75	1.8262	1.8665	1.8287	.0403	+ .0025
2.00	1.8647	1.8999	1.8552	.0352	− .0095
2.25	1.8946	1.9249	1.9011	.0303	+ .0065
2.50	1.9179	1.9437	1.9047	.0258	− .0132
2.75	1.9361	1.9578	1.9488	.0217	+ .0127
3.00	1.9502	1.9684	1.9303	.0182	− .0199
3.25	1.9612	1.9763	1.9837	.0151	+ .0225
3.50	1.9698	1.9822	1.9385	.0124	− .0313
3.75	1.9765	1.9867	2.0145	.0102	+ .0380
4.00	1.9817	1.9900	1.9313	.0083	− .0504
4.25	1.9857	1.9925	2.0489	.0068	+ .0632
4.50	1.9889	1.9944	1.9069	.0055	− .0820
4.75	1.9913	1.9958	2.0955	.0045	+ .1042
5.00	1.9933	1.9969	1.8952	.0036	− .0981

where λ_1 and λ_2 are the roots of the auxiliary equation

$$\lambda^2 + 2h\lambda - 1 = 0.$$

Thus,

$$\lambda_1 = \frac{-2h + \sqrt{4h^2 + 4}}{2} = -h + \sqrt{1 + h^2} \quad \text{and}$$

$$\lambda_2 = -h - \sqrt{1 + h^2}. \tag{8.48}$$

By the binomial theorem (see Exercise 25 of Section 5.1),

$$(1 + h^2)^{1/2} = 1 + \tfrac{1}{2}h^2 - \tfrac{1}{8}h^4 + \tfrac{1}{16}h^6 - \cdots,$$

where the omitted terms are higher powers of h. Hence the roots (8.48) of the auxiliary equation can be written as

$$\lambda_1 = 1 - h + \alpha(h) \quad \text{and} \quad \lambda_2 = -1 - h - \alpha(h), \tag{8.49}$$

where

$$\alpha(h) = \frac{h^2}{2} - \frac{h^4}{8} + \frac{h^6}{16} - \cdots.$$

Substituting (8.49) into (8.47) yields

$$y_n = c_1[1 - h + \alpha(h)]^n + c_2(-1)^n[1 + h + \alpha(h)]^n + 2. \tag{8.50}$$

From calculus we know that

$$\lim_{h \to \infty} \left(1 + \frac{1}{h}\right)^h = \lim_{\varepsilon \to 0} (1 + \varepsilon)^{1/\varepsilon} = e.$$

Therefore, since $x_n = 0 + nh = nh$, we have

$$\lim_{h \to 0} (1 - h)^n = \lim_{h \to 0} (1 - h)^{x_n/h} = e^{-x_n}, \qquad \lim_{h \to 0} (1 + h)^n = e^{x_n}.$$

Hence as $h \to 0$, we may ignore the higher order terms $\alpha(h)$ in (8.50) to obtain

$$y_n = c_1 e^{-x_n} + 2 + c_2(-1)^n e^{x_n}. \tag{8.51}$$

Here lies the problem. The exact solution of the problem requires that $c_1 = -1$ and $c_2 = 0$. However, even a small round-off error may cause c_2 to be nonzero and this error will grow exponentially while the real solution is approaching the constant two. This is the phenomenon we observed in Table 8.4. Note that the $(-1)^n$ in (8.51) causes the errors to oscillate (as we also observed).

The problem arose because we approximated a *first* order differential equation by a *second* order difference equation. Such approximations do not always lead to this kind of instability, but it is a possibility that cannot be ignored. In general, to analyze the effectiveness of a given method, we must not only estimate the discretization error, but also show that the method is not numerically unstable (that is, *it is numerically stable*), for the given problem.

8.6 SYSTEMS AND BOUNDARY VALUE PROBLEMS

The methods developed in Sections 8.1, 8.3, and 8.4 can be extended very easily to apply to higher order equations and systems of equations.

Euler Methods

For this technique and the improved Euler method, it is necessary only to re-interpret the formulas

$$y_{n+1} = y_n + hf(x_n, y_n) \tag{8.52}$$

and

$$y_{n+1} = y_n + \frac{h}{2}[f(x_n, y_n) + f(x_{n+1}, y_{n+1})], \tag{8.53}$$

with y_n a vector with as many entries as there are dependent variables. In this case, the function f consists of a vector of functions also. The methods are best illustrated by examples.

Example 8.6.1 Consider the initial value problem

$$\frac{dx}{dt} = -3x + 4y, \qquad x(0) = 1,$$

$$\frac{dy}{dt} = -2x + 3y, \qquad y(0) = 2.$$

(8.54)

Suppose we are seeking the values $x(1)$ and $y(1)$. In this problem t is the independent variable, while x and y are the dependent variables. If we wish to use Euler's method, formula (8.52) translates into the equations

$$x_{n+1} = x_n + hx'_n = x_n + h(-3x_n + 4y_n),$$

$$y_{n+1} = y_n + hy'_n = y_n + h(-2x_n + 3y_n).$$

(8.55)

The initial values are $x_0 = 1$, $y_0 = 2$, and $h = 0.2$, and the procedure is essentially the same as before (see Table 8.5).

Table 8.5

t_n	x_n	y_n	x'_n	y'_n	$x_{n+1} = x_n + hx'_n$	$y_{n+1} = y_n + hy'_n$
0.0	1.00	2.00	5.00	4.00	2.00	2.80
0.2	2.00	2.80	5.20	4.40	3.04	3.68
0.4	3.04	3.68	5.60	4.96	4.16	4.67
0.6	4.16	4.67	6.20	5.69	5.40	5.81
0.8	5.40	5.81	7.04	6.63	6.81	7.14
1.0	6.81	7.14				

The solution of (8.54) is given by

$$x(t) = 3e^t - 2e^{-t}, \qquad y(t) = 3e^t - e^{-t},$$

so $x(1) = 3e - 2e^{-1} \approx 7.419$ and $y(1) = 3e - e^{-1} \approx 7.787$, implying that our method has an error of about ten percent. The accuracy may be improved by selecting smaller values of h.

No additional difficulty is caused by having a nonhomogeneous or even non-linear system of equations.

Example 8.6.2 Consider the nonlinear system

$$\frac{dx}{dt} = x - y^2 + t, \qquad x(0) = 1,$$

$$\frac{dy}{dt} = x^2 + y - t, \qquad y(0) = 2.$$

(8.56)

We wish to find $x(1)$ and $y(1)$ with $h = 0.25$ by means of the improved Euler method. For this problem we obtain

$$\bar{x}_{n+1} = x_n + hx_n' = x_n + h(x_n - y_n^2 + t_n),$$

$$\bar{y}_{n+1} = y_n + hy_n' = y_n + h(x_n^2 + y_n - t_n),$$

$$x_{n+1} = x_n + \frac{h}{2}(x_n' + \bar{x}_{n+1}')$$

$$= x_n + \frac{h}{2}[(x_n - y_n^2 + t_n) + (\bar{x}_{n+1} - \bar{y}_{n+1}^2 + t_{n+1})],$$

$$y_{n+1} = y_n + \frac{h}{2}(y_n' + \bar{y}_{n+1}')$$

$$= y_n + \frac{h}{2}[(x_n^2 + y_n - t_n) + (\bar{x}_{n+1}^2 + \bar{y}_{n+1} - t_{n+1})],$$

with initial values $x_0 = 1$, $y_0 = 2$, and $h = 0.25$. The results of the calculation are shown in Table 8.6.

Table 8.6

t_n	x_n	y_n	$\bar{x}_{n+1}$	$\bar{y}_{n+1}$	x_{n+1}	y_{n+1}
0.00	1.00	2.00	0.25	2.75	−0.258	2.695
0.25	−0.258	2.695	−2.076	3.323	−2.744	3.884
0.50	−2.744	3.884	−7.076	6.612	−11.17	12.24
0.75	−11.17	12.24	−51.23	46.31	−301.8	363.0
1.00	−301.8	363.0				

Higher order equations can be treated as systems of first order equations.

Example 8.6.3 Let

$$y'' = x + y + y', \qquad y(0) = 1, \qquad y'(0) = 2. \tag{8.57}$$

Suppose that we wish to find $y(1)$ by Euler's method and with $h = 0.2$. Equation (8.57) may be rewritten as a system by letting $z = y'$. Then $z' = y''$, and we have the initial value problem

$$y' = z, \qquad\qquad y(0) = 1,$$

$$z' = x + y + z, \qquad z(0) = 2, \tag{8.58}$$

where x is the independent variable and y and z are the dependent variables. The calculation for $y(1)$ is shown in Table 8.7.

Table 8.7

x_n	y_n	z_n	y'_n	z'_n	y_{n+1}	z_{n+1}
0.0	1.00	2.00	2.00	3.00	1.40	2.60
0.2	1.40	2.60	2.60	4.20	1.92	3.44
0.4	1.92	3.44	3.44	5.76	2.61	4.59
0.6	2.61	4.59	4.59	7.80	3.53	6.15
0.8	3.53	6.15	6.15	10.48	4.76	8.25
1.0	4.76	8.25				

The solution to (8.57) is

$$y(x) = \frac{3}{\sqrt{5}} \left[\exp\left(\frac{1 + \sqrt{5}}{2} x \right) - \exp\left(\frac{1 - \sqrt{5}}{2} x \right) \right] - x + 1,$$

so the value of $y(1) \approx 6.043$ (with an error of twenty-seven percent).

Since the improved Euler method is a predictor-corrector procedure, it is clear that it can be adapted to apply to systems also.

Example 8.6.4 Adapt the predictor-corrector formulas (8.42) and (8.43) to the system in Example 8.6.2.

We will assume from the calculations in Example 8.6.2 that $x_0 = 1$, $x'_0 = -3$, $x_1 = -0.258$, $x'_1 = -7.27$, $y_0 = 2$, $y'_0 = 3$, $y_1 = 2.69$, $y'_1 = 2.51$, and $h = 0.25$. With these values we first predict

$$x_{n+2} = x_{n+1} + \frac{h}{2}(-x'_n + 3x'_{n+1}),$$

$$y_{n+2} = y_{n+1} + \frac{h}{2}(-y'_n + 3y'_{n+1}),$$

(8.59)

and then use these values to compute x'_{n+2}, y'_{n+2}, using the system of differential equations (8.56). Then, to correct these values, we employ Simpson's rule twice:

$$x_{n+2} = x_n + \frac{h}{3}(x'_n + 4x'_{n+1} + x'_{n+2}),$$

$$y_{n+2} = y_n + \frac{h}{3}(y'_n + 4y'_{n+1} + y'_{n+2}),$$

(8.60)

and recalculate x'_{n+2}, y'_{n+2} with (8.56).

It is clear that such methods are very laborious, but they are easily carried out on a computer.

Runge-Kutta

The Runge-Kutta formula for a system of differential equations is a direct generalization of Eqs. (8.8) and (8.9). Suppose that we are given the system

$$\frac{dx}{dt} = f(t,x,y), \qquad x(t_0) = x_0,$$

$$\frac{dy}{dt} = g(t,x,y), \qquad y(t_0) = y_0.$$

(8.61)

Then the rule becomes

$$x_1 = x_0 + \tfrac{1}{6}(m_1 + 2m_2 + 2m_3 + m_4),$$

$$y_1 = y_0 + \tfrac{1}{6}(n_1 + 2n_2 + 2n_3 + n_4),$$

(8.62)

where

$$m_1 = hf(t_0,x_0,y_0) \qquad\qquad n_1 = hg(t_0,x_0,y_0),$$

$$m_2 = hf\left(t_0 + \frac{h}{2}, x_0 + \frac{m_1}{2}, y_0 + \frac{n_1}{2}\right), \qquad n_2 = hg\left(t_0 + \frac{h}{2}, x_0 + \frac{m_1}{2}, y_0 + \frac{n_1}{2}\right),$$

$$m_3 = hf\left(t_0 + \frac{h}{2}, x_0 + \frac{m_2}{2}, y_0 + \frac{n_2}{2}\right), \qquad n_3 = hg\left(t_0 + \frac{h}{2}, x_0 + \frac{m_2}{2}, y_0 + \frac{n_2}{2}\right),$$

$$m_4 = hf(t_0 + h,x_0 + m_3,y_0 + n_3), \qquad n_4 = hg(t_0 + h,x_0 + m_3,y_0 + n_3).$$

(8.63)

It should now be apparent how this procedure is generalized for systems involving more dependent variables. We apply the formulas above to the system (8.56) in Example 8.6.2 with $h = 1$:

$$m_1 = -3, \qquad\qquad n_1 = 3,$$

$$m_2 = -\frac{49}{4}, \qquad\qquad n_2 = \frac{13}{4},$$

$$m_3 = -\frac{1137}{64}, \qquad\qquad n_3 = \frac{1881}{64},$$

$$m_4 = -\frac{4,100,657}{(64)^2}, \qquad n_4 = \frac{1,275,809}{(64)^2}.$$

We obtain

$$x_1 \approx -176.4, \qquad y_1 \approx 65.3.$$

Even though this process involves more complicated computations at each step, it involves less work than predictor-corrector methods. Since it is also quite accurate, it is thus the preferred method for hand calculations.

Boundary Value Problems

Consider the differential equation

$$y'' = f(x,y,y'), \tag{8.64}$$

with boundary conditions $y(a) = y_a$ and $y(b) = y_b$. We now describe a procedure, sometimes called the *shooting method*, that is often used to solve such problems. The idea is to convert (8.64) into an initial value problem, with initial conditions $y(a) = y_a$ and $y'(a) = M_0$, where the number M_0 is arbitrarily selected. Using one of the previously described techniques (Euler method, Runge-Kutta, predictor-corrector method), we now calculate the value $y(b) = N_0$ for the initial value problem. This number will undoubtedly be different from the required value y_b, so we again solve (8.64) as an initial value problem with $y(a) = y_a$ and $y'(a) = M_1$, obtaining $y(b) = N_1$. The assumption is now made that $y(b)$ varies linearly with the values $y'(a)$, so the next value we choose for $y'(a)$ ($= M_2$) is selected by solving the equation

$$\frac{M_2 - M_1}{M_0 - M_1} = \frac{y_b - N_1}{N_0 - N_1} \tag{8.65}$$

or

$$M_2 = M_1 + \frac{y_b - N_1}{N_0 - N_1}(M_0 - M_1). \tag{8.66}$$

Notice that (8.66) is very similar to the "regula falsi" method discussed in Exercise 7 of Section 2.7. It will yield very accurate information if the user is prepared to repeat the process several times. Again, it is clear that since this process is very laborious, it is best performed by employing a computer.

Finally, it should be noted that while shooting methods are useful in finding approximate solutions for many boundary value problems, they do not work, for a variety of reasons, for all such problems. Other methods, some very complex, are available to deal with these situations.

EXERCISES 8.6

1. Apply Euler's method to the initial value problem

$$\frac{dx}{dt} = 4x - 2y, \qquad x(0) = 1,$$

$$\frac{dy}{dt} = 5x + 2y, \qquad y(0) = 2.$$

Find $x(1)$ and $y(1)$ with $h = 0.2$. Check your accuracy by finding the exact value.

2. Write a BASIC (or FORTRAN) program for solving Exercise 1 with $h = 0.01$. Compare your answer to the exact value.

3. Do Exercise 1 using the improved Euler method.

4. Solve the initial value problem

$$y'' = y' + xy^2, \qquad y(0) = 1, \qquad y'(0) = 0,$$

for $y(1)$ with $h = 0.2$ by Euler's method.

5. Write a BASIC (or FORTRAN) program for solving Exercise 4 with $h = 0.01$ by Euler's method.

6. Modify the predictor-corrector formulas (8.44) for the system in Example 8.6.2. How many known values are necessary to initiate the method?

7. Write a BASIC (or FORTRAN) program for the predictor-corrector formulas (8.59) and (8.60). Use it to solve the problem in Example 8.6.2 with $h = 0.01$.

8. Use the "shooting method" to determine the value of $y'(0)$ for the boundary value problem

$$y'' = y^2, \qquad y(0) = y(1) = 0,$$

in such a way that y is positive over the interval $0 \leqslant x \leqslant 1$. Use the Runge-Kutta method (8.8), (8.9) with $h = 1$.

9. Find the maximum value of y over the interval $0 \leqslant x \leqslant 1$ for the boundary value problem

$$y'' + yy' + 1 = 0, \qquad y(0) = y(1) = 0.$$

10. Develop a BASIC (or FORTRAN) program to perform the "shooting method" using the improved Euler method for an equation of the form

$$y'' = f(x,y,y'), \qquad y(a) = y_a, \qquad y(b) = y_b.$$

Use it to find the maximum value of y over the interval $0 \leqslant x \leqslant 1$ for the boundary value problem

$$y'' + \sin y = 0, \qquad y(0) = y(1) = 0.$$

FOURIER
SERIES
AND
BOUNDARY
VALUE
PROBLEMS

9

9.1 INTRODUCTION

Until now we have considered differential equations whose solutions are determined by initial conditions. In this chapter we shall discuss another equally important way to specify a particular solution of a differential equation, namely, specifying certain values of the function and its derivatives at two or more points. The differential equation together with the given conditions at two or more points is called a *boundary value problem*. (This was defined earlier in Section 1.2.)

In this section we shall present some examples of boundary value problems, together with a discussion of some of the powerful methods that are used to solve them. We shall see how the concepts of eigenvalues and eigenfunctions arise naturally in these situations and discuss the importance of Fourier series techniques in solving nonhomogeneous boundary value problems.

Since a boundary value problem requires that conditions be given at two or more points, we can assume the differential equation involved to be at least of second order, because a first order equation is usually completely specified by a single condition. (This follows from the existence-uniqueness Theorems 11.3 and 11.5.)

Example 9.1.1 Consider the simple harmonic motion of a mass m attached to a coiled spring with spring constant λ. Applying Newton's law of motion and Hooke's law to this situation, we obtained in Section 3.9 [see Eq. (3.51)], the differential equation

$$x'' + k^2 x = 0, \qquad k^2 = \frac{\lambda}{m}, \tag{9.1}$$

where $x = x(t)$ denotes the displacement at time t of the mass from its equilibrium position. Suppose that the mass is initially at its equilibrium position when it is given an unknown initial velocity $v_0 > 0$. How can we guarantee that the mass will again be at its equilibrium position after precisely one second? Obviously, such a mechanism provides a simple, though crude, timepiece. (See Fig. 9.1.)

Our purpose now is to discover some of the properties of the unknown solution $x(t)$. If the time is measured in seconds, then we have the boundary conditions

$$x(0) = x(1) = 0. \tag{9.2}$$

The boundary conditions together with the differential equation (9.1) specify the boundary value problem we must solve.

We recall from Chapter 3 that to find the general solution of the differential equation (9.1) we need only solve the auxiliary equation

$$\lambda^2 + k^2 = 0.$$

Since the roots of the auxiliary equation are $\lambda = \pm ik$, the general solution of (9.1)

Equilibrium
position
$t = 0, 1, \ldots$

m

v_0

Figure 9.1

is given by

$$x(t) = A \cos kt + B \sin kt. \tag{9.3}$$

Setting $t = 0$ in (9.3) and using the boundary condition $x(0) = 0$, we observe that $A = 0$, so that the solution must consist only of multiples of $\sin kt$. For the boundary condition $x(1) = 0$ to hold we must also have

$$B \sin k = x(1) = 0.$$

There are two cases to consider: either $B = 0$ or $\sin k = 0$. If $B = 0$, then the solution (9.3) is constantly zero, which means that the mass remains at its equilibrium position at all times t. This is impossible, since the mass was given an initial velocity $v_0 > 0$. Therefore,

$$\sin k = 0,$$

which is possible only if k is a nonzero multiple of π:

$$k = \pm \pi, \pm 2\pi, \pm 3\pi, \ldots.$$

[If $k = 0$, we see that (9.3) is again constantly zero.] Thus we have arrived at a surprising conclusion: there are infinitely many solutions of the form

$$x(t) = B \sin (n\pi)t, \tag{9.4}$$

where $k = n\pi$ for some nonzero integer n, for the boundary value problem (9.1),

(9.2). Furthermore, the constant B cannot be specified without additional information. Indeed, B depends on v_0 and n, since

$$v_0 = x'(0) = n\pi B$$

or

$$B = v_0/n\pi.$$

Actually, in doing the above calculations, we used one fact that was not part of the boundary value problem (9.1), (9.2). This was the assumption, given in the statement of the problem, that $v_0 > 0$. This assumption was used to disallow the *trivial solutions* $x(t) \equiv 0$ for all t of the boundary value problem (9.1), (9.2). However, the results we have obtained illustrate one of the basic facts about boundary value problems: *nontrivial solutions will exist only for certain values of the parameter k*. All such nontrivial solutions (9.4) are called *eigenfunctions* of the problem, and the corresponding values of k^2 that yield these eigenfunctions are called *eigenvalues* of the problem. Rephrasing our work in this context, we see that

$$x(t) = \left(\frac{v_0}{n\pi} \right) \sin n\pi t \tag{9.5}$$

is an eigenfunction of this problem corresponding to the eigenvalue

$$\frac{\lambda}{m} = k^2 = (n\pi)^2. \tag{9.6}$$

It is interesting to discover the physical significance of this solution. Suppose we are given a spring with (fixed) spring constant λ. Solving (9.6) for m, we obtain the masses

$$m_n = \frac{\lambda}{n^2\pi^2}, \qquad n = 1, 2, 3, \ldots, \tag{9.7}$$

each corresponding to a different value of n. Since λ and π are fixed quantities, it is *only* for these masses that a solution of the boundary value problem will exist. For any other choice, no matter what initial velocity is given, the mass will not be back at its equilibrium position after exactly one second. Thus the construction of our clock depends on the selection of one of the masses (9.7). Furthermore, (9.5) indicates that if the mass m_n is chosen, it will be at its equilibrium position n times during the time period $0 < t \leqslant 1$. Since increasing n decreases both the weight m_n and the amplitude $(v_0/n\pi)$ of the oscillations, we arrive at the conclusion that *the lighter the mass the more rapid the oscillation*. This fact is easy to observe physically.

Example 9.1.2 A population is subject to an influenza epidemic. Let $s(t)$ and $i(t)$ be, respectively, the numbers of susceptible and infected individuals at time t. Suppose that the rate of change of the number of susceptible individuals is proportional to the number of infected individuals, while the rate of change of the number of infected individuals is proportional to the number of susceptible individuals. (The difference between the change in the numbers of infected and susceptible individuals consists of those who have gained immunity or died.) The epidemic lasts a certain period of time, say two months. [With this assumption, we can set $i(0) = i(2) = 0$.] We wish to discover the number of infected individuals at all time t (in months).

From the hypotheses above, we obtain the system of equations

$$\frac{ds}{dt} = -pi,$$

$$\frac{di}{dt} = qs,$$

(9.8)

where p and q are unknown positive constants of proportionality. Differentiating the second of the equations in (9.8) with respect to t, and substituting the first equation for ds/dt, we obtain the second order equation

$$\frac{d^2i}{dt^2} = q\frac{ds}{dt} = -qpi$$

or

$$i'' + k^2 i = 0, \qquad k^2 = pq,$$

(9.9)

with the boundary conditions

$$i(0) = i(2) = 0.$$

(9.10)

This boundary value problem is very similar to that of Example 9.1.1. We again find the general solution

$$i(t) = A\cos kt + B\sin kt.$$

Setting $t = 0$ and using the first boundary condition, we find that $0 = i(0) = A$. The second boundary condition $t = 2$ implies that

$$B\sin 2k = i(2) = 0.$$

Since we are not interested in the trivial solution $i(t) \equiv 0$, it follows that $B \neq 0$, so that $2k$ must be a nonzero multiple of π. Hence the eigenvalues of this problem are

$$k^2 = \left(\frac{\pi}{2}\right)^2, (\pi)^2, \left(\frac{3\pi}{2}\right)^2, \ldots, \left(\frac{n\pi}{2}\right)^2, \ldots,$$

and the corresponding eigenfunctions $i_k(t)$ are given by

$$i_k(t) = B\sin\frac{n\pi t}{2}, \qquad k^2 = \left(\frac{n\pi}{2}\right)^2.$$

(9.11)

If the number of infected individuals is assumed to be positive during the entire epidemic $0 < t < 2$, then $n = 1$. But if we allow for the possibility of a sequence of several consecutive epidemics, then other values of n could have physical meaning.

The boundary value problems in these two examples will also arise in several different problems in Chapter 10 (see Examples 10.3.1 and 10.4.1).

Example 9.1.3 We return to Example 9.1.1, but assume that the spring-weight system is also subject to a periodic external force $K \sin \omega t$, where $K, \omega > 0$. Using the theory of forced vibrations developed in Section 3.9, we obtain the nonhomogeneous second order equation

$$m \frac{d^2 x}{dt^2} = -\lambda x + K \sin \omega t.$$

Using the same boundary conditions as before, we are led to the nonhomogeneous boundary value problem

$$x'' + \frac{\lambda}{m} x = \frac{K}{m} \sin \omega t, \qquad x(0) = x(1) = 0. \tag{9.12}$$

There are two cases to consider: either $\omega = n\pi$ for some integer n or $\omega \neq n\pi$ for every integer n.

First, we use the method of undetermined coefficients (Section 3.6) to find a particular solution. We suppose that

$$x(t) = A \sin \omega t + B \cos \omega t. \tag{9.13}$$

Since $x(0) = 0$, it follows that $B = 0$. Differentiating (9.13) twice and substituting into Eq. (9.12), we have

$$-A\omega^2 \sin \omega t + \frac{\lambda}{m} A \sin \omega t = \frac{K}{m} \sin \omega t.$$

Canceling the $\sin \omega t$ on both sides of this equation and solving for A, we obtain

$$A = \frac{K/m}{(\lambda/m) - \omega^2} = \frac{K}{\lambda - m\omega^2}.$$

CASE 1. $(\omega = n_0\pi)$ Then

$$x_p(t) = \frac{K}{\lambda - mn_0^2\pi^2} \sin n_0\pi t \tag{9.14}$$

is a particular solution of (9.12) provided that

$$m \neq \frac{\lambda}{n_0^2\pi^2}. \tag{9.15}$$

From Example 9.1.1 we know that the homogeneous boundary value problem

$$x'' + \frac{\lambda}{m} x = 0, \qquad x(0) = x(1) = 0, \tag{9.16}$$

has the eigenfunction

$$x_n(t) = A \sin n\pi t,$$

whenever

$$m = m_n = \frac{\lambda}{n^2 \pi^2},$$

and the trivial solution for all other values of m. If $m = m_n$, $n \neq n_0$, then we have the general solution

$$
\begin{aligned}
x(t) &= A \sin n\pi t + \frac{K}{\lambda - m_n n_0^2 \pi^2} \sin n_0 \pi t \\
&= A \sin n\pi t + \frac{K n^2}{\lambda(n^2 - n_0^2)} \sin n_0 \pi t.
\end{aligned}
$$

For all other masses m satisfying (9.15), the solution of (9.12) is given by Eq. (9.14). There is no solution if $m = \lambda/n_0^2 \pi^2$ (see Section 9.7).

CASE 2. If $\omega \neq n\pi$, we see that the function we have obtained by the method of undetermined coefficients,

$$x_p(t) = \frac{K}{\lambda - m\omega^2} \sin \omega t, \tag{9.17}$$

does not satisfy the boundary condition $x_p(1) = 0$. Since only linear combinations of periodic functions of the form

$$x_n(t) = \sin n\pi t \tag{9.18}$$

satisfy the boundary conditions $x(0) = x(1) = 0$, we try to approximate (9.17) as closely as possible by a sum of such functions. We shall see in Section 9.3 that the trigonometric series

$$2\pi \sin \omega \sum_{n=1}^{\infty} \frac{(-1)^n n}{\omega^2 - (n\pi)^2} \sin n\pi t, \tag{9.19}$$

called a *Fourier series*, converges to the function $\sin \omega t$ at all points in $0 \leqslant t < 1$, and to zero at $t = 1$. Making use of this fact, we seek a solution of the form

$$x(t) = 2\pi \sin \omega \sum_{n=1}^{\infty} \frac{(-1)^n n c_n}{\omega^2 - (n\pi)^2} \sin n\pi t. \tag{9.20}$$

Substituting (9.20) into Eq. (9.12) and using the Fourier series (9.19), we can easily show that $c_n = k/(\lambda - mn^2\pi^2)$. Using these values in Eq. (9.20), we obtain the general solution of the boundary value problem (9.12).

At this point one may wonder whether very many functions can be represented as a Fourier series. The question is significant, since forcing functions can be quite arbitrary. Actually, we shall find that the class of functions that have Fourier series representations includes almost every practical forcing function. Although the Fourier series (9.19) may appear complicated, we shall discover that it is really quite easy to work with once we have become more familiar with it. We shall see that the nth term in the series is independent of the other terms and behaves essentially like the nth coordinate of a certain vector. This phenomenon, called *orthogonality*, will be one of the central themes of this chapter.

Example 9.1.4 Consider the ideal pendulum consisting of a weightless rod of length l supported at one end and attached to a particle of positive weight at the other (Fig. 9.2). In Exercise 3 of Section 7.7 we found that the angular motion $\theta(t)$ of the pendulum is given by the nonlinear differential equation

$$\frac{d^2\theta}{dt^2} + \frac{g}{l}\sin\theta(t) = 0.$$

Suppose that initially $\theta(0) = 0$ (the rod is vertical) and that we give the pendulum an unknown initial velocity $\theta'(0) = v_0 > 0$ (see Fig. 9.2). How can we guarantee that the mass will return to its initial position after precisely one second (i.e.,

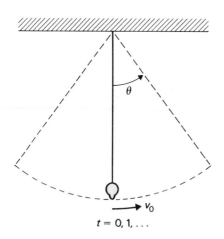

$t = 0, 1, \ldots$

Figure 9.2

$\theta(1) = 0$)? The situation, which is essentially that of constructing a pendulum clock, reduces to the solution of the nonlinear boundary value problem

$$\theta'' + \frac{g}{l} \sin \theta = 0, \qquad \theta(0) = \theta(1) = 0.$$

In practice, the situation is much more complicated, since it involves damping effects due to friction and a forcing term (such as tension on a spring) to prevent the pendulum from stopping. We shall not investigate nonlinear boundary value problems in this book; instead, we refer the interested reader to more advanced books on the subject.*

Finally, it should be noted that boundary value problems can involve differential equations of order greater than two. For example, in studying the deflection of a beam one will arrive at a fourth order boundary value problem†

$$\frac{d^4 y}{dx^4} = k^2 y,$$

with specified boundary condition $y(0)$, $y'(0)$, $y(L)$, and $y'(L)$, where L is the length of the beam.

EXERCISES 9.1

1. What eigenfunctions and eigenvalues will result in Example 9.1.1 if the boundary condition $x(1) = 0$ is replaced by $x(T) = 0$, where T is some fixed constant?

2. Consider the boundary value problem

$$\frac{d^2 x}{dt^2} - \lambda x(t) = 0, \qquad x(0) = x(1) = 0,$$

where λ is any real number. What are the eigenvalues and eigenfunctions of this problem?

3. Answer the question of Exercise 2 given that the boundary condition $x(1) = 0$ is replaced by $x(T) = 0$ for some fixed constant T.

4. Answer the question of Exercise 2 given that the boundary condition $x(0) = 0$ is replaced by $x(-1) = 0$, all other conditions remaining the same.

5. Find the eigenfunctions and eigenvalues of the boundary value problem

$$x''(t) + k^2 x(t) = 0, \qquad x(-T) = x(T) = 0,$$

where T is a fixed constant.

---

* See for example, P. B. Bailey, L. F. Shampine, and P. E. Waltman, *Nonlinear Two-Point Boundary Value Problems*, Academic Press, New York, 1968.

† This equation is derived in C. R. Wylie, *Advanced Engineering Mathematics*, McGraw-Hill, New York, 1966, p. 323.

9.2 ORTHOGONAL SETS OF FUNCTIONS AND FOURIER SERIES

Let $f_1(x)$ and $f_2(x)$ be two real-valued functions defined on an interval $a \leqslant x \leqslant b$, and suppose that the integral

$$(f_1, f_2) = \int_a^b f_1(x) f_2(x) \, dx \tag{9.21}$$

exists. We say that the functions f_1 and f_2 are *orthogonal on the interval* $a \leqslant x \leqslant b$ if the integral (9.21) vanishes, that is, if $(f_1, f_2) = 0$.* A set of real-valued functions $f_1(x), f_2(x), f_3(x), \ldots$ defined on $a \leqslant x \leqslant b$ is called an *orthogonal set of functions on the interval* $a \leqslant x \leqslant b$, if each integral

$$(f_n, f_m) = \int_a^b f_n(x) f_m(x) \, dx \tag{9.22}$$

exists and $(f_n, f_m) = 0$ whenever $n \neq m$. The nonnegative square root of the integral

$$(f_n, f_n) = \int_a^b f_n^2(x) \, dx$$

is called the *norm* of the function $f_n(x)$ and is often denoted by $\|f_n\|$ (see Exercise 10):

$$\|f_n\| = \sqrt{(f_n, f_n)} = \left[\int_a^b f_n^2(x) \, dx \right]^{1/2}. \tag{9.23}$$

An orthogonal set of functions $\{f_n(x)\}$ is called an *orthonormal set of functions on the interval* $a \leqslant x \leqslant b$ if $\|f_n\| = 1$ for all n.† Thus, if $\{f_n(x)\}$ is an orthonormal set of functions

$$(f_n, f_m) = \int_a^b f_n(x) f_m(x) \, dx = \begin{cases} 0 & \text{whenever } n \neq m, \\ 1 & \text{whenever } n = m. \end{cases} \tag{9.24}$$

---

* For the benefit of those readers familiar with the concept of the orthogonality of vectors in Euclidean n-space, we note that (9.21) is a generalization of the usual inner product

$$(\mathbf{f}_1, \mathbf{f}_2) = f_{11} f_{21} + f_{12} f_{22} + \cdots + f_{1n} f_{2n} = \sum_{j=1}^n f_{1j} f_{2j}.$$

Two vectors $\mathbf{f}_1$ and $\mathbf{f}_2$ are orthogonal in Euclidean n-space if $(\mathbf{f}_1, \mathbf{f}_2) = 0$. Also the *norm* of a vector is its Euclidean length

$$\|\mathbf{f}_1\| = \sqrt{(\mathbf{f}_1, \mathbf{f}_1)} = \left(\sum_{j=1}^n f_{1j}^2 \right)^{1/2}.$$

† This is equivalent to taking an orthonormal basis in Euclidean n-space, such as the *unit vectors*

$$e_1 = (1, 0, \ldots, 0), \quad e_2 = (0, 1, 0, \ldots, 0), \quad \ldots, \quad e_n = (0, \ldots, 0, 1).$$

Given any orthogonal set of functions $\{f_n(x)\}$ whose norms are nonzero, it is always possible to construct an orthonormal set of functions $\{g_n(x)\}$ by defining

$$g_n(x) = \frac{f_n(x)}{\|f_n\|}, \qquad a \leqslant x \leqslant b.$$

Then

$$(g_n, g_m) = \int_a^b \frac{f_n(x) \, f_m(x)}{\|f_n\| \, \|f_m\|} \, dx = \frac{1}{\|f_n\| \, \|f_m\|} (f_n, f_m), \tag{9.25}$$

which vanishes whenever $m \neq n$ since $\{f_n\}$ is an orthogonal set of functions. If $m = n$, then (9.25) yields

$$\|g_n\|^2 = (g_n, g_n) = \frac{1}{\|f_n\|^2} (f_n, f_n) = 1,$$

so that $\|g_n\| = 1$ for all n.

Example 9.2.1 The functions $f_n(x) = \sin nx$, $n = 1, 2, 3, \ldots$, form an orthogonal set of functions on the interval $-\pi \leqslant x \leqslant \pi$. To see this, we use the trigonometric identities

$$\cos (m - n)x = \cos mx \cos nx + \sin mx \sin nx,$$
$$\cos (m + n)x = \cos mx \cos nx - \sin mx \sin nx,$$

which imply (after subtracting and dividing by 2) that

$$\sin nx \sin mx = \tfrac{1}{2}[\cos (m - n)x - \cos (m + n)x].$$

Then

$$\begin{aligned}
(f_n, f_m) &= \int_{-\pi}^{\pi} \sin nx \sin mx \, dx \\
&= \frac{1}{2} \int_{-\pi}^{\pi} [\cos (m - n)x - \cos (m + n)x] \, dx \\
&= \frac{1}{2} \left[\frac{\sin (m - n)x}{m - n} - \frac{\sin (m + n)x}{m + n} \right] \Bigg|_{-\pi}^{\pi} = 0, \tag{9.26}
\end{aligned}$$

whenever $m \neq n$. If $m = n$, then $\cos (m - n)x = 1$ in the second integral of (9.26), so that

$$\begin{aligned}
(f_n, f_n) &= \frac{1}{2} \int_{-\pi}^{\pi} (1 - \cos 2nx) \, dx \\
&= \frac{1}{2} \left(x - \frac{\sin 2nx}{2n} \right) \Bigg|_{-\pi}^{\pi} = \pi. \tag{9.27}
\end{aligned}$$

Hence $\|f_n\| = \sqrt{\pi}$, and we may construct the orthonormal set of functions

$$g_n(x) = \frac{\sin nx}{\sqrt{\pi}}, \qquad n = 1, 2, \ldots.$$

Example 9.2.2 The functions

$$\frac{1}{\sqrt{2\pi}}, \quad \frac{\cos x}{\sqrt{\pi}}, \quad \frac{\sin x}{\sqrt{\pi}}, \quad \frac{\cos 2x}{\sqrt{\pi}}, \quad \frac{\sin 2x}{\sqrt{\pi}}, \ldots$$

form an orthonormal set of functions on the interval $-\pi \leqslant x \leqslant \pi$. We have already demonstrated the orthonormality of the $\sin nx/\sqrt{\pi}$ terms, and clearly

$$\int_{-\pi}^{\pi} \frac{\cos nx}{\sqrt{\pi}} \frac{\cos mx}{\sqrt{\pi}} \, dx = \frac{1}{2\pi} \int_{-\pi}^{\pi} [\cos (m - n)x + \cos (m + n)x] \, dx = 0 \tag{9.28}$$

whenever $m \neq n$ and equals

$$\frac{1}{2\pi} \int_{-\pi}^{\pi} (1 + \cos 2nx) \, dx = \frac{1}{2\pi} \left(x + \frac{\sin 2nx}{2n} \right) \Bigg|_{-\pi}^{\pi} = 1 \tag{9.29}$$

if $m = n$. The identities

$$\sin (m + n)x = \sin mx \cos nx + \cos mx \sin nx,$$
$$\sin (m - n)x = \sin mx \cos nx - \cos mx \sin nx$$

imply (after addition and division by 2) that

$$\int_{-\pi}^{\pi} \frac{\sin mx}{\sqrt{\pi}} \frac{\cos nx}{\sqrt{\pi}} \, dx = \frac{1}{2\pi} \int_{-\pi}^{\pi} [\sin (m - n)x + \sin (m + n)x] \, dx$$

$$= \frac{-1}{2\pi} \left[\frac{\cos (m - n)x}{m - n} + \frac{\cos (m + n)x}{m + n} \right] \Bigg|_{-\pi}^{\pi} = 0$$

for all m, n since $\sin (m - n)x = 0$ if $m = n$, and $\cos (-\theta) = \cos \theta$. Finally, the integrals

$$\int_{-\pi}^{\pi} \frac{\sin mx}{\sqrt{\pi}} \frac{1}{\sqrt{2\pi}} \, dx = \frac{-1}{\pi\sqrt{2}} \frac{\cos mx}{m} \Bigg|_{-\pi}^{\pi} = 0,$$

$$\int_{-\pi}^{\pi} \frac{\cos mx}{\sqrt{\pi}} \frac{1}{\sqrt{2\pi}} \, dx = \frac{1}{\pi\sqrt{2}} \frac{\sin mx}{m} \Bigg|_{-\pi}^{\pi} = 0,$$

$$\int_{-\pi}^{\pi} \left(\frac{1}{\sqrt{2\pi}} \right)^2 dx = \frac{1}{2\pi} x \Bigg|_{-\pi}^{\pi} = 1$$

complete the verification of the orthonormality.

Orthogonal sets of functions often generate useful series expansions. Let $\{f_n(x)\}$ be an orthogonal set of functions on an interval $a \leqslant x \leqslant b$ and suppose that the function $F(x)$ can be represented in terms of the functions $f_n(x)$ by a convergent series

$$F(x) = \sum_{n=1}^{\infty} c_n f_n(x) = c_1 f_1(x) + c_2 f_2(x) + \cdots. \tag{9.30}$$

The series (9.30) is called a *generalized Fourier series* of $F(x)$ and the coefficients $c_1, c_2, \ldots$ are called the *generalized Fourier coefficients of $F(x)$ with respect to the orthogonal set of functions* $\{f_n(x)\}$. It is easy to determine the constants $c_1, c_2, \ldots$ by the following procedure. Fixing m, we multiply both sides of (9.30) by $f_m(x)$ and integrate the result over the interval $a \leqslant x \leqslant b$:

$$(F, f_m) = \int_a^b F(x) f_m(x)\, dx = \sum_{n=1}^{\infty} c_n \int_a^b f_n(x) f_m(x)\, dx = \sum_{n=1}^{\infty} c_n (f_n, f_m).$$

Note that we have assumed term-by-term integration to be permissible—an assumption which sometimes is not valid.* Since $(f_n, f_m) = 0$ whenever $n \neq m$, we have

$$(F, f_m) = c_m \|f_m\|^2,$$

so that

$$c_m = \frac{(F, f_m)}{\|f_m\|^2} = \frac{1}{\|f_m\|^2} \int_a^b F(x) f_m(x)\, dx, \qquad m = 1, 2, \ldots. \tag{9.31}$$

Example 9.2.3 An infinite series of the form

$$\frac{a_0}{2} + \sum_{n=1}^{\infty} [a_n \cos nx + b_n \sin nx], \tag{9.32}$$

where a_n and b_n are constants, is generally referred to as a *trigonometric* or *Fourier series*. [Note that Eq. (9.19) is a special case of this formula with $a_n = 0$ for all n.]

If the series (9.32) converges to a function $F(x)$ on the interval $[-\pi, \pi]$, then it converges for all real x since

$$F(x) = \frac{a_0}{2} + \sum_{n=1}^{\infty} [a_n \cos nx + b_n \sin nx]$$

is a *periodic function of period 2π*:

$$F(x + 2\pi) = F(x) \qquad \text{for all } x.$$

---

* Most advanced calculus textbooks contain proofs that term-by-term integration is permissible if the series (9.30) converges *uniformly* on $a \leqslant x \leqslant b$.

This observation is obvious since the functions $\cos nx$ and $\sin nx$ are periodic of period 2π. The constants a_n and b_n are called the *Fourier coefficients* of $F(x)$ and are given by the *Euler formulas*:

$$a_n = \frac{1}{\pi} \int_{-\pi}^{\pi} F(x) \cos nx \, dx, \qquad b_n = \frac{1}{\pi} \int_{-\pi}^{\pi} F(x) \sin nx \, dx. \qquad (9.33)$$

These equalities follow from Eq. (9.31) and Example 9.2.2.

Example 9.2.4 Let $P_n(x)$ be the Legendre polynomial of degree n, $n = 0, 1, 2, \ldots$ (see Section 5.5). Then $P_n(x)$ is a solution of Legendre's equation

$$(1 - x^2)y'' - 2xy' + n(n + 1)y = 0$$

in the interval $-1 < x < 1$. Since $P_n(x)$ satisfies the differential equation on this interval, we consider the integral

$$(P_n, P_m) = \int_{-1}^{1} P_n(x)P_m(x) \, dx \qquad (9.34)$$

and suppose, without loss of generality, that $m < n$. By Rodrigues' formula (5.132),

$$P_n(x) = \frac{1}{2^n n!} \frac{d^n}{dx^n} [(x^2 - 1)^n], \qquad n = 0, 1, 2, \ldots,$$

so we may rewrite (9.34) as

$$(P_n, P_m) = \frac{1}{2^n n!} \int_{-1}^{1} P_m(x) \frac{d^n}{dx^n} [(x^2 - 1)^n] \, dx. \qquad (9.35)$$

Integrating (9.35) by parts, we have

$$(P_n, P_m) = \frac{P_m(x)}{2^n n!} \frac{d^{n-1}}{dx^{n-1}} [(x^2 - 1)^n] \Big|_{-1}^{1} - \frac{1}{2^n n!} \int_{-1}^{1} P'_m(x) \frac{d^{n-1}}{dx^{n-1}} [(x^2 - 1)^n] \, dx.$$

But $(x^2 - 1)^n = (x - 1)^n(x + 1)^n$, since $x^2 - 1 = (x - 1)(x + 1)$, and $(n - 1)$ repeated differentiations of this function yield a finite sum of terms of the form $(x - 1)^j(x + 1)^{n-j+1}$, with $1 \leqslant j \leqslant n$:

$$\frac{d}{dx} (x - 1)^n(x + 1)^n = n(x - 1)^{n-1}(x + 1)^n + n(x - 1)^n(x + 1)^{n-1},$$

$$\frac{d^2}{dx^2} (x - 1)^n(x + 1)^n = n(n - 1)(x - 1)^{n-2}(x + 1)^n + n^2(x - 1)^{n-1}(x + 1)^{n-1}$$

$$+ n(n - 1)(x - 1)^n(x + 1)^{n-2},$$

and so on. Clearly, one of the terms $(x - 1)^j$ or $(x + 1)^{n-j+1}$ vanishes at each endpoint of the interval, so that the first term vanishes and

$$(P_n,P_m) = \frac{-1}{2^n n!} \int_{-1}^{1} P'_m(x) \frac{d^{n-1}}{dx^{n-1}} [(x^2 - 1)^n] \, dx.$$

We again integrate by parts and continue doing so until we arrive at the integral

$$(P_n,P_m) = \frac{(-1)^n}{2^n n!} \int_{-1}^{1} P_m^{(n)}(x)(x^2 - 1)^n \, dx. \tag{9.36}$$

Since $P_m(x)$ is a polynomial of degree $m \, (<n)$, its nth derivative $P_m^{(n)}$ is zero. Thus $(P_n,P_m) = 0$ whenever $n \neq m$, since one of the subscripts must be smaller than the other.

If $m = n$, then

$$\|P_n\|^2 = (P_n,P_n) = \int_{-1}^{1} P_n^2(x) \, dx = \frac{(-1)^n}{2^n n!} \int_{-1}^{1} P_n^{(n)}(x)(x^2 - 1)^n \, dx$$

by Eq. (9.36). However, according to Eq. (5.131) the nth degree term of $P_n(x)$ has the coefficient $(2n)!/2^n(n!)^2$. Thus $P_n^{(n)}(x) = (2n)!/2^n n!$ (the other $n!$ term cancels with the term $d^n(x^n)/dx^n = n!$) so that

$$\|P_n\|^2 = \frac{(2n)!}{2^{2n}(n!)^2} \int_{-1}^{1} (1 - x^2)^n \, dx.$$

Setting $x = \sin \theta$ and $dx = \cos \theta \, d\theta$, we obtain

$$\begin{aligned}
\|P_n\|^2 &= \frac{(2n)!}{2^{2n}(n!)^2} \int_{-\pi/2}^{\pi/2} \cos^{2n+1} \theta \, d\theta \\
&= \frac{2(2n)!}{2^{2n}(n!)^2} \int_{0}^{\pi/2} \cos^{2n+1} \theta \, d\theta, \tag{9.37}
\end{aligned}$$

since the integrand is symmetric with respect to the y-axis. Finally, by formula 57 of Appendix 1,

$$\|P_n\|^2 = \frac{2(2n)!}{2^{2n}(n!)^2} \cdot \frac{2^{2n}(n!)^2}{(2n + 1)!} = \frac{2}{2n + 1}. \tag{9.38}$$

Thus the Legendre polynomials are an orthogonal set of functions, and the polynomials

$$\sqrt{\frac{2n + 1}{2}} \, P_n(x), \qquad n = 0, 1, 2, \ldots,$$

are an orthonormal set of functions on the interval $-1 \leqslant x \leqslant 1$. An easier proof of this result will be obtained in Example 9.5.2, after we have developed some additional facts.

A generalized Fourier series of Legendre polynomials,

$$\sum_{n=0}^{\infty} c_n P_n(x), \tag{9.39}$$

is called a *Legendre series*. Such series are extremely important for numerical approximations, even when only finitely many coefficients c_n are nonzero, since they provide the "best" *least squares approximation* for any integrable function $F(x)$ on the interval $-1 \leqslant x \leqslant 1$, as we shall shortly see.

Since the Legendre polynomial $P_n(x)$ has degree n, it is possible to express x^n as a linear combination of Legendre polynomials of degree $\leqslant n$. To see this, note that

$$x^0 = 1 = P_0(x), \quad x = P_1(x), \quad x^2 = \tfrac{1}{3}P_0(x) + \tfrac{2}{3}P_2(x), \ldots,$$

and if such expressions are known for all powers x^k, $k < n$, then, by Eq. (5.131),

$$\frac{(2n)!x^n}{2^{2n}(n!)^2} = P_n(x) - \sum_{k=1}^{M} \frac{(-1)^k}{2^n k!} \frac{(2n - 2k)!}{(n-k)!(n-2k)!} x^{n-2k},$$

(M is the greatest integer $\leqslant n/2$) so that x^n can also be expressed as a linear combination of Legendre polynomials of degree $\leqslant n$.

Now suppose we wish to find a polynomial $p(x)$ of degree n that best approximates an integrable function $F(x)$ on the interval $-1 \leqslant x \leqslant 1$, in the *least squares* sense. By this we mean that we want to choose a polynomial $p(x)$ that minimizes the definite integral

$$I = \int_{-1}^{1} [F(x) - p(x)]^2 \, dx. \tag{9.40}$$

Since $[F(x) - p(x)]^2 \geqslant 0$, this problem amounts to finding the polynomial $p(x)$ that minimizes the area under the curve $y = [F(x) - p(x)]^2$ over the interval $-1 \leqslant x \leqslant 1$.

Using (9.31), we calculate the coefficients

$$c_m = \frac{(F, P_m)}{\|P_m\|^2} = \left(\frac{2m + 1}{2}\right) \int_{-1}^{1} F(x) P_m(x) \, dx \tag{9.41}$$

and define the polynomial

$$p(x) = \sum_{k=0}^{n} c_k P_k(x). \tag{9.42}$$

Suppose $q(x)$ is any polynomial of degree n. Since every power x^k in $q(x)$ can be expressed as a linear combination of Legendre polynomials of degree $\leqslant k$, $q(x)$

can also be expressed as a linear combination of Legendre polynomials of degree $\leqslant n$:

$$q(x) = \sum_{k=0}^{n} b_k P_k(x).$$

(Of course, some of the coefficients b_k might be zero.) Then for this polynomial $q(x)$,

$$I = \int_{-1}^{1} [F(x) - q(x)]^2 \, dx$$

$$= \int_{-1}^{1} F(x)^2 \, dx + \int_{-1}^{1} \left[\sum_{k=0}^{n} b_k P_k(x) \right]^2 \, dx$$

$$- 2 \int_{-1}^{1} F(x) \sum_{k=0}^{n} b_k P_k(x) \, dx. \tag{9.43}$$

The second integral in Eq. (9.43) can be written as

$$\sum_{k=0}^{n} \sum_{j=0}^{n} b_k b_j \int_{-1}^{1} P_k(x) P_j(x) \, dx = \sum_{k=0}^{n} b_k^2 \, \|P_k\|^2,$$

since all the terms

$$(P_k, P_j) = \int_{-1}^{1} P_k(x) P_j(x) \, dx = 0$$

if $k \neq j$. Interchanging the finite sum and integral in the last term of (9.43), and noting that by (9.41)

$$b_k^2 \, \|P_k\|^2 - 2b_k(F, P_k) = \|P_k\|^2 (b_k^2 - 2b_k c_k) = \|P_k\|^2 (b_k - c_k)^2 - \|P_k\|^2 c_k^2,$$

we have

$$I = \int_{-1}^{1} F(x)^2 \, dx + \sum_{k=0}^{n} b_k^2 \, \|P_k\|^2 - 2 \sum_{k=0}^{n} b_k(F, P_k)$$

$$= \int_{-1}^{1} F(x)^2 \, dx + \sum_{k=0}^{n} \frac{2}{2k+1} (b_k - c_k)^2 - \sum_{k=0}^{n} \frac{2}{2k+1} c_k^2. \tag{9.44}$$

Since the c_k terms are fixed [by (9.41)] and we have no control over the integral in (9.44), we can affect the outcome only by suitably choosing the values b_k. Clearly I is minimized by letting $b_k = c_k$ for $k = 0, 1, \ldots, n$, that is, by choosing $p(x)$ as the minimizing polynomial. It is often incorrectly assumed by students that the "best" least squares approximation by polynomials of a function $F(x)$ that has a power series expansion, is given by the partial sums of that power series. The following example demonstrates this error.

Example 9.2.5 Suppose we wish to find the "best" least squares approximation of e^x on $-1 \leqslant x \leqslant 1$ by a straight line, that is by a polynomial of degree $\leqslant 1$. Then

$$c_0 = \frac{1}{\|P_0\|^2}(e^x, P_0) = \frac{1}{2}\int_{-1}^{1} e^x \, dx = \frac{e - e^{-1}}{2} = \sinh(1),$$

$$c_1 = \frac{1}{\|P_1\|^2}(e^x, P_1) = \frac{3}{2}\int_{-1}^{1} xe^x \, dx = \frac{3}{2}(x - 1)e^x \Big|_{-1}^{1} = \frac{3}{e},$$

implying that

$$p(x) = \sinh(1)P_0(x) + \frac{3}{e}P_1(x) = \sinh(1) + \frac{3x}{e}.$$

Clearly $p(x)$ is *not* the first two terms of the Maclaurin expansion of e^x ($= 1 + x + x^2/2 + x^3/3 + \cdots$).

Some important sets of functions $\{f_n(x)\}$ that occur in applications are not orthogonal, but have the property that for some function $w(x)$ the integral

$$(f_n, f_m)_w = \int_a^b w(x)f_n(x)f_m(x) \, dx = 0 \tag{9.45}$$

whenever $n \neq m$. If this happens, the set $\{f_n(x)\}$ is said to be *orthogonal with respect to the weight function $w(x)$ on the interval $a \leqslant x \leqslant b$*. The *weighted norm* of f_n is defined as

$$\|f_n\|_w = \left[\int_a^b w(x)f_n(x)^2 \, dx\right]^{1/2}, \tag{9.46}$$

and $\{f_n(x)\}$ is said to be *orthonormal with respect to the weight function $w(x)$*, if in addition to being orthogonal with respect to $w(x)$, it satisfies the condition $\|f_n\|_w = 1$ for all n. If $F(x)$ can be represented by a generalized Fourier series

$$F(x) = \sum_{n=0}^{\infty} c_n f_n(x), \tag{9.47}$$

then the general Fourier coefficients c_k can be determined by multiplying both sides of (9.47) by $w(x)f_k(x)$ and integrating over the interval $a \leqslant x \leqslant b$:

$$\int_a^b w(x)F(x)f_k(x) \, dx = \sum_{n=0}^{\infty} c_n \int_a^b w(x)f_n(x)f_k(x) \, dx = c_k \|f_k\|_w^2$$

or

$$c_k = \frac{(F, f_k)_w}{\|f_k\|_w^2}. \tag{9.48}$$

Example 9.2.6   Let $y_n(x)$ be a polynomial solution of *Laguerre's equation*[*]

$$(xe^{-x}y')' + ne^{-x}y = 0. \tag{9.49}$$

[Equation (9.49) is obtained by multiplying Eq. (5.143) by e^{-x}.] Then y_n is called a *Laguerre polynomial*. Consider the set of Laguerre polynomials $\{y_n(x)\}$, $n = 1, 2, 3, \ldots$, and suppose $m \neq n$. Then we have the equations

$$(xe^{-x}y_n')' + ne^{-x}y_n = 0,$$
$$(xe^{-x}y_m')' + me^{-x}y_m = 0.$$

We multiply the first equation by y_m and the second by $-y_n$ and add them to obtain

$$y_m(xe^{-x}y_n')' - y_n(xe^{-x}y_m')' = e^{-x}(m - n)y_n y_m. \tag{9.50}$$

Integrating both sides of (9.50) on the interval $0 \leqslant x \leqslant \infty$, we have

$$\int_0^\infty [y_m(xe^{-x}y_n')' - y_n(xe^{-x}y_m')']\,dx = (m - n)\int_0^\infty e^{-x}y_n y_m\,dx. \tag{9.51}$$

But by the product formula for differentiation,

$$[xe^{-x}(y_m y_n' - y_n y_m')]' = (xe^{-x}y_n')'y_m + xe^{-x}y_n'y_m' - (xe^{-x}y_m')'y_n - xe^{-x}y_m'y_n'$$
$$= y_m(xe^{-x}y_n')' - y_n(xe^{-x}y_m')'.$$

Thus (9.51) becomes

$$(m - n)\int_0^\infty e^{-x}y_n y_m\,dx = xe^{-x}(y_m y_n' - y_n y_m')\Big|_0^\infty = 0, \qquad m \neq n,$$

because $f(x) = x(y_m y_n' - y_n y_m')$ is a polynomial (since each term is a polynomial) so that the quotient $f(x)/e^x$ tends to zero as $x \to \infty$. This may be verified by applying L'Hôpital's rule as many times as the degree of f to the quotient $f(x)/e^x$:

$$\lim_{x \to \infty} \frac{f(x)}{e^x} = \lim_{x \to \infty} \frac{f'(x)}{e^x} = \cdots = \lim_{x \to \infty} \frac{c}{e^x} = 0.$$

Therefore, the set of functions $\{y_n\}$, $n = 1, 2, \ldots$, is orthogonal with respect to the weight function e^{-x} on $0 \leqslant x < \infty$.

EXERCISES 9.2

In Exercises 1 through 5 show that each given set of functions is orthogonal on the given interval and determine the corresponding orthonormal set.

1. $\{\cos nx\}$, $n = 0, 1, 2, \ldots$; $0 \leqslant x \leqslant 2\pi$

---

  [*]   Edmond Laguerre (1834–1886) was a French mathematician who made many contributions to geometry and the theory of infinite series.

2. $\left\{\sin\dfrac{n\pi x}{T}\right\}$, $n = 1, 2, 3, \ldots$; $-T \leqslant x \leqslant T$

3. $\left\{\cos\dfrac{2n\pi x}{T}\right\}$, $n = 0, 1, 2, \ldots$; $0 \leqslant x \leqslant T$

4. $\{\sin 2nx\}$, $n = 1, 2, 3, \ldots$; $0 \leqslant x \leqslant \pi$

5. $\{\cos 2nx\}$, $n = 0, 1, 2, \ldots$; $0 \leqslant x \leqslant \pi$

6. Represent each of the following polynomials as a linear combination of Legendre polynomials. [*Hint*: Use (9.41).]

a) x^4
b) x^5
c) $x^4 - x^3 + 7x^2 - 8x + 2$
d) $x^5 - x + 3$
e) $12x^4 - 8x^2 + 7$
f) $9x^3 - 8x^2 + 7x - 6$

7. Find the first three terms of the Legendre series of the functions:

a) e^x
b) $\sin x$
c) $\cos x$

d) $\sinh x$
e) $\cosh x$
f) $\dfrac{1}{1 + x^2}$

Compare your result with the Maclaurin expansions of the functions.

8. Given that $p(x)$ is a polynomial of degree n $(\geqslant 1)$ and

$$\int_{-1}^{1} x^k p(x)\, dx = 0 \qquad \text{for } k = 0, 1, 2, \ldots, n - 1,$$

show that $p(x) = cP_n(x)$ for some constant c.

9. (*Hermite polynomials**) The functions

$$H_0 = 1, \qquad H_n(x) = (-1)^n e^{x^2/2}\frac{d^n}{dx^n}e^{-x^2/2}, \qquad n = 1, 2, 3, \ldots,$$

are called Hermite polynomials (see Exercise 9 of Section 5.5). Prove that:

a) $H_1(x) = x$, $H_2(x) = x^2 - 1$, $H_3(x) = x^3 - 3x$.

b) The Hermite polynomials satisfy the relation

$$H_{n+1}(x) = xH_n(x) - H_n'(x).$$

c) $H_n(x)$ is a solution of *Hermite's equation*

$$y'' - 2xy + 2ny = 0.$$

d) The set of functions $\{H_n(x)\}$ is orthogonal with respect to the weight function $\exp(-x^2)$ on the interval $-\infty < x < \infty$.

*10. The norm of a function $f(x)$ on an interval $a \leqslant x \leqslant b$ is defined in Eq. (9.23) by the integral

$$\|f\| = \left(\int_a^b f^2(x)\, dx\right)^{1/2}.$$

---

* Charles Hermite (1822–1901) was a French mathematician known for his work in number theory.

This concept is a generalization of the notion of the distance of a point (x,y) in the plane from the origin:

$$d(x,y) = \sqrt{x^2 + y^2}.$$

Assume that f and g are continuous functions on $a \leqslant x \leqslant b$. Prove that the norm satisfies the following three properties:

a) $\|f\| \geqslant 0$, where the equality holds if and only if $f(x) \equiv 0$.

b) $\|af\| = |a| \cdot \|f\|$ for any constant a.

c) $\|f + g\| \leqslant \|f\| + \|g\|$ (triangle inequality).

[*Hint*: (a) Use the fact that if $f(x_0) \neq 0$, then by continuity $f^2(x) \geqslant \varepsilon > 0$ for all x in some interval $x_0 - \delta \leqslant x \leqslant x_0 + \delta$, where $\delta > 0$. (c) Show that for any two real numbers α and β

$$0 \leqslant \|\alpha f - \beta g\|^2 = \alpha^2 \|f\|^2 - 2\alpha\beta(f,g) + \beta^2 \|g\|^2.$$

Then let $\alpha = \|g\|$ and $\beta = \|f\|$, showing that

$$(f,g) \leqslant \|f\| \cdot \|g\|.$$

Finally, use the last inequality to show that

$$\|f + g\|^2 \leqslant (\|f\| + \|g\|)^2.]$$

9.3 FOURIER SERIES*

In the last section we developed some of the basic properties of Fourier series (see Example 9.2.3), but avoided discussing what kind of functions can be represented by a convergent Fourier series. Actually, the class of functions that can be represented in this fashion is surprisingly large, so large in fact, that it originally aroused a big controversy. The following theorem gives sufficient conditions for almost all conceivable practical applications.

Theorem 9.1 Let $F(x)$ be a periodic function with period $2T$ and such that $F(x)$ and $F'(x)$ are piecewise continuous† on the interval $-T \leqslant x \leqslant T$. Then $F(x)$ has a Fourier series

$$F(x) = \frac{a_0}{2} + \sum_{n=1}^{\infty} \left(a_n \cos \frac{n\pi x}{T} + b_n \sin \frac{n\pi x}{T} \right) \qquad (9.52)$$

whose coefficients are given by the *Euler formulas*

$$a_n = \frac{1}{T} \int_{-T}^{T} F(x) \cos \frac{n\pi x}{T} \, dx, \qquad n = 0, 1, 2, \ldots, \qquad (9.53)$$

---

* Named after the French physicist Joseph B. Fourier (1768–1830).
† See Section 6.1.

$$b_n = \frac{1}{T} \int_{-T}^{T} F(x) \sin \frac{n\pi x}{T} \, dx, \qquad n = 1, 2, \dots . \tag{9.54}$$

The Fourier series (9.52) converges to $F(x)$ at all points where F is continuous, and to $[F(x + 0) + F(x - 0)]/2$ at all points of jump discontinuity of F^*.

Remarks. Note that $[F(x + 0) + F(x - 0)]/2$ is the average of the right and left hand limits at the point x. At any point of continuity both of these values will coincide, so that (9.52) converges to $[F(x + 0) + F(x - 0)]/2$ for all x in the interval $-T \leqslant x \leqslant T$.

Although the conditions given in this theorem guarantee the existence and convergence of a Fourier series for the function $F(x)$, they are *not* the most general of such conditions. Moreover, the convergence of a Fourier series to a function $F(x)$ does not imply that F satisfies the conditions given in this theorem. In summary, the conditions given in this theorem are neither necessary nor the most general sufficient conditions. Furthermore, even with these limitations, the proof of the theorem as stated is still too complicated to be presented here.† Instead, we shall verify the theorem under the additional hypothesis that $F(x)$ has a continuous second derivative.

Proof. Integrating (9.53) by parts, we obtain for $n > 0$

$$a_n = \frac{F(x) \sin (n\pi x/T)}{n\pi} \Big|_{-T}^{T} - \frac{1}{n\pi} \int_{-T}^{T} F'(x) \sin \frac{n\pi x}{T} \, dx.$$

The first term on the right-hand side vanishes, and integrating again by parts, we have

$$a_n = \frac{TF'(x) \cos (n\pi x/T)}{(n\pi)^2} \Big|_{-T}^{T} - \frac{T}{(n\pi)^2} \int_{-T}^{T} F''(x) \cos \frac{n\pi x}{T} \, dx. \tag{9.55}$$

The periodicity of F implies that

$$F'(x + 2T) = \lim_{h \to 0} \frac{F(x + 2T + h) - F(x + 2T)}{h}$$

$$= \lim_{h \to 0} \frac{F(x + h) - F(x)}{h} = F'(x),$$

---

* $F(x + 0) = \lim_{h \to 0+} F(x + h)$, $F(x - 0) = \lim_{h \to 0+} F(x - h)$ for $h > 0$.

† Proofs of this theorem can be found in most advanced calculus or complex variables books. See, for example, W. Kaplan, *Advanced Calculus*, Addison-Wesley, Reading, Mass., 1973, p. 484; or W. R. Derrick, *Introductory Complex Analysis and Applications*, Academic Press, New York, 1972, p. 166.

and since $\cos(-t) = \cos t$, the first term in (9.55) is zero. By hypothesis, $F''(x)$ is continuous on $-T \leqslant x \leqslant T$, so that it attains its maximum and minimum values on that interval and some positive number M exists such that $|F''(x)| < M$. Hence the inequality $|\cos(n\pi x/T)| \leqslant 1$ implies that

$$|a_n| \leqslant \frac{T}{(n\pi)^2} \left| \int_{-T}^{T} F''(x) \cos\frac{n\pi x}{T} \, dx \right| < \frac{TM}{(n\pi)^2} \left| \int_{-T}^{T} dx \right| = \frac{2MT^2}{(n\pi)^2}.$$

A similar inequality also holds for $|b_n|$, so that the series of absolute values of the terms on the right-hand side of Eq. (9.52) is bounded by the convergent series

$$\frac{|a_0|}{2} + \frac{2MT^2}{\pi^2}\left(1 + 1 + \frac{1}{2^2} + \frac{1}{2^2} + \frac{1}{3^2} + \frac{1}{3^2} + \cdots\right).$$

Thus the Fourier series (9.52) converges absolutely for all values x. Readers familiar with the notion of uniform convergence will see that the series (9.52) converges uniformly. Since uniform convergence of a series permits term by term integration, the derivation of the Euler formulas (9.53) and (9.54), carried out in Example 9.2.3, can be completely justified.

Although the theory of Fourier series is complicated, the application of these series is very easy. It should be clear from Theorem 9.1 that Fourier series apply to a much wider class of functions than Taylor series (even under the assumption

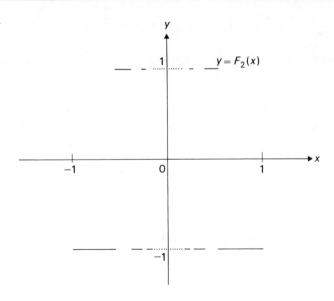

Figure 9.3

we have used in our proof), since discontinuous functions cannot have a Taylor series representation. It is also useful to consider some functions for which Theorem 9.1 *does not* apply.

Example 9.3.1 The functions $F_1(x) = 1/x$ and

$$F_2(x) = (-1)^n, \qquad \frac{1}{n+1} < |x| \leqslant \frac{1}{n}, \qquad n = 1, 2, 3, \ldots,$$

do not satisfy the hypotheses of Theorem 9.1 in the interval $-1 \leqslant x \leqslant 1$, since neither function is piecewise continuous. The function F_1 has an infinite jump discontinuity at $x = 0$, while F_2 has an infinite number of discontinuities in $-1 \leqslant x \leqslant 1$ (see Fig. 9.3).

Let us now illustrate the procedure involved in obtaining the Fourier series of a function by some examples.

Example 9.3.2 Find the Fourier series of the periodic function (see Fig. 9.4)

$$f(x) = (-1)^n k, \qquad n < x < n + 1.$$

Functions of this type occur as off-on controls in mechanical systems. Observe that $f(x)$ has period $2T = 2$, so that by (9.53),

$$a_n = \int_{-1}^{1} f(x) \cos n\pi x \, dx = (-k) \int_{-1}^{0} \cos n\pi x \, dx + k \int_{0}^{1} \cos n\pi x \, dx.$$

If $n = 0$, we obtain $a_0 = -k + k = 0$; and if $n \neq 0$, then

$$a_n = \frac{-k}{n\pi} \sin n\pi x \Big|_{-1}^{0} + \frac{k}{n\pi} \sin n\pi x \Big|_{0}^{1} = 0.$$

Similarly, since $\cos(-t) = \cos t$,

$$b_n = (-k) \int_{-1}^{0} \sin n\pi x \, dx + k \int_{0}^{1} \sin n\pi x \, dx$$

$$= \frac{k}{n\pi} \cos n\pi x \Big|_{-1}^{0} - \frac{k}{n\pi} \cos n\pi x \Big|_{0}^{1}$$

$$= \frac{2k}{n\pi} (1 - \cos n\pi),$$

which vanishes for even n and equals $4k/n\pi$ for odd n. Thus

$$f(x) = \frac{4k}{\pi} \left(\sin \pi x + \frac{1}{3} \sin 3\pi x + \frac{1}{5} \sin 5\pi x + \cdots \right). \tag{9.56}$$

The graphs of $f(x)$ and the first three partial sums of (9.56) are shown in Fig. 9.5.

Figure 9.4

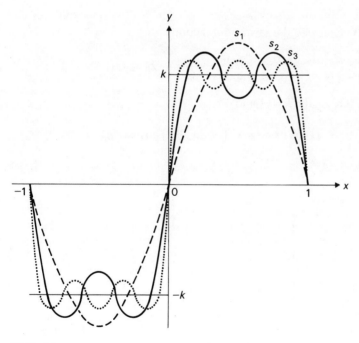

Figure 9.5

As a bonus for our work above, note that if we set $x = \frac{1}{2}$ in (9.56), we obtain the series

$$k = \frac{4k}{\pi}\left(1 - \frac{1}{3} + \frac{1}{5} - \frac{1}{7} + \cdots\right)$$

or

$$\frac{\pi}{4} = 1 - \frac{1}{3} + \frac{1}{5} - \frac{1}{7} + \cdots, \tag{9.57}$$

since $\sin \pi/2 = 1$, $\sin 3\pi/2 = -1$, and so on. This is a famous result which Leibniz obtained by means of a complicated geometrical construction.

As a final note, observe that the function $f(x)$ in this example satisfies the condition $f(-x) = -f(x)$. Functions having this property are said to be *odd* functions. In particular, observe that all functions of the form $\sin n\pi x$ are odd functions.

Example 9.3.3 Consider the sawtooth function

$$f(x) = \begin{cases} x + 1, & -1 \leqslant x \leqslant 0, \\ -x + 1, & 0 \leqslant x \leqslant 1, \end{cases} \qquad f(x + 2) = f(x),$$

shown in Fig. 9.6. Again $T = 1$ and

$$a_n = \int_{-1}^{0} (x + 1) \cos n\pi x \, dx + \int_{0}^{1} (-x + 1) \cos n\pi x \, dx.$$

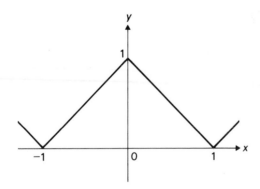

Figure 9.6

If $n = 0$, we have

$$a_0 = \left(\frac{x^2}{2} + x\right)\Big|_{-1}^{0} + \left(x - \frac{x^2}{2}\right)\Big|_{0}^{1} = 1,$$

and integrating by parts for $n \neq 0$, we obtain

$$a_n = \frac{(x + 1)}{n\pi} \sin n\pi x \Big|_{-1}^{0} - \frac{1}{n\pi} \int_{-1}^{0} \sin n\pi x \, dx$$

$$+ \frac{(1 - x)}{n\pi} \sin n\pi x \Big|_{0}^{1} + \frac{1}{n\pi} \int_{0}^{1} \sin n\pi x \, dx$$

$$= \frac{1}{(n\pi)^2} \cos n\pi x \Big|_{-1}^{0} - \frac{1}{(n\pi)^2} \cos n\pi x \Big|_{0}^{1} = \frac{2}{(n\pi)^2} (1 - \cos n\pi).$$

Therefore, a_n vanishes for even $n \neq 0$ and equals $4/(n\pi)^2$ for odd n. Similarly,

$$b_n = \frac{-(x + 1)}{n\pi} \cos n\pi x \Big|_{-1}^{0} + \frac{1}{n\pi} \int_{-1}^{0} \cos n\pi x \, dx$$

$$+ \frac{(x - 1)}{n\pi} \cos n\pi x \Big|_{0}^{1} - \frac{1}{n\pi} \int_{0}^{1} \cos n\pi x \, dx$$

$$= \frac{-1}{n\pi} + \frac{1}{(n\pi)^2} \sin n\pi x \Big|_{-1}^{0} + \frac{1}{n\pi} - \frac{1}{(n\pi)^2} \sin n\pi x \Big|_{0}^{1} = 0.$$

Thus

$$f(x) = \frac{1}{2} + \frac{4}{\pi^2}\left(\cos \pi x + \frac{1}{3^2}\cos 3\pi x + \frac{1}{5^2}\cos 5\pi x + \frac{1}{7^2}\cos 7\pi x + \cdots\right). \quad (9.58)$$

Incidentally, note that if we set $x = 1$, then

$$0 = \frac{1}{2} - \frac{4}{\pi^2}\left(1 + \frac{1}{3^2} + \frac{1}{5^2} + \frac{1}{7^2} + \cdots\right)$$

or

$$\frac{\pi^2}{8} = 1 + \frac{1}{3^2} + \frac{1}{5^2} + \frac{1}{7^2} + \cdots.$$

We observe that the term by term derivative of (9.58) yields

$$f'(x) = \frac{-4}{\pi}\left(\sin \pi x + \frac{1}{3}\sin 3\pi x + \frac{1}{5}\sin 5\pi x + \cdots\right),$$

which is the same series as (9.56) for $k = -1$. An explanation for this fact is found by noting that $f'(x) = (-1)^n(-1)$ for $n < x < n + 1$.

Finally, note that this function satisfies the condition $f(-x) = f(x)$. All such functions are called *even* functions. In particular, $\cos n\pi x$ is an even function for every integer n.

Example 9.3.4 Consider the clipped sine wave

$$f(x) = \begin{cases} k \sin Tx, & 0 < x < \pi/T, \\ 0, & -\pi/T < x < 0, \end{cases} \qquad f\left(x + \frac{2\pi}{T}\right) = f(x),$$

which is obtained by passing a sinusoidal voltage $k \sin Tx$ through a half-wave rectifier (see Fig. 9.7). Here the period is $2\pi/T$, so that

$$a_n = \frac{T}{\pi} \int_0^{\pi/T} k \sin Tx \cos nTx \, dx, \qquad a_0 = \frac{2k}{\pi}.$$

Using the substitution

$$2 \sin Tx \cos nTx = \sin (n + 1)Tx - \sin (n - 1)Tx$$

as in Example 9.2.1, we have

$$a_n = \frac{kT}{2\pi} \left[\frac{\cos (n - 1)Tx}{(n - 1)T} - \frac{\cos (n + 1)Tx}{(n + 1)T} \right]\Bigg|_0^{\pi/T}$$

$$= \frac{k}{2\pi} \left[\frac{\cos (n - 1)\pi - 1}{n - 1} - \frac{\cos (n + 1)\pi - 1}{n + 1} \right].$$

When n is odd, $a_n = 0$, and when n is even

$$a_n = \frac{k}{\pi} \left(\frac{1}{n + 1} - \frac{1}{n - 1} \right) = \frac{-2k}{\pi(n^2 - 1)}.$$

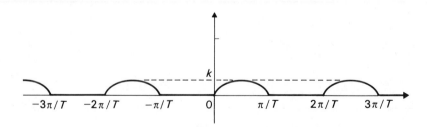

Figure 9.7

Similarly, for $n \neq 1$,

$$
\begin{aligned}
b_n &= \frac{T}{\pi} \int_0^{\pi/T} k \sin Tx \sin nTx \, dx \\
&= \frac{2kT}{\pi} \int_0^{\pi/T} [\cos(n-1)Tx - \cos(n+1)Tx] \, dx \\
&= \frac{2k}{\pi} \left[\frac{\sin(n-1)Tx}{n-1} - \frac{\sin(n+1)Tx}{n+1} \right] \Bigg|_0^{\pi/T} = 0;
\end{aligned}
$$

and if $n = 1$, by formula 18 of Appendix 1,

$$
b_1 = \frac{k}{2\pi}(Tx - \sin Tx \cos Tx) \Bigg|_0^{\pi/T} = \frac{k}{2}.
$$

Thus, since only b_1 and the evenly subscripted terms $a_{2n} = -2k/\pi(4n^2 - 1)$ are nonzero,

$$
f(x) = \frac{k}{\pi} + \frac{k}{2} \sin Tx - \frac{2k}{\pi} \sum_{n=1}^{\infty} \frac{\cos 2nTx}{4n^2 - 1}. \tag{9.59}
$$

Incidentally, if we set $x = \pi/2T$, then another intriguing identity results:

$$
k = \frac{k}{\pi} + \frac{k}{2} - \frac{2k}{\pi} \sum_{n=1}^{\infty} \frac{\cos n\pi}{4n^2 - 1}
$$

or

$$
\frac{\pi}{2} - 1 = 2\left(\frac{1}{1 \cdot 3} - \frac{1}{3 \cdot 5} + \frac{1}{5 \cdot 7} - \frac{1}{7 \cdot 9} + \cdots \right).
$$

Note that in this example the Fourier series (9.59) contains both sine and cosine terms. Since $f(x)$ is neither even nor odd, Examples 9.3.2 and 9.3.3 provide the necessary clues for the formulation of the following theorem.

Theorem 9.2 Let $F(x)$ be a periodic function of period $2T$ which has a Fourier series. If $F(x)$ is an even function, then all the Fourier coefficients b_n are zero, while if $F(x)$ is an odd function, then the Fourier coefficients a_n all vanish.

Remark. The Fourier series of an even function,

$$
F(x) = \frac{a_0}{2} + \sum_{n=1}^{\infty} a_n \cos \frac{n\pi x}{T}, \tag{9.60}
$$

is called a *Fourier cosine series*, while that of an odd function,

$$
F(x) = \sum_{n=1}^{\infty} b_n \sin \frac{n\pi x}{T}, \tag{9.61}
$$

is said to be a *Fourier sine series*. Observe that Theorem 9.2 implies that for even and odd functions, we need to calculate only half as many coefficients as are generally required.

Proof of Theorem 9.2 Let $g(x)$ be an odd function and let $x = -t$ so that $dx = -dt$. Then

$$\int_{-T}^{0} g(x)\, dx = -\int_{T}^{0} g(-t)\, dt = \int_{0}^{T} g(-t)\, dt = -\int_{0}^{T} g(t)\, dt,$$

so that we have

$$\int_{-T}^{T} g(x)\, dx = \int_{-T}^{0} g(x)\, dx + \int_{0}^{T} g(x)\, dx = 0.$$

Note now that if we multiply $g(x)$ by an even function $h(x)$, then the product $f(x) = g(x)h(x)$ is again an odd function:

$$f(-x) = g(-x)h(-x) = -g(x)h(x) = -f(x).$$

Now let $F(x)$ be an even function. Since $\sin(n\pi x/T)$ is an odd function,

$$F(x) \sin\left(\frac{n\pi x}{T}\right)$$

is odd and

$$Tb_n = \int_{-T}^{T} F(x) \sin\frac{n\pi x}{T}\, dx = 0,$$

implying that all the coefficients b_n vanish. Similarly, if $F(x)$ is odd, then

$$F(x) \cos\left(\frac{n\pi x}{T}\right)$$

is odd [since $\cos(n\pi x/T)$ is even] and

$$Ta_n = \int_{-T}^{T} F(x) \cos\frac{n\pi x}{T}\, dx = 0.$$

EXERCISES 9.3

1. Find the smallest positive period of the functions $\cos 2x$, $\sin \pi x$, $\cos(2\pi nx/T)$, $\sin 2k\pi x$.
2. Show that a constant function is periodic with any period $2T > 0$.
3. Suppose that $f(x)$ has period $2T$. What is the period of $f(ax/b)$?
4. Prove that a convergent infinite series of functions of period T is periodic of period T.

In Exercises 5 through 13 find the Fourier series of each function $f(x)$ of period 2π, where one period is defined, and accurately plot the first three partial sums

$$\frac{a_0}{2} + \sum_{n=1}^{k} (a_n \cos nx + b_n \sin nx), \qquad k = 1, 2, 3.$$

5. $f(x) = x, |x| < \pi$

6. $f(x) = \begin{cases} 0, & -\pi < x < 0 \\ 1, & 0 < x < \pi \end{cases}$

7. $f(x) = x^2, |x| < \pi$

8. $f(x) = \begin{cases} 0, & -\pi < x < 0 \\ x, & 0 < x < \pi \end{cases}$

9. $f(x) = |x|, |x| < \pi$

10. $f(x) = \begin{cases} x, & -\pi < x < 0 \\ x - \pi, & 0 < x < \pi \end{cases}$

11. $f(x) = \begin{cases} -1, & -\pi < x < -1 \\ x, & -1 < x < 1 \\ 1, & 1 < x < \pi \end{cases}$

12. $f(x) = \begin{cases} \pi + x, & -\pi < x < 0 \\ \pi - x, & 0 < x < \pi \end{cases}$

13. $f(x) = e^x, |x| < \pi$

In Exercises 14 through 21 find the Fourier series of each function $f(x)$ of period T, where one of the periods is defined.

14. $f(x) = x, |x| < 1, T = 2$

15. $f(x) = x, 0 < x < 2, T = 2$

16. $f(x) = x, 0 < x < 3, T = 3$

17. $f(x) = x^2, |x| < 1, T = 2$

18. $f(x) = x^2, 0 < x < 2, T = 2$

19. $f(x) = \begin{cases} 0, & 0 < x < 1 \\ 1, & 1 < x < 2, T = 2 \end{cases}$

20. $f(x) = \begin{cases} 0, & 0 < x < 1 \\ x - 1, & 1 < x < 2, T = 2 \end{cases}$

21. $f(x) = \begin{cases} x, & 0 < x < 1 \\ 1, & 1 < x < 2, T = 2 \end{cases}$

22. Find the Fourier series of the periodic function of period 2π:

$$f(x) = \frac{x^2}{4}, \qquad |x| < \pi,$$

and use this series to verify the identities:

$$\frac{\pi^2}{6} = 1 + \frac{1}{2^2} + \frac{1}{3^2} + \frac{1}{4^2} + \frac{1}{5^2} + \cdots,$$

$$\frac{\pi^2}{12} = 1 - \frac{1}{2^2} + \frac{1}{3^2} - \frac{1}{4^2} + \frac{1}{5^2} - \cdots,$$

$$\frac{\pi^2}{8} = 1 + \frac{1}{3^2} + \frac{1}{5^2} + \frac{1}{7^2} + \frac{1}{9^2} + \cdots.$$

9.4 STURM-LIOUVILLE PROBLEMS

Examples 9.1.1 and 9.1.2, as well as several of the homogeneous boundary value problems we shall encounter in Chapter 10, belong to a wide class of problems whose eigenfunctions and eigenvalues have particularly nice properties. We shall develop a small part of the fascinating theory of this class of problems in this and the next two sections.

Consider the differential equation

$$[r(x)y']' + [p(x) + \lambda q(x)]y = 0, \tag{9.62}$$

where $r(x)$, $p(x)$, and $q(x)$ are continuous functions on some interval $a \leqslant x \leqslant b$ and λ is a real parameter. We shall see that many of the differential equations of applied mathematics, such as Legendre's equation and Bessel's equation, can be written in the form (9.62). Suppose that we impose the boundary conditions

$$a_1 y(a) - a_2 y'(a) = 0,$$
$$b_1 y(b) - b_2 y'(b) = 0, \tag{9.63}$$

at the endpoints of the interval, and require that at least one coefficient in each equation in (9.63) be nonzero. Equation (9.62), together with the boundary conditions (9.63), is known as a *Sturm-Liouville* problem.*

Observe, as in Example 9.1.1, that any Sturm-Liouville problem has the trivial solution $y \equiv 0$. For certain values of the parameter λ, nontrivial solutions of the Sturm-Liouville problem may exist. All such nontrivial solutions are called *eigenfunctions* of the problem, and the corresponding values of λ that yield these solutions are called *eigenvalues* of the problem. In Example 9.1.1 the eigenvalues of Eq. (9.1) were given by $\lambda = k^2 = \pi^2, 4\pi^2, 9\pi^2, \ldots, (n\pi)^2, \ldots$ and the eigenfunctions corresponding to the eigenvalue $(n\pi)^2$ $(n \neq 0)$ had the form (9.4):

$$x(t) = B \sin n\pi t,$$

where B was any nonzero real number.

As we shall see in the next section, Sturm-Liouville problems have the following four basic properties:

i) All eigenvalues are real. (This tells us that we need not worry about complex eigenvalues, which may occur for other kinds of boundary value problems. See Exercise 9.5.9.)

ii) Corresponding to each eigenvalue there is only one linearly independent eigenfunction. (This simplifies the task of finding the eigenfunctions.)

iii) There are an infinite number of eigenvalues and eigenfunctions.

---

* The Swiss mathematician J. C. F. Sturm (1803–1855) made many significant contributions to the theory of differential equations. Joseph Liouville (1809–1882) was a French mathematician noted for his work in complex analysis.

iv)   The eigenfunctions can be chosen so as to be orthogonal with respect to the weight function $q(x)$. (The last two facts allow us to represent wide classes of functions by generalized Fourier series in terms of the eigenfunctions.)

Example 9.4.1   Find the eigenvalues and eigenfunctions of the Sturm-Liouville problem

$$y'' + \lambda y = 0, \qquad y(0) = y'(1) = 0. \qquad (9.64)$$

Setting $r(x) \equiv q(x) \equiv 1$ and $p(x) \equiv 0$, we see that (9.64) has the form (9.62), and letting $a = 0, b = 1, a_1 = -b_2 = 1, a_2 = b_1 = 0$, we have the right type of boundary conditions. Thus (9.64) is a Sturm-Liouville problem.

The roots of the auxiliary equation are $\pm\sqrt{-\lambda}$ so that we obtain the general solutions

$$y(x) = \begin{cases} c_1 e^{\sqrt{-\lambda}x} + c_2 e^{-\sqrt{-\lambda}x} & \text{if } \lambda < 0, \\ c_1 + c_2 x & \text{if } \lambda = 0, \\ c_1 \cos \sqrt{\lambda}x + c_2 \sin \sqrt{\lambda}x & \text{if } \lambda > 0. \end{cases} \qquad (9.65)$$

Thus we need to examine three cases:

1.   If $\lambda < 0$, then the boundary conditions $y(0) = 0$ and $y'(1) = 0$ yield, after a short computation, the homogeneous system of equations

$$c_1 + c_2 = 0,$$
$$\sqrt{-\lambda}c_1 - \sqrt{-\lambda}c_2 = 0.$$

It is easy to verify that the only solution to this system is $c_1 = c_2 = 0$, which indicates that problem (9.64) has only trivial solutions if $\lambda < 0$.

2.   If $\lambda = 0$, the boundary condition $y(0) = 0$ implies that $c_1 = 0$, and since $y' \equiv c_2$, the second condition forces c_2 to vanish. Again (9.64) has only a trivial solution.

3.   If $\lambda > 0$, the condition $y(0) = 0$ implies that $c_1 = 0$ so that the solution to the problem must have the form

$$y(x) = c_2 \sin \sqrt{\lambda}x. \qquad (9.66)$$

Differentiating (9.66), we have

$$y'(x) = c_2\sqrt{\lambda} \cos \sqrt{\lambda}x,$$

and setting $x = 1$, we obtain the equation

$$c_2\sqrt{\lambda} \cos \sqrt{\lambda} = 0. \qquad (9.67)$$

Hence in this case we will have nontrivial solutions whenever $\cos \sqrt{\lambda}$ is zero. This occurs whenever the constant $\sqrt{\lambda}$ is one of the numbers

$$\frac{\pi}{2}, \frac{3\pi}{2}, \frac{5\pi}{2}, \frac{7\pi}{2}, \ldots,$$

that is, whenever λ takes on one of the values

$$\frac{\pi^2}{4}, \frac{9\pi^2}{4}, \frac{25\pi^2}{4}, \frac{49\pi^2}{4}, \dots \qquad (9.68)$$

The values (9.68) are the eigenvalues of (9.64), and the eigenfunctions corresponding to the eigenvalues $\lambda_k = (2k - 1)^2\pi^2/4$, $k = 1, 2, 3, \dots$, have the form

$$y_k(x) = A \sin \frac{(2k - 1)\pi x}{2}, \qquad A \neq 0.$$

Example 9.4.2 Consider the Sturm-Liouville problem

$$(xy')' + \frac{\lambda}{x} y = 0, \qquad y(1) = y(e) = 0. \qquad (9.69)$$

If we write $(xy')'$ as $xy'' + y'$ and multiply the differential equation on both sides by x, we obtain the Euler equation considered in Section 3.8. We therefore assume that $y = x^r$ is a solution of (9.69) and find, after a short calculation, that

$$(rx^r)' + \lambda x^{r-1} = (r^2 + \lambda)x^{r-1} = 0.$$

Since x does not vanish on $1 \leqslant x \leqslant e$, it follows that $r = \pm\sqrt{-\lambda}$. Again we must deal with three cases:

1. If $\lambda < 0$, we have a general solution of the form

    $$y(x) = c_1 x^{\sqrt{-\lambda}} + c_2 x^{-\sqrt{-\lambda}}. \qquad (9.70)$$

    Using the boundary conditions, we obtain the homogeneous equations

    $$c_1 + c_2 = 0,$$
    $$e^{\sqrt{-\lambda}}c_1 + e^{-\sqrt{-\lambda}}c_2 = 0.$$

    Setting $c_2 = -c_1$ in the second equation, we obtain

    $$2c_1 \sinh \sqrt{-\lambda} = c_1(e^{\sqrt{-\lambda}} - e^{-\sqrt{-\lambda}}) = 0,$$

    and since $\sinh \sqrt{-\lambda} \neq 0$ if $\lambda \neq 0$, it follows that $c_1 = c_2 = 0$. Thus (9.69) has only trivial solutions when $\lambda < 0$.

2. If $\lambda = 0$, (9.69) reduces to the equation

    $$(xy')' = 0.$$

    Integrating both sides, we obtain

    $$xy' = c_1,$$

    so that

    $$y' = \frac{c_1}{x}$$

and

$y(x) = c_1 \ln x + c_2.$

The condition $y(1) = 0$ implies that $c_2 = 0$. Then $y(e) = c_1 \ln e = c_1 = 0$. Again we have only trivial solutions.

3. If $\lambda > 0$, we use the identity $e^{i\theta} = \cos \theta + i \sin \theta$ [see (3.28)] to write

$y_1(x) = x^{\sqrt{-\lambda}} = (e^{\ln x})^{i\sqrt{\lambda}} = \cos (\sqrt{\lambda} \ln x) + i \sin (\sqrt{\lambda} \ln x),$

$y_2(x) = x^{-\sqrt{-\lambda}} = (e^{\ln x})^{-i\sqrt{\lambda}} = \cos (\sqrt{\lambda} \ln x) - i \sin (\sqrt{\lambda} \ln x).$

Letting $y_1^* = (y_1 + y_2)/2$ and $y_2^* = (y_1 - y_2)/2i$ [see (3.31)] we may write the general solution of (9.69) in the form

$y(x) = c_1 \cos (\sqrt{\lambda} \ln x) + c_2 \sin (\sqrt{\lambda} \ln x), \qquad \lambda > 0.$

Setting $x = 1$, we find that the condition $y(1) = 0$ implies that $c_1 = 0$, so that

$y(x) = c_2 \sin (\sqrt{\lambda} \ln x).$

Finally, setting $x = e$, we have

$0 = y(e) = c_2 \sin \sqrt{\lambda}.$

Therefore (9.69) has a nontrivial solution whenever

$\sqrt{\lambda} = k\pi, \qquad k = 1, 2, 3, \ldots.$

Thus the numbers

$\lambda_k = k^2\pi^2, \qquad k = 1, 2, 3, \ldots,$

are all eigenvalues of (9.69), and the eigenfunctions corresponding to λ_k have the form

$y_k(x) = A \sin (k\pi \ln x).$

Example 9.4.3 Consider the Sturm-Liouville problem

$y'' + \lambda y = 0, \qquad y(0) + y'(0) = 0, \qquad y(1) = 0.$

Then, as in Example 9.4.1,

$$y(x) = \begin{cases} c_1 e^{\sqrt{-\lambda} x} + c_2 e^{-\sqrt{-\lambda} x} & \text{if } \lambda < 0, \\ c_1 + c_2 x & \text{if } \lambda = 0, \\ c_1 \cos \sqrt{\lambda} x + c_2 \sin \sqrt{\lambda} x & \text{if } \lambda > 0. \end{cases}$$

It is easy to verify that the boundary conditions imply that $c_1 = c_2 = 0$ if $\lambda < 0$. If $\lambda = 0$, then we have

$y(0) = c_1, \qquad y'(0) = c_2, \qquad y(1) = c_1 + c_2,$

so that both boundary conditions imply that $c_1 + c_2 = 0$. We therefore obtain the eigenfunctions

$$y_0(x) = c_1(1 - x).$$

If $\lambda > 0$, then

$$y(0) = c_1, \qquad y'(0) = \sqrt{\lambda}c_2, \qquad y(1) = c_1 \cos \sqrt{\lambda} + c_2 \sin \sqrt{\lambda}.$$

The boundary conditions imply that

$$c_1 + \sqrt{\lambda}c_2 = 0,$$
$$c_1 \cos \sqrt{\lambda} + c_2 \sin \sqrt{\lambda} = 0.$$

Setting $c_1 = -\sqrt{\lambda}c_2$, we find that the second equation becomes

$$-c_2\sqrt{\lambda} \cos \sqrt{\lambda} + c_2 \sin \sqrt{\lambda} = 0$$

or

$$\sqrt{\lambda} \cos \sqrt{\lambda} = \sin \sqrt{\lambda}$$

and (dividing by $\cos \sqrt{\lambda}$)

$$\tan \sqrt{\lambda} = \sqrt{\lambda}.$$

Unlike the previous examples, there is no explicit way to obtain these eigenvalues. We can easily see that there are an infinite number of real eigenvalues by plotting the curves $y = \tan x$ and $y = x$. The eigenvalues are the squares of the x-values of the points of intersection (see Fig. 9.8). From Fig. 9.8 it seems clear that the square roots of the eigenvalues get closer and closer to the vertical asymptotes of $\tan x$. That is,

$$\sqrt{\lambda_n} \approx \left(\frac{2n + 1}{2}\right) \pi \qquad \text{or} \qquad \lambda_n \approx \frac{(2n + 1)^2}{4} \pi^2.$$

The eigenfunctions are

$$y_n(x) = A(\sin \sqrt{\lambda_n}x - \sqrt{\lambda_n} \cos \sqrt{\lambda_n}x), \qquad n \geqslant 1, \qquad A \neq 0.$$

EXERCISES 9.4

In the following exercises find the eigenvalues and the corresponding eigenfunctions of the given Sturm-Liouville problems.

1. $y'' + \lambda y = 0$, $y(0) = y(\pi) = 0$

2. $y'' + \lambda y = 0$, $y(0) = y'(\pi) = 0$

3. $y'' + \lambda y = 0$, $y(-\pi/2) = y(\pi/2) = 0$

4. $y'' + \lambda y = 0$, $y'(0) = y'(\pi) = 0$

5. $(xy')' + \dfrac{\lambda}{x}y = 0$, $y(1) = y(e^2) = 0$

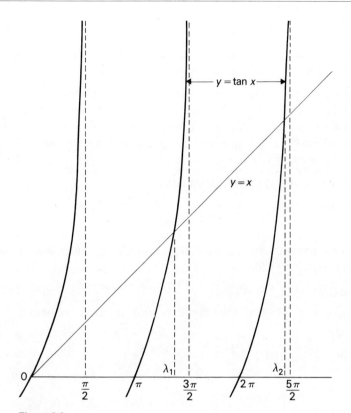

Figure 9.8

6. $(x^2y')' + \lambda y = 0$, $y(1) = y(e) = 0$

7. $(x^2y')' + \dfrac{\lambda}{x^2} y = 0$, $y(\frac{1}{2}) = y(1) = 0$ [*Hint*: Try $y_1 = \sin(\lambda/x)$.]

8. $\left(\dfrac{1}{2x} y'\right)' - 2x\lambda y = 0$, $y(0) = y(1) = 0$

9.5 PROPERTIES OF STURM-LIOUVILLE PROBLEMS

In this section we shall investigate the four basic properties of the eigenvalues and eigenfunctions of the Sturm-Liouville problem mentioned in the last section:

$$[r(x)y']' + [p(x) + \lambda q(x)]y = 0, \tag{9.71}$$

with the boundary conditions

$$a_1 y(a) - a_2 y'(a) = 0, \qquad b_1 y(b) - b_2 y'(b) = 0. \tag{9.72}$$

We begin by showing that the eigenfunctions of this problem are orthogonal with respect to the weight function $q(x)$.

Theorem 9.3. Orthogonality Theorem Let the real-valued functions $r(x)$, $p(x)$, and $q(x)$ of Eq. (9.71) be continuous on the interval $a \leqslant x \leqslant b$, and let $y_m(x)$ and $y_n(x)$ be eigenfunctions corresponding to distinct eigenvalues λ_m and λ_n of the Sturm-Liouville problem (9.71), (9.72). Then y_m and y_n are orthogonal with respect to the weight function $q(x)$.

Proof. The functions y_m and y_n satisfy the equations

$$(r y_m')' + (p + \lambda_m q) y_m = 0,$$
$$(r y_n')' + (p + \lambda_n q) y_n = 0.$$

Multiplying the first equation by y_n and the second by $-y_m$ and adding the resulting equations together yields

$$y_n (r y_m')' - y_m (r y_n')' = (\lambda_n - \lambda_m) q y_m y_n.$$

Integrating all terms from a to b and the first two terms by parts, we obtain

$$\left(y_n r y_m' \Big|_a^b - \int_a^b r y_m' y_n' \, dx \right) - \left(y_m r y_n' \Big|_a^b - \int_a^b r y_n' y_m' \, dx \right)$$
$$= (\lambda_n - \lambda_m) \int_a^b q y_m y_n \, dx. \tag{9.73}$$

Canceling the two identical integrals on the left-hand side, we obtain the equation

$$r(b)[y_n(b) y_m'(b) - y_m(b) y_n'(b)] - r(a)[y_n(a) y_m'(a) - y_m(a) y_n'(a)]$$
$$= (\lambda_n - \lambda_m) \int_a^b q y_m y_n \, dx. \tag{9.74}$$

By hypothesis, at least one of the constants a_1, a_2 in (9.72) is nonzero [see Eq. (9.63)], so that we have either

$$y(a) = \frac{a_2}{a_1} y'(a) \qquad \text{or} \qquad y'(a) = \frac{a_1}{a_2} y(a). \tag{9.75}$$

If the first of these equations is permissible, then it is satisfied by both $y_m(a)$ and $y_n(a)$, so the second term in brackets in Eq. (9.74) becomes

$$y_n(a) y_m'(a) - y_m(a) y_n'(a) = \frac{a_2}{a_1} y_n'(a) y_m'(a) - \frac{a_2}{a_1} y_m'(a) y_n'(a) = 0.$$

Similarly, if the second equation in (9.75) is permissible, then the second term in brackets will vanish. In the same way, the second boundary condition in (9.72) causes the first term in brackets in (9.74) to vanish. Thus,

$$(\lambda_n - \lambda_m) \int_a^b q y_m y_n \, dx = 0,$$

and since $\lambda_n \neq \lambda_m$, the proof is complete.

Remark. Observe that if $r(a) = 0$, we do not require the first boundary condition (9.72) to prove the orthogonality theorem. Similarly, the second boundary condition is not required if $r(b) = 0$. Finally, if $r(a) = r(b)$, we can also obtain the conclusion of the orthogonality theorem by assuming the conditions

$$y(a) = y(b), \qquad y'(a) = y'(b) \tag{9.76}$$

instead of (9.72). The proof of this fact is obvious, since the quantities in brackets in Eq. (9.74) are now identical, so that the left-hand side of (9.74) vanishes. Furthermore, *if the functions $r(x)$, $p(x)$, and $q(x)$ are periodic with period $(b - a)$, then any eigenfunction of this problem will also be periodic with period $(b - a)$*. (See Exercise 6.)

Example 9.5.1 Consider the Sturm-Liouville problem

$$y'' + \lambda y = 0, \qquad y(0) = y'(1) = 0.$$

As we saw in Example 9.4.1, the eigenfunctions of this problem have the form

$$\sin \frac{\pi}{2} x, \quad \sin \frac{3\pi}{2} x, \quad \sin \frac{5\pi}{2} x, \quad \ldots, \quad \sin \frac{(2k - 1)}{2} \pi x.$$

Theorem 9.3 implies that these functions are orthogonal, a fact which we established directly in Section 9.2.

Example 9.5.2 Using the orthogonality theorem, it is easy to verify that the Legendre polynomials form an orthogonal set of functions. Since $[(1 - x^2)y']' = (1 - x^2)y'' - 2xy'$, we can write Legendre's equation (see Example 9.2.4) in the Sturm-Liouville form

$$[(1 - x^2)y']' + \lambda y = 0, \qquad \lambda = n(n + 1).$$

Since $r(x) = 1 - x^2$ vanishes at $x = \pm 1$, no boundary conditions are needed for the theorem to apply; and since $q(x) \equiv 1$, we immediately have

$$(P_m, P_n) = \int_{-1}^{1} P_m(x) P_n(x) \, dx = 0 \qquad \text{if} \qquad m \neq n.$$

The endpoints a and b in the orthogonality theorem need not be finite provided the improper integrals in (9.73) all exist. We illustrate this situation with the following example.

Example 9.5.3 Laguerre's equation (Example 9.2.6)

$$(xe^{-x}y')' + ne^{-x}y = 0$$

is of the Sturm-Liouville type and $r(x) = xe^{-x}$ vanishes at $x = 0$. By L'Hôpital's rule

$$\lim_{x \to \infty} r(x) = \lim_{x \to \infty} \frac{x}{e^x} = \lim_{x \to \infty} \frac{1}{e^x} = 0,$$

so no boundary conditions are required to apply the orthogonality theorem. Thus the Laguerre polynomials $\{y_n(x)\}$ form an orthogonal set of functions with respect to the weight function $q(x) = e^{-x}$:

$$\int_0^\infty e^{-x}y_m(x)y_n(x)\,dx = 0 \qquad \text{if} \qquad m \neq n.$$

In every example we have seen so far, the eigenvalues of the given Sturm-Liouville problem are real, not because the examples were carefully chosen, but because the situation holds for every Sturm-Liouville problem. The proof of this fact, stated below, is very similar to the proof of Theorem 9.3 and is therefore left as an exercise (see Exercise 8).

Theorem 9.4 The eigenvalues of the Sturm-Liouville problem (9.71), (9.72) are all real.

Finally, we have observed that in every example there are an infinite number of eigenvalues, and that corresponding to each eigenvalue there is only one linearly independent eigenfunction. That is, if y_1 and y_2 are two eigenfunctions corresponding to the eigenvalue λ, then there is a constant c such that $y_1 = cy_2$. When the latter condition holds, the eigenvalue is said to be *simple*. These results are summarized in the theorem below, whose proof, however, is beyond the scope of this text.* (The simplicity of the eigenvalues can easily be proved; see Exercise 11.)

Theorem 9.5 The eigenvalues of the Sturm-Liouville problem (9.71), (9.72) are simple. Moreover, there are an infinite number of them, which can be arranged in increasing order

$$\lambda_1 < \lambda_2 < \lambda_3 < \cdots < \lambda_k < \cdots,$$

where λ_k tends to ∞ as k tends to ∞.

All three theorems of this section will be needed for the solution of nonhomogeneous boundary value problems in Section 9.7.

---

* See, for example, the excellent book *Methods of Mathematical Physics* by R. Courant and D. Hilbert, Wiley, New York, 1953, Vol. I, Chapter 6.

EXERCISES 9.5

In Exercises 1 through 5 verify the implications of Theorems 9.3, 9.4, and 9.5 for each given
Sturm-Liouville problem.

1. $y'' + \lambda y = 0$, $y(0) = y(\pi) = 0$

2. $y'' + \lambda y = 0$, $y(0) = y'(\pi) = 0$

3. $y'' + \lambda y = 0$, $y(-\pi/2) = y(\pi/2) = 0$

4. $(xy')' + (\lambda/x)y = 0$, $y(1) = y(e^2) = 0$

5. $(x^2 y')' + (\lambda/x^2)y = 0$, $y(\frac{1}{2}) = y(1) = 0$

6. Prove that if $r(x)$, $p(x)$, and $q(x)$ are continuous periodic functions of period $(b - a)$,
   then all eigenfunctions of the boundary value problem (7.71), (7.72) are periodic with
   period $(b - a)$.

7. a) Show that Hermite's equation (see Exercise 9(c) of Section 9.2) can be written in
   the form
   $$(e^{-x^2}y')' + 2ne^{-x^2}y = 0.$$

   b) Using the orthogonality theorem, prove that the Hermite polynomials are orthogonal
   with respect to the weight function $\exp(-x^2)$ on $-\infty < x < \infty$.

8. Prove that the eigenvalues of the Sturm-Liouville problem (9.71), (9.72) are all real by
   carrying out the following steps:

   a) Show that if $\lambda = \alpha + i\beta$ is an eigenvalue with the corresponding eigenfunction
   $y_\lambda(x) = u(x) + iv(x)$, then $\lambda = \alpha - i\beta$ is an eigenvalue with the corresponding
   eigenfunction $\bar{y}_\lambda(x) = u(x) - iv(x)$.

   b) Using the same proof as that of Theorem 9.3, show that
   $$(\lambda - \bar{\lambda}) \int_a^b y_\lambda(x)\bar{y}_\lambda(x)q(x)\, dx = 0.$$

   c) Use the result of part (b) of this exercise and the fact that $y_\lambda \bar{y}_\lambda = u^2 + v^2$ to show
   that $\beta = 0$.

9. The boundary value problem
   $$y'' - \frac{\lambda}{x}y' + \frac{\lambda}{x^2}y = 0, \qquad y(1) = y(2) = 0,$$

   is not a Sturm-Liouville problem. Show that none of the eigenvalues are real.

10. Consider the boundary value problem
    $$y'' + \lambda y = 0, \qquad y(-1) = y(1), \qquad y'(-1) = y'(1).$$

    a) Explain why it is not a Sturm-Liouville problem.
    b) Calculate the eigenvalues of this problem.
    c) Show that corresponding to each eigenvalue there are two linearly independent
       eigenfunctions. That is, show that the eigenvalues are not simple.

11. Prove that the eigenvalues of a Sturm-Liouville problem are simple. [*Hint*: Assume that y_1 and y_2 are two eigenfunctions corresponding to the eigenvalue λ. Use the boundary conditions (9.72) to show that the Wronskian $W(y_1,y_2)(x) = 0$ for $a \leqslant x \leqslant b$.]

*9.6 THE STURM SEPARATION THEOREM AND THE ZEROS OF BESSEL FUNCTIONS

In this section we will establish some further properties of the solutions of the second order differential equation (9.71), which, as we saw earlier, arises in a number of physical problems. We begin by analyzing the behavior of the zeros of two solutions of Eq. (9.71). The procedure we will use is similar to that employed in the orthogonality theorem (9.3). *Note that we are not assuming the boundary conditions* (9.72) *in this section.*

Theorem 9.6. Sturm Separation Theorem Let y_1 and y_2 be two linearly independent solutions of Eq. (9.71) on the interval $a \leqslant x \leqslant b$. Suppose that $r(x) > 0$ for all $a \leqslant x \leqslant b$. Then between any two consecutive zeros of y_1 there will be precisely one zero of y_2.

Proof. Since y_1 and y_2 are linearly independent, neither solution is the zero solution. Let x_1 and x_2 be two consecutive zeros of y_1, with $a \leqslant x_1 < x_2 \leqslant b$ (see Fig. 9.9). We can assume that $y_1(x) > 0$ for $x_1 < x < x_2$, since $-y_1$ is also

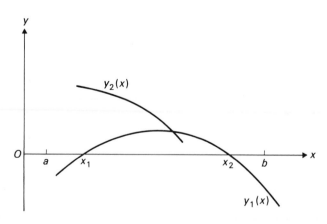

Figure 9.9

a solution of (9.71) and one of the two functions y_1 and $-y_1$ is positive in this interval. Similarly, we can assume that $y_2(x_1) \geq 0$. Since both y_1 and y_2 satisfy (9.71), we have

$$(ry_1')' + (p + \lambda q)y_1 = 0,$$
$$(ry_2')' + (p + \lambda q)y_2 = 0.$$

Multiplying the first of these equations by y_2 and the second by $-y_1$, and adding the resulting equations together, we obtain

$$y_2(ry_1')' - y_1(ry_2')' = 0.$$

Moving the second term to the right-hand side of the equation and integrating both sides from x_1 to x_2, we have

$$\int_{x_1}^{x_2} y_2(ry_1')' \, dx = \int_{x_1}^{x_2} y_1(ry_2')' \, dx. \tag{9.77}$$

Integrating both sides of (9.77) by parts yields the equation

$$y_2(ry_1') \Big|_{x_1}^{x_2} - \int_{x_1}^{x_2} ry_1' y_2' \, dx = y_1(ry_2') \Big|_{x_1}^{x_2} - \int_{x_1}^{x_2} ry_2' y_1' \, dx.$$

Both integrals are the same; so canceling them and gathering all terms involving x_2 on the right-hand side and all terms in x_1 on the left-hand side, we have

$$r(x_1)[y_1(x_1)y_2'(x_1) - y_1'(x_1)y_2(x_1)]$$
$$= r(x_2)[y_1(x_2)y_2'(x_2) - y_1'(x_2)y_2(x_2)]. \tag{9.78}$$

Note that the quantities in brackets are the Wronskians of y_1 and y_2 evaluated at the points x_1 and x_2, respectively. Since y_1 and y_2 are linearly independent on $a \leq x \leq b$, the Wronskian $W(x)$ is never zero, by Lemma 3.3. However, by hypothesis, $y_1(x_1) = y_1(x_2) = 0$, so that

$$W(x_1) = -y_1'(x_1)y_2(x_1) \neq 0 \tag{9.79}$$

and

$$W(x_2) = -y_1'(x_2)y_2(x_2) \neq 0. \tag{9.80}$$

Recall now that we have assumed $y_2(x_1) \geq 0$. By (9.79), we must have $y_2(x_1) > 0$ and $y_1'(x_1) \neq 0$, since $W(x_1) \neq 0$. Since $y_1(x) > 0$ in $x_1 < x < x_2$, we necessarily have $y_1'(x_1) > 0$, and hence $W(x_1) < 0$. Similarly, $y_1'(x_2) < 0$, and since $r(x)$ is always positive on $a \leq x \leq b$, Eq. (9.78) implies that $W(x_2)$ is negative. Thus, by (9.80) and the fact that $y_1'(x_2) < 0$, we have $y_2(x_2) < 0$. Therefore, y_2 changes sign between x_1 and x_2. Since y_2 must be continuous, the intermediate value theorem of calculus implies that y_2 must have a zero between x_1 and x_2.

Finally, we must make sure y_2 has only one zero between x_1 and x_2. If it had more than one zero, we could reverse the roles of y_1 and y_2, and the proof above, applied to y_1, would imply the existence of a zero of y_1 between x_1 and x_2. Since x_1 and x_2 are *consecutive* zeros of y_1, no such zero can exist. Thus y_2 has exactly one zero between x_1 and x_2.

Remark. If two solutions of (9.71) have a common zero, then they must be linearly dependent, since their Wronskian will vanish at this point.

The Sturm separation theorem indicates that the number of zeros of any two solutions of (9.71) on any interval is approximately the same. However, it does not guarantee the existence of *any* zeros. We shall now prove a theorem which compares the number of zeros (that is, the rate of *oscillation*) of solutions of two differential equations

$$[r(x)y']' + p_1(x)y = 0, \tag{9.81}$$

and

$$[r(x)w']' + p_2(x)w = 0, \tag{9.82}$$

where r, p_1, and p_2 are continuous functions on $a \leqslant x \leqslant b$. The motivation behind this theorem can be seen by examining the solutions $y = \sin x$ and $w = \sin 3x$ of the two equations

$$y'' + y = 0 \quad \text{and} \quad w'' + 9w = 0.$$

The solutions of the second equation ($\sin 3x, \cos 3x$) oscillate more rapidly than those of the first equation ($\sin x, \cos x$). That is, $\sin 3x$ oscillates three times in $[0, 2\pi]$ while $\sin x$ oscillates only once. It is reasonable to suspect that whenever p_2 exceeds p_1, there will be an increase in the rate of oscillation.

Theorem 9.7. Sturm Comparison Theorem Let $r(x)$ be positive and $y(x)$ and $w(x)$ be solutions of Eqs. (9.81) and (9.82), respectively, on the interval $a \leqslant x \leqslant b$. Suppose that x_1 and x_2 are consecutive zeros of y with $a \leqslant x_1 < x_2 \leqslant b$, $p_2(x) \geqslant p_1(x)$ on $x_1 \leqslant x \leqslant x_2$, and that $p_2(x_0) > p_1(x_0)$ for some point $x_1 \leqslant x_0 \leqslant x_2$. Then, if w vanishes at x_1, it will vanish again in the interval $x_1 < x < x_2$. Moreover, in that case, every solution of (9.82) will vanish at some point in the interval $x_1 < x < x_2$.

Proof. As in Theorem 9.6 we can assume $y > 0$ on $x_1 < x < x_2$ and that $y'(x_1) > 0$ and $y'(x_2) < 0$. Multiplying (9.81) by $w(x)$ and (9.82) by $-y(x)$ and adding the resulting equations together, we obtain

$$w(ry')' - y(rw')' + (p_1 - p_2)yw = 0.$$

Integrating all the terms in this equation from x_1 to x_2 and the first two terms by parts, we have

$$\left(wry' \Big|_{x_1}^{x_2} - \int_{x_1}^{x_2} ry'w' \, dx \right) - \left(yrw' \Big|_{x_1}^{x_2} - \int_{x_1}^{x_2} rw'y' \, dx \right) = \int_{x_1}^{x_2} (p_2 - p_1)yw \, dx.$$

The two integrals on the left-hand side of this equation cancel out, so that gathering like terms yields

$$r(x_2)[w(x_2)y'(x_2) - y(x_2)w'(x_2)] - r(x_1)[w(x_1)y'(x_1) - y(x_1)w'(x_1)]$$
$$= \int_{x_1}^{x_2} (p_2 - p_1)yw \, dx.$$

The second term in brackets is zero since $w(x_1) = y(x_1) = 0$; hence the equation reduces to

$$r(x_2)w(x_2)y'(x_2) = \int_{x_1}^{x_2} (p_2 - p_1)yw \, dx. \qquad (9.83)$$

If $w(x) > 0$ on $x_1 < x < x_2$, then the integrand on the right-hand side of (9.83) is nonnegative on $x_1 \leqslant x \leqslant x_2$ and *positive* in some smaller interval containing the point x_0. Hence the right-hand side of (9.83) is positive. Since $r(x_2) > 0$ and $y'(x_2) < 0$, it must follow that $w(x_2) < 0$. Since w is continuous, the last conclusion contradicts the assumption that $w > 0$ on $x_1 < x < x_2$. Hence w must vanish in the interval $x_1 < x < x_2$. Finally, if w_1 is another linearly independent solution of (9.82), then the Sturm separation theorem 9.6 implies that w_1 must have a zero between the zeros of w, and therefore in the interval $x_1 < x < x_2$.

Example 9.6.1 Let $p(x) < 0$ on the interval $a \leqslant x \leqslant b$ and suppose the equation

$$y'' + p(x)y = 0 \qquad (9.84)$$

has a nontrivial solution y with zeros at x_1 and x_2, with $a \leqslant x_1 < x_2 \leqslant b$. Since $w = 1$ is a solution of $w'' = 0$, the Sturm comparison theorem 9.7 implies that w must vanish in $x_1 < x < x_2$. But this is impossible. So every nontrivial solution of (9.84) has at most one zero in $a \leqslant x \leqslant b$.

Example 9.6.2 Bessel's equation of order p,

$$x^2 y'' + xy' + (x^2 - p^2)y = 0,$$

may be transformed, after an easy calculation, into the equation

$$w'' + \left(1 + \frac{1 - 4p^2}{4x^2} \right) w = 0 \qquad (9.85)$$

by means of the substitution $y = w/\sqrt{x}$, which does not change the zeros. Comparing (9.85) with the equation $y'' + y = 0$, the Sturm comparison theorem yields the following results:

1. If $0 \leqslant p < \frac{1}{2}$, then

$$1 + \frac{1 - 4p^2}{4x^2} > 1,$$

so every solution of (9.85), and hence of Bessel's equation, will vanish at least once between the zeros of any nontrivial solution of $y'' + y = 0$. But the general solution of $y'' + y = 0$ has the form

$$y = c_1 \cos x + c_2 \sin x = \sqrt{c_1^2 + c_2^2} \sin (x + \theta),$$

where $\theta = \tan^{-1} (c_1/c_2)$. Thus every nontrivial solution of $y'' + y = 0$ has zeros separated by an interval of length π. Hence *every solution of Bessel's equation of order p, $0 \leqslant p < \frac{1}{2}$, has at least one zero in every interval of length π, for $x > 0$.*

2. If $p = \frac{1}{2}$, then Eq. (9.85) reduces to $w'' + w = 0$, with the general solution $w = c_1 \cos x + c_2 \sin x$. Then *every nontrivial solution of Bessel's equation of order $\frac{1}{2}$,*

$$y = \frac{w}{\sqrt{x}} = \frac{1}{\sqrt{x}} (c_1 \cos x + c_2 \sin x),$$

has zeros separated by an interval of length π, for $x > 0$.

3. If $p > \frac{1}{2}$, then

$$1 + \frac{1 - 4p^2}{4x^2} < 1,$$

so every solution of Bessel's equation can have *at most* one zero in every interval of length π. We leave the proof of the fact that each solution of Bessel's equation of order $p > \frac{1}{2}$ has infinitely many zeros in $x > 0$ as an exercise (see Exercise 8).

Let us now return to the Sturm-Liouville problem (9.71) with the boundary conditions $y(a) = y(b) = 0$, and ask the following question: *How many eigenvalues does this problem have?* If $q(x)$ is positive on the interval $a \leqslant x \leqslant b$, then a suitably large value of λ will guarantee that $p(x) + \lambda q(x)$ is positive on $a \leqslant x \leqslant b$. Assuming also that $r(x)$ is positive on $a \leqslant x \leqslant b$, we can make the nontrivial solutions of (9.71) oscillate as rapidly as we wish by selecting appropriate values for λ. Since it is always possible to pick a solution y_λ of (9.71) that vanishes at $x = a$ for each λ, it is reasonable to expect that some number λ_0 exists such that the next zero of y_{λ_0} will occur at $x = b$. Similarly, we can expect to find a λ_k such that $y_{\lambda_k}(b) = 0$ and y_{λ_k} has k zeros between a and b, $k = 1, 2, 3, \ldots$. The above discussion, when made precise, leads to a proof of the existence of an infinite number of eigenvalues for the Sturm-Liouville problem (9.71), (9.72) (see Theorem 9.5).

EXERCISES 9.6

1. Prove that the substitution $y = w/\sqrt{x}$ will transform Bessel's equation of order p into Eq. (9.85).

2. Extend Example 9.6.1 to equations of the form (9.81).

3. Use the Sturm separation theorem to show that the functions $f(x) = a \sin x + b \cos x$ and $g(x) = c \sin x + d \cos x$ have alternating zeros whenever $ad - bc \neq 0$.

4. Show that the zeros of the functions $f(x) = \sin \ln x$ and $g(x) = \cos \ln x$ alternate. [*Hint*: Find a differential equation for which f and g are both solutions.]

5. Does the Sturm separation theorem apply to the differential equation $y'' - y = 0$?

6. Which of the equations

    $$y'' + (x^2 + 1)y = 0,$$
    $$w'' + 2xw = 0$$

    has the more rapidly oscillating solution in the interval $1 \leqslant x \leqslant 10$?

7. How many zeros does every solution of

    $$y'' + xy = 0$$

    have on the interval $0 < x < \infty$?

8. Prove that every solution of Bessel's equation of order $p \geqslant 0$ has infinitely many zeros in the interval $x > 0$.

9. Prove that the distance between successive zeros of any nontrivial solution of Bessel's equation of order $p \geqslant 0$ approaches π as x tends to ∞.

* 10. Prove that the conclusion of the Sturm comparison theorem will remain valid if the differential equations (9.81) and (9.82) are replaced by the equations

    $$[r_1(x)y']' + p_1(x)y = 0,$$
    $$[r_2(x)w']' + p_2(x)w = 0,$$

    with $r_2(x) \leqslant r_1(x)$ on $a \leqslant x \leqslant b$. This result is sometimes called *Picone's theorem*. [*Hint*: Use the identity

    $$\left[\frac{y}{w}(r_1 y'w - r_2 w'y) \right]' = (p_2 - p_1)y^2 + (r_1 - r_2)y'^2 + r_2 \left(y' - \frac{w'y}{w} \right)^2. \right]$$

* 11. Consider Eq. (9.81) with $r(x) > 0$ and $p_1(x)$ having continuous derivatives on the interval $a \leqslant x \leqslant b$. Suppose that $(rp_1)' \geqslant 0$. Prove that the absolute values of the relative extremums of every nontrivial solution of (9.81) are nonincreasing as x increases. This result, due to *Sonin*, indicates that the solutions are damped as x increases. [*Hint*: Differentiate the function $f(x) = y^2 + (py'^2/q)$ and show that $f' \leqslant 0$ on $a \leqslant x \leqslant b$. Then evaluate f at the points where y' vanishes.]

* 12. Let $r(x) > 0$ and $r(x)$ and $p_1(x)$ be continuous on the finite interval $a \leqslant x \leqslant b$. Prove that the only solution of (9.81) that will vanish infinitely often in this interval is the trivial solution $y = 0$.

9.7 NONHOMOGENEOUS BOUNDARY VALUE PROBLEMS

In Example 9.1.3 we considered the nonhomogeneous boundary value problem (9.12). At that time we stated that the solution of (9.12), for $\omega \neq n\pi$, could be expressed by the Fourier sine series (9.20). In this section we shall justify that statement and unite the seemingly unrelated concepts of Fourier series and Sturm-Liouville problems. We shall discover that the solutions of certain nonhomogeneous boundary value problems can be obtained as generalized Fourier series in the eigenfunctions of the associated Sturm-Liouville problem.

This method of *eigenfunction expansions* is one of the main techniques used in solving nonhomogeneous boundary value problems. Furthermore, it provides the motivation for the very powerful method of *Green's functions*, developed in Exercises 8 through 12, which is extremely useful for solving such problems.

Suppose that we wish to solve the *nonhomogeneous* differential equation

$$[r(x)w']' + p(x)w = f(x) \tag{9.86}$$

on $a \leqslant x \leqslant b$ with the boundary conditions

$$a_1 w(a) - a_2 w'(a) = 0, \qquad b_1 w(b) - b_2 w'(b) = 0. \tag{9.87}$$

We shall consider a different boundary value problem

$$[r(x)y']' + [p(x) + \lambda q(x)]y = 0 \tag{9.88}$$

with identical boundary conditions

$$a_1 y(a) - a_2 y'(a) = 0, \qquad b_1 y(b) - b_2 y'(b) = 0, \tag{9.89}$$

and assume that its eigenvalues and eigenfunctions are known. We are free to choose $q(x)$ in any way we wish; however, the choice is limited, since we must be able to determine the eigenfunctions of (9.88), (9.89). Furthermore, different choices of $q(x)$ will yield different sets of eigenfunctions, and some sets may be easier to work with than others. We assume that $r(x)$, $p(x)$, and $q(x)$ are continuous and r and q are positive on $a \leqslant x \leqslant b$. By Theorem 9.5 and the orthogonality theorem, we know that the problem (9.88), (9.89) has an infinite orthogonal set $\{y_k\}$ of eigenfunctions with respect to the weight function $q(x)$.

The object now is to express the solution $w(x)$ of (9.86), (9.87) (if one exists) as a generalized Fourier series [see Eq. 9.30)] of the form

$$w(x) = \sum_{n=1}^{\infty} c_n y_n, \tag{9.90}$$

where the functions y_n are eigenfunctions of (9.88), (9.89). Of course, since we are proceeding formally, when the process is complete we shall have to check that our result is indeed a solution. As we saw in Section 9.3, the problem of determining which functions can be represented in this way is not easy. Although the proof is beyond the scope of this book, *any piecewise continuous function $w(x)$ with piecewise continuous derivative $w'(x)$ has a representation* (9.90) *at all points of continuity, provided that a and b are finite.** (At points of discontinuity the generalized Fourier series converges to $[w(x + 0) + w(x - 0)]/2$.)

To obtain the formal representation (9.90), we proceed in a manner similar to the proof of the orthogonality theorem. Multiplying (9.86) by y and (9.88) by $-w$, we obtain the equation

$$y(rw')' - w(ry')' = fy + \lambda qyw.$$

Integrating both sides from a to b and the left-hand side by parts, we have

$$\left(yrw' \Big|_a^b - \int_a^b rw'y'\, dx\right) - \left(wry' \Big|_a^b - \int_a^b ry'w'\, dx\right) = \int_a^b (fy + \lambda qyw)\, dx.$$

The integrals on the left-hand side cancel out, yielding the equation

$$r(b)[y(b)w'(b) - w(b)y'(b)] - r(a)[y(a)w'(a) - w(a)y'(a)]$$
$$= \int_a^b (fy + \lambda qyw)\, dx. \qquad (9.91)$$

Letting $\lambda = \lambda_k$ and $y = y_k$, we see as in the proof of the orthogonality theorem that the left-hand side of Eq. (9.91) vanishes. Thus,

$$\lambda_k \int_a^b qwy_k\, dx = -\int_a^b fy_k\, dx,$$

and since by Eq. (9.48),

$$c_k = \frac{(w, y_k)_q}{\|y_k\|_q^2} = \frac{\int_a^b qwy_k\, dx}{\int_a^b qy_k^2\, dx},$$

we finally obtain

$$c_k = -\frac{\int_a^b fy_k\, dx}{\lambda_k \int_a^b qy_k^2\, dx}. \qquad (9.92)$$

In particular, if the eigenfunctions y_k have been chosen so that they are orthonormal with respect to q, then the denominator in the right-hand side of (9.92) reduces to

---

*   Much more general theorems are known. See, for example, E. A. Coddington and N. Levinson, *Theory of Ordinary Differential Equations*, McGraw-Hill, New York, 1955, p. 199.

just the eigenvalue λ_k, and we have the representation

$$w(x) = -\sum_{n=1}^{\infty} \left(\lambda_n^{-1} \int_a^b fy_n \, dx \right) y_n(x) \tag{9.93}$$

for the solution of the boundary value problem (9.86), (9.87). We now apply this result to justify the solution (9.20) of Example 9.1.3.

Example 9.7.1 Suppose that we again consider the forced vibrations of a spring-weight system, with spring constant λ_0, attached to an external periodic force $K \sin \omega t$. For simplicity, we ignore the damping term in Eq. (3.61) and merely consider the nonhomogeneous equation

$$x'' + \frac{\lambda_0}{m} x = \frac{K}{m} \sin \omega t. \tag{9.94}$$

We again suppose that the mass is at its equilibrium position initially and when $t = 1$ second, and seek a solution to the nonhomogeneous boundary value problem.

To apply the eigenfunction expansion (9.93), we consider the Sturm-Liouville problem

$$y'' + (\lambda + 1) \frac{\lambda_0}{m} y = 0, \qquad y(0) = y(1) = 0. \tag{9.95}$$

The eigenvalues of (9.95) are obtained by setting $\sqrt{(\lambda + 1)\lambda_0/m}$ equal to a multiple of π; hence

$$\lambda_k = \frac{mk^2\pi^2}{\lambda_0} - 1, \qquad k = 1, 2, 3, \ldots.$$

The corresponding eigenfunctions have the form

$$y_k = A \sin k\pi t, \qquad k = 1, 2, 3, \ldots,$$

and are orthonormal with respect to $q(x) = \lambda_0/m$ if $A = \sqrt{2m/\lambda_0}$. By (9.92), the Fourier coefficients c_k are given by

$$c_k = -\sqrt{\frac{2m}{\lambda_0}} \cdot \frac{K}{m\lambda_k} \int_0^1 \sin \omega t \sin k\pi t \, dt.$$

Using formula 19 of Appendix 1 and the identity

$$\sin (\omega \pm k\pi)t = \sin \omega t \cos k\pi t \pm \cos \omega t \sin k\pi t,$$

we have

$$c_k = -\sqrt{\frac{2m}{\lambda_0}} \cdot \frac{K}{2m\lambda_k} \left[\frac{\sin (\omega - k\pi)t}{\omega - k\pi} - \frac{\sin (\omega + k\pi)t}{\omega + k\pi} \right] \Big|_0^1$$

$$= -\sqrt{\frac{2m}{\lambda_0}} \cdot \frac{K}{m\lambda_k} \sin \omega \cos k\pi \frac{k\pi}{\omega^2 - k^2\pi^2}, \qquad \omega \neq k\pi.$$

If $\omega = k\pi$, then

$$c_k = -\sqrt{\frac{2m}{\lambda_0}} \cdot \frac{K}{2m\lambda_k},$$

and all the other Fourier coefficients vanish by the orthogonality of the eigen-functions. Thus the solution $x(t)$ must be stated for two cases:

1. If $\omega \neq k\pi, k = 1, 2, 3, \ldots$, then by (9.93) and the identity $\cos n\pi = (-1)^n$,

$$x(t) = -\frac{2K\pi \sin \omega}{\lambda_0} \sum_{n=1}^{\infty} \frac{(-1)^n n}{\lambda_n(\omega^2 - n^2\pi^2)} \sin n\pi t.$$

   It is easy to show that this solution is identical to Eq. (9.20).
2. Suppose that $\omega = k\pi$ for some positive integer k. Then using (9.93) and the value of λ_k, we obtain the particular solution

$$x_p(t) = -\frac{K}{\lambda_0 \lambda_k} \sin k\pi t = \frac{K \sin k\pi t}{\lambda_0 - mk^2\pi^2}.$$

   This solution is valid provided that the spring constant $\lambda_0 \neq mk^2\pi^2$.
   The general solution is now obtained as in Case 1 of Example 9.1.3. When $\lambda_0 = mk^2\pi^2$, there is *no* solution. Observe that in this case the eigenvalue

$$\lambda_k = \frac{mk^2\pi^2}{\lambda_0} - 1 = 0.$$

The following result is beyond the scope of this book, but provides the necessary information to decide whether or not there is a solution in the case above.

Theorem 9.8* Let one of the eigenvalues λ_k of the problem (9.88), (9.89) be zero. Then the nonhomogeneous boundary value problem (9.86), (9.87) has a solution if and only if

$$\int_a^b f(t)y_k(t) \, dt = 0.$$

Example 9.7.2 Consider the boundary value problem

$$(xw')' + \frac{w}{x} = \frac{1}{x}, \qquad w(1) = w(e) = 0. \tag{9.96}$$

To apply the method of eigenfunction expansion (9.93), we let $p(x) = q(x) = f(x) = 1/x$ and consider the Sturm-Liouville problem

$$(xy')' + \frac{(1 + \lambda)}{x} y = 0, \qquad y(1) = y(e) = 0. \tag{9.97}$$

---

* See, for example, E. A. Coddington and N. Levinson, *Theory of Ordinary Differential Equations*, McGraw-Hill, New York, 1955, p. 294.

By Example 9.4.2, we know that the eigenvalues $\lambda^* = 1 + \lambda$ of (9.97) are given by

$$1 + \lambda_k = \lambda_k^* = k^2\pi^2, \quad k = 1, 2, 3, \ldots,$$

so that $\lambda_k = k^2\pi^2 - 1$, and the eigenfunctions are of the form

$$y_k = A \sin (k\pi \ln x).$$

Using the substitution $u = k\pi \ln x$, $du = (k\pi/x) \, dx$, we have

$$\int_1^e \frac{1}{x} \sin^2 (k\pi \ln x) \, dx = \frac{1}{k\pi} \int_0^{k\pi} \sin^2 u \, du$$

$$= \frac{u - \cos u \sin u}{2k\pi} \Big|_0^{k\pi} = \frac{1}{2}.$$

Letting $A = \sqrt{2}$, we obtain an orthonormal set of eigenfunctions. By (9.92), the coefficients c_k of the generalized Fourier series are given by

$$c_k = -\lambda_k^{-1} \int_1^e \frac{\sqrt{2}}{x} \sin (k\pi \ln x) \, dx = \frac{\sqrt{2}\lambda_k^{-1}}{k\pi} \cos (k\pi \ln x) \Big|_1^e$$

$$= \frac{\sqrt{2}(\cos k\pi - 1)}{k\pi\lambda_k} = \begin{cases} 0, & k \text{ even}, \\ \dfrac{-2\sqrt{2}}{k\pi(k^2\pi^2 - 1)}, & k \text{ odd}. \end{cases}$$

Hence the solution of the problem (9.96) is given by the series

$$w(x) = -4 \sum_{n=0}^{\infty} \frac{\sin [(2n + 1)\pi \ln x]}{[(2n + 1)\pi]^3 - [(2n + 1)\pi]}.$$

A number of further examples of the use of eigenfunction expansions will be found in Chapter 10, where they are used in the solution of certain partial differential equations.

The methods we have presented in this chapter are not the only techniques that can be used to solve boundary value problems. A very powerful procedure, attributed to George Green (1793–1841), involves the representation of the solution of the boundary value problems (9.86), (9.87) in the form

$$w(x) = \int_a^b K(x,t)f(t) \, dt. \tag{9.98}$$

Actually, in light of the eigenfunction expansion (9.93), a representation of the form (9.98) is not surprising, for if we use t as the variable of integration and interchange the sum and integral in (9.93), we will obtain the expression

$$\int_a^b \left(-\sum_{n=1}^{\infty} \frac{y_n(t)y_n(x)}{\lambda_n} \right) f(t) \, dt.$$

Hence *Green's function* $K(x,t)$ may be presumed to equal

$$K(x,t) = -\sum_{n=1}^{\infty} \frac{y_n(t)y_n(x)}{\lambda_n}. \tag{9.99}$$

All the assumptions we made above can be justified.* Exercises 8 through 12 at the end of this section will indicate another method of obtaining the Green's function $K(x,t)$.

EXERCISES 9.7

In Exercises 1 through 5 use the method of eigenfunction expansions to solve the given nonhomogeneous boundary value problems.

1. $y'' + y = x$, $y(0) = y(\pi) = 0$
2. $y'' + 2y = \cos x$, $y(0) = y(1) = 0$
3. $y'' + 4y = x^2$, $y(0) = y'(1) = 0$
4. $(xy')' + \dfrac{1}{x}y = \ln x$, $y'(1) = y'(e^{2\pi}) = 0$
5. $(xy')' + \dfrac{3}{x}y = \dfrac{1}{x}\sin(\ln x)$, $y(1) = y(2) = 0$

* 6. Consider the nonhomogeneous problem

$$y'' + \lambda y = f(x), \qquad y(0) = y(1) = 0.$$

Show that if $f(x)$ is continuous, then there is a unique solution to this problem if and only if λ is *not* an eigenvalue of the associated homogeneous equation $y'' + \lambda y = 0$, $y(0) = y(1) = 0$.

* 7. Prove the same claim as in Exercise 6 for the problem

$$y'' + \lambda y = f(x), \qquad y(0) = y(1), \qquad y'(0) = y'(1).$$

Consider the differential equation

$$(rw')' + qw = 0.$$

It is easy to find a nontrivial solution w_1 of this equation with $w_1(a) = 0$, and a nontrivial solution w_2 with $w_2(b) = 0$. Define the function

$$k(x,t) = \begin{cases} w_1(x)w_2(t), & a < x < t, \\ w_1(t)w_2(x), & t < x < b. \end{cases}$$

---

* See, for example, E. A. Coddington and N. Levinson, *Theory of Ordinary Differential Equations*, McGraw-Hill, New York, 1955.

8. Ignoring any difficulties at $t = x$, show that

$$\frac{d}{dx}\int_a^b k(x,t)f(t)\,dt = \int_a^b \frac{\partial}{\partial x}k(x,t)f(t)\,dt.$$

9. Show that

$$\frac{d}{dx}\left[r(x)\frac{d}{dx}\int_a^b k(x,t)f(t)\,dt\right] = r(x)f(x)W(x) - q(x)\int_a^b k(x,t)f(t)\,dt,$$

where W is the Wronskian of the functions w_1 and w_2.

10. Using the result of Exercise 9, conclude that

$$w_3(x) = \int_a^b k(x,t)f(t)\,dt$$

satisfies the equation

$$[r(x)w_3']' + q(x)w_3 = f(x)r(x)W(x).$$

11. Show that

$$\frac{d}{dx}(r(x)W(x)) = 0,$$

and conclude that $r(x)W(x) = C$, a constant.

12. Prove that

$$w(x) = C^{-1}w_3(x)$$

is a solution of the boundary value problem

$$(r(x)w')' + q(x)w = f(x), \qquad w(a) = w(b) = 0.$$

Hence conclude that the Green's function is given by $K(x,t) = C^{-1}k(x,t)$.

13. Using the ideas of Exercises 8 through 12, find the Green's function for the boundary value problem

$$w''(x) - w(x) = f(x), \qquad w(0) = w(1) = 0.$$

14. Find the Green's function for the boundary value problem

$$w''(x) + k^2w(x) = f(x), \qquad w(0) = w(a) = 0.$$

15. Generalize the procedure used in Exercises 8 through 12 to obtain the Green's function for the boundary value problem

$$(rw')' + qw = f, \qquad w(0) = w'(1) = 0.$$

Apply the generalized procedure of Exercise 15 to Exercises 16 through 18.

16. $w''(x) = f(x), w(0) = w'(1) = 0$

17. $w''(x) + 4w(x) = f(x), w(0) = w'(1) = 0$

18. $(xw')' = f(x), w(1) = w'(e) = 0$

PARTIAL
DIFFERENTIAL
EQUATIONS

10

As was indicated in Example 1.1.6, a *partial differential equation* is an equation involving a function of two or more variables and some of its partial derivatives. Thus the crucial difference between partial and ordinary differential equations is the number of independent variables involved in the equation.

In this chapter we shall present a brief introduction to partial differential equations. Since the general theory is much too difficult to be presented here, we shall be content merely to examine a few simple examples. Since these examples have a number of important practical applications, even this cursory treatment will be of significant value.

10.1 PRELIMINARIES

In this chapter we shall be concerned only with first or second order linear partial differential equations with constant coefficients and in two independent variables; that is, equations of the form

$$a \frac{\partial^2 w}{\partial x^2}(x,y) + 2b \frac{\partial^2 w}{\partial x \, \partial y}(x,y) + c \frac{\partial^2 w}{\partial y^2}(x,y) + k \frac{\partial w}{\partial x}(x,y)$$

$$+ m \frac{\partial w}{\partial y}(x,y) + nw(x,y) = f(x,y), \quad (10.1)$$

where a, b, c, k, m, and n are constants. (The number 2 in front of the coefficient b will simplify later computations.) Equation (10.1) is *linear* since it does not involve any nonlinear functions of the dependent variable w and its partial derivatives. That is, (10.1) is of first degree in w and its partial derivatives.

If $a = b = c = 0$, we say that (10.1) is a *first order* equation. The first order equation

$$k \frac{\partial w}{\partial x} + m \frac{\partial w}{\partial y} + nw = f(x,y)$$

with constant coefficients k, m, and n is easy to solve, because the rotation of axes

$$u = x \cos \alpha + y \sin \alpha,$$
$$v = -x \sin \alpha + y \cos \alpha, \quad (10.2)$$

with $\tan \alpha = m/k$, will transform all such equations into an ordinary differential equation by eliminating the partial derivative with respect to v. We illustrate this technique below.

Example 10.1.1 Consider the partial differential equation

$$\frac{\partial w}{\partial x} + \sqrt{3} \frac{\partial w}{\partial y} - w = 4x + 2y. \quad (10.3)$$

Since $\tan \alpha = \sqrt{3}$, $\alpha = \pi/3$ radians, so that $\cos \alpha = \frac{1}{2}$ and $\sin \alpha = \sqrt{3}/2$. Thus, by (10.2),

$$u = \frac{x}{2} + \frac{\sqrt{3}y}{2}, \qquad v = -\frac{\sqrt{3}}{2}x + \frac{y}{2}, \tag{10.4}$$

and we have

$$\frac{\partial w}{\partial x} = \frac{\partial w}{\partial u}\frac{\partial u}{\partial x} + \frac{\partial w}{\partial v}\frac{\partial v}{\partial x} = \frac{1}{2}\frac{\partial w}{\partial u} - \frac{\sqrt{3}}{2}\frac{\partial w}{\partial v},$$

$$\frac{\partial w}{\partial y} = \frac{\partial w}{\partial u}\frac{\partial u}{\partial y} + \frac{\partial w}{\partial v}\frac{\partial v}{\partial y} = \frac{\sqrt{3}}{2}\frac{\partial w}{\partial u} + \frac{1}{2}\frac{\partial w}{\partial v}. \tag{10.5}$$

Solving Eq. (10.4) for x and y, we obtain

$$x = \frac{1}{2}u - \frac{\sqrt{3}}{2}v, \qquad y = \frac{\sqrt{3}}{2}u + \frac{1}{2}v. \tag{10.6}$$

Substituting (10.5) and (10.6) into Eq. (10.3) and gathering like terms, we have

$$2\frac{\partial w}{\partial u} - w = (2 + \sqrt{3})u + (1 - 2\sqrt{3})v. \tag{10.7}$$

We now treat v as a parameter and solve (10.7) as a first order linear ordinary differential equation in w with u as the independent variable:

$$2w' - w = (2 + \sqrt{3})u + (1 - 2\sqrt{3})v, \qquad w' = \frac{dw}{du}.$$

Using the methods of Section 2.3, we can multiply each side of this equation by $e^{-u/2}/2$ and write

$$(e^{-u/2}w)' = e^{-u/2}\left[\left(1 + \frac{\sqrt{3}}{2}\right)u + \left(\frac{1}{2} - \sqrt{3}\right)v\right].$$

Integrating both sides, we have

$$e^{-u/2}w = \left(1 + \frac{\sqrt{3}}{2}\right)\int ue^{-u/2}\,du + \left(\frac{1}{2} - \sqrt{3}\right)v\int e^{-u/2}\,du + g(v),$$

where the last term is an arbitrary differentiable function of the parameter v (which will vanish whenever we take its partial derivative with respect to u). Integrating the right-hand side and gathering terms, we have

$$w(u,v) = -(2 + \sqrt{3})u + (2\sqrt{3} - 1)v - (4 + 2\sqrt{3}) + g(v)e^{u/2}.$$

Replacing u and v by the substitution (10.4), we finally obtain

$$w(x,y) = g\left(\frac{y - \sqrt{3}x}{2}\right)e^{(x+\sqrt{3}y)/4} - 4x - 2y - (4 + 2\sqrt{3}). \tag{10.8}$$

That this is, indeed, a solution of (10.3) is easily checked by performing the operations indicated in that equation.

Up to now we have stressed the similarity between first order ordinary and partial differential equations. We shall encounter a substantial difference when we consider initial value problems. The solution of an ordinary differential equation is completely determined by prescribing a value for it at a single point. This is generally not true for partial differential equations. In order to obtain a *unique* solution, it is usually necessary to prescribe values for the solution on an entire *line*. A suitable initial condition is

$$w(x,0) = w_0(x)$$

for all x, where w_0 is a differentiable function. A solution satisfying this initial condition is easily obtained by setting $y = 0$ in Eq. (10.8). We have:

$$w_0(x) = g\left(-\frac{\sqrt{3}}{2}x\right)e^{x/4} - 4x - (4 + 2\sqrt{3})$$

or

$$g\left(-\frac{\sqrt{3}}{2}x\right) = [w_0(x) + 4x + 4 + 2\sqrt{3}]e^{-x/4}.$$

Setting $z = -\sqrt{3}x/2$ now produces the exact form of the function g:

$$g(z) = \left[w_0\left(-\frac{2z}{\sqrt{3}}\right) - \frac{8z}{\sqrt{3}} + 4 + 2\sqrt{3}\right]e^{z/(2\sqrt{3})}.$$

For example, if $w_0(x) = \sin x$, then

$$g(z) = \left[\sin\left(-\frac{2z}{\sqrt{3}}\right) - \frac{8z}{\sqrt{3}} + 4 + 2\sqrt{3}\right]e^{z/(2\sqrt{3})},$$

so that when z is replaced by $(y - \sqrt{3}x)/2$ in (10.8) we obtain, after simplification,

$$w(x,y) = \left[\sin\left(x - \frac{y}{\sqrt{3}}\right) - \frac{4y}{\sqrt{3}} + 4x + 4 + 2\sqrt{3}\right]e^{y/\sqrt{3}} - 4x - 2y - (4 + 2\sqrt{3}).$$

Returning now to Eq. (10.1), we assume that at least one of the coefficients a, b, and c is nonzero. If so, (10.1) is called a *second order* equation. The procedure for solving second order equations is much more complicated than that for first order equations. Initially, it is convenient to simplify Eq. (10.1) by another rotation of axes. Using the substitution (10.2) with $\tan 2\alpha = 2b/(a - c)$, we can eliminate the mixed second partial $\partial^2 w/\partial u\,\partial v$ and obtain an equation of the form

$$A\frac{\partial^2 w}{\partial u^2} + C\frac{\partial^2 w}{\partial v^2} + K\frac{\partial w}{\partial u} + M\frac{\partial w}{\partial v} + nw$$

$$= f(u\cos\alpha - v\sin\alpha, u\sin\alpha + v\cos\alpha) \qquad (10.9)$$

(see Exercise 6), where

$$A = a \cos^2 \alpha + 2b \sin \alpha \cos \alpha + c \sin^2 \alpha,$$
$$C = a \sin^2 \alpha - 2b \sin \alpha \cos \alpha + c \cos^2 \alpha,$$
$$K = k \cos \alpha + m \sin \alpha,$$
$$M = m \cos \alpha - k \sin \alpha. \tag{10.10}$$

Define

$$B = (c - a) \sin \alpha \cos \alpha + b(\cos^2 \alpha - \sin^2 \alpha). \tag{10.11}$$

Then by the double angle formulas of trigonometry,

$$B = \frac{(c - a)}{2} \sin 2\alpha + b \cos 2\alpha,$$

or

$$B = b \cos 2\alpha \left[\frac{(c - a)}{2b} \tan 2\alpha + 1 \right].$$

Since

$$\tan 2\alpha = \frac{2b}{a - c},$$

it follows that $B = 0$. An easy but messy calculation (see Exercise 7) will show that

$$b^2 - ac = B^2 - AC = -AC. \tag{10.12}$$

By analogy with analytic geometry, we call the partial differential equation (10.9) [or (10.1)] an *elliptic* equation if $b^2 - ac < 0$, that is, if A and C have the same sign. If A and C have opposite signs, or equivalently, if $b^2 - ac > 0$, we say that Eq. (10.9) [or (10.1)] is a *hyperbolic* equation. Finally, if $b^2 = ac = 0$, Eq. (10.9) [or (10.1)] is called a *parabolic* equation. In summary, Eq. (10.1) is called

$$\left.\begin{array}{l} \text{hyperbolic} \\ \text{parabolic} \\ \text{elliptic} \end{array}\right\} \quad \text{if} \quad b^2 - ac \begin{cases} > 0, \\ = 0, \\ < 0. \end{cases}$$

As a final simplification (see Exercises 8 and 9), it is possible by changing the *dependent* variable and the scale to transform Eq. (10.1) into one of the following four standard forms:

$$\frac{\partial^2 \omega}{\partial u^2} + \frac{\partial^2 \omega}{\partial v^2} + \lambda \omega = F(u,v) \quad \text{(elliptic)}, \tag{10.13}$$

$$\frac{\partial^2 \omega}{\partial u^2} - \frac{\partial^2 \omega}{\partial v^2} + \lambda \omega = F(u,v) \quad \text{(hyperbolic)}, \tag{10.14}$$

$$\frac{\partial^2 \omega}{\partial u^2} - \lambda \frac{\partial \omega}{\partial v} = F(u,v) \qquad \text{(parabolic)}, \tag{10.15}$$

$$\frac{\partial^2 \omega}{\partial u^2} + \lambda \omega = F(u,v) \qquad \text{(degenerate)}. \tag{10.16}$$

The degenerate case may be treated as an ordinary differential equation with parameter v, in a manner very similar to Example 10.1.1. For this reason, we shall not consider this case any further. In the next three sections we shall consider an example of each of these equations, showing how each arises in practice and indicating what steps can be taken to obtain a solution.

Example 10.1.2 To transform the equation

$$\frac{\partial^2 w}{\partial x^2} + 4\frac{\partial^2 w}{\partial x\, \partial y} + \frac{\partial^2 w}{\partial y^2} + \frac{\partial w}{\partial x} = 0$$

into standard form, observe that $b^2 - ac = 3$, since $a = c = 1$ and $b = 2$, and therefore the equation is hyperbolic. Since $\tan 2\alpha$ is infinite, we have $\alpha = 45° = \pi/4$. Substituting this value into (10.10), we obtain

$$3\frac{\partial^2 w}{\partial u^2} - \frac{\partial^2 w}{\partial v^2} + \frac{1}{\sqrt{2}}\frac{\partial w}{\partial u} - \frac{1}{\sqrt{2}}\frac{\partial w}{\partial v} = 0.$$

Setting

$$w = \omega \exp\left[-\frac{\sqrt{2}}{12}(u + 3v)\right]$$

and using Eq. (10.17) in the exercises, we find that the last equation reduces to

$$3\frac{\partial^2 \omega}{\partial u^2} - \frac{\partial^2 \omega}{\partial v^2} + \frac{\omega}{12} = 0.$$

Finally, letting $u = \sqrt{3}x$ and $v = y$ yields the equation

$$\frac{\partial^2 \omega}{\partial x^2} - \frac{\partial^2 \omega}{\partial y^2} + \frac{\omega}{12} = 0.$$

EXERCISES 10.1

In Exercises 1 through 4 find the solution $w(x,y)$ for each partial differential equation that satisfies the given condition.

1. $3\dfrac{\partial w}{\partial x} - 4\dfrac{\partial w}{\partial y} + 2w = 7,\ w(x,0) = e^x$

2. $\dfrac{\partial w}{\partial x} + \dfrac{\partial w}{\partial y} + w = 2,\ w(x,0) = \sin x$

3. $\dfrac{\partial w}{\partial x} + \dfrac{\partial w}{\partial y} - w = e^x,\ w(x,0) = 0$

4. $\dfrac{\partial w}{\partial x} - \dfrac{\partial w}{\partial y} + w = y,\ w(x,0) = x^2$

5. Using the chain rule and the substitution (10.2), prove that

$$\frac{\partial^2 w}{\partial x^2} = \cos^2 \alpha \frac{\partial^2 w}{\partial u^2} - 2 \sin \alpha \cos \alpha \frac{\partial^2 w}{\partial u\, \partial v} + \sin^2 \alpha \frac{\partial^2 w}{\partial v^2},$$

$$\frac{\partial^2 w}{\partial x\, \partial y} = \sin \alpha \cos \alpha \frac{\partial^2 w}{\partial u^2} + (\cos^2 \alpha - \sin^2 \alpha) \frac{\partial^2 w}{\partial u\, \partial v} - \sin \alpha \cos \alpha \frac{\partial^2 w}{\partial v^2},$$

$$\frac{\partial^2 w}{\partial y^2} = \sin^2 \alpha \frac{\partial^2 w}{\partial u^2} + 2 \sin \alpha \cos \alpha \frac{\partial^2 w}{\partial u\, \partial v} + \cos^2 \alpha \frac{\partial^2 w}{\partial v^2}.$$

6. Substitute the results of Exercise 5 into Eq. (10.1) to verify Eq. (10.9), and show that the mixed second partial $\partial^2 w/\partial u\, \partial v$ vanishes if $\tan 2\alpha = 2b/(a - c)$.

7. Obtain Eq. (10.12) by a direct calculation.

8. Verify that for a suitable choice of α and β the substitution

$$w = e^{\alpha u + \beta v}\omega$$

will transform the equation

$$A \frac{\partial^2 w}{\partial u^2} + C \frac{\partial^2 w}{\partial v^2} + K \frac{\partial w}{\partial u} + M \frac{\partial w}{\partial v} + Nw = F$$

into an equation of the form

$$A \frac{\partial^2 \omega}{\partial u^2} + C \frac{\partial^2 \omega}{\partial v^2} + \left(N - \frac{K^2 + AM^2}{4AC}\right)\omega = G. \tag{10.17}$$

What can be said about the situation if $C = 0$? Write the equation in one of the forms (10.15) or (10.16).

9. Show that the change of scale $u = \sqrt{|A|}\, x,\ v = \sqrt{|C|}\, y$ will transform Eq. (10.17), with $A, C \neq 0$, into one of the standard forms (10.13), (10.14).

10. Transform the following equations into standard form:

a) $4\dfrac{\partial^2 w}{\partial x^2} + 3\dfrac{\partial^2 w}{\partial y^2} - w = 0$

b) $\dfrac{\partial^2 w}{\partial x^2} - \dfrac{\partial^2 w}{\partial y^2} + 2\dfrac{\partial w}{\partial x} - 5\dfrac{\partial w}{\partial y} + 7w = 0$

c) $\dfrac{\partial^2 w}{\partial x^2} + 2\dfrac{\partial^2 w}{\partial x\, \partial y} + \dfrac{\partial^2 w}{\partial y^2} + w = 0$

d) $\dfrac{\partial^2 w}{\partial x^2} + 2\dfrac{\partial^2 w}{\partial x\, \partial y} + 4\dfrac{\partial^2 w}{\partial y^2} + 5w = 0$

10.2 THE VIBRATING STRING

As our first physical application, consider a string which is tightly stretched between two fixed points 0 and L on the x-axis (see Fig. 10.1). Suppose that the string is pulled back and released at time $t = 0$, causing it to vibrate. Our problem is to determine the displacement $y(x,t)$ of the point on the string which is x units away from the end 0, at any time t.

To avoid making our equation too complicated, we make two simplifying assumptions:

i) Our "ideal" string has uniform mass m per unit length and offers no resistance to bending.
ii) The tension T in the string is so large that the gravitational force on the string may be neglected.

Our procedure will be similar to that of Example 1.3.3. Consider a small segment of length Δx. By Newton's second law of motion (see Example 2.1.5), the total force acting on this piece of string is equal to the mass of the string multiplied by its acceleration:

$$F = ma = (m\Delta x)\frac{\partial^2 y}{\partial t^2}. \tag{10.18}$$

We assumed in this equation that the string is moving only in the xy-plane and that each particle in the string moves only vertically.

Let T_x and $T_{x+\Delta x}$ be the tension vectors at the endpoints of the given segment. These forces are applied tangentially (see Fig. 10.1) since the string offers no resistance to bending. Since there is no motion in the x-direction, the x-components of the tension vectors must coincide:

$$T_{x+\Delta x} \cos \beta = T_x \cos \alpha \equiv T. \tag{10.19}$$

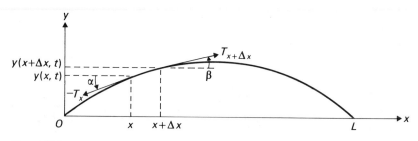

Figure 10.1

Thus T is constant, since x and Δx are arbitrary. Similarly, the difference in the y-components of the tension vectors must equal the total force acting on the string. Then by (10.18),

$$T_{x+\Delta x} \sin \beta - T_x \sin \alpha = m\Delta x \frac{\partial^2 y}{\partial t^2}. \tag{10.20}$$

Dividing each term in (10.20) by the corresponding term in (10.19), we have

$$\frac{T_{x+\Delta x} \sin \beta}{T_{x+\Delta x} \cos \beta} - \frac{T_x \sin \alpha}{T_x \cos \alpha} = \frac{m}{T} \Delta x \frac{\partial^2 y}{\partial t^2}$$

or

$$\tan \beta - \tan \alpha = \frac{m}{T} \Delta x \frac{\partial^2 y}{\partial t^2}. \tag{10.21}$$

Since

$$\tan \alpha = \frac{\partial y}{\partial x}\bigg|_x \qquad \text{and} \qquad \tan \beta = \frac{\partial y}{\partial x}\bigg|_{x+\Delta x},$$

we may rewrite (10.21) in the form

$$\frac{1}{\Delta x}\left[\frac{\partial y}{\partial x}\bigg|_{x+\Delta x} - \frac{\partial y}{\partial x}\bigg|_x\right] = \frac{m}{T}\frac{\partial^2 y}{\partial t^2}.$$

Letting Δx tend to zero, we obtain in the limit the equation

$$\frac{\partial^2 y}{\partial x^2} = \frac{m}{T}\frac{\partial^2 y}{\partial t^2}.$$

Since m/T is positive, it is clear that this is a hyperbolic partial differential equation. The equation, written in the form

$$\frac{\partial^2 y}{\partial t^2} = c^2 \frac{\partial^2 y}{\partial x^2}, \qquad c^2 = \frac{T}{m}, \tag{10.22}$$

is often called the one dimensional *wave equation*, where the constant c^2 indicates that the coefficient is positive.

As yet we have not made use of the fact that the string is fixed at its endpoints. We may write these boundary conditions as

$$y(0,t) = y(L,t) = 0, \qquad t \geqslant 0. \tag{10.23}$$

In addition, we have not taken into account the initial distortion of the string and the fact that it was at rest when released. These initial conditions can be written as

$$y(x,0) = f(x), \qquad 0 \leqslant x \leqslant L, \tag{10.24}$$

$$\frac{\partial y}{\partial t}\bigg|_{t=0} = 0, \tag{10.25}$$

where $f(x)$ is a function depicting the original distortion (see Examples 10.2.1 and 10.2.2), and $(\partial y/\partial t)(x,t)$ is the velocity of the point x units away from the origin at time t.

The direct method we shall use to solve this problem is due to D'Alembert.* Since it is only rarely possible to apply this technique, we shall develop two more applicable methods in the next three sections.

We begin by defining the variables

$$u = x + ct, \qquad v = x - ct, \tag{10.26}$$

and transforming the wave equation (10.22) into one involving variables u and v. By the chain rule, we have

$$\frac{\partial y}{\partial t} = \frac{\partial y}{\partial u}\frac{\partial u}{\partial t} + \frac{\partial y}{\partial v}\frac{\partial v}{\partial t} = c\left(\frac{\partial y}{\partial u} - \frac{\partial y}{\partial v}\right), \tag{10.27}$$

where we used Eqs. (10.26) to obtain the partial derivatives $\partial u/\partial t = c = \partial v/\partial t$. Taking the partial derivative with respect to t of Eq. (10.27), we obtain

$$\frac{\partial^2 y}{\partial t^2} = c\frac{\partial}{\partial t}\left(\frac{\partial y}{\partial u} - \frac{\partial y}{\partial v}\right) = c\left[\frac{\partial}{\partial u}\left(\frac{\partial y}{\partial u} - \frac{\partial y}{\partial v}\right)\frac{\partial u}{\partial t} + \frac{\partial}{\partial v}\left(\frac{\partial y}{\partial u} - \frac{\partial y}{\partial v}\right)\frac{\partial v}{\partial t}\right]$$

$$= c^2\left[\left(\frac{\partial^2 y}{\partial u^2} - \frac{\partial^2 y}{\partial u\,\partial v}\right) - \left(\frac{\partial^2 y}{\partial v\,\partial u} - \frac{\partial^2 y}{\partial v^2}\right)\right].$$

Since the mixed second partials are equal, we finally have

$$\frac{\partial^2 y}{\partial t^2} = c^2\left[\frac{\partial^2 y}{\partial u^2} - 2\frac{\partial^2 y}{\partial u\,\partial v} + \frac{\partial^2 y}{\partial v^2}\right].$$

Similarly, we find that

$$\frac{\partial^2 y}{\partial x^2} = \left[\frac{\partial^2 y}{\partial u^2} + 2\frac{\partial^2 y}{\partial u\,\partial v} + \frac{\partial^2 y}{\partial v^2}\right].$$

Substituting these results into Eq. (10.22) and canceling like terms yield the

---

* The French mathematician J. D'Alembert (1717–1783) is known for his contributions in mechanics.

equation

$$\frac{\partial^2 y}{\partial u \, \partial v} = 0,$$ (10.28)

since $4c^2 \neq 0$. It is now easy to solve (10.28) by performing two successive integrations. Integrating with respect to u, we obtain

$$\frac{\partial y}{\partial v} = g'(v),$$ (10.29)

where g' is an unknown function in v. From (10.29), it is evident that g' is the partial derivative, with respect to v, of the solution y—which is the reason for using the prime in our notation. Integrating Eq. (10.29) with respect to v yields

$$y = g(v) + h(u),$$ (10.30)

where h is an unknown function in u. Substituting (10.26) into Eq. (10.30) yields D'Alembert's solution

$$y(x,t) = g(x - ct) + h(x + ct)$$ (10.31)

of the wave equation (10.22).

Since h is arbitrary, if we ignore (10.24), we will see that one possible solution of the wave equation is $g(x - ct)$. If we set $t = 0$, then the initial distortion of the string is given by $y(x,0) = g(x)$. At a fixed time later, its distortion is $y(x,t) = g(x - ct)$. Thus the shape of the string can be obtained from the graph of the function $g(x)$ by moving an interval of length L to the right with velocity c. For this reason, $g(x - ct)$ is called a *traveling wave*. Similarly $h(x + ct)$ is a traveling wave, which moves to the left with velocity c. Hence D'Alembert's solution is a superposition of two traveling waves, one moving to the right, the other to the left, with the same velocity c.

The functions g and h can be determined from the initial conditions. Setting $t = 0$, we find that

$$f(x) = y(x,0) = g(x) + h(x).$$ (10.32)

Noting that g and h are functions of one variable, we may use the chain rule to differentiate (10.31) with respect to t. And setting $t = 0$, we obtain

$$0 = \frac{\partial y}{\partial t}\bigg|_{t=0} = -cg'(x - ct) + ch'(x + ct)\bigg|_{t=0} = c[h'(x) - g'(x)].$$

From the last equation we know that $h' = g'$, so that

$$h(x) = g(x) + k,$$

where k is a constant. Replacing this function for h in (10.32) yields $f(x) = 2g(x) + k$ or $g = (f - k)/2$. Hence $h = (f + k)/2$, so that the solution of the problem finally becomes

$$y(x,t) = \tfrac{1}{2}[f(x - ct) + f(x + ct)]. \tag{10.33}$$

Note that since the boundary conditions (10.23) must be satisfied by Eq. (10.33), setting $x = 0$ in (10.33) will yield

$$0 = \tfrac{1}{2}[f(-ct) + f(ct)]$$

or

$$f(-ct) = -f(ct).$$

Thus $f(x)$ is odd. This result may seem curious, since $f(x)$ has been defined only on the interval $0 \leqslant x \leqslant L$. However, it causes no difficulty, since we may extend f to the interval $-L \leqslant x \leqslant L$ by defining

$$f(-x) = -f(x) \qquad \text{for} \qquad -L \leqslant x \leqslant 0.$$

This extension is always possible as $f(0) = 0$. Similarly, setting $x = L$ and using the fact that f is odd, we find that

$$f(L + ct) = -f(L - ct) = f(ct - L),$$

so that f has period $2L$. Again, it is clear that we can extend $f(x)$ to the entire real line by letting

$$f(x + 2L) = f(x),$$

for all x. This extension of $f(x)$ to the entire real line will seem very reasonable if we return to the notion of traveling waves. It is clear, from Eq. (10.33), that we must immediately leave the interval $0 \leqslant x \leqslant L$ when $t > 0$. Now suppose we move $f(x)$ gradually to the right. Since $y(0,t) = y(L,t) = 0$ for all t, any positive contribution at 0 or L from the traveling wave $f(x - ct)$ must be offset by an equal negative contribution from $f(x + ct)$ [which is obtained by moving $f(x)$ the same distance to the left], and vice versa. In effect, this procedure amounts to a reflection of wave forms at the boundary.

By Theorem 9.2, the function $f(x)$ has a Fourier sine series—a fact we shall refer to in the next section. (The continuity of f is guaranteed by the fact that the string is unbroken. We assume that f' is piecewise continuous, since that is what it will be in most practical situations, such as in the example below.)

Example 10.2.1 Suppose that we pluck the string at its center (see Fig. 10.2) a distance $y(L/2,0) = y_0$ meters (y_0 is assumed to be small). We can assume that the distortion consists of two straight lines from $(0,0)$ to $(L/2,y_0)$ and from $(L/2,y_0)$

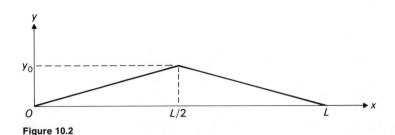

Figure 10.2

to $(L,0)$. Hence

$$f(x) = \begin{cases} \dfrac{2y_0}{L}\, x, & 0 \leqslant x \leqslant \dfrac{L}{2}, \\[2mm] \dfrac{2y_0}{L}\, (L - x), & \dfrac{L}{2} \leqslant x \leqslant L, \end{cases}$$

which has the piecewise continuous derivative

$$f'(x) = \begin{cases} \dfrac{2y_0}{L}, & 0 \leqslant x < \dfrac{L}{2}, \\[2mm] \dfrac{-2y_0}{L}, & \dfrac{L}{2} < x \leqslant L. \end{cases}$$

If the string is four meters long, weighs 0.08 kilogram, and is subject to a constant tension of two kilograms, then we obtain the solution

$$y(x,t) = \tfrac{1}{2}[f(x - 10t) + f(x + 10t)],$$

since

$$c^2 = \frac{T}{m} = \frac{2}{0.08/4} = 100.$$

To discover the exact location of a point one meter from the end 0 after exactly one second, we set $x = t = 1$. Then

$$y(1,1) = \tfrac{1}{2}[f(-9) + f(11)],$$

and since f has period $2L = 8$ and is odd, we get

$$y(1,1) = \frac{1}{2}[f(-1) + f(3)] = \frac{1}{2}[f(3) - f(1)]$$

$$= \frac{1}{2}\left(\frac{2y_0}{L} - \frac{2y_0}{L}\right) = 0.$$

Example 10.2.2 Suppose that $f(x) = y_0 \sin(\pi x/L)$. Then by (10.33)

$$y(x,t) = \frac{y_0}{2}\left[\sin\frac{\pi}{L}(x - ct) + \sin\frac{\pi}{L}(x + ct)\right],$$

and by the addition formulas of trigonometry, this reduces to

$$y(x,t) = y_0 \sin\frac{\pi x}{L}\cos\frac{\pi ct}{L}.$$

Thus, if we fix our attention on the point x_0 units from the left, we see that it oscillates with a period $2L/c$ and an amplitude $y_0 \sin(\pi x_0/L)$.

EXERCISES 10.2

In Exercises 1 through 6 use D'Alembert's solution (10.33) to find the displacement $y(x,t)$ for each given function $f(x)$, point x, and time t. Assume that $c = 5$ and $L = 6$ meters.

1. $f(x) = 0.01 \sin(\pi x/L)$, $x = 2$, $t = 1$
2. $f(x) = 0.1 \sin(\pi x/L)$, $x = 3$, $t = 2$
3. $f(x) = 0.01 \sin(2\pi x/L)$, $x = 3$, $t = 1$
4. $f(x) = 0.1 \sin(2\pi x/L)$, $x = 2$, $t = 2$
5. $f(x) = \begin{cases} 0.1x, & 0 \leqslant x \leqslant 3, x = 2, t = 2 \\ 0.1(6 - x), & 3 \leqslant x \leqslant 6 \end{cases}$
6. $f(x) = \begin{cases} 0.1x, & 0 \leqslant x \leqslant 2 \\ 0.2(3 - x), & 2 \leqslant x \leqslant 4, x = 3, t = 5 \\ 0.1(x - 6), & 4 \leqslant x \leqslant 6 \end{cases}$
7. Assume, instead of the initial conditions (10.24) and (10.25), that the string is initially in its equilibrium position, and is set vibrating by each of its points x being given an initial velocity $f'(x)$, where f is a differentiable function. Prove that the solution is given by

$$y(x,t) = \frac{1}{2c}[f(x + ct) - f(x - ct)].$$

Using the result of Exercise 7, $c = 5$, and $L = 6$ meters, do Exercises 8 through 11 for the given functions $f'(x)$ and values x and t.

8. $f'(x) = 0.01 \sin(\pi x/L)$, $x = 2$, $t = 3$
9. $f'(x) = 0.1 \sin(2\pi x/L)$, $x = 1$, $t = 1$
10. $f'(x) = 3(6x - x^2)$, $x = 3$, $t = 0.01$ second
11. $f'(x) = 0.1 \sin^2(\pi x/L)$, $x = 1$, $t = 0.1$ second
12. Suppose that we subject a vibrating string to a damping force that is proportional at each instance to the velocity at each point. Show that the resulting hyperbolic differential equation is of the form

$$\frac{\partial^2 y}{\partial t^2} - k\frac{\partial y}{\partial t} = c^2\frac{\partial^2 y}{\partial x^2}.$$

10.3 SEPARATION OF VARIABLES

One of the simplest techniques for solving ordinary differential equations consists of separating the variables. Although we now have *two* independent variables, we can nevertheless adapt the technique to all partial differential equations of the forms (10.13) through (10.16), whenever the function $F(u,v) \equiv 0$. The method is best explained by means of an example.

Suppose that we again consider the problem of the vibrating string given by the wave equation

$$\frac{\partial^2 y}{\partial t^2} = c^2 \frac{\partial^2 y}{\partial x^2}, \qquad c^2 = \frac{T}{m}, \tag{10.34}$$

with boundary conditions

$$y(0,t) = y(L,t) = 0, \qquad t \geq 0, \tag{10.35}$$

and initial conditions

$$y(x,0) = f(x), \qquad \frac{\partial y}{\partial t}\bigg|_{t=0} = 0, \qquad 0 \leq x \leq L. \tag{10.36}$$

We begin by seeking a solution of the form

$$y(x,t) = X(x)T(t), \tag{10.37}$$

where X is a function of x alone, and T is a function involving only the independent variable t. Of course, there is no guarantee that such a solution exists. In fact, as we shall see later, separation is often not possible. Nevertheless, we shall try it out and see if it works. Using primes to denote differentiation, we observe that

$$\frac{\partial^2 y}{\partial t^2} = \frac{\partial^2}{\partial t^2}[XT] = XT'' \qquad \text{and} \qquad \frac{\partial^2 y}{\partial x^2} = X''T,$$

so that Eq. (10.34) becomes

$$XT'' = c^2 X''T. \tag{10.38}$$

It is now possible to separate the variables and write Eq. (10.38) as

$$\frac{T''}{T} = c^2 \frac{X''}{X}. \tag{10.39}$$

Initially we had no assurance that it would be possible to separate the variables as we have done. Should such a separation prove impossible, the method would not apply and we would have to turn to other methods of solving the given problem.

Now suppose we fix t and allow x to vary. Since the left-hand side of (10.39) is constant, the right-hand side must be constant for all values of x. Consequently, the right-hand side is constant *regardless* of the value of t, and we have

$$\frac{T''}{T} = c^2 \frac{X''}{X} = k,$$

where k is constant. From this set of equalities we obtain two ordinary differential equations

$$T'' - kT = 0, \tag{10.40}$$

$$X'' - \frac{k}{c^2} X = 0. \tag{10.41}$$

Thus the problem of solving the partial differential equation (10.34) has been reduced to solving two ordinary differential equations (10.40) and (10.41). The constant k is arbitrary, but its value must be the same for both equations. Using the auxiliary equation (see Chapter 3) $m^2 - k = 0$, we can easily obtain the general solution

$$T(t) = \begin{cases} a_1 e^{\sqrt{k}t} + a_2 e^{-\sqrt{k}t}, & \text{if } k > 0, \\ a_1 t + a_2, & \text{if } k = 0, \\ a_1 \cos \sqrt{-k}t + a_2 \sin \sqrt{-k}t, & \text{if } k < 0. \end{cases}$$

Similarly, we also have the general solution

$$X(x) = \begin{cases} b_1 e^{\sqrt{k}x/c} + b_2 e^{-\sqrt{k}x/c}, & \text{if } k > 0, \\ b_1 x + b_2, & \text{if } k = 0, \\ b_1 \cos \dfrac{\sqrt{-k}x}{c} + b_2 \sin \dfrac{\sqrt{-k}x}{c}, & \text{if } k < 0. \end{cases}$$

Since our solution must satisfy the boundary conditions (10.35), if a solution of the form (10.37) exists, we must have the equations

$$X(0)T(t) = X(L)T(t) = 0 \qquad \text{for all } t.$$

If $T(t) \equiv 0$, then $y \equiv 0$ and the string is constantly at rest, which is impossible if $f(x) \not\equiv 0$. For $f \not\equiv 0$, T must be nonzero for at least one value of t, which implies that $X(0) = X(L) = 0$. Thus (10.41) becomes a Sturm-Liouville problem with these boundary conditions.

Setting $x = 0$ and $x = L$ in the general solution yields, for $k > 0$, the homogeneous system of equations

$$X(0) = b_1 + b_2 = 0,$$
$$X(L) = b_1 e^{\sqrt{k}L/c} + b_2 e^{-\sqrt{k}L/c} = 0.$$

This system has a nonzero solution if and only if the determinant of its coefficients is zero. But

$$\begin{vmatrix} 1 & 1 \\ e^{\sqrt{kL}/c} & e^{-\sqrt{kL}/c} \end{vmatrix} = e^{-\sqrt{kL}/c} - e^{\sqrt{kL}/c} = -2 \sinh \frac{\sqrt{kL}}{c} \neq 0,$$

so that $X(x) \equiv 0$ for $k > 0$. Then $y \equiv 0$, which is impossible if $f(x) \not\equiv 0$. For $k = 0$, the condition $X(0) = 0$ implies that $b_2 = 0$, while $X(L) = 0$ requires that $b_1 = 0$, again leading to the zero solution. So we are left with only the possibility that k is negative, and we set $k = -r^2$. This yields the general solution

$$X(x) = b_1 \cos \frac{rx}{c} + b_2 \sin \frac{rx}{c}.$$

Letting $x = 0$ and $x = L$, we have

$$X(0) = b_1 = 0 \qquad \text{and} \qquad X(L) = b_2 \sin \frac{rL}{c} = 0.$$

To prevent our again having the zero solution, we must choose an r such that rL/c is a positive multiple of π:

$$\frac{rL}{c} = n\pi \qquad \text{or} \qquad r = \frac{n\pi c}{L}.$$

Thus we obtain an infinite set of solutions

$$X_n(x) = b_2 \sin \frac{n\pi x}{L}, \tag{10.42}$$

each associated with the choice

$$k = -\frac{n^2\pi^2 c^2}{L^2}. \tag{10.43}$$

Hence the corresponding solutions for T are

$$T_n(t) = a_1 \cos \frac{n\pi ct}{L} + a_2 \sin \frac{n\pi ct}{L}.$$

Finally, we obtain an infinite set of solutions

$$y_n(x,t) = X_n(x)T_n(t) = \left(B_n \cos \frac{n\pi ct}{L} + B_n^* \sin \frac{n\pi ct}{L} \right) \sin \frac{n\pi x}{L}, \tag{10.44}$$

where B_n and B_n^* are constants, that satisfy the partial differential equation (10.34) and the boundary conditions (10.35). In the language of Chapter 9, we see that we have found the *eigenvalues* (10.43) and the corresponding *eigenfunctions* (10.44) of the boundary value problem (10.34), (10.35).

We now try to select a solution that satisfies the initial conditions (10.36). It is extremely likely that no one solution (10.44) will satisfy Eqs. (10.36). However, any finite sum of solutions (10.44) is again a solution of the boundary value problem (by the *principle of superposition*), so we can try to find a finite linear combination of these solutions that satisfies the initial conditions (10.36). Generally, even this will fail, so as a last resort we try an infinite series of the solutions (10.44):

$$y(x,t) = \sum_{n=1}^{\infty} y_n(x,t) = \sum_{n=1}^{\infty} \left(B_n \cos \frac{n\pi ct}{L} + B_n^* \sin \frac{n\pi ct}{L} \right) \sin \frac{n\pi x}{L}. \qquad (10.45)$$

There is no guarantee at this point that such a series will converge, but since its form is very similar to that of the Fourier series (9.32), there is every reason to be optimistic. Furthermore, if (10.45) converges and the series obtained by the formal term by term partial differentiations $\partial/\partial t$, $\partial^2/\partial t^2$, $\partial/\partial x$, and $\partial^2/\partial x^2$ all converge uniformly, then (10.45) will again be a solution of the boundary value problem (10.34), (10.35). If, in addition, we can also satisfy the initial conditions (10.36), then we shall have solved the vibrating string problem.

Setting $t = 0$ in (10.45) and using the first of Eqs. (10.36), we have

$$y(x,0) = \sum_{n=1}^{\infty} B_n \sin \frac{n\pi x}{L} = f(x). \qquad (10.46)$$

We again assume that $f(x)$ and $f'(x)$ are piecewise continuous (see Section 10.2). The infinite series (10.46) is a Fourier sine series, which requires that the function $f(x)$ be odd $[f(-x) = -f(x)]$, and periodic with period $2L$. As in Section 10.2, it is easy to extend the function $f(x)$, defined on the interval $0 \leqslant x \leqslant L$, in such a way that it is odd and periodic with period $2L$. Since $f(0) = 0$, we let $f(-x) = -f(x)$, on $-L \leqslant x \leqslant 0$ and require that $f(x + 2L) = f(x)$ for all real numbers x. Then the coefficients B_n are given by the Euler formula

$$B_n = \frac{1}{L} \int_{-L}^{L} f(x) \sin \frac{n\pi x}{L} \, dx, \qquad n = 1, 2, 3, \ldots. \qquad (10.47)$$

Taking the partial derivative of (10.45) with respect to t and using the second of the initial conditions (10.36), we have

$$\frac{\partial y}{\partial t} = \frac{n\pi c}{L} \sum_{n=1}^{\infty} \left(B_n^* \cos \frac{n\pi ct}{L} - B_n \sin \frac{n\pi ct}{L} \right) \sin \frac{n\pi x}{L} \bigg|_{t=0} = 0$$

or

$$\sum_{n=1}^{\infty} B_n^* \sin \frac{n\pi x}{L} = 0.$$

Again we obtain an expression of the form (10.47) for B_n^*, but this time the function involved is zero. Hence all the coefficients B_n^* vanish. Thus the solution (10.45) reduces to

$$y(x,t) = \sum_{n=1}^{\infty} B_n \cos \frac{n\pi ct}{L} \sin \frac{n\pi x}{L}, \tag{10.48}$$

where the coefficients B_n are given by Eq. (10.47). It is possible to write the series (10.48) in closed form by using the trigonometric identity

$$2 \sin A \cos B = \sin (A + B) + \sin (A - B).$$

Then

$$y(x,t) = \frac{1}{2} \sum_{n=1}^{\infty} B_n \left[\sin \frac{n\pi}{L} (x + ct) + \sin \frac{n\pi}{L} (x - ct) \right]$$

$$= \frac{1}{2} \left[\sum_{n=1}^{\infty} B_n \sin \frac{n\pi}{L} (x + ct) + \sum_{n=1}^{\infty} B_n \sin \frac{n\pi}{L} (x - ct) \right],$$

and the series can be evaluated by substituting $x + ct$ and $x - ct$, respectively, for the variable x in Eq. (10.46). Hence

$$y(x,t) = \tfrac{1}{2}[f(x + ct) + f(x - ct)],$$

which is precisely the same result as we obtained directly in Section 10.2.

Example 10.3.1 Let a vibrating string have length π and let $c^2 = 1$. Suppose that the initial velocity is zero and the initial distortion is $f(x) = x(\pi - x)$. Instead of using D'Alembert's solution, we shall obtain an expansion of the form (10.48). We need only calculate the coefficients B_n. Integrating by parts, we obtain

$$B_n = \frac{1}{\pi} \int_{-\pi}^{\pi} x(\pi - x) \sin nx \, dx$$

$$= \frac{1}{\pi} \left[-x(\pi - x) \frac{\cos nx}{n} \bigg|_{-\pi}^{\pi} + \frac{1}{n} \int_{-\pi}^{\pi} (\pi - 2x) \cos nx \, dx \right]$$

$$= \frac{1}{\pi} \left[-\frac{2\pi^2}{n} \cos n\pi + \frac{(\pi - 2x)}{n^2} \sin nx \bigg|_{-\pi}^{\pi} - \frac{2}{n^3} \cos nx \bigg|_{-\pi}^{\pi} \right]$$

$$= -\frac{2\pi}{n} \cos n\pi = (-1)^{n+1} \frac{2\pi}{n}.$$

Hence the solution is

$$y(x,t) = 2\pi[\cos t \sin x - \tfrac{1}{2} \cos 2t \sin 2t + \tfrac{1}{3} \cos 3t \sin 3t - \cdots].$$

We now apply the method of separation of variables to the one dimensional heat flow problem we discussed in Example 1.1.6. Recall that we have a rod of length L whose lateral surface is completely insulated. The temperature at any point x units from a fixed end of a uniform rod, at time t, is given by the function $T(x,t)$ that satisfies the partial differential equation

$$\frac{\partial T}{\partial t} = \delta \frac{\partial^2 T}{\partial x^2}. \tag{10.49}$$

Equation (10.49) is called the *heat equation* and is clearly parabolic. The constant δ is positive and measures the *diffusivity* of the material of the rod. To completely specify our problem, we select the case where the *boundary conditions* are

$$T(0,t) = T(L,t) = 0, \qquad t \geqslant 0, \tag{10.50}$$

and the *initial condition* is

$$T(x,0) = f(x), \qquad 0 \leqslant x \leqslant L, \tag{10.51}$$

where f is a given function. In physical terms, we are keeping the ends of the rod at zero temperature and letting $f(x)$ denote the initial temperature at any point x of the rod. It is easy to modify the problem by selecting other, possibly time dependent, boundary conditions (see Exercises 7, 8, and 9).

We begin by asuming the existence of a solution to the problem of the form

$$T(x,t) = X(x)\mathcal{T}(t).$$

Substituting this equation into (10.49) yields

$$X\mathcal{T}' = \delta X''\mathcal{T}$$

or

$$\frac{\mathcal{T}'}{\delta\mathcal{T}} = \frac{X''}{X}. \tag{10.52}$$

Since the left-hand side is constant for fixed t and arbitrary x, the right-hand side is a constant, say k. Thus we again obtain a pair of ordinary differential equations

$$\mathcal{T}' - k\,\delta\mathcal{T} = 0, \tag{10.53}$$

$$X'' - kX = 0. \tag{10.54}$$

The boundary conditions (10.50) may be written as

$$X(0)\mathcal{T}(t) = X(L)\mathcal{T}(t) = 0,$$

implying that $X(0) = X(L) = 0$ unless the rod has zero initial temperature at every point. If we ignore this uninteresting case, we will note that the boundary value problem (10.54), $X(0) = X(L) = 0$, is almost identical with the boundary value problem for X in the case of the vibrating string, where we discovered that nonzero solutions exist only for negative k. Setting $k = -r^2$, we find that k is an eigenvalue of the problem only if r is a multiple of π/L, and the eigenfunctions corresponding to $-(n\pi/L)^2$ have the form

$$X_n(x) = A \sin \frac{n\pi x}{L}. \tag{10.55}$$

Using $k = -(n\pi/L)^2$, we find that Eq. (10.53) yields the general solution

$$\mathcal{T}_n(t) = B e^{-(n\pi/L)^2 \delta t}. \tag{10.56}$$

Hence we consider an infinite series of the form

$$T(x,t) = \sum_{n=1}^{\infty} X_n(x)\mathcal{T}_n(t) = \sum_{n=1}^{\infty} B_n \sin \frac{n\pi x}{L} e^{-(n\pi/L)^2 \delta t}. \tag{10.57}$$

Setting $t = 0$ and making use of the initial condition (10.51), we have

$$T(x,0) = \sum_{n=1}^{\infty} B_n \sin \frac{n\pi x}{L} = f(x),$$

so that for Theorem 9.1 to be applicable, f must again be an odd, periodic, piecewise continuous function of period $2L$ with piecewise continuous derivative; the coefficients B_n are given by Eq. (10.47). Thus, the solution (10.57) is completely determined and the series must converge, since $T(x,0)$ converges and the exponential factors are less than 1 for all $t > 0$. Observe that the exponential factors in (10.57) will cause $T(x,t)$ to approach zero as t tends to infinity.

Example 10.3.2 Suppose that the rod has length π, $\delta = 1$, and initial temperature $f(x) = \sin 2x$. By the orthogonality of the set of functions $\{\sin nx\}$ (see Example 9.2.1), the coefficients B_n are all zero for $n \neq 2$. For $n = 2$, we have from (10.47) and by Eq. (9.27),

$$B_2 = \frac{1}{\pi} \int_{-\pi}^{\pi} \sin^2 2x \, dx = 1.$$

Thus the solution of this problem is given by

$$T(x,t) = (\sin 2x)e^{-4t}.$$

EXERCISES 10.3

In Exercises 1 through 6 find the temperature $T(x,t)$ in an insulated rod π units long whose ends are kept at $0\,°C$, where the initial temperature is given by:

1. $f(x) = x(\pi - x)$
2. $f(x) = x(\pi^2 - x^2)$

3. $f(x) = x^2(\pi - x)$

4. $f(x) = x \cos \dfrac{x}{2}$

5. $f(x) = \begin{cases} x, & 0 \leqslant x \leqslant \pi/2 \\ 0, & \pi/2 < x \leqslant \pi \end{cases}$

6. $f(x) = \begin{cases} 0, & 0 \leqslant x < \pi/3 \\ 1, & \pi/3 \leqslant x < 2\pi/3 \\ 0, & 2\pi/3 \leqslant x \leqslant \pi \end{cases}$

7. Suppose that the ends of the rod are kept at different constant temperatures

$$T(0,t) = T_1, \qquad T(L,t) = T_2.$$

Find the temperature at x as t tends to ∞; that is, find the *steady state* temperature. [*Hint*: The steady state temperature is *not* time dependent.]

8. Denote the steady state temperature by $T_s(x)$ and define the *transient* temperature in the rod by

$$T_t(x,t) = T(x,t) - T_s(x).$$

Verify that $T_t(x,t)$ is given by Eq. (10.57). This shows that we need only superimpose the steady state temperature on the solution (10.57) to obtain the solution for nonzero boundary conditions.

9. Suppose that the ends of the rod are kept at the temperatures

$$T(0,t) = T_1 e^{-c^2 \delta t}, \qquad T(L,t) = T_2 e^{-c^2 \delta t}.$$

For what initial conditions $f(x)$ can a solution be found by the separation of variables? What happens if cL is a multiple of π?

Find the solutions $y(x,t)$ of the equations in Exercises 10 through 15 by the separation of variables.

10. $\dfrac{\partial^2 y}{\partial x^2} + \dfrac{1}{x^2} \dfrac{\partial^2 y}{\partial t} = 0$

11. $\dfrac{\partial^2 y}{\partial x \, \partial t} = y$

12. $\dfrac{\partial^2 y}{\partial x^2} + \dfrac{\partial^2 y}{\partial t^2} = 0$

13. $\dfrac{\partial^2 y}{\partial x^2} - 2\dfrac{\partial y}{\partial x} = -\dfrac{\partial y}{\partial t}$

14. $\dfrac{\partial^2 y}{\partial t^2} = \dfrac{\partial^2 y}{\partial x^2} + y$

15. $\dfrac{\partial^2 y}{\partial t^2} = x^2 \dfrac{\partial^2 y}{\partial x^2}$

10.4 TWO-DIMENSIONAL HEAT FLOW AND LAPLACE'S EQUATION

Consider a rectangular sheet of heat conducting material of uniform thickness θ, density ρ, specific heat c, and heat conductivity κ. We may suppose that the set of points (x,y) with $0 \leqslant x \leqslant L$ and $0 \leqslant y \leqslant M$ is a face of the sheet. We assume that the faces of the sheet are insulated and that heat can enter and leave only through the edges of the sheet.

Select any interior point (x,y) on the face and consider a rectangular region $ABCD$ whose corner coordinates are given in Fig. 10.3. It is reasonable to assume the heat flow in each plane to be identical, so that the flow is two dimensional. In the discussion below we shall assume that the distances Δx and Δy are very small.

The rate of change of heat energy in the sheet $ABCD$ at any time t is approximately [see Eq. (1.12)]

$$c\rho\theta\Delta x\Delta y\,\frac{\partial T}{\partial t},\tag{10.58}$$

where $T = T(x,y,t)$ is the temperature at any point (x,y) at time t. Since Δx and Δy are small, we assume that the heat flux along each edge is constant. Then the heat energy passing through $ABCD$ in the vertical direction is

$$\theta\Delta x[Q_F(x,y,t) - Q_F(x,y + \Delta y,t)],\tag{10.59}$$

where Q_F is the heat flux. In the horizontal direction the heat energy is

$$\theta\Delta y[Q_F(x,y,t) - Q_F(x + \Delta x,y,t)].\tag{10.60}$$

Using Eq. (1.14) and adding (10.59) to (10.60), we obtain another approximation of the rate of change of heat energy in the sheet $ABCD$ [see Eq. (1.13)]:

$$\kappa\theta\Delta y\left[\frac{\partial T}{\partial x}(x + \Delta x,y,t) - \frac{\partial T}{\partial x}(x,y,t)\right]$$
$$+ \kappa\theta\Delta x\left[\frac{\partial T}{\partial y}(x,y + \Delta y,t) - \frac{\partial T}{\partial y}(x,y,t)\right].\tag{10.61}$$

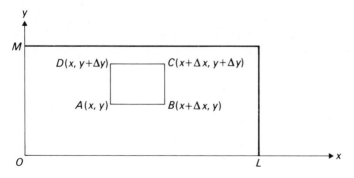

Figure 10.3

Equating (10.61) and (10.58) and dividing by $\Delta x \Delta y$ yield

$$c\rho\theta \frac{\partial T}{\partial t} = \kappa\theta \frac{\left[\dfrac{\partial T}{\partial x}(x + \Delta x, y, t) - \dfrac{\partial T}{\partial x}(x, y, t)\right]}{\Delta x}$$

$$+ \kappa\theta \frac{\left[\dfrac{\partial T}{\partial y}(x, y + \Delta y, t) - \dfrac{\partial T}{\partial y}(x, y, t)\right]}{\Delta y}.$$

Passing to the limit as Δx and Δy both approach zero, we obtain

$$c\rho\theta \frac{\partial T}{\partial t} = \kappa\theta \left[\frac{\partial^2 T}{\partial x^2} + \frac{\partial^2 T}{\partial y^2}\right],$$

which may be written in the form

$$\frac{\partial^2 T}{\partial x^2} + \frac{\partial^2 T}{\partial y^2} = \frac{1}{\delta}\frac{\partial T}{\partial t}, \tag{10.62}$$

where $\delta = \kappa/c\rho$ is the diffusivity. Equation (10.62) is called the *two-dimensional heat equation*.

If the edges of the sheet are kept at constant temperatures, and an extremely long time is allowed to pass, then the temperature at any given point will stabilize and remain almost constant for all larger values of t; that is, steady state conditions will be attained. The temperature T at a point on the sheet will then depend only on its position and not on time. Thus $\partial T/\partial t$ will be zero throughout the sheet, and Eq. (10.62) becomes

$$\frac{\partial^2 T}{\partial x^2} + \frac{\partial^2 T}{\partial y^2} = 0. \tag{10.63}$$

This elliptic partial differential equation, commonly referred to as *Laplace's equation*, arises in many other problems of applied mathematics. We shall now present three examples of the solution of Laplace's equation by the method of separation of variables.

Example 10.4.1 We begin by studying the steady state heat flow problem in the rectangular sheet of size $L \times M$ in which Laplace's equation arose. Suppose that the upper horizontal edge is kept at $100°C$, while the other three edges are kept at $0°C$. We may write these boundary conditions for $T(x,y)$ as follows:

$$T(0,y) = T(L,y) = 0, \qquad 0 < y < M, \tag{10.64}$$

$$T(x,0) = 0, \qquad T(x,M) = 100, \qquad 0 < x < L. \tag{10.65}$$

Since we are seeking the steady state solution, time will not be a factor in this problem, and therefore no initial condition is needed.

Setting $T(x,y) = X(x)Y(y)$, we find that Eq. (10.63) becomes

$$X''Y + XY'' = 0,$$

which may be written in separated form as

$$\frac{X''}{X} = -\frac{Y''}{Y}. \tag{10.66}$$

The left-hand side of (10.66) is constant for fixed x and arbitrary y. Hence Y''/Y must be constant. Letting k be this constant, we obtain the pair of ordinary differential equations

$$X'' - kX = 0,$$
$$Y'' + kY = 0. \tag{10.67}$$

From (10.64) we have

$$X(0)Y(y) = X(L)Y(y) = 0,$$

and $Y(y) \not\equiv 0$ [since otherwise $T(x,y) \equiv 0$, contradicting the second equation of (10.65)], so that $X(0) = X(L) = 0$. Thus again we have the same situation that we encountered in Section 10.3, and the eigenvalues are all negative. Setting $k = -r^2$, we find that rL must be a multiple of π in order that we may have a nonzero solution of the boundary value problem for X. Hence $r = n\pi/L$, and the functions

$$X_n(x) = A \sin \frac{n\pi x}{L}, \qquad n = 1, 2, 3, \ldots, \tag{10.68}$$

are the eigenfunctions of the problem. Setting $k = -(n\pi/L)^2$ in the second equation of (10.67) yields the general solution

$$Y_n(y) = B_1 e^{n\pi y/L} + B_2 e^{-n\pi y/L}.$$

The first condition of (10.65) implies that $Y(0) = 0$, thus $B_1 = -B_2$, and we can rewrite Y_n as

$$Y_n(y) = B \sinh \frac{n\pi y}{L}. \tag{10.69}$$

As in Section 10.3, to enlarge the class of possible solutions, we consider an infinite sum of products of the terms (10.68) and (10.69):

$$T(x,y) = \sum_{n=1}^{\infty} c_n \sin \frac{n\pi x}{L} \sinh \frac{n\pi y}{L}. \tag{10.70}$$

Evaluating (10.70) at any point (x,M), we obtain by (10.65)

$$100 = \sum_{n=1}^{\infty} \left(c_n \sinh \frac{n\pi M}{L} \right) \sin \frac{n\pi x}{L}, \qquad 0 < x < L,$$

which again is a Fourier sine series, requiring that we extend the boundary condition $T(x,M)$ to the interval $-L \leqslant x \leqslant L$ as an odd piecewise continuous periodic function of period $2L$ with a piecewise continuous derivative. Clearly the function

$$f(x) = \begin{cases} 100, & 0 < x < L, \\ -100, & -L < x < 0, \end{cases}$$

with $f(x + 2L) = f(x)$ satisfies these conditions. Hence $c_n \sinh (n\pi M/L)$ must equal the nth Fourier (sine) coefficient of $f(x)$:

$$\begin{aligned} c_n \sinh \frac{n\pi M}{L} &= \frac{1}{L} \int_{-L}^{L} f(x) \sin \frac{n\pi x}{L} \, dx \\ &= -\frac{100}{L} \int_{-L}^{0} \sin \frac{n\pi x}{L} \, dx + \frac{100}{L} \int_{0}^{L} \sin \frac{n\pi x}{L} \, dx \\ &= \frac{200}{n\pi} (1 - \cos n\pi) = \frac{200}{n\pi} [1 - (-1)^n]. \end{aligned}$$

Hence $c_{2k} = 0$, while

$$c_{2k+1} = \frac{400}{\pi(2k+1) \sinh (2k+1)\pi M/L},$$

so that the solution is given by

$$\begin{aligned} T(x,y) &= \frac{400}{\pi} \left[\frac{\sin (\pi x/L) \sinh (\pi y/L)}{\sinh (\pi M/L)} + \frac{\sin (3\pi x/L) \sinh (3\pi y/L)}{3 \sinh (3\pi M/L)} + \cdots \right] \\ &= \frac{400}{\pi} \sum_{n=1}^{\infty} \frac{\sin [(2n-1)\pi x/L] \sinh [(2n-1)\pi y/L]}{(2n-1) \sinh [(2n-1)\pi M/L]}. \end{aligned}$$

We may use this formula to compute the steady state temperature at any point in the rectangle. For example,

$$T\left(\frac{L}{2}, \frac{M}{2} \right) = \frac{400}{\pi} \sum_{n=1}^{\infty} \frac{\sin [(2n-1)\pi/2] \sinh [(2n-1)\pi M/2L]}{(2n-1) \sinh [(2n-1)\pi M/L]}.$$

And since

$$\frac{\sinh (A/2)}{\sinh A} = \frac{e^{A/2} - e^{-A/2}}{e^A - e^{-A}} = \frac{1}{e^{A/2} + e^{-A/2}} = \frac{1}{2 \cosh (A/2)},$$

we obtain

$$T\left(\frac{L}{2}, \frac{M}{2}\right) = \frac{200}{\pi}\left[\frac{1}{\cosh\left(\pi M/2L\right)} - \frac{1}{3\cosh\left(3\pi M/2L\right)} + \frac{1}{5\cosh\left(5\pi M/2L\right)} - \cdots\right],$$

which, for given L and M, can be approximated to any desired degree of accuracy.

Example 10.4.2 We now consider the steady state heat flow problem in the disk D of radius 1 centered at the origin. It should be clear from the nature of the situation that the problem will be simplified if we can use polar coordinates (r,θ) instead of rectangular coordinates (x,y). Assume that the boundary of the disk is kept at the temperature

$$T(1,\theta) = f(\theta), \qquad 0 \leqslant \theta \leqslant 2\pi, \tag{10.71}$$

where the function $f(\theta)$ is periodic with period 2π and has a continuous derivative. Now we need only write Laplace's equation (10.63) in polar coordinates. Let us recall the transformation formulas

$$r = \sqrt{x^2 + y^2}, \qquad x = r\cos\theta,$$

$$\theta = \tan^{-1}\frac{y}{x}, \qquad y = r\sin\theta.$$

By the chain rule, we have

$$\frac{\partial T}{\partial x} = \frac{\partial T}{\partial r}\frac{\partial r}{\partial x} + \frac{\partial T}{\partial \theta}\frac{\partial \theta}{\partial x} = \frac{\partial T}{\partial r}\cos\theta - \frac{\partial T}{\partial \theta}\frac{\sin\theta}{r},$$

since

$$\frac{\partial r}{\partial x} = \frac{\partial}{\partial x}\sqrt{x^2 + y^2} = \frac{x}{\sqrt{x^2 + y^2}} = \frac{x}{r} = \cos\theta,$$

$$\frac{\partial \theta}{\partial x} = \frac{-y/x^2}{1 + (y/x)^2} = \frac{-y}{r^2} = \frac{-\sin\theta}{r},$$

and

$$\frac{\partial^2 T}{\partial x^2} = \frac{\partial}{\partial r}\left(\frac{\partial T}{\partial r}\frac{\partial r}{\partial x} + \frac{\partial T}{\partial \theta}\frac{\partial \theta}{\partial x}\right)\frac{\partial r}{\partial x} + \frac{\partial}{\partial \theta}\left(\frac{\partial T}{\partial r}\frac{\partial r}{\partial x} + \frac{\partial T}{\partial \theta}\frac{\partial \theta}{\partial x}\right)\frac{\partial \theta}{\partial x}$$

$$= \frac{\partial^2 T}{\partial r^2}\cos^2\theta - \frac{\partial^2 T}{\partial r\,\partial \theta}\frac{\sin\theta\cos\theta}{r} + \frac{\partial T}{\partial \theta}\frac{\sin\theta\cos\theta}{r^2}$$

$$- \frac{\partial^2 T}{\partial \theta\,\partial r}\frac{\sin\theta\cos\theta}{r} + \frac{\partial T}{\partial r}\frac{\sin^2\theta}{r} + \frac{\partial^2 T}{\partial \theta^2}\frac{\sin^2\theta}{r^2} + \frac{\partial T}{\partial \theta}\frac{\cos\theta\sin\theta}{r^2}.$$

Similarly,

$$\frac{\partial^2 T}{\partial y^2} = \frac{\partial}{\partial r}\left(\frac{\partial T}{\partial r}\sin\theta + \frac{\partial T}{\partial \theta}\frac{\cos\theta}{r}\right)\frac{\partial r}{\partial y} + \frac{\partial}{\partial \theta}\left(\frac{\partial T}{\partial r}\sin\theta + \frac{\partial T}{\partial \theta}\frac{\cos\theta}{r}\right)\frac{\partial \theta}{\partial y}$$

$$= \frac{\partial^2 T}{\partial r^2}\sin^2\theta + \frac{\partial^2 T}{\partial r\,\partial\theta}\frac{\sin\theta\cos\theta}{r} - \frac{\partial T}{\partial\theta}\frac{\cos\theta\sin\theta}{r^2}$$

$$+ \frac{\partial^2 T}{\partial\theta\,\partial r}\frac{\sin\theta\cos\theta}{r} + \frac{\partial T}{\partial r}\frac{\cos^2\theta}{r} + \frac{\partial^2 T}{\partial\theta^2}\frac{\cos^2\theta}{r^2} - \frac{\partial T}{\partial\theta}\frac{\sin\theta\cos\theta}{r^2}.$$

Adding the two equations, we obtain

$$\frac{\partial^2 T}{\partial x^2} + \frac{\partial^2 T}{\partial y^2} = \frac{\partial^2 T}{\partial r^2} + \frac{1}{r^2}\frac{\partial^2 T}{\partial\theta^2} + \frac{1}{r}\frac{\partial T}{\partial r} = 0, \tag{10.72}$$

which is Laplace's equation in polar coordinates.

Now we let $T(r,\theta) = R(r)\Theta(\theta)$ and transform Eq. (10.72) into

$$R''\Theta + \frac{1}{r^2}R\Theta'' + \frac{1}{r}R'\Theta = 0$$

or

$$\left(R'' + \frac{1}{r}R'\right)\Theta = -\frac{R}{r^2}\Theta''.$$

Separating the variables, we have

$$\frac{r^2 R'' + rR'}{R} = -\frac{\Theta''}{\Theta} = k,$$

since each side of the first equality is a function of only one variable. Hence we obtain two ordinary differential equations:

$$\Theta'' + k\Theta = 0, \tag{10.73}$$

and

$$r^2 R'' + rR' - kR = 0. \tag{10.74}$$

Since $f(\theta) = T(1,\theta) = R(1)\Theta(\theta)$, it is clear that $\Theta(\theta)$ is periodic with period 2π. We assume that f is not identically zero [see Exercise 12 for the case where $f(\theta) \equiv 0$] so that $R(1) \neq 0$. Observe that $\Theta(0) = \Theta(2\pi)$ and $\Theta'(0) = \Theta'(2\pi)$. By the remark following the orthogonality theorem and Exercise 6 of Section 9.5, the boundary value problem in Θ has periodic eigenfunctions of period 2π. Thus the eigenvalues must be $k = n^2$ with the corresponding eigenfunctions

$$\Theta_n(\theta) = A_1 \cos n\theta + A_2 \sin n\theta, \qquad n = 0, 1, 2, 3, \ldots.$$

Equation (10.74) is an Euler equation (see Section 3.8) and can be solved by letting $R = r^\lambda$. Letting $k = n^2$, we see that (10.74) becomes

$$r^\lambda[\lambda(\lambda - 1) + \lambda - n^2] = r^\lambda(\lambda^2 - n^2) = 0,$$

and the roots of the equation $\lambda^2 - n^2 = 0$ are $\lambda = \pm n$. Hence the general solution of (10.74) with $k = n^2$ is

$$R_n(r) = B_1 r^n + B_2 r^{-n}, \qquad n = 0, 1, 2, 3, \dots.$$

Since we want R_n to exist for all values in the range $0 \leqslant r \leqslant 1$, B_2 must vanish and $R_n(r) = Br^n$, $n = 0, 1, 2, 3, \dots$. Thus we consider the infinite series

$$T(r,\theta) = \frac{A_0}{2} + \sum_{n=1}^{\infty} r^n(A_n \cos n\theta + B_n \sin n\theta), \tag{10.75}$$

and seek a solution to our problem in this form. To evaluate the constants A_n, B_n, we note that

$$f(\theta) = T(1,\theta) = \frac{A_0}{2} + \sum_{n=1}^{\infty} A_n \cos n\theta + B_n \sin n\theta. \tag{10.76}$$

Hence (10.76) is the Fourier series of f, and the coefficients A_n and B_n are the Fourier coefficients of f, given by the Euler formulas

$$A_n = \frac{1}{\pi} \int_{-\pi}^{\pi} f(\varphi) \cos n\varphi \, d\varphi \qquad B_n = \frac{1}{\pi} \int_{-\pi}^{\pi} f(\varphi) \sin n\varphi \, d\varphi.$$

Since (10.76) converges and $r \leqslant 1$, the series (10.75) will also converge. For example, if $f(\theta) = \cos \theta$, then all the coefficients except A_1 will vanish, and $A_1 = 1$. Hence the temperature is given by

$$T(r,\theta) = r \cos \theta$$

at all points in the unit disk.

It is interesting to note what will occur if we replace the coefficients A_n, B_n in (10.75) by the Euler formulas and interchange the sums and integrals:

$$T(r,\theta) = \frac{1}{2\pi} \int_{-\pi}^{\pi} f(\varphi) \, d\varphi$$

$$+ \sum_{n=1}^{\infty} r^n \left[\frac{\cos n\theta}{\pi} \int_{-\pi}^{\pi} f(\varphi) \cos n\varphi \, d\varphi + \frac{\sin n\theta}{\pi} \int_{-\pi}^{\pi} f(\varphi) \sin n\varphi \, d\varphi \right]$$

$$= \frac{1}{\pi} \int_{-\pi}^{\pi} f(\varphi) \left[\frac{1}{2} + \sum_{n=1}^{\infty} r^n(\cos n\theta \cos n\varphi + \sin n\theta \sin n\varphi) \right] d\varphi.$$

Since

$$\cos n(\theta - \varphi) = \cos n\theta \cos n\varphi + \sin n\theta \sin n\varphi,$$

we obtain

$$T(r,\theta) = \frac{1}{\pi} \int_{-\pi}^{\pi} f(\varphi) \left[\frac{1}{2} + \sum_{n=1}^{\infty} r^n \cos n(\theta - \varphi) \right] d\varphi.$$

But

$$\cos n(\theta - \varphi) = \frac{e^{in(\theta - \varphi)} + e^{-in(\theta - \varphi)}}{2},$$

so that

$$\sum_{n=1}^{\infty} r^n \cos n(\theta - \varphi) = \frac{1}{2} \sum_{n=1}^{\infty} r^n [e^{in(\theta - \varphi)} + e^{-in(\theta - \varphi)}]$$

$$= \frac{1}{2} \left[\sum_{n=1}^{\infty} [re^{i(\theta - \varphi)}]^n + \sum_{n=1}^{\infty} [re^{-i(\theta - \varphi)}]^n \right].$$

The complex number $e^{it} = \cos t + i \sin t$ can be represented as a vector of length $\sqrt{(\cos t)^2 + (\sin t)^2} = 1$ (see Section 3.5), so that $re^{i(\theta - \varphi)}$ and $re^{i(\varphi - \theta)}$ both have length equal to r (<1). Using the geometric series (6.17), we have

$$\sum_{n=1}^{\infty} r^n \cos n(\theta - \varphi) = \frac{1}{2} \left[\frac{re^{i(\theta - \varphi)}}{1 - re^{i(\theta - \varphi)}} + \frac{re^{-i(\theta - \varphi)}}{1 - re^{-i(\theta - \varphi)}} \right]$$

$$= \frac{1}{2} \left[\frac{r[e^{i(\theta - \varphi)} + e^{-i(\theta - \varphi)}] - 2r^2}{r^2 - r[e^{i(\theta - \varphi)} + e^{-i(\theta - \varphi)}] + 1} \right]$$

$$= \frac{r \cos(\theta - \varphi) - r^2}{r^2 - 2r \cos(\theta - \varphi) + 1}.$$

Thus

$$\frac{1}{2} + \sum_{n=1}^{\infty} r^n \cos n(\theta - \varphi) = \frac{1}{2} + \frac{r \cos(\theta - \varphi) - r^2}{r^2 + 1 - 2r \cos(\theta - \varphi)}$$

$$= \frac{1 - r^2}{2[r^2 + 1 - 2r \cos(\theta - \varphi)]}$$

and

$$T(r,\theta) = \frac{1 - r^2}{2\pi} \int_{-\pi}^{\pi} \frac{f(\varphi)}{r^2 + 1 - 2r \cos(\theta - \varphi)} d\varphi. \tag{10.77}$$

Equation (10.77) is called the *Poisson integral formula** and is valid for all values $r < 1$ (see Exercise 9). It indicates that the temperature at any interior point (r,θ) of the unit disk may be obtained by integrating the boundary temperatures according to the formula (10.77). In particular, if $r = 0$, then the temperature at the center of the disk is

$$T(0,\theta) = \frac{1}{2\pi} \int_{-\pi}^{\pi} f(\varphi) \, d\varphi,$$

that is, *the temperature at the center is the integral average of the boundary temperatures*. This fact is often called the *mean value theorem* and holds for all functions that satisfy Laplace's equation on the unit disk $r \leqslant 1$.

Example 10.4.3 In this example we shall consider the three-dimensional Laplace equation

$$\frac{\partial^2 T}{\partial x^2} + \frac{\partial^2 T}{\partial y^2} + \frac{\partial^2 T}{\partial z^2} = 0 \tag{10.78}$$

associated with the steady state heat flow problem in a solid.

Suppose the solid is a rectangular cylinder of height π and radius 1 (Fig. 10.4). Since such a solid is easily described in cylindrical coordinates, it will be convenient to obtain Laplace's equation (10.78) in terms of the coordinates (r,θ,z), where

$$x = r \cos \theta, \qquad y = r \sin \theta, \qquad z = z.$$

Obviously, we are simply writing the xy-coordinates in polar form. Thus the calculations we did in Example 10.4.2 still apply, and Laplace's equation becomes [from (10.72)]

$$\frac{\partial^2 T}{\partial r^2} + \frac{1}{r} \frac{\partial T}{\partial r} + \frac{1}{r^2} \frac{\partial^2 T}{\partial \theta^2} + \frac{\partial^2 T}{\partial z^2} = 0. \tag{10.79}$$

We now apply the method of separation of variables to the function T and, accordingly, set $T(r,\theta,z) = R(r)\Theta(\theta)Z(z)$ and write Eq. (10.79) as

$$R''\Theta Z + \frac{1}{r} R'\Theta Z + \frac{1}{r^2} R\Theta''Z + R\Theta Z'' = 0$$

or

$$[(R'' + r^{-1}R')\Theta + r^{-2}R\Theta'']Z = -R\Theta Z''.$$

---

* Named in honor of S. D. Poisson (1781–1840), a French mathematician and physicist.

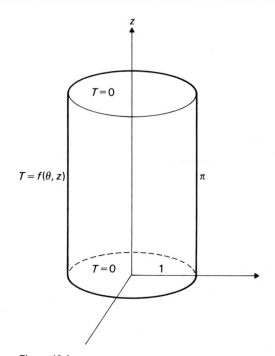

$T = 0$

$T = f(\theta, z)$

π

$T = 0$

1

Figure 10.4

Separating out the z-variables, we have

$$\frac{R'' + r^{-1}R'}{R} + r^{-2}\frac{\Theta''}{\Theta} = -\frac{Z''}{Z} = k_1,$$

since the left-hand side of the equation is a function of r and θ, while the right-hand side is a function of z alone. Thus

$$Z'' + k_1 Z = 0,$$

and

$$\frac{R'' + r^{-1}R'}{R} = k_1 - r^{-2}\frac{\Theta''}{\Theta}.$$

Multiplying both sides of the last equation by r^2, we have

$$\frac{r^2 R'' + rR'}{R} = k_1 r^2 - \frac{\Theta''}{\Theta},$$

or

$$\frac{r^2 R'' + rR'}{R} - k_1 r^2 = -\frac{\Theta''}{\Theta} = k_2,$$

since each side is a function of only one variable. Hence we also obtain the ordinary differential equations

$$\Theta'' + k_2 \Theta = 0$$

and

$$r^2 R'' + rR' - (k_2 + k_1 r^2)R = 0. \tag{10.80}$$

If we now assume the boundary conditions

$$T(r,\theta,0) = T(r,\theta,\pi) = 0, \qquad T(1,\theta,z) = f(\theta,z),$$

where f is a continuously differentiable function in θ and z, we can translate the conditions into terms of R, Θ, Z:

$$R(r)\Theta(\theta)Z(0) = R(r)\Theta(\theta)Z(\pi) = 0,$$
$$R(1)\Theta(\theta)Z(z) = f(\theta,z).$$

Since R and Θ are not constantly zero [assuming $f(\theta,z) \neq 0$], it follows that $Z(0) = Z(\pi) = 0$. Also, since f is continuous, it must be periodic with period 2π in θ (since f is the temperature on the surface of the cylinder). Thus $\Theta(\theta)$ is periodic with period 2π, so that $\Theta(0) = \Theta(2\pi)$ and $\Theta'(0) = \Theta'(2\pi)$. Finally, we know that $R(1) \neq 0$ and $R(0)$ is finite. As in Example 10.4.2, the eigenvalues of the boundary value problem

$$\Theta'' + k_2 \Theta = 0, \qquad \Theta(0) = \Theta(2\pi), \qquad \Theta'(0) = \Theta'(2\pi),$$

are $k_2 = m^2$, $m = 1, 2, 3, \ldots$, with the corresponding eigenfunctions

$$\Theta_m(\theta) = A_1 \cos m\theta + A_2 \sin m\theta.$$

By the result of Exercise 1 of Section 9.4, the boundary value problem

$$Z'' + k_1 Z = 0, \qquad Z(0) = Z(\pi) = 0,$$

has the eigenvalues $k_1 = n^2$, $n = 1, 2, 3, \ldots$, and the corresponding eigenfunctions

$$Z_n(z) = B \sin nz.$$

Setting $k_1 = n^2$ and $k_2 = m^2$ in Eq. (10.80), we obtain

$$r^2 R'' + rR' - (n^2 r^2 + m^2)R = 0,$$

which may be transformed into Bessel's equation by the substitution $\rho = inr$. To verify this fact, we note that

$$\frac{dR}{d\rho} = \frac{dR}{dr}\frac{dr}{d\rho} = \frac{1}{in} R'(r),$$

and

$$\frac{d^2R}{d\rho^2} = \frac{d}{d\rho}\left(\frac{1}{in} R'\right) = \frac{1}{(in)^2} R''(r) = -\frac{1}{n^2} R''(r).$$

Substituting these values into the equation, we obtain

$$-n^2r^2 \frac{d^2R}{d\rho^2} + inr \frac{dR}{d\rho} - (n^2r^2 + m^2)R = 0$$

or

$$\rho^2 \frac{d^2R}{d\rho^2} + \rho \frac{dR}{d\rho} + (\rho^2 - m^2)R = 0. \tag{10.81}$$

By Theorem 5.5, Eq. (10.81) has the general solution

$$R(\rho) = C_1 J_m(\rho) + C_2 Y_m(\rho), \qquad \rho \neq 0.$$

And since $R(0)$ is finite, the constant $C_2 = 0$, since the Bessel function of the second kind, Y_m, involves a $\ln \rho$ term which approaches $-\infty$ as ρ approaches 0. We are, therefore, led to the double series

$$T(r,\theta,z) = \sum_{m=1}^{\infty} \sum_{n=1}^{\infty} (a_{mn} \cos m\theta + b_{mn} \sin m\theta) \sin nz \, J_m(inr). \tag{10.82}$$

The coefficients a_{mn} and b_{mn} can be determined by setting $r = 1$ and requiring that

$$f(\theta,z) = \sum_{m=1}^{\infty} \left[\left(\sum_{n=1}^{\infty} a_{mn} J_m(in) \sin nz \right) \cos m\theta \right.$$

$$\left. + \left(\sum_{n=1}^{\infty} b_{mn} J_m(in) \sin nz \right) \sin m\theta \right]. \tag{10.83}$$

If we define

$$a_m(z) = \sum_{n=1}^{\infty} a_{mn} J_m(in) \sin nz,$$

$$b_m(z) = \sum_{n=1}^{\infty} b_{mn} J_m(in) \sin nz, \tag{10.84}$$

we see that (10.83) becomes

$$f(\theta,z) = \sum_{m=1}^{\infty} a_m(z) \cos m\theta + b_m(z) \sin m\theta,$$

which is a Fourier series in θ, where z is an arbitrary parameter. By Euler's formulas, it follows that

$$a_m(z) = \frac{1}{\pi} \int_{-\pi}^{\pi} f(\theta,z) \cos m\theta \, d\theta, \qquad b_m(z) = \frac{1}{\pi} \int_{-\pi}^{\pi} f(\theta,z) \sin m\theta \, d\theta.$$

If the functions $a_m(z)$ and $b_m(z)$ can be calculated, we can then use the Fourier sine series (10.84) in z to calculate the coefficients a_{mn} and b_{mn}:

$$a_{mn} = \frac{2}{\pi J_m(in)} \int_0^{\pi} a_m(z) \sin nz \, dz,$$

$$b_{mn} = \frac{2}{\pi J_m(in)} \int_0^{\pi} b_m(z) \sin nz \, dz.$$

Although the formal procedure delineated above is straightforward, the calculations involved are, to say the least, very tedious. In Exercise 8, the reader will be asked to developed a similar procedure that leads to the use of series of Legendre polynomials in place of Bessel functions.

EXERCISES 10.4

1. A square plate with sides of length L has both faces insulated. The upper horizontal edge is kept at 50 °C, while all the other edges are at 0 °C. Find the steady state temperatures at the points $(L/2,L/2)$ and $(L/4,L/4)$.

2. If in Exercise 1 the temperatures along the edge $y = L$ are given by

    $$T(x,L) = x(L - x),$$

    and all the other conditions remain the same, what is $T(x,y)$?

3. If in Exercise 1 the temperatures along the upper horizontal edge are

    $$T(x,L) = 100 \sin (\pi x/L),$$

    with all the other conditions remaining unchanged, what is $T(x,y)$?

4. Suppose that the temperature along the edge $y = L$ in Exercise 2 is changed to

    $$T(x,L) = x(L^2 - x^2),$$

    with all the other conditions remaining the same. Find the temperature at the points $(L/4,L/4)$ and $(L/4,3L/4)$.

5. Consider the steady state heat flow problem on the unit disk (see Example 10.4.2) and

suppose that $f(\theta) = \sin^2 \theta$. Find the temperature at the point $r = \frac{1}{2}$, $\theta = 0°$ by:

a) the eigenfunction expansion method,

b) Poisson's integral formula.

6. Repeat Exercise 5 with $f(\theta) = \sin \theta - \cos \theta$.

7. Repeat Exercise 5 with $f(\theta) = \theta(2\pi - \theta)$.

* 8. a) Show that the three-dimensional Laplace equation (10.78) may be written in spherical coordinates r, θ, φ ($x = r \cos \theta \sin \varphi$, $y = r \sin \theta \sin \varphi$, $z = r \cos \varphi$) in the form

$$\frac{\partial^2 T}{\partial r^2} + \frac{2}{r}\frac{\partial T}{\partial r} + \frac{1}{r^2}\frac{\partial^2 T}{\partial \varphi^2} + \frac{\cot \varphi}{r^2}\frac{\partial T}{\partial \varphi} + \frac{1}{r^2 \sin^2 \varphi}\frac{\partial^2 T}{\partial \theta^2} = 0. \tag{10.85}$$

b) Separate the variables in (10.85) to obtain the ordinary differential equations

$$\Theta'' + k_1\Theta = 0,$$

$$R'' + \left(\frac{2}{r}\right)R' + \left(\frac{k_2}{r^2}\right)R = 0,$$

$$[(\sin \varphi)\Phi']' - \left(\frac{k_2 \sin \varphi + k_1}{\sin \varphi}\right)\Phi = 0.$$

c) Assume the boundary condition $T(1,\theta,\varphi) = f(\theta,\varphi)$ for a sphere of radius 1, where f is continuously differentiable in θ and φ, and periodic with period 2π in θ. Find the eigenfunctions of the three boundary value problems thus defined. [*Hint*: Use Legendre polynomials.]

d) Obtain the eigenfunction expansion for $T(r,\theta,\varphi)$.

* 9. Justify the steps involved in obtaining Poisson's integral formula (10.77) from the solution (10.75) of the steady state heat equation in the disk $r \leq 1$. [*Hint*: Show that the series (10.75) converges uniformly in the disk $r \leq r_0 < 1$.]

* 10. Generalize Example 10.4.2 to obtain the steady state temperature in a disk of radius R, centered at the origin, and prove that the corresponding Poisson integral formula is

$$T(r,\theta) = \frac{R^2 - r^2}{2\pi} \int_{-\pi}^{\pi} \frac{f(\varphi)\, d\varphi}{R^2 + r^2 - 2rR \cos(\theta - \varphi)}, \qquad r < R,$$

where the continuously differentiable function $f(\varphi)$ describes the boundary temperature in the direction $0 \leq \varphi < 2\pi$.

* 11. Let D be any bounded polygon in the xy-plane, and let (x_0,y_0) be a point interior to D. Let u be a real valued function defined on D that has continuous derivatives of at least order 2.

a) Show that if u attains its maximal value M at (x_0,y_0), then

$$\frac{\partial^2 u}{\partial x^2} + \frac{\partial^2 u}{\partial y^2}\bigg|_{(x_0,y_0)} \leq 0.$$

b) Show that if

$$\frac{\partial^2 u}{\partial x^2} + \frac{\partial^2 u}{\partial y^2} > 0$$

at all interior points of D, then u does not attain its maximal value at any interior point of D.

c) Suppose that

$$\frac{\partial^2 u}{\partial x^2} + \frac{\partial^2 u}{\partial y^2} \geq 0 \tag{10.86}$$

at all interior points of D. Show that the function $w(x,y) = u(x,y) + \varepsilon[(x - x_0)^2 + (y - y_0)^2]$ does not attain its maximal value in D at (x_0,y_0), for any value $\varepsilon > 0$.

d) Conclude that any function satisfying (10.86) at all interior points of D must attain its maximal value on the boundary of D. (This is called the *maximum principle*.)

* 12. Use the maximum principle obtained in Exercise 11 to show that any function u that satisfies Laplace's equation attains both its maximum and its minimum on the boundary of D. (Physically, this means that in a steady state heat flow problem, the maximum and minimum temperatures will occur on the boundary of the region.)

10.5 LAPLACE TRANSFORM METHODS FOR PARTIAL DIFFERENTIAL EQUATIONS

The one-dimensional heat equation

$$\frac{\partial T}{\partial t} = \delta \frac{\partial^2 T}{\partial x^2}$$

is sometimes called the *thermal diffusion equation,* since it describes the flow of heat along a uniform rod. The same equation may also be used to describe one-dimensional fluid flow, since we can replace temperature by mass (or concentration) in the considerations of Example 1.1.6. By doing this, we can employ the heat equation to study such diverse problems as the dispersal of pollutants in the atmosphere or the flow of a chemical across a membrane. In some cases, the fluid flow is aided (or retarded) by other forces. For example, the motion of a pollutant emitted from a smoke stack involves both a mixing (*diffusion*) of the pollutant in the air and the transport (*convection*) of the pollutant by air currents. Thus the change in the concentration T in the interval from x_0 to x_1, due to convection, is

$$v[T(x_0,t) - T(x_1,t)],$$

where v is the velocity of the current. Adding this term to Eq. (1.15), we obtain

$$\frac{dQ}{dt} = \int_{x_0}^{x_1} \frac{\partial}{\partial x}\left(\kappa \frac{\partial T}{\partial x} - vT\right) A\, dx.$$

Differentiating (1.12), we finally have

$$c\rho \frac{\partial T}{\partial t} = \frac{\partial}{\partial x}\left(\kappa \frac{\partial T}{\partial x}\right) - v \frac{\partial T}{\partial x}. \tag{10.87}$$

Equation (10.87) is generally called the *diffusion equation*. If κ is zero, we call (10.87) a *pure convection*; if $v = 0$, a *pure diffusion*.

The method of separation of variables is again very useful in solving diffusion equations. However, in some situations the boundary and initial conditions are such that it is not possible to obtain a solution by this technique. Consider the following example of a pure diffusion.

Example 10.5.1 Suppose that κ is constant and $v = 0$ in Eq. (10.87). Then Eq. (10.87) reduces to the thermal diffusion equation

$$\frac{\partial T}{\partial t} = \delta \frac{\partial^2 T}{\partial x^2}. \tag{10.88}$$

Assume that we are given the initial condition

$$T(x,0) = 0, \qquad 0 \leqslant x < \infty, \tag{10.89}$$

and the boundary conditions

$$\lim_{x \to +\infty} T(x,t) = \lim_{x \to +\infty} \frac{\partial T}{\partial x}(x,t) = 0, \tag{10.90}$$

$$T(0,t) = T_0(\neq 0), \qquad 0 < t. \tag{10.91}$$

These conditions could be used to describe the heat flow in an infinitely long rod when a constant source of heat is applied at $x = 0$, or to describe the concentration of a pollutant at varying distances from a smoke stack, in the absence of winds, with the stack constantly emitting pollutants, for all time $t > 0$. If we let $T(x,t) = X(x)\mathcal{T}(t)$, we have

$$X\mathcal{T}' = \delta X''\mathcal{T} \qquad \text{or} \qquad \frac{\mathcal{T}'}{\delta\mathcal{T}} = \frac{X''}{X} = k,$$

from which we obtain the two differential equations

$$\mathcal{T}' = k\,\delta\mathcal{T} \qquad \text{and} \qquad X'' - \mathcal{T}X \doteq 0. \tag{10.92}$$

The first differential equation in (10.92) has the general solution $\mathcal{T}(t) = ce^{k\,\delta t}$. Using the initial condition (10.89), we observe that

$$0 = T(x,0) = X(x)\mathcal{T}(0) = cX(x).$$

Hence either $c = 0$, implying that $\mathcal{T} \equiv 0$, or $X \equiv 0$. In either case $T \equiv 0$, contradicting the second boundary condition (10.91). Clearly, the method of separation of variables fails for this problem.

It is therefore apparent that other techniques must be developed to treat problems of this type. The effectiveness of Laplace transform methods for ordinary differential equations suggests that they might be useful in this context. The rest of this section will be devoted to a discussion of how transformation techniques may be applied to partial differential equations. It should be noted, however, that the effectiveness of Laplace transforms is subject to several limitations. The equation *must* be linear, should have constant coefficients, and there must be appropriate initial conditions. Even when these conditions are met, there is no guarantee of success; and even if a solution can be obtained, it may be easier to obtain by other methods.

To apply Laplace transforms to a partial differential equation, we must begin by considering the ranges of the independent variables. We shall illustrate the entire procedure as it relates to Example 10.5.1. In this problem, the independent variables x and t may both assume all values in the range 0 to ∞. This property is very desirable, since the definition of the Laplace transform of a function requires its integration over this range [see Eq. (6.1)]. At this point both variables look promising. We recall, however, that the differentiation property (Theorem 6.5) of Laplace transforms requires an additional initial condition for each order of the derivative involved. In this case, Eq. (10.88) contains first derivatives in t and second derivatives in x. Since $(\partial T/\partial x)(0,t)$ is not known, there are not enough data for the x variable. Thus we shall treat x as a parameter and consider the Eq. (10.88) as an ordinary differential equation with t as the only independent variable. Let

$$\mathcal{L}\{T(x,t)\} = \int_0^\infty e^{-st}T(x,t)\,dt.$$

Then $\mathcal{L}\{T\}$ is a function of s and the parameter x. Using the differentiation property (Theorem 6.5) and the initial condition (10.89), we may transform Eq. (10.88) into the ordinary differential equation

$$s\mathcal{L}\{T\} - 0 = \mathcal{L}\left\{\frac{\partial T}{\partial t}\right\} = \delta\mathcal{L}\left\{\frac{\partial^2 T}{\partial x^2}\right\} = \delta\frac{\partial}{\partial x^2}\mathcal{L}\{T\}. \tag{10.93}$$

Note that we have interchanged the operations of taking the Laplace transform and differentiating with respect to x. This exchange may not be valid, but the objection can be ignored if the method succeeds in producing a solution to the problem. Hence it is essential that we verify any result obtained by this method.

It is important now to interpret the boundary conditions (10.90) and (10.91) for $\mathcal{L}\{T\}$ whenever possible. Note that (10.91) yields

$$\mathcal{L}\{T\}(0,s) = \int_0^\infty e^{-st}T(0,t)\,dt = \frac{T_0}{s}. \tag{10.94}$$

Also, after warning that the indicated operations may not always be valid, we obtain

$$\lim_{x \to +\infty} \mathcal{L}\{T\}(x,s) = \lim_{x \to +\infty} \int_0^\infty e^{-st}T(x,t)\,dt$$

$$= \int_0^\infty e^{-st}\left[\lim_{x \to +\infty} T(x,t)\right]\,dt = 0. \tag{10.95}$$

Now we treat s as a parameter and x as the independent variable. Setting $z = \mathscr{L}\{T\}$, we may rewrite (10.93) as

$$\frac{d^2z}{dx^2} = \frac{s}{\delta}\,z, \tag{10.96}$$

where s is fixed. The auxiliary equation for the differential equation (10.96) has the roots $\pm\sqrt{s/\delta}$, so that z has the general solution

$$z(x) = c_1 e^{\sqrt{s/\delta}\,x} + c_2 e^{-\sqrt{s/\delta}\,x}.$$

Since $s > 0$, the coefficient c_1 must be zero, because the term $e^{\sqrt{s/\delta}\,x}$ tends to infinity as $x \to +\infty$, violating the "identity" (10.95). Setting $x = 0$, we find that $c_2 = T_0/s$, by (10.94), so that we finally have

$$\mathscr{L}\{T\} = z(x) = \frac{T_0}{s}\,e^{-\sqrt{s/\delta}\,x}. \tag{10.97}$$

Again we treat x as a parameter and seek the inverse Laplace transform of Eq. (10.97). By formula 43 of Appendix 2, with $r = x/\sqrt{\delta}$, we see that

$$T(x,t) = T_0\left(1 - \frac{2}{\sqrt{\pi}}\int_0^{x/2\sqrt{\delta t}} e^{-u^2}\,du\right). \tag{10.98}$$

To check that (10.98) is indeed the solution of Example 10.5.1, we note from formula 58 of Appendix 1 that

$$\int_0^\infty e^{-u^2}\,du = \frac{\sqrt{\pi}}{2}. \tag{10.99}$$

The integral in (10.98) equals (10.99) if $t \to 0+$ or $x \to +\infty$. Thus (10.89) and the first limit of (10.90) hold. If $x = 0$, we obviously obtain (10.91). Differentiating (10.98) with respect to x, we have

$$\frac{\partial T}{\partial x} = -\frac{T_0}{\sqrt{\pi\,\delta t}}\,e^{-(x^2/4\,\delta t)},$$

which vanishes as $x \to +\infty$, and differentiating again, we obtain

$$\delta\,\frac{\partial^2 T}{\partial x^2} = \frac{T_0 x}{2\sqrt{\pi\delta}}\,\frac{e^{-(x^2/4\,\delta t)}}{t^{3/2}} = \frac{\partial T}{\partial t}.$$

Thus in this case the Laplace transform method does yield a solution.

Example 10.5.2 We shall consider the vibrating string problem of Section 10.2, given by the wave equation

$$\frac{\partial^2 y}{\partial t^2} = c^2\,\frac{\partial^2 y}{\partial x^2}, \tag{10.100}$$

with the boundary conditions

$$y(0,t) = y(L,t) = 0 \tag{10.101}$$

and initial conditions

$$y(x,0) = f(x), \quad \left.\frac{\partial y}{\partial t}\right|_{(x,0)} = 0. \tag{10.102}$$

Since $0 \leqslant x \leqslant L$, we select t as our independent variable for the Laplace transform. By the differentiation property [see Eq. (6.23)], we transform (10.100) into

$$s^2 \mathscr{L}\{y\} - sf(x) - 0 = c^2 \frac{\partial^2}{\partial x^2} \mathscr{L}\{y\}.$$

Thus we have the nonhomogeneous second order ordinary differential equation

$$c^2 Y'' - s^2 Y = -sf(x), \tag{10.103}$$

where $Y(x) = \mathscr{L}\{y\}$ and s is a parameter. By Eq. (10.101), we have $Y(0) = Y(L) = 0$. Our problem thus reduces to a nonhomogeneous boundary value problem. It may then be solved by the methods of Chapter 3 or Section 9.6. To simplify matters, we assume that $f(x) = x(L - x)$, so that we can avoid having to use eigenfunction expansions to find the solution Y. A particular solution of (10.103) will be of the form

$$Y_p = Ax^2 + Bx + C.$$

Hence

$$2c^2 A - s^2(Ax^2 + Bx + C) = -sx(L - x).$$

Equating like powers of x, we find that

$$A = -\frac{1}{s}, \quad B = \frac{L}{s}, \quad C = -\frac{2c^2}{s^3}.$$

Thus the general solution of (10.103) is

$$Y = \alpha_1 e^{sx/c} + \alpha_2 e^{-sx/c} - \frac{x^2}{s} + \frac{Lx}{s} - \frac{2c^2}{s^3}.$$

Setting $x = 0$ and $x = L$ in the equation, we obtain the simultaneous equations

$$\alpha_1 + \alpha_2 = \frac{2c^2}{s^3},$$

$$\alpha_1 e^{sL/c} + \alpha_2 e^{-sL/c} = \frac{2c^2}{s^3}.$$

Therefore,

$$\alpha_1 = \frac{(2c^2/s^3)(1 - e^{-sL/c})}{e^{sL/c} - e^{-sL/c}} = \frac{c^2(1 - e^{-sL/c})}{s^3 \sinh (sL/c)},$$

$$\alpha_2 = \frac{(2c^2/s^3)(e^{sL/c} - 1)}{e^{sL/c} - e^{-sL/c}} = \frac{c^2(e^{sL/c} - 1)}{s^3 \sinh (sL/c)}.$$

We now have

$$\mathcal{L}\{y\} = Y(x) = \frac{c^2[e^{sx/c}(1 - e^{-sL/c}) + e^{-sx/c}(e^{sL/c} - 1)]}{s^3 \sinh (sL/c)} - \frac{x^2}{s} + \frac{Lx}{s} - \frac{2c^2}{s^3}$$

$$= \frac{2c^2}{s^3} \frac{\{\sinh (sx/c) + \sinh [s(L - x)/c]\}}{\sinh (sL/c)} + \frac{x(L - x)}{s} - \frac{2c^2}{s^3}. \qquad (10.104)$$

To find y, we must find the inverse Laplace transform of the expressions

$$\frac{\sinh (sx/c)}{s^3 \sinh (sL/c)} \qquad \text{and} \qquad \frac{\sinh [s(L - x)/c]}{s^3 \sinh [sL/c]}.$$

Now using the geometric series

$$\frac{1}{1 - x} = 1 + x + x^2 + \cdots$$

(see Example 5.1.1), we have

$$\frac{\sinh (sx/c)}{\sinh (sL/c)} = \frac{e^{sx/c} - e^{-sx/c}}{e^{sL/c} - e^{-sL/c}}$$

$$= \frac{e^{sx/c} - e^{-sx/c}}{e^{sL/c}} \cdot \frac{1}{1 - e^{-2sL/c}}$$

$$= [e^{s(x - L)/c} - e^{-s(x + L)/c}] \sum_{n=0}^{\infty} e^{-2nsL/c}$$

$$= \sum_{n=0}^{\infty} [e^{s(x - (2n + 1)L)/c} - e^{-s(x + (2n + 1)L)/c}].$$

Since $\mathcal{L}\{H(t - a)\} = e^{-as}/s$ and $\mathcal{L}\{t\} = 1/s^2$, we apply the convolution theorem (6.11) to each term of the series

$$\frac{1}{s^3} \frac{\sinh (sx/c)}{\sinh (sL/c)}$$

$$= \sum_{n=0}^{\infty} \left[\frac{e^{-s[(2n + 1)L - x]/c}}{s^3} - \frac{e^{-s[(2n + 1)L + x]/c}}{s^3} \right]$$

$$= \sum_{n=0}^{\infty} \mathcal{L}\{t*H[t - ((2n + 1)L - x)/c]\} - \mathcal{L}\{t*H[t - ((2n + 1)L + x)/c]\}.$$

Similarly,

$$\frac{1}{s^3} \frac{\sinh \left[s(L - x)/c \right]}{\sinh (sL/c)}$$

$$= \sum_{n=0}^{\infty} \left[\frac{e^{-s(2nL + x)/c}}{s^3} - \frac{e^{-s(2(n+1)L - x)/c}}{s^3} \right]$$

$$= \sum_{n=0}^{\infty} \mathscr{L} \{t*H[t - (2nL + x)/c]\} - \mathscr{L} \{t*H[t - (2(n + 1)L - x)/c]\}.$$

Taking the inverse Laplace transform of (10.104) term by term, we have

$$y(x,t) = 2c^2 \sum_{n=0}^{\infty} \left\{ t*H \left[t - \frac{(2n + 1)L - x}{c} \right] - t*H \left[t - \frac{(2n + 1)L + x}{c} \right] \right.$$

$$\left. + t*H \left(t - \frac{2nL + x}{c} \right) - t*H \left[t - \frac{2(n + 1)L - x}{c} \right] \right\}$$

$$+ x(L - x) - c^2 t^2.$$

Since we intend to verify our solution, we need not justify interchanging the operations of taking the inverse Laplace transforms, convolution, and the infinite series. Simplifying the expression above, we obtain, after some algebra,

$$y(x,t) = 2c^2 t* \sum_{n=0}^{\infty} (-1)^n \left[H \left(t - \frac{nL + x}{c} \right) + H \left(t - \frac{nL + (L - x)}{c} \right) \right]$$

$$+ x(L - x) - c^2 t^2.$$

But

$$t*H(t - a) = \int_0^t (t - u)H(u - a) \, du = \begin{cases} 0, & \text{if } t \leqslant a, \\ \dfrac{(t - a)^2}{2} & \text{if } a \leqslant t. \end{cases}$$

Finally, interchanging the sums and integrals yields

$$y(x,t) = c^2 \sum_{n=0}^{\infty} (-1)^n \left[H \left(t - \frac{nL + x}{c} \right) \left(t - \frac{nL + x}{c} \right)^2 \right.$$

$$\left. + H \left(t - \frac{nL + (L - x)}{c} \right) \cdot \left(t - \frac{nL + (L - x)}{c} \right)^2 \right]$$

$$+ x(L - x) - c^2 t^2. \tag{10.105}$$

We leave the task of verifying that (10.105) is indeed the solution of the boundary and initial condition problem as an exercise (see Exercise 1).

EXERCISES 10.5

1. Verify that (10.105) is a solution of Example 10.5.2.

In Exercises 2 through 7 obtain the solution of the given boundary and initial condition problem by the method of Laplace transforms. Be sure to verify your solution.

2. $\dfrac{\partial^2 y}{\partial x^2} = 4\dfrac{\partial^2 y}{\partial t^2}, t > 0, x > 0$

$y(x,0) = 0, \dfrac{\partial y}{\partial t}\bigg|_{(x,0)} = 1, x \geqslant 0$

$y(0,t) = t, \lim_{x \to \infty} y(x,t)$ is finite

3. $\dfrac{\partial^2 y}{\partial x^2} = 16\dfrac{\partial^2 y}{\partial t^2}, t > 0, x > 0$

$y(x,0) = 0, \dfrac{\partial y}{\partial t}\bigg|_{(x,0)} = 1, x \geqslant 0$

$y(0,t) = \sin t, \lim_{x \to \infty} y(x,t)$ is finite

4. $\dfrac{\partial T}{\partial t} = \delta\dfrac{\partial^2 T}{\partial x^2}, t > 0, x > 0$

$T(x,0) = 0, 0 \leqslant x < \infty, T(0, t) = 1, t > 0$

$\lim_{x \to \infty} T(x,t) = \lim_{x \to \infty} \dfrac{\partial T}{\partial x}(x,t) = 0, t > 0$

5. $\dfrac{\partial T}{\partial t} = \delta\dfrac{\partial^2 T}{\partial x^2}, t > 0, x > 0$

$T(x,0) = e^{-x}, x \geqslant 0,$

$T(0,t) = T_0 (\neq 0), \lim_{x \to \infty} T(x,t) = \lim_{x \to \infty} \dfrac{\partial T}{\partial x}(x,t) = 0, t > 0$

6. $\dfrac{\partial T}{\partial t} = \delta\dfrac{\partial^2 T}{\partial x^2}, t > 0, x > 0$

$T(x,0) = 0, x \geqslant 0,$

$T(0,t) = \sin \omega t, \lim_{x \to \infty} T(x,t) = \lim_{x \to \infty} \dfrac{\partial T}{\partial x}(x,t) = 0, t > 0$

7. $\dfrac{\partial T}{\partial t} = \delta\dfrac{\partial^2 T}{\partial x^2} + \mu\dfrac{\partial T}{\partial x}, t > 0, x > 0$

$T(x,0) = 0, x \geqslant 0$

$T(0,t) = T_0 (\neq 0), \lim_{x \to \infty} T(x,t) = \lim_{x \to \infty} \dfrac{\partial T}{\partial x}(x,t) = 0, t > 0$

THE
EXISTENCE
AND
UNIQUENESS
OF
SOLUTIONS

11

In most of the preceding material in this book, we were seeking techniques for finding solutions to differential equations, assuming the *existence* of *unique* solutions for ordinary differential equations with specified initial conditions. The aim of this chapter is to prove some basic existence and uniqueness theorems for solutions of initial value problems and to show that the solution depends continuously on the initial conditions. In particular, we will prove the existence-uniqueness Theorems 3.1, 7.2, and 7.11 stated earlier in the text. At this point many readers may ask the following three questions:

1. Why do we need to prove the existence of a solution, particularly if we know that the differential equation arises from a physical problem that has a solution?
2. Why must we worry about uniqueness?
3. Why is continuous dependence on initial conditions important?

To answer the first question, it is important to remember that a differential equation is only a model of a physical problem. It is possible that the differential equation is a very bad model, so bad in fact that it has no solution. Countless hours could be spent using the techniques we have developed looking for a solution that may not even exist. Thus existence theorems are not only of theoretical value in telling us which equations have solutions, but also useful in developing mathematical models of physical problems.

Similarly, uniqueness theorems have both theoretical and practical implications. If we know that a problem has a unique solution, then once we have found one solution, we are done. If the solution is not unique, then we cannot talk about "the" solution but must, instead, worry about *which* solution is being discussed. In practice, if the physical problem has a unique solution, so should any mathematical model of the problem.

Finally, continuous dependence on the initial conditions is very important, since some inaccuracy is always present in practical situations. We need to know that if the initial conditions are slightly changed, the solution of the differential equation will change only slightly. Otherwise, slight inaccuracies could yield very different solutions. This property will be discussed in Exercise 18.

Before continuing with our discussion, we advise the reader that in this chapter we shall use theoretical tools from calculus that have not been widely employed earlier in the text. In particular, we shall need the following facts about continuity and convergence of functions. They are discussed in most intermediate and advanced calculus texts*:

---

* See, for example, R. C. Buck, *Advanced Calculus*, McGraw-Hill, New York, 1956.

A) Let $f(t,x)$ be a continuous function of the two variables t and x and let the closed, bounded region D be defined by

$$D = \{(t,x): a \leqslant t \leqslant b, c \leqslant x \leqslant d\},$$

where a, b, c, and d are finite real numbers. Then $f(t,x)$ is bounded for (t,x) in D. That is, there is a number $M > 0$ such that $|f(t,x)| \leqslant M$ for every pair (t,x) in D.

B) Let $f(x)$ be continuous on the closed interval $a \leqslant x \leqslant b$ and differentiable on the open interval $a < x < b$. Then the *mean value theorem* of differential calculus states that there is a number ξ between a and b $(a < \xi < b)$ such that

$$f(b) - f(a) = f'(\xi)(b - a).$$

This equation can be written as

$$\frac{f(b) - f(a)}{b - a} = f'(\xi),$$

which says, geometrically, that the slope of the tangent to the curve $y = f(x)$ at the point ξ between a and b is equal to the slope of the secant line passing through the points $(a,f(a))$ and $(b,f(b))$ (see Fig. 11.1).

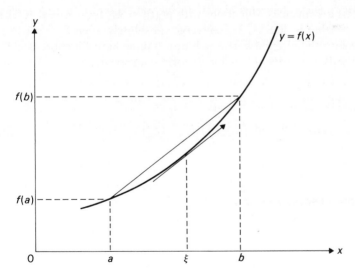

Figure 11.1

C) Let $\{x_n(t)\}$ be a sequence of functions. Then $x_n(t)$ is said to *converge uniformly* to a (limit) function $x(t)$ on the interval $a \leqslant t \leqslant b$ if for every real number $\varepsilon > 0$, there exists an integer $N > 0$ such that whenever $n \geqslant N$, we have

$$|x_n(t) - x(t)| < \varepsilon$$

for every t, $a \leqslant t \leqslant b$.

D) If the functions $\{x_n(t)\}$ of statement (C) are continuous on the interval $a \leqslant t \leqslant b$, then the limit function $x(t)$ is also continuous there. This fact is often stated as "the uniform limit of continuous functions is continuous."

E) Let $f(t,x)$ be a continuous function in the variable x and suppose that $x_n(t)$ converges to $x(t)$ uniformly as $n \to \infty$. Then

$$\lim_{n \to \infty} f(t,x_n(t)) = f(t,x(t)).$$

F) Let $f(t)$ be an integrable function on the interval $a \leqslant t \leqslant b$. Then

$$\left| \int_a^b f(t)\, dt \right| \leqslant \int_a^b |f(t)|\, dt,$$

and if $|f(t)| \leqslant M$, then

$$\int_a^b |f(t)|\, dt \leqslant M \int_a^b dt = M(b - a).$$

G) Let $\{x_n(t)\}$ be a sequence of functions with $|x_n(t)| \leqslant M_n$ for $a \leqslant t \leqslant b$. Then, if $\sum_{n=0}^{\infty} |M_n| < \infty$ (that is, if $\sum_{n=0}^{\infty} M_n$ converges absolutely) then $\sum_{n=0}^{\infty} x_n(t)$ converges uniformly on the interval $a \leqslant t \leqslant b$ to a unique limit function $x(t)$. This is often called the *Weierstrass M-test* for uniform convergence.

H) Let $\{x_n(t)\}$ converge uniformly to $x(t)$ on the interval $a \leqslant t \leqslant b$ and let $f(t,x)$ be a continuous function of t and x in the region D defined in statement (A). Then

$$\lim_{n \to \infty} \int_a^b f(s,x_n(s))\, ds = \int_a^b \lim_{n \to \infty} f(s,x_n(s))\, ds = \int_a^b f(s,x(s))\, ds.$$

11.1 SUCCESSIVE APPROXIMATIONS: A LOCAL EXISTENCE-UNIQUENESS THEOREM

In this section we shall prove a general theorem about existence and uniqueness of solutions of the first order initial value problem

$$x'(t) = f(t,x(t)), \qquad x(t_0) = x_0, \tag{11.1}$$

where t_0 and x_0 are real numbers. Equation (11.1) includes all the first order equations we have discussed in this book. For example, for the linear nonhomogeneous equation $x' + a(t)x = b(t)$,

$$f(t,x) = -a(t)x + b(t).$$

We shall show that if $f(t,x)$ and $(\partial f/\partial x)(t,x)$ are continuous in some region containing the point (t_0,x_0), then there is an interval (containing t_0) on which a unique solution of Eq. (11.1) exists. First, we need some preliminary results.

Theorem 11.1   Let $f(t,x)$ be continuous for all values t and x. Then the initial value problem (11.1) is equivalent to the integral equation

$$x(t) = x_0 + \int_{t_0}^{t} f(s,x(s))\, ds \tag{11.2}$$

in the sense that $x(t)$ is a solution of (11.1) if and only if $x(t)$ is a solution of (11.2).

Proof.   If $x(t)$ satisfies (11.1), then

$$\int_{t_0}^{t} f(s,x(s))\, ds = \int_{t_0}^{t} x'(s)\, ds = x(s)\Big|_{t_0}^{t} = x(t) - x_0,$$

which shows that $x(t)$ satisfies (11.2). Conversely, if $x(t)$ satisfies (11.2), then differentiating (11.2), we have

$$x'(t) = \frac{d}{dt} \int_{t_0}^{t} f(s,x(s))\, ds = f(t,x(t))$$

and

$$x(t_0) = x_0 + \int_{t_0}^{t_0} f(s,x(s))\, ds = x_0.$$

Hence $x(t)$ also satisfies (11.1).

Let D denote the rectangular region in the tx-plane defined by

$$D: a \leqslant t \leqslant b,\, c \leqslant x \leqslant d, \tag{11.3}$$

where $-\infty < a < b < +\infty$ and $-\infty < c < d < +\infty$. See Fig. 11.2. We say that the function $f(t,x)$ is *Lipschitz continuous in x over D* if there exists a constant $k, 0 < k < \infty$, such that

$$|f(t,x_1) - f(t,x_2)| \leqslant k|x_1 - x_2| \tag{11.4}$$

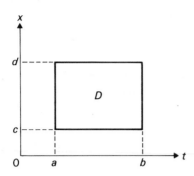

Figure 11.2

whenever (t,x_1) and (t,x_2) belong to D. The constant k is called a *Lipschitz constant*. Clearly, according to (11.4), every Lipschitz continuous function is continuous in x for each fixed t. However, *not every continuous function is Lipschitz continuous*.

Example 11.1.1 Let $f(t,x) = \sqrt{x}$ on the set $0 \leqslant t \leqslant 1, 0 \leqslant x \leqslant 1$. Then $f(t,x)$ is certainly continuous on this region. But

$$|f(t,x) - f(t,0)| = |\sqrt{x} - 0| = \frac{1}{\sqrt{x}}|x - 0|$$

for all $0 < x < 1$, and $x^{-1/2}$ tends to ∞ as x approaches zero. Thus no finite Lipschitz constant can be found to satisfy Eq. (11.4).

However, Lipschitz continuity is not a rare occurrence, as shown by the following theorem.

Theorem 11.2 Let $f(t,x)$ and $(\partial f/\partial x)(t,x)$ be continuous on D. Then $f(t,x)$ is Lipschitz continuous in x over D.

Proof. Let (t,x_1) and (t,x_2) be points in D. For fixed t, $(\partial f/\partial x)(t,x)$ is a function of x, and so we may apply the mean value theorem of differential calculus [statement (**B**)] to obtain

$$|f(t,x_1) - f(t,x_2)| = \left|\frac{\partial f}{\partial x}(t,\xi)\right||x_1 - x_2|$$

where $x_1 < \xi < x_2$. But since $\partial f/\partial x$ is continuous in D, it is bounded there [according to statement (A)]. Hence there is a constant k, $0 < k < \infty$, such that

$$\left| \frac{\partial f}{\partial x} (t,x) \right| \leqslant k$$

for all (t,x) in D, and the proof is complete.

Example 11.1.2   If $f(t,x) = tx^2$ on $0 \leqslant t \leqslant 1, 0 \leqslant x \leqslant 1$, then

$$\left| \frac{\partial f}{\partial x} \right| = |2tx| \leqslant 2,$$

so that

$$|f(t,x_1) - f(t,x_2)| \leqslant 2 |x_1 - x_2|.$$

We now define a sequence of functions $\{x_n(t)\}$, called *Picard* iterations*, by the successive formulas

$$x_0(t) = x_0,$$

$$x_1(t) = x_0 + \int_{t_0}^{t} f(s,x_0(s))\, ds,$$

$$x_2(t) = x_0 + \int_{t_0}^{t} f(s,x_1(s))\, ds, \tag{11.5}$$

$$\cdot$$
$$\cdot$$
$$\cdot$$

$$x_n(t) = x_0 + \int_{t_0}^{t} f(s,x_{n-1}(s))\, ds,$$

$$\cdot$$
$$\cdot$$
$$\cdot$$

We will show that under certain conditions the Picard iterations defined by (11.5) converge uniformly to a solution of (11.2). First we illustrate the process of this iteration by a simple example.

Example 11.1.3   Consider the initial value problem

$$x'(t) = x(t), \qquad x(0) = 1. \tag{11.6}$$

As we know, (11.6) has the unique solution $x(t) = e^t$. In this case, the function $f(t,x)$ of (11.1) is given by $f(t,x(t)) = x(t)$, so that the Picard iterations defined by

---

*   Emile Picard (1856–1941) was one of the most eminent French mathematicians of the past century who made several outstanding contributions to mathematical analysis.

(11.5) yield successively

$$x_0(t) = x_0 = 1,$$

$$x_1(t) = 1 + \int_0^t (1) \, ds = 1 + t,$$

$$x_2(t) = 1 + \int_0^t (1 + s) \, ds = 1 + t + \frac{t^2}{2},$$

$$x_3(t) = 1 + \int_0^t \left(1 + s + \frac{s^2}{2}\right) ds = 1 + t + \frac{t^2}{2!} + \frac{t^3}{3!},$$

and clearly,

$$x_n(t) = 1 + t + \frac{t^2}{2!} + \cdots + \frac{t^n}{n!} = \sum_{k=0}^{n} \frac{t^k}{k!}.$$

Hence

$$\lim_{n \to \infty} x_n(t) = \sum_{k=0}^{\infty} \frac{t^k}{k!} = e^t$$

by formula (5.16).

Theorem 11.3. Existence Theorem Let $f(t,x)$ be Lipschitz continuous in x with the Lipschitz constant k on the region D of all points (t,x) satisfying the inequalities

$$|t - t_0| \leqslant a, \qquad |x - x_0| \leqslant b.$$

(See Fig. 11.3.) Then there exists a number $\delta > 0$ with the property that the initial

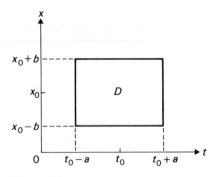

Figure 11.3

value problem

$$x' = f(t,x), \qquad x(t_0) = x_0,$$

has a solution $x = x(t)$ on the interval $|t - t_0| \leqslant \delta$.

Proof. The proof of this theorem is complicated and will be done in several stages. However, the basic idea is simple: we need only justify that the Picard iterations converge uniformly and yield, in the limit, the solution of the integral equation (11.2).

Since f is continuous on D, it is bounded there [statement (A)] and we may begin by letting M be a finite upper bound for $|f(t,x)|$ on D. We then define

$$\delta = \min \{a,b/M\}. \tag{11.7}$$

1. We first show that the iterations $\{x_n(t)\}$ are continuous and satisfy the inequality

$$|x_n(t) - x_0| \leqslant b. \tag{11.8}$$

Inequality (11.8) is necessary in order that $f(t,x_n(t))$ be defined for $n = 0, 1, 2, \ldots$. To show the continuity of $x_n(t)$, we first note that $x_0(t) = x_0$ is continuous (a constant function is always continuous). Then

$$x_1(t) = x_0 + \int_{t_0}^{t} f(t,x_0(s))\, ds.$$

But $f(t,x_0)$ is continuous [since $f(t,x)$ is continuous in t and x], and the integral of a continuous function is continuous. Thus $x_1(t)$ is continuous. In a similar fashion, we can show that

$$x_2(t) = x_1(t) + \int_{t_0}^{t} f(t,x_1(s))\, ds$$

is continuous and so on for $n = 3, 4, \ldots$.

Obviously the inequality (11.8) holds when $n = 0$, because $x_0(t) = x_0$. For $n \neq 0$, we use the definition (11.5) and Eq. (11.7) to obtain

$$|x_n(t) - x_0| = \left| \int_{t_0}^{t} f(s,x_{n-1}(s))\, ds \right| \leqslant \left| \int_{t_0}^{t} |f(s,x_{n-1}(s))|\, ds \right|$$

$$\leqslant M \left| \int_{t_0}^{t} ds \right| = M\,|t - t_0| \leqslant M\delta \leqslant b.$$

These inequalities follow from statement (F). Note that the last inequality helps explain the choice of δ in (11.7).

2. Next, we show by induction that

$$|x_n(t) - x_{n-1}(t)| \leqslant Mk^{n-1} \frac{|t - t_0|^n}{n!} \leqslant \frac{Mk^{n-1}\delta^n}{n!}. \tag{11.9}$$

If $n = 1$, we obtain

$$\left|x_1(t) - x_0(t)\right| \leq \left|\int_{t_0}^t f(s, x_0(s))\, ds\right| \leq M \left|\int_{t_0}^t ds\right|$$

$$= M \left|t - t_0\right| \leq M\delta.$$

Thus the result is true for $n = 1$.

We assume that the result is true for $n = m$ and prove that it holds for $n = m + 1$. That is, we assume that

$$\left|x_m(t) - x_{m-1}(t)\right| \leq \frac{Mk^{m-1}\left|t - t_0\right|^m}{m!} \leq \frac{Mk^{m-1}\delta^m}{m!}.$$

Then, since $f(t, x)$ is Lipschitz continuous in x over D,

$$\left|x_{m+1}(t) - x_m(t)\right| = \left|\int_{t_0}^t f(s, x_m(s))\, ds - \int_{t_0}^t f(s, x_{m-1}(s))\, ds\right|$$

$$\leq \left|\int_{t_0}^t \left|f(s, x_m(s)) - f(s, x_{m-1}(s))\right|\, ds\right|$$

$$\leq k \left|\int_{t_0}^t \left|x_m(s) - x_{m-1}(s)\right|\, ds\right|$$

$$\leq \frac{Mk^m}{m!} \left|\int_{t_0}^t (s - t_0)^m\, ds\right|^* = \frac{Mk^m\left|t - t_0\right|^{m+1}}{(m+1)!} \leq \frac{Mk^m\delta^{m+1}}{(m+1)!},$$

which is what we wanted to show.

3. We will now show that $x_n(t)$ converges uniformly to a limit function $x(t)$ on the interval $\left|t - t_0\right| \leq \delta$. By statement (D), this will show that $x(t)$ is continuous. We first note that

$$x_n(t) - x_0(t) = x_n(t) - x_{n-1}(t) + x_{n-1}(t) - x_{n-2}(t) + \cdots + x_1(t) - x_0(t)$$

$$= \sum_{k=0}^n \left[x_m(t) - x_{m-1}(t)\right]. \tag{11.10}$$

But by (11.9),

$$\left|x_m(t) - x_{m-1}(t)\right| \leq \frac{Mk^{m-1}\delta^m}{m!} = \frac{M}{k}\frac{k^m\delta^m}{m!},$$

---

* This inequality follows from the induction assumption that (11.9) holds for $n = m$.

so that

$$\sum_{m=1}^{\infty} |x_m(t) - x_{m-1}(t)| \leq \frac{M}{k} \sum_{m=1}^{\infty} \frac{(k\delta)^m}{m!} = \frac{M}{k}(e^{k\delta} - 1),$$

by Eq. (5.16) since

$$e^{k\delta} = \sum_{m=0}^{\infty} \frac{(k\delta)^m}{m!} = 1 + \sum_{m=1}^{\infty} \frac{(k\delta)^m}{m!}.$$

By the Weierstrass M-test [statement (G)], we conclude that the series

$$\sum_{m=1}^{\infty} [x_m(t) - x_{m-1}(t)]$$

converges absolutely and uniformly on $|t - t_0| \leq \delta$ to a unique limit function $y(t)$. But

$$y(t) = \lim_{n \to \infty} \sum_{m=1}^{n} [x_m(t) - x_{m-1}(t)]$$

$$= \lim_{n \to \infty} [x_n(t) - x_0(t)] = \lim_{n \to \infty} x_n(t) - x_0(t)$$

or

$$\lim_{n \to \infty} x_n(t) = y(t) + x_0(t).$$

We denote the right-hand side of this equation by $x(t)$. Thus the limit of the Picard iterations $x_n(t)$ exists and the convergence $x_n(t) \to x(t)$ is uniform for all t in the interval $|t - t_0| \leq \delta$.

4. It remains to be shown that $x(t)$ is a solution to (11.2) for $|t - t_0| < \delta$. Since $f(t,x)$ is a continuous function of x and $x_n(t) \to x(t)$ as $n \to \infty$, we have by statement (E)

$$\lim_{n \to \infty} f(t,x_n(t)) = f(t,x(t)).$$

Hence by (11.5),

$$x(t) = \lim_{n \to \infty} x_{n+1}(t) = x_0 + \lim_{n \to \infty} \int_{t_0}^{t} f(s,x_n(s)) \, ds$$

$$= x_0 + \int_{t_0}^{t} \lim_{n \to \infty} f(s,x_n(s)) \, ds = x_0 + \int_{t_0}^{t} f(s,x(s)) \, ds.$$

The step where we interchange the limit and integral is justified by statement (H). Thus $x(t)$ solves (11.2) and therefore it solves the initial value problem (11.1).

It turns out that the solution obtained in Theorem 11.3 is unique. Before proving this, however, we shall derive a simple version of a very useful result known as *Gronwall's inequality*.

Theorem 11.4. Gronwall's Inequality Let $x(t)$ be a continuous nonnegative function and suppose that

$$x(t) \leqslant A + B \left| \int_{t_0}^{t} x(s) \, ds \right|, \tag{11.11}$$

where A and B are positive constants, for all values $|t - t_0| \leqslant \delta$. Then

$$x(t) \leqslant A e^{B|t - t_0|} \tag{11.12}$$

for all t in the interval $|t - t_0| \leqslant \delta$.

Proof. We shall prove this result for $t_0 \leqslant t \leqslant t_0 + \delta$. The proof for $t_0 - \delta \leqslant t \leqslant t_0$ is similar (see Exercise 14). We define

$$y(t) = B \int_{t_0}^{t} x(s) \, ds.$$

Then

$$y'(t) = Bx(t) \leqslant B \left[A + B \int_{t_0}^{t} x(s) \, ds \right] = AB + By$$

or

$$y'(t) - By(t) \leqslant AB. \tag{11.13}$$

We note that

$$\frac{d}{dt} \left[y(t) e^{-B(t-t_0)} \right] = e^{-B(t-t_0)} \left[y'(t) - By(t) \right].$$

Therefore, multiplying both sides of (11.13) by the integrating factor $e^{-B(t-t_0)}$ (which is greater than zero), we have

$$\frac{d}{dt} \left[y(t) e^{-B(t-t_0)} \right] \leqslant AB e^{-B(t-t_0)}.$$

An integration of both sides of the inequality from t_0 to t yields

$$y(s) e^{-B(s-t_0)} \Big|_{t_0}^{t} \leqslant AB \int_{t_0}^{t} e^{-B(s-t_0)} \, ds = -A e^{-B(s-t_0)} \Big|_{t_0}^{t}.$$

But $y(t_0) = 0$, so that

$$y(t) e^{-B(t-t_0)} \leqslant A(1 - e^{-B(t-t_0)}),$$

from which, after multiplying both sides by $e^{B(t-t_0)}$, we obtain

$$y(t) \leqslant A[e^{B(t-t_0)} - 1].$$

Then by (11.11),

$$x(t) \leqslant A + y(t) \leqslant A e^{B(t-t_0)},$$

and the theorem is proved.

Theorem 11.5. Uniqueness Theorem Let the conditions of Theorem 11.3 (existence theorem) hold. Then $x(t) = \lim_{n \to \infty} x_n(t)$ is the only continuous solution of the initial value problem (11.1) in $|t - t_0| \leq \delta$.

Proof. Let $x(t)$ and $y(t)$ be two continuous solutions of (11.2) in the interval $|t - t_0| \leq \delta$ and suppose that $(t, y(t))$ belongs to the region D for all t in that interval.* Define $v(t) = |x(t) - y(t)|$. Then $v(t) \geq 0$ and $v(t)$ is continuous. Since $f(t,x)$ is Lipschitz continuous in x over D,

$$v(t) = \left\| \left[x_0 + \int_{t_0}^{t} f(s,x(s))\, ds \right] - \left[x_0 + \int_{t_0}^{t} f(s,y(s))\, ds \right] \right\|$$

$$\leq k \left\| \int_{t_0}^{t} |x(s) - y(s)|\, ds \right\| = k \left\| \int_{t_0}^{t} v(s)\, ds \right\|$$

$$\leq \varepsilon + k \left\| \int_{t_0}^{t} v(s)\, ds \right\|$$

for every $\varepsilon > 0$. By Gronwall's inequality, we have

$$v(t) \leq \varepsilon\, e^{k|t - t_0|}.$$

But $\varepsilon > 0$ can be chosen arbitrarily close to zero, so that $v(t) \leq 0$. Since $v(t) \geq 0$, it follows that $v(t) \equiv 0$, implying that $x(t)$ and $y(t)$ are identical. Hence the limit of the Picard iterations is the only continuous solution.

Theorem 11.6   Let $f(t,x)$ and $\dfrac{\partial f}{\partial x}(t,x)$ be continuous on D. Then there exists a constant $\delta > 0$ such that the Picard iteration $\{x_n(t)\}$ converge to a unique continuous solution of the initial value problem (11.1) on $|t - t_0| \leq \delta$.

Proof. This theorem follows directly from Theorem 11.2 and the existence and uniqueness theorems.

We note that Theorems 11.3, 11.5, and 11.6 are *local* results. By this we mean that unique solutions are guaranteed to exist only "near" the initial point (t_0, x_0).

Example 11.1.4 Let

$$x'(t) = x^2(t), \qquad x(1) = 2.$$

Without solving this equation, we can show that there is a unique solution in some interval $|t - t_0| = |t - 1| \leq \delta$. Let $a = b = 1$. Then $|f(t,x)| = x^2 \leq 9\ (= M)$ for all $|x - x_0| \leq 1$, $x_0 = x(1) = 2$. Therefore, $\delta = \min\{a, b/M\} = \frac{1}{9}$, and Theorem 6 guarantees the existence of a unique solution on the interval $|t - 1| \leq \frac{1}{9}$. The solution of this initial value problem is easily found by a separation of variables

---

* Note that without this assumption, the function $f(t, y(t))$ may not even be defined at points where $(t, y(t))$ is not in D.

to be $x(t) = 2/(3 - 2t)$. This solution exists so long as $t \neq \frac{3}{2}$. Starting at $t_0 = 1$, we see that the maximum interval of existences is $|t - t_0| < \frac{1}{2}$. Hence the value $\delta = \frac{1}{9}$ is not the best possible.

Example 11.1.5 Consider the initial value problem

$$x' = \sqrt{x}, \qquad x(0) = 0.$$

As we saw in Example 11.1.1, $f(t,x) = \sqrt{x}$ does *not* satisfy a Lipschitz condition in any region containing the point (0,0). By a separation of variables, it is easy to calculate the solution

$$x(t) = \left(\frac{t}{2}\right)^2.$$

However $y(t) = 0$ is also a solution. Hence without a Lipschitz condition, the solution to an initial value problem (if one exists) may fail to be unique.

The last two examples illustrate the local nature of our existence-uniqueness theorem. We shall show in Section 11.2 that it is possible to derive *global* existence-uniqueness results for certain linear differential equations. That is, we shall show that a unique solution exists for every real number t (see Exercise 17).

EXERCISES 11.1

For each initial value problem of Exercises 1 through 10, determine whether a unique solution can be guaranteed. If so, let $a = b = 1$ if possible, and find the number δ as given by Eq. (11.7). When possible, solve the equation and find a better value for δ, as in Example 11.1.4.

1. $x' = x^3$, $x(2) = 5$

2. $x' = x^3$, $x(1) = 2$.

3. $x' = \dfrac{x}{t - x}$, $x(0) = 1$

4. $x' = x^{1/3}$, $x(1) = 0$

5. $x' = \sin x$, $x(1) = \pi/2$

6. $x' = \sqrt{x(x - 1)}$, $x(1) = 2$

7. $x' = \ln \sin x$, $x(\pi/2) = 1$

8. $x' = \sqrt{x(x - 1)}$, $x(2) = 3$

9. $x' = |x|$, $x(0) = 1$

10. $x' = tx$, $x(5) = 10$

11. Compute a Lipschitz constant for each of the following functions on the indicated region D:

a) $f(t,x) = te^{-2x/t}$, $t > 0$, $x > 0$

b) $f(t,x) = \sin tx$, $|t| \leqslant 1$, $|x| \leqslant 2$

c) $f(t,x) = e^{-t^2}x^2 \sin \dfrac{1}{x}$, all t, $-1 \leqslant x < 1$, $x \neq 0$

(This exercise shows that a Lipschitz constant may exist even when $\partial f/\partial x$ is not bounded in D.)

d) $f(t,x) = (t^2 x^3)^{3/2}$, $|t| \leqslant 2$, $|x| \leqslant 3$

12. Consider the initial value problem

    $$x' = x^2, \qquad x(1) = 3.$$

    Show that the Picard iterations converge to the unique solution of this problem.

13. Construct the sequence $\{x_n(t)\}$ of Picard iterations for the initial value problem

    $$x' = -x, \qquad x(0) = 3,$$

    and show that it converges to the unique solution $x(t) = 3e^{-t}$.

14. Prove Gronwall's inequality (Theorem 11.4) for $t_0 - \delta \leqslant t \leqslant t_0$. [*Hint:* For $t < t_0$, $y(t) \leqslant 0$.]

15. Let $v(t)$ be a positive function that satisfies the inequality

    $$v(t) \leqslant A + \int_{t_0}^{t} r(s)v(s)\, ds, \tag{11.14}$$

    where $A \geqslant 0$, $r(t)$ is a continuous, positive function, and $t \geqslant t_0$. Prove that

    $$v(t) \leqslant A \exp\left[\int_{t_0}^{t} r(s)\, ds\right] \tag{11.15}$$

    for all $t \geqslant t_0$. What kind of result holds for $t \leqslant t_0$? This is a more general form of Gronwall's inequality. [*Hint:* Define $y(t) \equiv \int_{t_0}^{t} r(s)v(s)\, ds$ and show that $y'(t) = r(t)v(t) \leqslant r(t)[A + y(t)]$. Finish the proof by following the steps of the proof of Theorem 11.4, using the integrating factor $\exp\left[-\int_{t_0}^{t} r(s)\, ds\right]$.]

16. Consider the initial value problem

    $$x'' = f(t,x), \qquad x(0) = x_0, \qquad x'(0) = x_1, \tag{11.16}$$

    where f is defined on the rectangle $D: |t| \leqslant a, |x - x_0| \leqslant b$. Prove under appropriate hypotheses that if a solution exists, then it must be unique. [*Hint:* Let $x(t)$ and $y(t)$ be continuous solutions of (11.16). Verify by differentiation that

    $$x(t) = x_0 + x_1 t - \int_0^t (t - s)f(s,x(s))\, ds,$$

    $$y(t) = x_0 + x_1 t - \int_0^t (t - s)f(s,y(s))\, ds$$

    in some interval $|t| \leqslant \delta, \delta > 0$. Then subtract these two expressions, use an appropriate Lipzchitz condition, and apply the Gronwall inequality of Exercise 15.]

17. Consider the first order linear problem

    $$x'(t) = a(t)x + b(t), \qquad x(t_0) = x_0, \tag{11.17}$$

    where $a(t)$ and $b(t)$ are continuous in the interval and $a \leqslant t_0 \leqslant b$.

    a) Using $f(t,x) = a(t)x + b(t)$, show that the Picard iterations $\{x_n(t)\}$ exist and are continuous for all t in the interval $a \leqslant t \leqslant b$.

    b) Show that $f(t,x)$ defined in part (a) is Lipschitz continuous in x over the region D of points (t,x) satisfying the conditions $a \leqslant t \leqslant b, |x| < \infty$.

    *c) Modify the proofs of Theorems 11.3 and 11.5 to show that (11.17) has a unique solution in the *entire* interval $[a,b]$.

d) Show that if $a(t)$ and $b(t)$ are continuous for all values t, then (11.17) has a unique solution which is defined for *all* values $-\infty < t < \infty$. (This is a *global* existence-uniqueness theorem.)

18. Consider the initial value problem

$$x' = f(t,x), \qquad x(t_0) = x_0. \tag{11.18}$$

Show that if $f(t,x)$ satisfies the conditions of Theorem 11.3, then the solution $x(t) \equiv x(t;x_0)$ depends continuously on the value of x_0. Thus a small change in x_0 produces a small change in the solution over the interval $|t - t_0| \leq \delta$. This result is often termed *continuous dependence on initial conditions*. [*Hint*: Define $x_0(t)$ to be the unique solution of (11.18) and $x_1(t)$ to be the unique solution of

$$x' = f(t,x), \qquad x(t_0) = x_1. \tag{11.19}$$

Rewrite the initial value problems (11.18) and (11.19) as integrals (11.2) and subtract one from the other to obtain

$$\left|x_1(t) - x_0(t)\right| \leq |x_1 - x_0| + \left|\int_{t_0}^{t} |f(s,x_1(s)) - f(s,x_0(s))| \, ds\right|.$$

Then apply the Lipschitz condition to show that

$$\left|x_1(t) - x_0(t)\right| \leq |x_1 - x_0| + k\left|\int_{t_0}^{t} |x_1(s) - x_0(s)| \, ds\right|.$$

Finally, apply Gronwall's inequality (Theorem 11.4) to show that

$$\left|x_1(t) - x_0(t)\right| \leq |x_1 - x_0|e^{k|t - t_0|} \leq |x_1 - x_0|e^{k\delta}$$

for $|t - t_0| \leq \delta$, from which the desired result is immediately obtained.]

---

*11.2 SYSTEMS OF LINEAR DIFFERENTIAL EQUATIONS

In this section we shall extend the local results of Section 11.1 in two ways. First, we will show that, with little modification, the existence and uniqueness theorems 11.3 and 11.5 are true when $\mathbf{x}(t)$ and $\mathbf{f}(t,\mathbf{x}(t))$ are vectors. Then we will show that *global* existence and uniqueness results can be proved as easy corollaries when the system is linear (see Exercise 17 of Section 11.1).

We consider the initial value system

$$
\begin{aligned}
x_1' &= f_1(t,x_1,x_2,\ldots,x_n), & x_1(t_0) &= x_{10}, \\
x_2' &= f_2(t,x_1,x_2,\ldots,x_n), & x_2(t_0) &= x_{20}, \\
&\ \ \vdots \\
x_n' &= f_n(t,x_1,x_2,\ldots,x_n), & x_n(t_0) &= x_{n0}.
\end{aligned}
\tag{11.20}
$$

As in Chapter 7, we write the vectors

$$\mathbf{x} = \begin{bmatrix} x_1 \\ \cdot \\ \cdot \\ \cdot \\ x_n \end{bmatrix}, \qquad \mathbf{f}(t,\mathbf{x}) = \begin{bmatrix} f_1(t,x_1, \ldots, x_n) \\ \cdot \\ \cdot \\ \cdot \\ f_n(t,x_1, \ldots, x_n) \end{bmatrix}, \qquad \mathbf{x}_0 = \begin{bmatrix} x_{10} \\ \cdot \\ \cdot \\ \cdot \\ x_{n0} \end{bmatrix}.$$

Then we may write (11.20) in the compact form

$$\mathbf{x}' = \mathbf{f}(t,\mathbf{x}), \qquad \mathbf{x}(t_0) = \mathbf{x}_0. \tag{11.21}$$

In order to generalize the results of Section 11.1, it is necessary to generalize the notion of the absolute value of a number to the absolute value or *norm* of a vector $\mathbf{x}$ or matrix A. If $\mathbf{x} = (x_1,x_2)$ is a two-vector, then the distance from (x_1,x_2) to the origin is given by the Pythagorean theorem (see Fig. 11.4):

$$\|\mathbf{x}\| = \sqrt{(x_1^2 + x_2^2)}. \tag{11.22}$$

Therefore, it is natural to define the length or *norm* of an n-vector $\mathbf{x} = (x_1,x_2,\ldots,x_n)$, by

$$\|\mathbf{x}\| = \sqrt{(x_1^2 + x_2^2 + \cdots + x_n^2)}. \tag{11.23}$$

If A is an $n \times n$ matrix, there are several ways to define its norm. For our purposes,

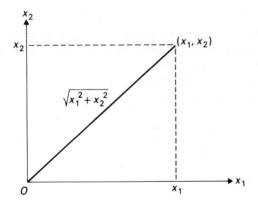

Figure 11.4

the simplest choice for the norm of A is

$$\|A\| = \sum_{i=1}^{n} \sum_{j=1}^{n} |a_{ij}|. \tag{11.24}$$

That is, the norm of A is the sum of the absolute values of the components of A. Using (11.23) and (11.24), we may prove that:

i) $\|\mathbf{x}\| \geqslant 0$, where the equality holds only if $\mathbf{x}$ is the zero vector. $\tag{11.25}$

ii) $\|\alpha\mathbf{x}\| = |\alpha| \, \|\mathbf{x}\|$ when α is a scalar. $\tag{11.26}$

iii) $\|\mathbf{x} + \mathbf{y}\| \leqslant \|\mathbf{x}\| + \|\mathbf{y}\|$. $\tag{11.27}$

Moreover, if A is an $n \times n$ matrix, we can show (see Exercises 1 through 4) that

$$\|A\mathbf{x}\| \leqslant \|A\| \, \|\mathbf{x}\|. \tag{11.28}$$

Now using this notation and a norm $\| \cdot \|$ for every absolute value $| \cdot |$ that appeared in the proofs of Theorems 11.3 and 11.5, we obtain the existence and uniqueness of a local vector solution $\mathbf{x}(t)$ of the initial value problem (11.21). The verification of these details is left as an exercise (see Exercise 8). We therefore have:

Theorem 11.7 Let D denote the region [in $(n + 1)$-dimensional space, one dimension for t and n dimensions for the vector $\mathbf{x}$]

$$|t - t_0| \leqslant a, \qquad \|\mathbf{x} - \mathbf{x}_0\| \leqslant b, \tag{11.29}$$

and suppose that $\mathbf{f}(t,\mathbf{x})$ satisfies the Lipschitz condition

$$\|\mathbf{f}(t,\mathbf{x}_1) - \mathbf{f}(t,\mathbf{x}_2)\| \leqslant k \, \|\mathbf{x}_1 - \mathbf{x}_2\| \tag{11.30}$$

whenever the pairs $(t,\mathbf{x}_1)$ and $(t,\mathbf{x}_2)$ belong to D, where k is a positive constant. Then there is a constant $\delta > 0$ such that there exists a unique continuous vector solution $\mathbf{x}(t)$ of the system (11.21) in the interval $|t - t_0| \leqslant \delta$.

Before leaving this theorem, we should briefly discuss the condition (11.30). It is easy to see that (11.30) is implied by the inequality

$$|f_i(t,x_{11}, \ldots, x_{1n}) - f_i(t,x_{21}, \ldots, x_{2n})| \leqslant k_1 \sum_{j=1}^{n} |x_{1j} - x_{2j}| \tag{11.31}$$

for some number k_1. This fact follows from the double inequality

$$\frac{1}{n} \sum_{j=1}^{n} |x_j| \leqslant \|\mathbf{x}\| \leqslant \sum_{j=1}^{n} |x_j|, \tag{11.32}$$

which is an immediate consequence of the definition of $\|\mathbf{x}\|$ (see Exercise 5).

Another condition that implies (11.30) is

$$|f_i(t,x_{11}, \ldots, x_{1n}) - f_i(t,x_{21}, \ldots, x_{2n})| \leqslant k_2 \max_j |x_{1j} - x_{2j}| \tag{11.33}$$

for $i = 1, 2, \ldots, n$. The two inequalities (11.31) and (11.33) are useful since it is often very difficult to verify inequality (11.30) directly.

Finally, if the partial derivatives

$$\frac{\partial f_i}{\partial x_j}, \quad i, j = 1, 2, \ldots, n,$$

are continuous in D, then they are bounded on D and conditions (11.31) and (11.33) both follow from the mean value theorem of differential calculus (see Exercise 7).

We now turn to the main purpose of this section which is to show the global existence and uniqueness of solutions of linear systems with continuous coefficients. Consider the system

$$\mathbf{x}' = A(t)\mathbf{x} + \mathbf{f}(t), \quad \mathbf{x}(t_0) = \mathbf{x}_0, \tag{11.34}$$

where t_0 is a point in the interval $\alpha \leqslant t \leqslant \beta$ and $A(t)$ is an $n \times n$ matrix.

Theorem 11.8 (Theorem 7.13) Let $A(t)$ and $\mathbf{f}(t)$ be a continuous matrix and vector function, respectively, on the interval $\alpha \leqslant t \leqslant \beta$. Then there exists a unique vector function $\mathbf{x}(t)$ which is a solution of (11.34) on the entire interval $\alpha \leqslant t \leqslant \beta$.

Proof. We define the Picard iterations

$$\mathbf{x}_0(t) = \mathbf{x}_0,$$

$$\mathbf{x}_1(t) = \mathbf{x}_0 + \int_{t_0}^{t} [A(s)\mathbf{x}_0(s) + \mathbf{f}(s)] \, ds,$$

.
.
.

$$\tag{11.35}$$

$$\mathbf{x}_{n+1}(t) = \mathbf{x}_0 + \int_{t_0}^{t} [A(s)\mathbf{x}_n(s) + \mathbf{f}(s)] \, ds,$$

.
.
.

Clearly these iterations are continuous on $\alpha \leqslant t \leqslant \beta$ for $n = 1, 2, \ldots$. Since $A(t)$ is a continuous matrix function, we have

$$\sup_{\alpha \leqslant t \leqslant \beta} \|A(t)\| = \sup_{\alpha \leqslant t \leqslant \beta} \sum_{i=1}^{n} \sum_{j=1}^{n} |a_{ij}(t)| = k < \infty.$$

Let

$$M = \sup_{\alpha \leqslant t \leqslant \beta} \|A(t)\mathbf{x}_0 + \mathbf{f}(t)\|,$$

which is finite [according to statement (A) of Section 11.1] since $A(t)$ and $\mathbf{f}(t)$ are continuous. Observe that we are unable to define the value δ of Eq. (11.7), since the vector functions $\mathbf{x}(t)$ are *unrestricted*. Thus a slight modification of the proof of Theorem 11.3 is necessary:

i) Note that for any vectors $\mathbf{x}_1$ and $\mathbf{x}_2$ and for $\alpha \leqslant t \leqslant \beta$, we have

$$\|[A(t)\mathbf{x}_1 + \mathbf{f}(t)] - [A(t)\mathbf{x}_2 + \mathbf{f}(t)]\| \leqslant \|A(t)\|\,\|\mathbf{x}_1 - \mathbf{x}_2\| \leqslant k\,\|\mathbf{x}_1 - \mathbf{x}_2\|.$$

Thus the vector function $A(t)\mathbf{x} + \mathbf{f}$ satisfies a Lipschitz condition for any $\mathbf{x}$ and $\alpha \leqslant t \leqslant \beta$.

ii) By induction, we now show that the Picard iterations (11.35) satisfy the inequality

$$\|\mathbf{x}_n(t) - \mathbf{x}_{n-1}(t)\| \leqslant Mk^{n-1}\frac{|t - t_0|^n}{n!} \leqslant Mk^{n-1}\frac{(\beta - \alpha)^n}{n!}. \tag{11.36}$$

We shall prove the first inequality. The second follows easily. First, if $n = 1$, then

$$\|\mathbf{x}_1(t) - \mathbf{x}_0(t)\| = \left\|\int_{t_0}^t [A(s)\mathbf{x}_0 + \mathbf{f}(s)]\,ds\right\|$$

$$\leqslant \left|\int_{t_0}^t \|A(s)\mathbf{x}_0 + \mathbf{f}(s)\|\,ds\right|$$

$$\leqslant M\left|\int_{t_0}^t ds\right| = M\,|t - t_0| \leqslant M(\beta - \alpha).*$$

Assume that the result holds for $n = m$. Then by part (i),

$$\|\mathbf{x}_{m+1}(t) - \mathbf{x}_m(t)\| \leqslant \left|\int_{t_0}^t \|[A(s)\mathbf{x}_m(s) + \mathbf{f}(s)] - [A(s)\mathbf{x}_{m-1}(s) + \mathbf{f}(s)]\|\,ds\right|$$

$$\leqslant k\left|\int_{t_0}^t \|\mathbf{x}_m(s) - \mathbf{x}_{m-1}(s)\|\,ds\right| \leqslant k\left|\int_{t_0}^t Mk^{m-1}\frac{|s - t_0|^m}{m!}\,ds\right|$$

$$= Mk^m\frac{|t - t_0|^{m+1}}{(m + 1)!} \leqslant Mk^m\frac{(\beta - \alpha)^{m+1}}{(m + 1)!},$$

which is the desired result.

The major difference between this proof and the proof of Theorem 11.3 is that, in the linear case, the bound M can be defined independently of x, whereas in Theorem 11.3 we must restrict x as well as t. This difference enables us to prove the global result that solutions exist and are unique on the entire interval $\alpha \leqslant t \leqslant \beta$.

---

* The first of these inequalities is called the Cauchy-Schwartz inequality. Its proof may be found in any advanced calculus text. See, for example, R. Buck, *Advanced Calculus*, McGraw-Hill, New York, 1956.

With the estimate (11.36) [which is identical to the estimate (11.9)] the proof proceeds exactly as those for Theorems 11.3 and 11.5. Note that Theorems 3.1 and 7.2 are special cases of Theorem 11.8.

Corollary 11.9 Let $A(t)$ and $\mathbf{f}(t)$ be continuous on the interval $-\infty < t < \infty$. Then there exists a unique continuous solution $\mathbf{x}(t)$ of (11.34) defined for *all* t on $-\infty < t < \infty$.

Proof. Suppose $|t_0| \leqslant n$. We define $\mathbf{x}_n(t)$ to be the unique solution of (11.34) on the interval $|t| \leqslant n$ which is guaranteed by Theorem 11.8. We observe that $\mathbf{x}_n(t)$ coincides with $\mathbf{x}_{n+k}(t)$ on the interval $|t| \leqslant n$ for $k = 1, 2, 3, \ldots$, since both are solutions of (11.34), and Theorem 11.8 requires any solution on $|t| \leqslant n$ to be unique. Thus the vector function

$$\mathbf{x}(t) = \lim_{n \to \infty} \mathbf{x}_n(t)$$

is defined for all real values t, since $|t| \leqslant n$ for sufficiently large n. Clearly $\mathbf{x}(t)$ is a solution of (11.34) on $-\infty < t < \infty$, and such a solution must be unique since it is uniquely defined on each finite interval containing t_0.

EXERCISES 11.2

1. Prove the inequality (11.25).

2. Prove Eq. (11.26).

* 3. Prove the inequality (11.27). [*Hint:* (a) First show that the *inner product*

   $$\mathbf{x} \cdot \mathbf{y} = x_1 y_1 + x_2 y_2 + \cdots + x_n y_n$$

   satisfies the condition $|\mathbf{x} \cdot \mathbf{y}| \leqslant \|\mathbf{x}\| \, \|\mathbf{y}\|$, where $\mathbf{x} = (x_1, x_2, \ldots, x_n)$ and $\mathbf{y} = (y_1, y_2, \ldots, y_n)$. (b) Show that $\mathbf{x} \cdot \mathbf{x} = \|\mathbf{x}\|^2$. (c) Apply (b) to $\|\mathbf{x} + \mathbf{y}\|^2$ and use the inequality in (a) to verify (11.27).]

* 4. Prove the inequality (11.28).

5. Prove the inequality (11.32).

6. Using (11.32), show that the inequalities (11.30), (11.31), and (11.33) are all equivalent.

7. Show that if the partial derivatives $\partial f_i / \partial x_j$ are continuous on the region D defined by (11.29), then there exists a constant $k_1 > 0$ such that the inequality (11.30) holds for $i = 1, \ldots, n$.

* 8. Prove Theorem 11.7 by carrying out the following steps:

   a) Show that the system (11.21) is equivalent to the integral equation

   $$\mathbf{x}(t) = \mathbf{x}_0 + \int_{t_0}^{t} \mathbf{f}(s, \mathbf{x}(s)) \, ds. \tag{11.37}$$

   b) Define the vector valued Picard iterates $\{\mathbf{x}_n\}$ as in (11.5).

c) Prove that

$$M = \sup_{(t,x) \in D} \|f(t,x)\|$$

is finite.

d) Show that if $\delta = \min \{a, b/M\}$, then the Picard iterates defined in (b) satisfy the condition $\|x_n(t) - x_0\| \leq b$, for $|t - t_0| \leq a$.

e) Show by induction that

$$\|x_n(t) - x_{n-1}(t)\| \leq Mk^{n-1} \frac{|t - t_0|^n}{n!} \leq Mk^{n-1} \frac{\delta^n}{n!}.$$

f) Prove that the series

$$\sum_{m=1}^{\alpha} \|x_m(t) - x_{m-1}(t)\|$$

converges uniformly for $|t - t_0| < \delta$ and conclude that $x_n(t)$ converges uniformly to a vector function $x(t)$ as n tends to ∞.

g) Show that the $x(t)$ defined in part (f) is a continuous solution to (11.37) on the interval $|t - t_0| \leq \delta$.

h) Use Gronwall's inequality to show that this solution is unique.

9. Consider the system

$$x_1' = tx_1 x_2 + t^2, \qquad x_1(0) = 1,$$
$$x_2' = x_1^2 + x_2^2 + t, \qquad x_2(0) = 2.$$

a) Write the system in the form (11.21).

b) Find a Lipschitz constant for f in the region

$$D: |t| \leq 1, (x_1 - 1)^2 + (x_2 - 2)^2 \leq 1.$$

c) Find $\delta = \min \{a, b/M\}$ as in Exercise 8.

10. Follow the steps of Exercise 9 for the system

$$x_1' = x_2^2 + 1, \qquad x_1(1) = 0,$$
$$x_2' = x_1^2 + t, \qquad x_2(1) = 1,$$

in the region $D: |t - 1| \leq 3, x_1^2 + (x_2 - 1)^2 \leq 4$.

11. Follow the steps of Exercise 9 for the system

$$x_1' = x_1^2 x_2,$$
$$x_2' = x_3 + t, \qquad x(0) = c,$$
$$x_3' = x_3^2,$$

in the region $D: |t| \leq a, \|x - c\| \leq b$.

12. Under the assumption of Theorem 11.7 consider the systems

$$x' = f(t,x), \qquad x(t_0) = x_{10},$$
$$x' = f(t,x), \qquad x(t_0) = x_{20}.$$

Let x_1 and x_2 be solutions of these two systems, respectively. Prove that

$$\|x_1(t) - x_2(t)\| \leqslant \|x_{10} - x_{20}\| e^{k\delta}.$$

This shows that the solutions of Theorem 11.7 vary continuously with the initial vector $x(t_0)$. [*Hint*: See Exercise 18 of Section 11.1]

* 13.   Consider the system

$$x' = f(t,x). \tag{11.38}$$

Let D denote the region $a \leqslant t \leqslant b$, $x \in R^n$, and assume that

$$\|f(t,x_1) - f(t,x_2)\| \leqslant k \|x_1 - x_2\|,$$

for any pair of points (t,x_1) and (t,x_2) in D. Prove that for any pair (t_0,x_0) in D there exists a unique solution $x(t)$ of Eq. (11.38) defined on the entire interval $a \leqslant t \leqslant b$, which satisfies $x(t_0) = x_0$. [*Hint*: Proceed as in the proof of Theorem 11.8. This is a more general global existence-uniqueness result.]

NONLINEAR
EQUATIONS
AND
STABILITY

12

12.1 INTRODUCTION

In Chapter 11 we saw that there is a large class of differential equations and systems having solutions defined in some interval. However, if an equation is nonlinear, then usually there is not any way to find this solution. For this reason, it is necessary to seek methods for describing the nature of a solution without explicitly solving the equation.

First, it is necessary to ask: what kind of information about a solution is it useful to have? We indicate a partial answer by considering our old standby: second order linear equations with constant coefficients.

Example 12.1.1 Consider the three equations:

a) $x'' + 3x' + 2x = 0$,

b) $x'' - 3x' + 2x = 0$,

c) $x'' + x = 0$.

The general solutions to these equations are

a) $x(t) = c_1 e^{-t} + c_2 e^{-2t}$,

b) $x(t) = c_1 e^t + c_2 e^{2t}$,

c) $x(t) = c_1 \cos t + c_2 \sin t$.

It is clear that all solutions of (a) approach zero as t tends to ∞, all solutions of (b) approach ∞ as t tends to ∞, and that all solutions of (c) remain bounded but do not approach any constant as t tends to ∞. Furthermore, the solutions of (c) are periodic of period 2π.

Nonlinear equations, too, may approach zero, become unbounded, or remain bounded as t becomes large. They also may be periodic. It is fair to say that a major portion of modern research in the theory of ordinary differential equations is concerned with finding conditions that will ensure that the solution of a nonlinear equation have one of these properties. There is no general method for analyzing *all* nonlinear equations. In this chapter we will discuss some of the oldest known and most elementary ways of obtaining this information.

One of these methods is to consider the nonlinear equation as a *perturbation* of some linear equation; that is, attempt to approximate a nonlinear equation by a "related" linear equation. We shall illustrate this method with some examples.

Example 12.1.2 Consider the freely swinging (frictionless) pendulum of length l depicted in Fig. 12.1. In Exercise 3 of Section 7.7 we indicated that Newton's law of motion yields the nonlinear second order equation

$$\frac{d^2\theta}{dt^2} + \omega^2 \sin \theta = 0, \tag{12.1}$$

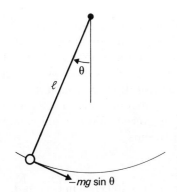

Figure 12.1

where $\omega^2 = g/l$. However, since

$$\lim_{\theta \to 0} \frac{\sin \theta}{\theta} = 1,$$

we may approximate (12.1) for small values of θ by the linear equation

$$\frac{d^2\theta}{dt^2} + \omega^2\theta = 0. \tag{12.2}$$

The general solution of Eq. (12.2) is periodic:

$$\theta(t) = c_1 \cos \omega t + c_2 \sin \omega t.$$

How similar in behavior are the solutions of these two equations? This question will be answered in Section 12.3.

Example 12.1.3 Consider the nonlinear first order scalar equation

$$x' = -x + x^2. \tag{12.3}$$

This equation has two constant solutions,

$$x = 0 \quad \text{and} \quad x = 1,$$

easily verified by substitution into Eq. (12.3). For x close to zero, the nonlinear term x^2 is relatively small compared to the linear term $-x$, since

$$\lim_{x \to 0} \frac{x^2}{x} = 0.$$

Thus we wish to compare the solutions of (12.3) to those of the linear equation

$$x' = -x, \tag{12.4}$$

whose general solution is

$$x(t) = x(0)e^{-t}.$$

The nonlinear equation (12.3) can be solved by a separation of variables:

$$\int \frac{dx}{-x + x^2} = \int dt = t + C.$$

By using partial fractions we have

$$\int \frac{dx}{-x + x^2} = \int \left(\frac{1}{x - 1} - \frac{1}{x} \right) dx = \ln (x - 1) - \ln (x) = \ln \left(\frac{x - 1}{x} \right),$$

which implies that

$$\frac{x - 1}{x} = Ce^t \tag{12.5}$$

for some new constant C.

We may assume that $x(0) \neq 0$, because if $x(0) = 0$, we already have the unique solution $x(t) \equiv 0$. Then, for $t = 0$, Eq. (12.5) yields

$$\frac{x(0) - 1}{x(0)} = C.$$

Thus

$$\frac{x(t) - 1}{x(t)} = \frac{x(0) - 1}{x(0)} e^t$$

or, after some simple algebra,

$$x(t) = \frac{x(0)}{x(0)(1 - e^t) + e^t}. \tag{12.6}$$

This solution is defined so long as the denominator is not zero; that is, so long as

$$x(0)(1 - e^t) + e^t \neq 0$$

or

$$e^t \neq \frac{x(0)}{x(0) - 1} = \frac{1}{C}$$

or

$$t \neq \ln \frac{x(0)}{x(0) - 1} = -\ln C.$$

Note that for small values of t, $1 - e^t$ is close to zero so that the solution $x(t)$, given by (12.6), is close to $x(0)e^{-t}$, which is the solution to the linear equation (12.4).

Now suppose that $0 < x(0) < 1$. Then by (12.5), $C < 0$. Thus $Ce^t \neq 1$ for all $t \geqslant 0$, and the solution of (12.3) [given by (12.6)] approaches zero as t tends to ∞. Hence for $0 < x(0) < 1$, Eq. (12.4) is a "good" approximation to (12.3) in the sense that the solutions to the two equations exhibit the same asymptotic behavior. When $x(0) \geqslant 1$, we are no longer "near" the zero solution, and Eq. (12.4) is not a good approximation. If $x(0) = 1$, we obtain the constant solution $x \equiv 1$. If $x(0) > 1$, then $0 < C < 1$ and $-\ln C > 0$, so that the solution (12.5) approaches ∞ as t tends to $-\ln C$. This situation is illustrated in Fig. 12.2. In the terminology to be introduced in the next section, we may say that the solution $x = 0$ of Eq. (12.3) is *asymptotically stable*, while the solution $x = 1$ is *unstable*.

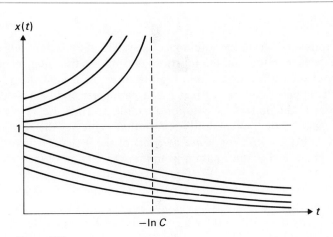

Figure 12.2

In earlier chapters we introduced a terminology for classifying certain types of differential equations. While nonlinear equations and systems can take many different forms, they can be roughly classified into two different categories. To illustrate this, we consider the system of two first order equations

$$x' = f(t,x,y),$$
$$y' = g(t,x,y),$$
(12.7)

where f and g are assumed to be continuously differentiable functions of t, x, and y over some region

$$D: a < t < b, c < x < d, e < y < h$$

in three-dimensional space. By Theorems 11.3 and 11.5, this is enough to guarantee that there is a unique solution (defined over some interval in t) that passes through any initial point (t_0, x_0, y_0) in D.

The system (12.7) is said to be *autonomous* (time independent) if the functions f and g do not depend on t. Otherwise, (12.7) is said to be *nonautonomous*. Hence (12.7) is autonomous if $f(t,x,y) = f(x,y)$ and $g(t,x,y) = g(x,y)$.

Example 12.1.4 The system

$$x' = -x^2 + y,$$
$$y' = -x + y^2$$

is autonomous, while the system

$$x' = ty,$$
$$y' = -x$$

is nonautonomous.

In previous chapters, we wrote the solutions of a system of two equations as a pair $(x(t), y(t))$, where each of the solution functions depends on the independent variable t. A graph of such a solution would require three dimensions for t, x, and y. However, it is often of interest to treat t as a parameter and express the solution function y as a function of the solution function x. Then the xy-plane is called the *phase plane* of the system, and a curve that expresses the relation between x and y is called an *orbit* (or *phase portrait*) of the system. It is often possible to derive a great deal of information from an examination of the orbits of the system.

For example, let $x(t)$ and $y(t)$ denote the respective populations of two species competing for the same limited resources. A model for the interaction between the two species is given by the Lotka-Volterra equations (see Example 7.1.4)

$$x' = \beta_1 x - \delta_{11} x^2 - \delta_{12} xy,$$
$$y' = \beta_2 y - \delta_{21} xy - \delta_{22} y^2.$$

We shall examine this system more closely in Section 12.3. At this point, we merely remark that an orbit of this system would give us very useful information. Increases or decreases in the population of one of the two species will affect the population of the other species. An orbit will graphically depict this effect. Consider, for example, the orbit shown in Fig. 12.3. We can derive a great deal of information from such an orbit. When $x = a$ units, an increase in x will cause a decrease in y until $x = b$ and $y = d$. Then both x and y will increase until x reaches its maximum sustainable population of c units. The second population will continue to rise until it reaches a population of e units, and so on. Moreover, both populations are periodic, that is, all population levels recur again and again. Note that all this

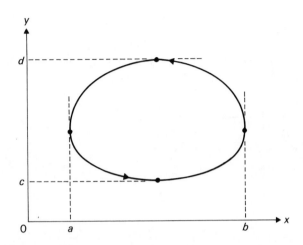

Figure 12.3

information can be obtained (if the orbits can be drawn) *without actually solving the system!*

We shall illustrate the calculation of orbits in the next few examples of linear systems and return to the discussion of nonlinear systems (such as the Lotka-Volterra equations) in Section 12.3.

Example 12.1.5   Consider the equation of the harmonic oscillator

$$x'' + x = 0,$$

with initial conditions $x(0) = 1$, $x'(0) = 0$. We may rewrite it as the autonomous system

$$x' = y, \qquad x(0) = 1,$$
$$y' = -x, \qquad y(0) = 0.$$

There are two ways of finding the orbit. First, we observe that the unique solution of the initial value problem is the solution pair $(\cos t, -\sin t)$. Since $\cos^2 t + \sin^2 t = 1$, the orbit satisfies the equation

$$x^2 + y^2 = 1,$$

which is the unit circle in the xy-plane (see Fig. 12.4). The arrows in the figure indicate the direction in which the solutions move about the orbit as t increases. Note that as t increases, $\cos t$ (the x-coordinate) moves from 1 (when $t = 0$) to 0 (when $t = \pi/2$), to -1 (when $t = \pi$), to 0 (when $t = 3\pi/2$), and back to 1 (at

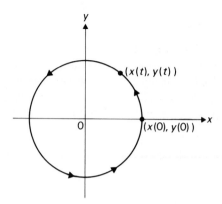

Figure 12.4

$t = 2\pi$). Similarly, $\sin t$ (the y-coordinate) moves from 0 to 1, to 0, to -1, and back to 0. This phenomenon explains the direction indicated by the arrows in Fig. 12.4. Therefore, starting at the point $(1,0)$ (corresponding to $t = 0$), x decreases while y increases.

We can also find this orbit without solving the system. By the chain rule, we have

$$\frac{dy}{dx} = \frac{dy/dt}{dx/dt} = -\frac{x}{y}.$$

Separating the variables, we find that

$$y \, dy = -x \, dx$$

or, after an integration,

$$x^2 + y^2 = C.$$

To evaluate the constant C, we note that at $t = 0$, $x = 1$ and $y = 0$, so that $C = x^2(0) + y^2(0) = 1$. (Thus the radius of the orbit depends on the initial conditions.)

Example 12.1.6 Again consider the harmonic oscillator with the new initial condition

$$x(t_0) = 1, \qquad y(t_0) = 0.$$

The unique solution pair is

$$(x(t), y(t)) = (\cos(t - t_0), -\sin(t - t_0)).$$

But then

$$x^2 + y^2 = \cos^2(t - t_0) + \sin^2(t - t_0) = 1,$$

which is the *same* orbit as in the previous example. In other words, *the orbit is independent of the initial value of t.* [But not, of course, of the initial values $x(t_0)$ and $y(t_0)$]. This is a property shared by *all autonomous systems.* This fact will be assumed for the remainder of this chapter. Its proof can be found in most advanced differential equations texts.* The property does not hold for nonautonomous systems, as is easily illustrated by the next example.

Example 12.1.7 Consider the nonautonomous system

$$x' = \frac{1}{t}x, \qquad x(t_0) = 1,$$

$$y' = y, \qquad y(t_0) = 2.$$

Since these equations are uncoupled, it is easy to calculate the solution pair

$$(x(t), y(t)) = \left(\frac{t}{t_0}, 2e^{t - t_0}\right).$$

Since $t = t_0 x$, the orbit is

$$y = 2e^{t_0(x - 1)}.$$

Thus different values of the initial value of t lead to different orbits. Figure 12.5 shows the orbits for $t_0 = 1, t_0 = 2$.

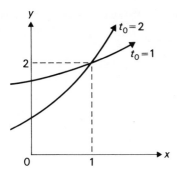

Figure 12.5

---

* See, for example, H. K. Wilson, *Ordinary Differential Equations*, Addison-Wesley, Reading, Mass., 1972.

In light of the properties of autonomous systems illustrated in Examples 12.1.5 and 12.1.6, we will deal with them exclusively in the remainder of this chapter. As we saw earlier (see Example 12.1.2), autonomous systems can and do arise naturally in applications. A discussion of the properties of nonautonomous systems is more complicated (one reason being the necessity to worry about the initial value of t). It can be found in many intermediate and advanced textbooks on differential equations.

EXERCISES 12.1

In each of Exercises 1 through 6 (a) find a related linear equation (as in Examples 12.1.2 and 12.1.3), (b) find all constant solutions of the nonlinear equation, (c) solve the nonlinear equation, and (d) determine the behavior of the solutions to the nonlinear equation for values of $x(0)$ in the indicated range.

1. $x' = x - x^2, -\infty < x(0) < \infty$

2. $x' = -2x + 3x^2, x(0) \geqslant 0$

3. $x' = 2x + 3x^2, x(0) \geqslant 0$

4. $x' = 2x - 3x^2, -\infty < x(0) < \infty$

5. $x' = x(x - 1)(x - 2), x(0) \geqslant 0$

6. $x' = -x(x - 1)(x - 2), x(0) \geqslant 0$

7. a) Draw the orbits for the initial value problem

$$x' = y, \qquad x(0) = a,$$
$$y' = -x, \qquad y(0) = b.$$

   b) Show that these orbits are identical to those for the same system with the initial conditions

$$x(t_0) = a, \qquad y(t_0) = b.$$

8. Show that the orbits for the equation

$$x'' + \omega^2 x = 0$$

   are ellipses centered at the origin.

9. Find the orbits for the system

$$x' = tx, \qquad x(t_0) = 1,$$
$$y' = -y, \qquad y(t_0) = 1,$$

   and graph these orbits for $t_0 = 0$, $t_0 = 1$, and $t_0 = 2$.

* 10. Suppose that $f(x,y)$ and $g(x,y)$ satisfy the conditions of the existence-uniqueness theorems

11.3 and 11.5. Then show that no two orbits of the autonomous system

$$x' = f(x,y), \qquad y' = g(x,y) \tag{12.8}$$

can ever intersect.

* 11.   Use the result of Exercise 10 to show that if (x_0,y_0) is a point having the property that $f(x_0,y_0) = g(x_0,y_0) = 0$ and if $(x(t), y(t))$ is a nonzero solution pair of the system (12.8), then there is no value of t for which $(x(t),y(t)) = (x_0,y_0)$.

12.2 CRITICAL POINTS, STABILITY, AND PHASE PORTRAITS FOR LINEAR SYSTEMS

The general autonomous system of two first order equations is given by

$$x' = f(x,y),$$
$$y' = g(x,y). \tag{12.9}$$

Since orbits are independent of t_0, we shall assume that $t_0 = 0$.

A point (x_0,y_0) is called a *critical point* of the system (12.9) if

$$f(x_0,y_0) = g(x_0,y_0) = 0.$$

Any critical point (x_0,y_0) is a constant solution pair of (12.9), since the derivative of a constant is zero:

$$x'_0 = 0 = f(x_0,y_0),$$
$$y'_0 = 0 = g(x_0,y_0).$$

A critical point (x_0,y_0) of the system (12.9) is a point of *equilibrium*, since once we reach this point we can never leave it, the derivatives of both $x(t)$ and $y(t)$ being zero there. Physically, a critical point is often a point at which the potential energy is at a minimum. For instance, in Example 12.1.2, if we use the substitution $\mu = d\theta/dt$, we can write the system (12.1) as

$$\theta' = \mu,$$
$$\mu' = -\omega^2 \sin \theta.$$

From Fig. 12.1, it is clear that the potential energy is a minimum when $\theta = 0$. The point $(0,0)$ is a point of equilibrium of the system. Of course, there are other critical points (see the next example). This situation will be discussed in great detail in Section 12.3 (Example 12.3.4).

Example 12.2.1 Consider the system

$$x' = y,$$
$$y' = -\omega^2 \sin x,$$

which was just obtained from Eq. (12.1). This system has infinitely many critical points, since $(k\pi, 0)$ is critical for all integers k.

Example 12.2.2 Consider the system

$$x' = -x^2 + y,$$
$$y' = x - y^2.$$

The two critical points are $(0,0)$ and $(1,1)$.

The notion of stability is central to any discussion of the behavior of differential equations. Roughly, a solution $\varphi(t)$ to a system of equations is *stable* if whenever we start "close" to $\varphi(t)$, we will stay close to $\varphi(t)$ for all future values of t. It is *asymptotically stable* if it is stable and the solutions that start close to $\varphi(t)$ approach $\varphi(t)$ as t tends to ∞. Finally, $\varphi(t)$ is *unstable* if it is not stable.

We already saw an example of a system that is stable. In Example 12.1.5 the orbits (see Fig. 12.4) were circles centered at the origin of radius $\sqrt{x^2(0) + y^2(0)}$. If the initial conditions are changed by a small amount, then the radius of the circular orbit is changed by a small amount, and thus the new orbit stays close to the original one.

To define "closeness" more precisely, we make use of the Pythagorean distance between two points in the plane. If (x_1, y_1) and (x_2, y_2) are the two points, then the distance between them is

$$d = [(x_1 - x_2)^2 + (y_1 - y_2)^2]^{1/2}.$$

We denote by $(x(t,x^*), y(t,y^*))$ the unique solution pair to the system (12.9) that satisfies the initial conditions

$$x(0) = x^*, \qquad y(0) = y^*.$$

Now we can give formal definitions of the above concepts. The constant solution (or critical point) (x_0, y_0) is said to be:

i) *stable* if for every number $\varepsilon > 0$ there is a number $\delta > 0$ such that whenever

$$[(x_0 - x^*)^2 + (y_0 - y^*)^2]^{1/2} < \delta,$$

we have

$$[(x_0 - x(t,x^*))^2 + (y_0 - y(t,y^*))^2]^{1/2} < \varepsilon,$$

for all $t \geq 0$ (i.e., if you start close, you stay close);

ii) *asymptotically stable* if it is stable and there exists a number $A > 0$ such that whenever

$$[(x_0 - x^*)^2 + (y_0 - y^*)^2]^{1/2} < A,$$

we have

$$\lim_{t \to \infty} \left[(x_0 - x(t,x^*))^2 + (y_0 - y(t,y^*))^2 \right] = 0$$

(i.e., if you start close enough, the solution will approach the critical point as t tends to ∞);

iii) *unstable* if it is not stable (i.e., no matter how close to the constant solution you start, there are solutions that will move away from the constant solution).

The requirement that the solution be stable is part of the definition of asymptotic stability. There are examples, which we shall not cite here, of systems where all solutions eventually tend to zero but each gets very large, no matter how close to zero it starts. The zero solution for such a system is unstable.

The foregoing ideas are sketched in Fig. 12.6, where the critical point is taken to be the origin.

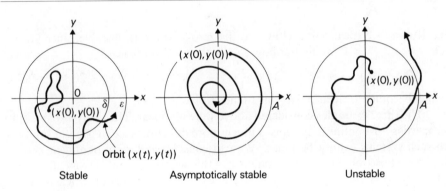

| Stable | Asymptotically stable | Unstable |

Figure 12.6

Example 12.2.3 Consider again the harmonic oscillator

$$x' = y, \qquad y' = -x.$$

Clearly $(0,0)$ is a constant solution. The solution pair with the initial conditions $x(0) = x^*$, $y(0) = y^*$ is

$$(x(t,x^*), y(t,y^*)) = (x^* \cos t + y^* \sin t, \ -x^* \sin t + y^* \cos t),$$

so that

$$(x(t,x^*) - 0)^2 + (y(t,y^*) - 0)^2 = (x^* \cos t + y^* \sin t)^2 + (-x^* \sin t + y^* \cos t)^2$$
$$= 2(x^{*2} + y^{*2}).$$

Thus $\delta = \varepsilon/2$ in the definition of stability and the zero solution is stable.

Example 12.2.4 Consider the system

$$x' = y,$$
$$y' = -2x - 3y.$$

Again (0,0) is a critical point. The solution pair that satisfies $x(0) = x^*$, $y(0) = y^*$ is given by

$$(x(t,x^*),y(t,y^*))$$
$$= ((2x^* + y^*)e^{-t} - (x^* + y^*)e^{-2t}, -(2x^* + y^*)e^{-t} + 2(x^* + y^*)e^{-2t}),$$

which tends to zero as t tends to ∞ no matter how large x^* and y^* are. Hence the zero solution is asymptotically stable. As in this example, any situation in which A can be arbitrarily large is called *global asymptotic stability*.

Example 12.2.5 The system

$$x' = y,$$
$$y' = -2x + 3y$$

has (0,0) as a critical point. However, the zero solution is unstable, since for any $x^* \neq 0$ and $y^* \neq 0$ (no matter how small), the solution pair is

$$(x(t,x^*),y(t,y^*)) = ((2x^* - y^*)e^t + (y^* - x^*)e^{2t}, (2x^* - y^*)e^t + 2(y^* - x^*)e^{2t}),$$

which becomes arbitrarily large as t tends to ∞.

The three examples considered above were all linear. The analysis of stability properties for nonlinear systems is much more difficult, since in general solutions to such systems cannot be found. We will consider a class of nonlinear systems in the next section, but in the remainder of this section we will classify all possible linear systems* so that we may have some basis of comparison when we get to the nonlinear ones.

The linear system we will consider is

$$x' = a_{11}x + a_{12}y,$$
$$y' = a_{21}x + a_{22}y, \tag{12.10}$$

where the coefficients a_{ij} are real constants. As in Section 7.5, we derive the auxiliary equation of the system

$$\lambda^2 - (a_{11} + a_{22})\lambda + (a_{11}a_{22} - a_{21}a_{12}) = 0, \tag{12.11}$$

with the roots λ_1 and λ_2. The orbits of the system (12.10) will depend on the nature of these two roots. We shall therefore consider each case separately. We note that in most cases (0,0) will be the only constant solution (critical point) of the system

---

* Except those in which at least one of the roots of the auxiliary equation (12.11) is zero.

(why?), so that we shall restrict our attention to the nature of the orbits around the origin. This restriction also means that none of the roots of the auxiliary equation (12.11) will be zero (see Exercise 11).

In the following discussion, we shall examine the different possibilities for the roots of (12.11) and draw representative orbits of the system for each case. Thus, merely knowing the roots of the auxiliary equation is enough to determine the nature of the orbits.

CASE 1. λ_1 and λ_2 are real, distinct, and of the same sign. We may assume, for simplicity, that $\lambda_1 > \lambda_2$. Then by Theorem 7.8 each solution pair has the form

$$(x(t),y(t)) = (c_1\alpha_1 e^{\lambda_1 t} + c_2\alpha_2 e^{\lambda_2 t}, c_1\beta_1 e^{\lambda_1 t} + c_2\beta_2 e^{\lambda_2 t}), \tag{12.12}$$

where c_1 and c_2 are arbitrary.

CASE 1(a). $\lambda_2 < \lambda_1 < 0$ (both roots are negative). Clearly all solutions tend to (0,0) as t tends to ∞. First, we assume that $c_1 = 0$ and $c_2 \neq 0$. Then $y = (\beta_2/\alpha_2)x$, which means that the orbit is a straight line with slope β_2/α_2. If $c_1 \neq 0$ and $c_2 = 0$, then the situation is similar and we obtain the line $y = (\beta_1/\alpha_1)x$. To obtain the other orbits, we assume that c_1 and c_2 are both nonzero. Then

$$\frac{y(t)}{x(t)} = \frac{c_1\beta_1 e^{\lambda_1 t} + c_2\beta_2 e^{\lambda_2 t}}{c_1\alpha_1 e^{\lambda_1 t} + c_2\alpha_2 e^{\lambda_2 t}},$$

and dividing the numerator and denominator by $e^{\lambda_1 t}$, we have

$$\frac{y(t)}{x(t)} = \frac{c_1\beta_1 + c_2\beta_2 e^{(\lambda_2 - \lambda_1)t}}{c_1\alpha_1 + c_2\alpha_2 e^{(\lambda_2 - \lambda_1)t}} \rightarrow \frac{c_1\beta_1}{c_1\alpha_1} = \frac{\beta_1}{\alpha_1}$$

as t tends to ∞. Thus all orbits approach the origin with the slope β_1/α_1. Similarly, as t tends to $-\infty$, all solutions become asymptotic to the line with the slope β_2/α_2. This situation is illustrated in Fig. 12.7 for three different values of the slopes β_1/α_1 and β_2/α_2. It is clear that in this case the zero solution is asymptotically stable. Here the origin is called a *stable node*.

CASE 1(b). $\lambda_1 > \lambda_2 > 0$ (both roots are positive). Then all solutions (except the zero solution) approach ∞ as t tends to ∞. Hence the zero solution is unstable. The orbits are the same as in the previous case except that the direction of motion is reversed. The origin here is called an *unstable node*. As t tends to $-\infty$, the orbits approach zero with the slope β_2/α_2 and as t tends to ∞, the orbits become asymptotic to the line with the slope β_1/α_1.

CASE 2. λ_1 and λ_2 are real with opposite signs. We assume that $\lambda_1 > 0 > \lambda_2$. The situation here is very different. If $c_1 = 0$ and $c_2 \neq 0$, we obtain, as before,

$$\frac{y}{x} = \frac{\beta_2}{\alpha_2} \quad \text{or} \quad y = \frac{\beta_2}{\alpha_2}x.$$

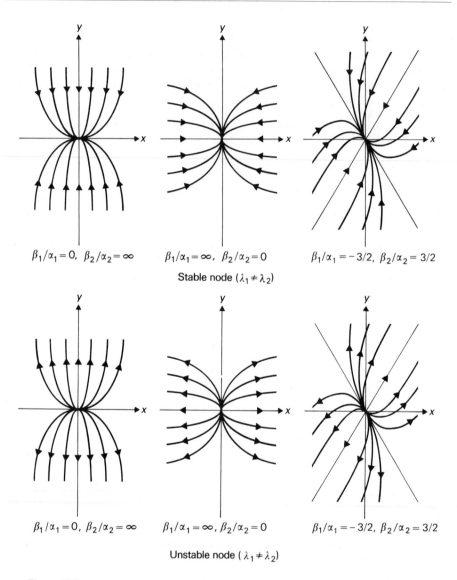

$\beta_1/\alpha_1 = 0,\ \beta_2/\alpha_2 = \infty$ $\beta_1/\alpha_1 = \infty,\ \beta_2/\alpha_2 = 0$ $\beta_1/\alpha_1 = -3/2,\ \beta_2/\alpha_2 = 3/2$

Stable node ($\lambda_1 \neq \lambda_2$)

$\beta_1/\alpha_1 = 0,\ \beta_2/\alpha_2 = \infty$ $\beta_1/\alpha_1 = \infty,\ \beta_2/\alpha_2 = 0$ $\beta_1/\alpha_1 = -3/2,\ \beta_2/\alpha_2 = 3/2$

Unstable node ($\lambda_1 \neq \lambda_2$)

Figure 12.7

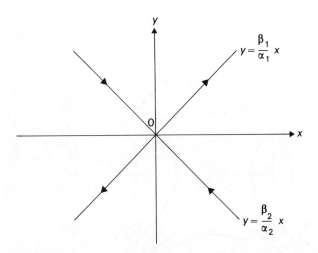

Figure 12.8

As t tends to ∞, $x(t)$ and $y(t)$ approach zero. If $c_1 \neq 0$ and $c_2 = 0$, then $y = (\beta_1/\alpha_1)x$ and both x and y approach ∞ as t tends to ∞, and approach zero as t tends to $-\infty$. These two orbits are sketched in Fig. 12.8. When both c_1 and c_2 are nonzero, the situation is more complicated. Again

$$\frac{y(t)}{x(t)} = \frac{c_1\beta_1 e^{\lambda_1 t} + c_2\beta_2 e^{\lambda_2 t}}{c_1\alpha_1 e^{\lambda_1 t} + c_2\alpha_2 e^{\lambda_2 t}} = \frac{c_1\beta_1 + c_2\beta_2 e^{(\lambda_2 - \lambda_1)t}}{c_1\alpha_1 + c_2\alpha_2 e^{(\lambda_2 - \lambda_1)t}},$$

which approaches β_1/α_1 as t tends to ∞. Hence all orbits are asymptotic to the line $y = (\beta_1/\alpha_1)x$ as t tends to ∞. Also

$$\frac{y(t)}{x(t)} = \frac{c_1\beta_1 e^{(\lambda_1 - \lambda_2)t} + c_2\beta_2}{c_1\alpha_1 e^{(\lambda_1 - \lambda_2)t} + c_2\alpha_2},$$

which approaches β_2/α_2 as t tends to $-\infty$. Hence all orbits are asymptotic to the line $y = (\beta_2/\alpha_2)x$ as t tends to $-\infty$. Finally, we observe that both $x(t)$ and $y(t)$ approach ∞ as t tends to $\pm\infty$, and by uniqueness, no orbit can pass through the origin.* The orbits are therefore as depicted in Fig. 12.9.

---

* By the uniqueness theorem 11.5, there is a unique solution to system (12.10) which satisfies $x(t_0) = y(t_0) = 0$. Certainly the zero solution satisfies these conditions for every t_0. Hence no other solution can pass through the origin at any time t_0.

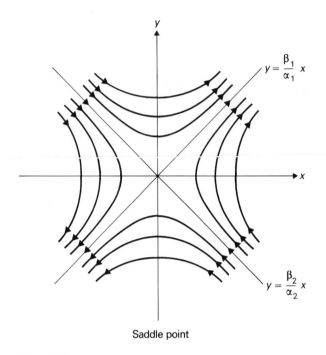

Saddle point

Figure 12.9

It is clear here that the origin is unstable. In this situation, the origin is called (for obvious reasons) a *saddle point*. We note that a saddle point has the property that exactly one orbit approaches the origin and all others are "repelled" by it. The physical behavior corresponding to a saddle point is illustrated in Example 12.3.4.

CASE 3. $\lambda_1 = \lambda_2 = \lambda$. Here either $\lambda < 0$ or $\lambda > 0$.

CASE 3(a). $\lambda < 0$. There are two ways in which the auxiliary Eq. (12.11) can yield a double root. One possibility is

$$a_{11} = a_{22} \neq 0, \qquad a_{21} = a_{12} = 0. \tag{12.13}$$

Then the auxiliary equation is

$$\lambda^2 - 2a_{11}\lambda + a_{11}^2 = 0,$$

or $\lambda = a_{11}$ is the double root. Then the system (12.10) becomes

$$x' = \lambda x,$$
$$y' = \lambda y.$$

The solutions are obviously of the form

$$(x(t),y(t)) = (c_1 e^{\lambda t}, c_2 e^{\lambda t}),$$

so that

$$\frac{y}{x} = \frac{c_2}{c_1} \quad \text{or} \quad y = \frac{c_2}{c_1} x.$$

Thus all orbits are straight lines with the slope c_2/c_1. Since $\lambda < 0$, all solutions approach zero as t tends to ∞, and the zero solution is asymptotically stable. The situation is graphed in Fig. 12.10. The origin in this case is also called a node. Sometimes the nodes depicted in Fig. 12.6 are called *improper nodes*, while the node in Fig. 12.10 is called a *proper node*. We will not use this terminology. Also, we should add that the node of Fig. 12.10 is sometimes called a *star-shaped node*.

If $\lambda < 0$ is a double root but the equalities (12.13) do not hold, then the equations are coupled and the general solution is, according to Theorem 7.9,

$$(x(t),y(t)) = ([c_1\alpha_1 + c_2(\alpha_2 + \alpha_3 t)]e^{\lambda t}, [c_1\beta_1 + c_2(\beta_2 + \beta_3 t)]e^{\lambda t}). \tag{12.14}$$

Then

$$\frac{y}{x} = \frac{c_1\beta_1 + c_2\beta_2 + c_2\beta_3 t}{c_1\alpha_1 + c_2\alpha_2 + c_2\alpha_3 t} = \frac{c_1\beta_1/t + c_2\beta_2/t + c_2\beta_3}{c_1\alpha_1/t + c_2\alpha_2/t + c_2\alpha_3},$$

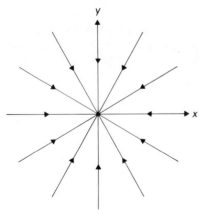

(Star-shaped) stable node $(\lambda_1 = \lambda_2)$

Figure 12.10

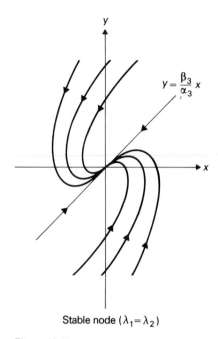

$$y = \frac{\beta_3}{\alpha_3} x$$

Stable node ($\lambda_1 = \lambda_2$)

Figure 12.11

which approaches β_3/α_3 as t tends to $\pm\infty$. Since both $x(t)$ and $y(t)$ approach zero as t tends to ∞, the zero solution is asymptotically stable. Also all orbits are asymptotic to the line $y = (\beta_3/\alpha_3)x$ as t tends to $\pm\infty$, as illustrated in Fig. 12.11.

CASE 3(b). $\lambda > 0$. The orbits are the same but with the arrows reversed, since now all solutions approach ∞ as t tends to ∞. (See Figs. 12.12 and 12.13). In these two cases the origin is unstable. Cases 3(a) and 3(b) provide another situation in which nodes arise.

CASE 4. λ_1 *and* λ_2 *are complex conjugates but not pure imaginary.* Then $\lambda_1 = a + ib$ and $\lambda_2 = a - ib$ where neither a nor b is zero.

CASE 4(a). $a < 0$. According to Theorem 7.10, all solutions have the form

$$(x(t), y(t))$$
$$= (e^{at}[c_1(A_1 \cos bt - A_2 \sin bt) + c_2(A_1 \sin bt + A_2 \cos bt)],$$
$$\quad e^{at}[c_1(B_1 \cos bt - B_2 \sin bt) + c_2(B_1 \sin bt + B_2 \cos bt)]). \tag{12.15}$$

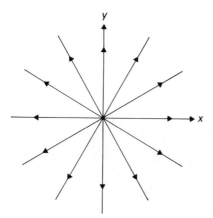

(Star-shaped) unstable node ($\lambda_1 = \lambda_2$)

Figure 12.12

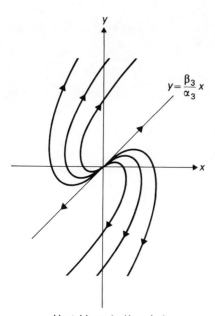

$$y = \frac{\beta_3}{\alpha_3} x$$

Unstable node ($\lambda_1 = \lambda_2$)

Figure 12.13

To simplify the notation, we define

$$k_1 = c_1A_1 + c_2A_2, \qquad k_2 = -c_1A_2 + c_2A_1,$$
$$k_3 = c_1B_1 + c_2B_2, \qquad k_4 = -c_1B_2 + c_2B_1,$$

so that (12.15) becomes

$$x(t) = e^{at}(k_1 \cos bt + k_2 \sin bt),$$
$$y(t) = e^{at}(k_3 \cos bt + k_4 \sin bt). \tag{12.16}$$

We now define

$$A = \sqrt{k_1^2 + k_2^2} \qquad \text{and} \qquad B = \sqrt{k_3^2 + k_4^2}.$$

Then

$$x(t) = A\left(\frac{k_1}{A}\cos bt + \frac{k_2}{A}\sin bt\right)e^{at},$$

and

$$y(t) = B\left(\frac{k_3}{B}\cos bt + \frac{k_4}{B}\sin bt\right)e^{at}.$$

Now, by the choice of A and B, k_1/A, k_2/A, k_3/B, and k_4/B all lie between -1 and $+1$. Thus we may define α_1 and α_2 by

$$\cos \alpha_1 = \frac{k_1}{A}, \qquad \cos \alpha_2 = \frac{k_3}{B},$$

$$\sin \alpha_1 = -\frac{k_2}{A}, \qquad \sin \alpha_2 = -\frac{k_4}{B}, \tag{12.17}$$

and α_1 and α_2 are between zero and 2π. Note that

$$\sin^2 \alpha_1 + \cos^2 \alpha_1 = \sin^2 \alpha_2 + \cos^2 \alpha_2 = 1.$$

Finally, we have

$$x(t) = Ae^{at}(\cos \alpha_1 \cos bt - \sin \alpha_1 \sin bt)$$
$$= Ae^{at} \cos (bt + \alpha_1),$$
$$y(t) = Be^{at}(\cos \alpha_2 \cos bt - \sin \alpha_2 \sin bt) \tag{12.18}$$
$$= Be^{at} \cos (bt + \alpha_2).$$

Then

$$\frac{y}{x} = \frac{B \cos (bt + \alpha_1)}{A \cos (bt + \alpha_2)},$$

which is defined whenever $\cos(bt + \alpha_2) \neq 0$. Since this expression is periodic, it is clear that as t tends to ∞, the ratio y/x does not approach a limit, but the orbits must circle around the origin. Since $a < 0$, $x(t)$ and $y(t)$ approach zero as t tends to ∞. Hence the orbits must spiral in toward the origin. The zero solution is asymptotically stable, and the origin is called a *stable focus* (or *spiral point*). See Fig. 12.14.

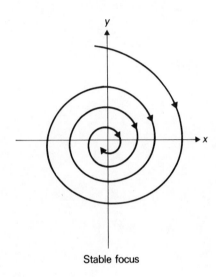

Stable focus

Figure 12.14

CASE 4(b). $a > 0$. Here the analysis is as before except that all solutions approach ∞ as t tends to ∞. See Fig. 12.15.

CASE 5. λ_1 and λ_2 *are pure imaginary*. Then $\lambda_1 = ib$ and $\lambda_2 = -ib$, and we can use the same analysis as above with $a = 0$: we have

$$(x(t), y(t)) = (A \cos(bt + \alpha_1), B \cos(bt + \alpha_2)). \tag{12.19}$$

Clearly $x(t)$ and $y(t)$ are periodic with period $2\pi/b$ so that every orbit beginning at the point (x^*, y^*) when $t = t^*$ will return to the *same point* when $t = t^* + 2\pi/b$. Thus the orbits are closed curves. To get a feeling for the nature of these curves, we set $k_2 = k_3 = 0$ in (12.16) so that

$$x(t) = k_1 \cos bt \quad \text{and} \quad y(t) = k_4 \sin bt.$$

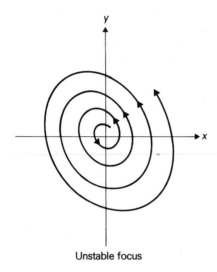

Unstable focus

Figure 12.15

Then

$$\frac{x^2}{k_1^2} + \frac{y^2}{k_4^2} = \cos^2 bt + \sin^2 bt = 1,$$

which is the equation of an ellipse centered about (0,0) with the x-axis as its axis of symmetry. If k_2 and k_3 are nonzero, then we obtain ellipses which have a rotated axis of symmetry. In this situation the zero solution is stable but not asymptotically stable, since solutions do not approach zero. Then the origin is called a *center*, which is illustrated in Fig. 12.16.

The entire preceding analysis is summarized in the theorem below.

Theorem 12.1 Consider the system

$$x' = a_{11}x + a_{12}y,$$
$$y' = a_{21}x + a_{22}y,$$

where the a_{ij} are real constants and $a_{11}a_{22} - a_{12}a_{21} \neq 0$ so that the origin (0,0) is the only critical point (see Exercise 11). Let λ_1 and λ_2 be the two roots of the auxiliary equation (12.11). Then

a) the origin is stable if λ_1 and λ_2 are pure imaginary,
b) the origin is asymptotically stable if Re $\lambda_1 < 0$ and Re $\lambda_2 < 0$,
c) the origin is unstable in all other cases. Moreover, the behavior of the orbits near the origin is as indicated in Table 12.1.

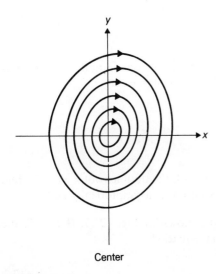

Center

Figure 12.16

Table 12.1

λ_1, λ_2	Type of critical point
real, distinct, negative	stable node
real, distinct, positive	unstable node
real, distinct, opposite signs	saddle point (unstable)
real, equal, negative	stable node
real, equal, positive	unstable node
complex conjugate, not pure imaginary, negative real parts	stable focus
complex conjugate, not pure imaginary, positive real parts	unstable focus
pure imaginary	center (stable)

Example 12.2.6   Consider the system

$$x' = x - 3y,$$
$$y' = x - y.$$

The auxiliary equation is $\lambda^2 + 2 = 0$ with the roots $\pm\sqrt{2}i$. Therefore, the zero solution is stable and the origin is a center.

Example 12.2.7 Consider the system

$$x' = 4x - y,$$
$$y' = 6x - 3y.$$

The auxiliary equation is $\lambda^2 - \lambda - 6 = 0$ with the roots 3 and -2. Hence the origin is a saddle point and the zero solution is unstable.

Example 12.2.8 Consider the system

$$x' = -4x - y,$$
$$y' = x - 2y.$$

The auxiliary equation is $\lambda^2 + 6\lambda + 9 = 0$ with the double root $\lambda = -3$. Hence the origin is a stable node and the zero solution is asymptotically stable.

Example 12.2.9 Consider the RLC circuit shown in Fig. 12.17. If $E = 0$, then from Section 3.10 we obtain the differential equation

$$\frac{d^2I}{dt^2} + \frac{R}{L}\frac{dI}{dt} + \frac{I}{CL} = 0 \tag{12.20}$$

for the description of this circuit. Writing the equation as a system, we obtain

$$I' = y,$$
$$y' = -\frac{1}{CL}I - \frac{R}{L}y.$$

The auxiliary equation is

$$\lambda^2 + \frac{R}{L}\lambda + \frac{1}{CL} = 0$$

with the roots

$$\lambda_1 = \frac{-R + \sqrt{R^2 - 4L/C}}{2L}, \qquad \lambda_2 = \frac{-R - \sqrt{R^2 - 4L/C}}{2L}. \tag{12.21}$$

There are three cases to consider, according to whether $R^2 - 4L/C$ is greater than zero, less than zero, or equal to zero.

i) If $R^2 - 4L/C > 0$, then λ_1 and λ_2 are real, distinct, and negative, so that the origin is a stable node.

ii) If $R^2 - 4L/C = 0$, then $\lambda_1 = \lambda_2 = -R/2L$, so that the zero solution is again a stable node.

iii) If $R^2 - 4L/C < 0$, then λ_1 and λ_2 are complex conjugates with negative real parts, so that the origin is a stable focus.

Figure 12.17

In all three cases the critical point $(0,0)$ is asymptotically stable, and the transient current tends to zero as $t \to \infty$. Of course, the emf E is generally nonzero, but it affects only the steady state current. By superposition, the current $I(t)$ is the sum of the steady state and transient currents.

EXERCISES 12.2

In Exercises 1 through 8 describe the nature of the critical point $(0,0)$ of each system and sketch the orbits.

1. $x' = 4x - 3y$
   $y' = 5x - 4y$

2. $x' = -2x + 3y$
   $y' = x - 3y$

3. $x' = -x + y$
   $y' = -5x + 3y$

4. $x' = x + y$
   $y' = -x - 3y$

5. $x' = -4x - 2y$
   $y' = 5x + 2y$

6. $x' = 4x - 3y$
   $y' = 8x + 5y$

7. $x' = 2x + y$
   $y' = -4x + 2y$

8. $x' = 2x$
   $y' = 2y$

9. Consider the system

$$x' = a_{11}x + a_{12}y,$$
$$y' = a_{21}x + a_{22}y.$$

Let $T = a_{11} + a_{22}$, $D = a_{11}a_{22} - a_{12}a_{21}$, and $C = T^2 - 4D$. Show that the origin is:

a) a node if $D > 0$ and $C \geqslant 0$;
b) a saddle point if $D < 0$;
c) a focus if $T \neq 0$ and $C < 0$;
d) a center if $T = 0$ and $D > 0$.

10. Show that the zero solution in Exercise 9 is:

a) asymptotically stable if $D > 0$ and $T < 0$;
b) stable if $T = 0$ and $D > 0$;
c) unstable if $T > 0$ or $D < 0$.

11. Show that (0,0) is the only critical point of the system (12.10) if and only if $\lambda = 0$ is *not* a root of the auxiliary equation (12.11). [*Hint*: See Appendix 4.]

12. Let $L = 2$ henries, $R = 50$ ohms, and $C = 0.0003$ farads. Locate graphically the position on the orbit of Eq. (12.20) at $t = 1$, if $I(0) = 0$, and $I'(0)$ is one of the values 10, 100, and 1000.

13. Let $L = 0.44$ henries, $R = 1200$ ohms, and $C = 10^{-6}$ farads. Locate graphically the position of the solution on the orbit at $t = 1$ if $I(0) = 0$ and $I'(0)$ is one of the values 10, 100, and 1000.

14. Consider the mass-spring system of Section 3.9 which is damped so that Eq. (3.56) holds. That is,

$$x'' + \frac{\mu}{m} x' + \frac{\lambda}{m} x = 0.$$

Following the analysis of Example 12.2.9, describe the behavior of the orbits near the origin for different positive values of m, λ, and μ.

15. Consider the system

$$x' = 0,$$
$$y' = -x + y.$$

a) Solve the system.
b) Show that $\lambda_1 = 0$ and $\lambda_2 = 1$ are the roots of the auxiliary equation (12.11).
c) Find all the critical points of the system.
d) Show that all the orbits are straight lines.
e) Show that the origin is an unstable equilibrium point.

16. Consider the system

$$x' = 0,$$
$$y' = -x - y.$$

a) Solve the system.
b) Show that the roots of (12.11) for this system are $\lambda_1 = 0$ and $\lambda_2 = -1$.
c) Graph the orbits.
d) Show that the zero solution is stable but *not* asymptotically stable.

12.3 STABILITY OF NONLINEAR SYSTEMS

In this section we shall discuss the stability properties of the autonomous nonlinear system

$$x' = f(x,y),$$
$$y' = g(x,y),$$
(12.22)

where $f(0,0) = g(0,0) = 0$ so that the origin is a critical point. We assume that $f(x,y)$ and $g(x,y)$ possess continuous third partial derivatives so that they can each be expanded in a Taylor series:

$$f(x,y) = f(0,0) + \frac{\partial f}{\partial x}(0,0)x + \frac{\partial f}{\partial y}(0,0)y + \frac{\partial^2 f}{\partial x^2}(0,0)x^2$$

$$+ \frac{\partial^2 f}{\partial x\, \partial y}(0,0)xy + \frac{\partial^2 f}{\partial y^2}(0,0)y^2 + \cdots,$$

$$g(x,y) = g(0,0) + \frac{\partial g}{\partial x}(0,0)x + \frac{\partial g}{\partial y}(0,0)y + \frac{\partial^2 g}{\partial x^2}(0,0)x^2$$

$$+ \frac{\partial^2 g}{\partial x\, \partial y}(0,0)xy + \frac{\partial^2 g}{\partial y^2}(0,0)y^2 + \cdots, \tag{12.23}$$

where the omitted terms all involve higher powers of x and y. Now we define

$$a_{11} = \frac{\partial f}{\partial x}(0,0), \quad a_{12} = \frac{\partial f}{\partial y}(0,0), \quad a_{21} = \frac{\partial g}{\partial x}(0,0), \quad a_{22} = \frac{\partial g}{\partial y}(0,0),$$

and assume that

$$a_{11}a_{22} - a_{12}a_{21} \neq 0 \tag{12.24}$$

(see Exercise 11 of Section 12.2). Then using the fact that $f(0,0) = g(0,0) = 0$, we can write $f(x,y)$ and $g(x,y)$ as

$$f(x,y) = a_{11}x + a_{12}y + f_1(x,y),$$
$$g(x,y) = a_{21}x + a_{22}y + g_1(x,y),$$

where

$$\lim_{x,y \to 0} \frac{f_1(x,y)}{\sqrt{x^2 + y^2}} = \lim_{x,y \to 0} \frac{g_1(x,y)}{\sqrt{x^2 + y^2}} = 0. \tag{12.25}$$

The last property simply says that the point $(f_1(x,y), g_1(x,y))$ approaches the point $(0,0)$ "faster" than the point (x,y) does. In one dimension, this is easily visualized by noting, for example, that the function $f(x) = x^2$ goes to zero faster than x, or

$$\lim_{x \to 0} \frac{x^2}{x} = 0.$$

We now show that condition (12.25) holds for the function x^2. Once this has been done, it will be clear that the condition holds for xy, y^2, and higher powers of x and y (such as x^3, $x^2 y$, xy^2, and y^3, etc.). Now,

$$\lim_{x,y \to 0} \frac{x^2}{\sqrt{x^2 + y^2}} = \lim_{x,y \to 0} \frac{x}{\sqrt{1 + y^2/x^2}}$$

(obtained by dividing the numerator and denominator by x). We may let x and y tend to zero in a variety of ways. For convenience, we let $y = x$ as they both tend to zero. Then $y^2/x^2 = 1$ and

$$\lim_{x,y \to 0} \frac{x}{\sqrt{1 + y^2/x^2}} = \lim_{x,y \to 0} \frac{x}{\sqrt{2}} = 0.$$

Note that condition (12.24) will be satisfied if the determinant

$$\begin{vmatrix} \partial f/\partial x & \partial f/\partial y \\ \partial g/\partial x & \partial g/\partial y \end{vmatrix}$$

is nonzero at (0,0).

Example 12.3.1 The system

$$x' = 2x + 3y + x^3,$$
$$y' = x - 2y - y^{3/2}$$

satisfies the conditions (12.24) and (12.25) since (prove this)

$$\lim_{x,y \to 0} \frac{x^3}{\sqrt{x^2 + y^2}} = \lim_{x,y \to 0} \frac{y^{3/2}}{\sqrt{x^2 + y^2}} = 0.$$

We shall consider in the rest of this section the system

$$x' = a_{11}x + a_{12}y + f_1(x,y),$$
$$y' = a_{21}x + a_{22}y + g_1(x,y),$$
$$\tag{12.26}$$

where (12.24) and (12.25) are satisfied, and the *associated linear system*

$$x' = a_{11}x + a_{12}y,$$
$$y' = a_{21}x + a_{22}y.$$
$$\tag{12.27}$$

The following theorem enables us to determine the nature of the unique critical point (0,0) of (12.26) by indicating the behavior of the solutions of the linear system (12.27) near the origin. The proof of this theorem is difficult and beyond the scope of this text.*

Theorem 12.2 Let λ_1 and λ_2 be the roots of the auxiliary equation of the linear system (12.27).

i) The nonlinear system (12.26) has the same type of critical point at the origin as the linear system (12.27) whenever:

(a) $\lambda_1 \neq \lambda_2$ and (0,0) is a node of (12.27),
(b) $\lambda_1 = \lambda_2$ and (0,0) is not a star-shaped node of (12.27),

---

* For a proof, see J. K. Hale, *Ordinary Differential Equations*, Wiley, New York, 1969.

   (c) (0,0) is a saddle point of (12.27),

   (d) (0,0) is a focus of (12.27).

 ii) The origin is not necessarily the same type of critical point for the two systems:

   (e) if $\lambda_1 = \lambda_2$ and (0,0) is a star-shaped node of (12.27); then (0,0) is either a node or a focus of (12.26).

   (f) if (0,0) is a center of (12.27); then (0,0) is either a center or a focus of (12.26).

The next theorem relates the stability of the nonlinear system to that of the associated linear system. As before, the proof is omitted. The proof of part (i), however, is suggested in Exercise 11 of Section 12.4.

Theorem 12.3
  i) If the zero solution of (12.27) is asymptotically stable, then the zero solution of (12.26) is asymptotically stable.
 ii) If the zero solution of (12.27) is unstable, then the zero solution of (12.26) is unstable.
iii) If the zero solution of (12.27) is stable but not asymptotically stable, then the zero solution of (12.26) may be asymptotically stable, stable, or unstable.

Remark. Part (ii) of this theorem holds even when (12.24) is not satisfied. We shall use this fact in Example 12.3.3.

    Theorems 12.2 and 12.3 are often called *perturbation* theorems, because we may consider the nonlinear system (12.26) as a small perturbation of the linear system (12.27) *near the origin.*

Example 12.3.2 In Chapter 7 (Example 7.1.4) we discussed the Lotka-Volterra equations as a model for the interaction of two competing species:

$$x' = \beta_1 x - \delta_{11}x^2 - \delta_{12}xy,$$
$$y' = \beta_2 y - \delta_{21}xy - \delta_{22}y^2. \tag{12.28}$$

We can think of this system as a perturbation of the associated uncoupled linear system

$$x' = \beta_1 x,$$
$$y' = \beta_2 y. \tag{12.29}$$

The system (12.29) has the obvious solution

$$(x(t),y(t)) = (c_1 e^{\beta_1 t}, c_2 e^{\beta_2 t}),$$

and

$$\lim_{x,y\to 0} \frac{x^2}{\sqrt{x^2 + y^2}} = \lim_{x,y\to 0} \frac{xy}{\sqrt{x^2 + y^2}} = \lim_{x,y\to 0} \frac{y^2}{\sqrt{x^2 + y^2}} = 0.$$

Hence Theorems 12.2 and 12.3 apply in the following cases:

1. If β_1 and β_2 are negative, then the critical point $(0,0)$ is asymptotically stable for both systems (12.28) and (12.29). In both systems the origin is a stable node. This means that both populations will become extinct, in the absence of other factors, if the initial populations are small [that is, we start near $(0,0)$]. Thus small initial populations cannot sustain themselves.
2. If β_1 and β_2 have opposite signs, then the origin in both cases is a saddle point, and the zero solution is unstable. This situation implies that one of the populations will become extinct while the other will grow without bound (since the asymptotes of the saddle point are the x and y axes).
3. If β_1 and β_2 are unequal and positive, then the origin in both cases is an unstable node. Hence both populations will increase if they start near $(0,0)$. However, the orbits of (12.28) need not increase without bound, since as we move away from the origin the orbits may approach other critical points of (12.28) exhibiting a different type of behavior.

This is not a complete analysis of the system (12.28), since we can expect different and perhaps more interesting types of behavior near other critical points (see Exercise 16). We shall illustrate a similar situation in the next example.

Example 12.3.3 Consider the system

$$x' = -2xy = f(x,y),$$
$$y' = -x + y + xy - y^3 = g(x,y). \tag{12.30}$$

Setting $-2xy = -x + y + xy - y^3 = 0$, we find the three critical points $(0,0)$, $(0,1)$, and $(0,-1)$. We shall treat each of these separately.

1. $(0,0)$. The associated linear system is

$$x' = 0,$$
$$y' = -x + y. \tag{12.31}$$

The auxiliary equation for (12.31) is

$$\lambda^2 - \lambda = 0,$$

with the roots $\lambda_1 = 0$, $\lambda_2 = 1$. Hence the origin of (12.31) is unstable (see Exercise 15 of Section 12.2), and therefore the zero solution of (12.30) is unstable by the remark following Theorem 12.3.
2. $(0,1)$. To obtain the associated linear system, we need to use Taylor's theorem to expand the right-hand sides of (12.30) around the point $(0,1)$. Then

$$-2xy = f(0,1) + \frac{\partial f}{\partial x}\bigg|_{(0,1)} (x - 0) + \frac{\partial f}{\partial y}\bigg|_{(0,1)} (y - 1) + \cdots$$
$$= -2x + \cdots [= -2x - 2x(y - 1)]$$

and

$$-x + y + xy - y^3 = g(0,1) + \frac{\partial g}{\partial x}\bigg|_{(0,1)} (x - 0) + \frac{\partial g}{\partial y}\bigg|_{(0,1)} (y - 1) + \cdots$$

$$= -2(y - 1) + \cdots$$

$$[= -2(y - 1) + x(y - 1) - 6(y - 1)^2 - 6(y - 1)^3].$$

Selecting only the first degree terms of these expansions, we obtain the associated linear system

$$x' = -2x,$$

$$y' = -2(y - 1).$$

(12.32)

Here we treat the "new" phase plane variables as x and $z = y - 1$. Then the roots of the auxiliary equation of (12.32) are $\lambda_1 = \lambda_2 = -2$, which means that the point $(0,1)$ is a stable star-shaped node and the critical point $(0,1)$ is an asymptotically stable solution of (12.30). By Theorem 12.2(e), the point $(0,1)$ is either a stable node or a stable focus of (12.30).

3. $(0, -1)$. Expanding $f(x,y)$ and $g(x,y)$ around $(0, -1)$, we obtain

$$-2xy = f(0, -1) + \frac{\partial f}{\partial x}\bigg|_{(0,-1)} (x - 0) + \frac{\partial f}{\partial y}\bigg|_{(0,-1)} (y + 1) + \cdots$$

$$= 2x + \cdots [= 2x - 2x(y + 1)]$$

and

$$-x + y + xy - y^3 = g(0, -1) + \frac{\partial g}{\partial x}\bigg|_{(0,-1)} (x - 0) + \frac{\partial g}{\partial y}\bigg|_{(0,-1)} (y + 1) + \cdots$$

$$= -2x + \cdots$$

$$[= -2x - 2(y + 1) + x(y + 1) + 6(y + 1)^2 - 6(y + 1)^3],$$

so that the associated linear system around the point $(0, -1)$ is

$$x' = 2x,$$

$$y' = -2x - 2(y + 1).$$

(12.33)

Letting $z = y + 1$, we find the roots of the auxiliary equation for the system

$$x' = 2x,$$

$$z' = -2x - 2z$$

to be $\lambda_1 = 2, \lambda_2 = -2$, so that the solution $(0, -1)$ is a saddle point (unstable) of (12.30). This situation is illustrated in Fig. 12.18, where the critical point $(0,1)$ is shown as a node, although the theory only asserts that it is a node or a focus.

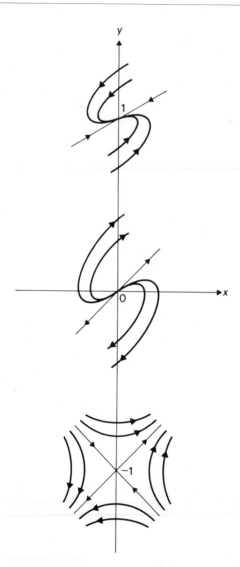

Figure 12.18

Example 12.3.4 In Example 12.1.2 we found that the equation of motion of a frictionless pendulum is given by

$$\theta'' + \omega^2 \sin \theta = 0.$$

Defining $\mu = \theta'$, we have the system

$$\theta' = \mu, \tag{12.34}$$
$$\mu' = -\omega^2 \sin \theta.$$

This system has an infinite number of critical points of the form $(k\pi, 0)$, $k = \pm 1$, $\pm 2, \pm 3, \ldots$. The associated linear system near the critical point $(0,0)$ is

$$\theta' = \mu, \tag{12.35}$$
$$\mu' = -\omega^2 \theta,$$

since the Taylor series of $\sin \theta$ is

$$\sin \theta = \theta - \frac{\theta^3}{3!} + \frac{\theta^5}{5!} - \cdots.$$

This is the system of the familiar harmonic oscillator. The origin of (12.35) is a center, so that we cannot conclude anything about the stability of the origin of (12.34). The results are more definitive for the critical point $(\pi, 0)$. Expanding $\sin \theta$ around $\theta = \pi$, we have

$$\sin \theta = \sin \pi + (\sin \theta)' \Big|_{\theta = \pi} (\theta - \pi) + \cdots$$

$$= (\pi - \theta) - \frac{1}{3!}(\pi - \theta)^3 + \frac{1}{5!}(\pi - \theta)^5 - \cdots.$$

Thus the associated linear system is

$$\theta' = \mu, \tag{12.36}$$
$$\mu = \omega^2(\theta - \pi).$$

Letting $z = \theta - \pi$, the roots of the auxiliary equation $\lambda^2 + \omega^2 = 0$ are $\lambda_1 = \omega$, $\lambda_2 = -\omega$, so that $(\pi, 0)$ is a saddle point. Hence we may conclude that the critical point $(\pi, 0)$ of (12.34) is an unstable saddle point. Figure 12.1 illustrates intuitively what is happening. When $\theta = \pi$ the pendulum is pointing vertically upward, which is a position at which the potential energy is at a maximum and the kinetic energy is zero. Clearly such a position is unstable, and a small displacement will cause large deviations from the initial position. On the other hand, it is just as clear that $\theta = 0$ (the vertically downward position) is a stable center. An initial displacement of θ_0 of a frictionless pendulum will lead to periodic oscillations with a maximum displacement of θ_0. It is evident that all critical points of the form $(k\pi,0)$ will be of the same type as $(0,0)$ if k is even and of the type $(\pi,0)$ if k is odd. Before drawing the orbits, we can show that the zero solution of (12.34) really is a center. By Theorem 12.2 it must be a center or a focus. But the system is *conservative* (there is no loss or gain of energy), whereas a stable focus would imply that the energy was decreasing to zero and an unstable focus would imply an increase in the total energy. Hence focus is ruled out, and the origin must be a center. This is all illustrated in Fig. 12.19.

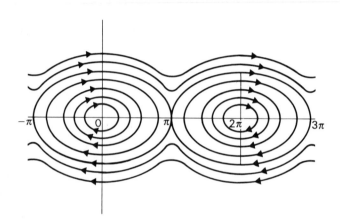

Figure 12.19

Example 12.3.5 We shall consider the pendulum of the previous example, this time including the effect of friction. We assume that the frictional force is proportional to the angular velocity θ'. Denoting this constant of proportionality by $\varepsilon > 0$, we obtain the equation

$$\theta'' + \varepsilon\theta' + \omega^2 \sin \theta = 0$$

or, equivalently,

$$\theta' = \mu,$$
$$\mu' = -\omega^2 \sin \theta - \varepsilon\mu. \qquad (12.37)$$

Again the origin is a critical point. However, in this case it is clear from physical considerations that the origin is asymptotically stable, since the friction will cause initial deviations from the vertical to damp out. We can prove this easily by considering the zero solution of the associated linear system

$$\theta' = \mu,$$
$$\mu' = -\omega^2\theta - \varepsilon\mu.$$

(12.38)

The auxiliary equation is $\lambda^2 + \varepsilon\lambda + \omega^2 = 0$ with the roots

$$\lambda_1 = \frac{-\varepsilon + \sqrt{\varepsilon^2 - 4\omega^2}}{2} \quad \text{and} \quad \lambda_2 = \frac{-\varepsilon - \sqrt{\varepsilon^2 - 4\omega^2}}{2}.$$

Hence the origin of (12.38) is a stable node (if $\varepsilon > 2\omega$), a star-shaped node (if $\varepsilon = 2\omega$), or a stable focus (if $\varepsilon < 2\omega$). In all cases, by Theorem 12.3, the origin of (12.37) is asymptotically stable. The last case is the most interesting. The "focal behavior" near the origin implies that the pendulum will continue to oscillate, but with constantly decreasing amplitudes and constantly increasing frequency.

Example 12.3.6 Consider the *RLC* circuit of Example 12.2.9. The Eq. (12.20) is an idealized description of the circuit, since there may very well be nonlinear terms present (such as induced currents from other sources). To take this factor into account, we have the equation

$$I'' + \frac{R}{L}I' + \frac{1}{CL}I + f(I,I') = 0,$$

(12.39)

where $f(x,y)$ is a nonlinear function such that

$$f(0,0) = \frac{\partial f}{\partial x}\bigg|_{(0,0)} = \frac{\partial f}{\partial y}\bigg|_{(0,0)} = 0.$$

(12.40)

Equation (12.39) can be written in the form

$$I' = y,$$
$$y' = -\frac{1}{CL}I - \frac{R}{L}y - f(I,y).$$

(12.41)

The associated linear system is

$$I' = y,$$
$$y' = -\frac{1}{CL}I - \frac{R}{L}y.$$

(12.42)

But as we showed in Example 12.2.9, the zero solution of (12.42) is asymptotically stable, and so by Theorem 12.3, the zero solution of (12.39) is also asymptotically stable. Therefore, in analyzing the circuit, we may safely ignore the nonlinear terms for small initial values of I and I'.

EXERCISES 12.3

In Exercises 1 through 6 verify that in each case $(0,0)$ is a critical point and determine the asymptotic behavior of solutions near that point.

1. $x' = 3 \sin x + e^y - 1$
   $y' = xy - y$

2. $x' = \ln (1 + y) + \cosh x - 1$
   $y' = \tan y + 2x$

3. $x' = -1 - y - e^{-x}$
   $y' = y - \sin x$

4. $x' = 2 \cos y - \frac{1}{2}y + e^x - 1 - \sin 2x$
   $y' = 2 \tan x - y$

5. $x' = (\sin x)(\cos x) + y^2$
   $y' = y^2 - x - x^3 + 3y$

6. $x' = -y + \varepsilon x(1 - y^2), \varepsilon > 0$
   $y' = x$

In Exercises 7 through 13 determine the critical points of each nonlinear equation; find the associated linear system for each of these critical points; determine, if possible, the nature of each critical point and its stability properties; and sketch the orbits near each such point.

7. $x' = x + x^3$
   $y' = y + y^3$

8. $x' = -\sin x + x^2$
   $y' = \sin y$

9. $x' = x - x^2 + xy$
   $y' = 2y - xy - 6y^2$

10. $x' = -e^y + 1$
    $y' = e^x - 1$

11. $x' = -xy^2 + y^2 - 7xy - x^2 - 6x$
    $y' = x^2 + y$

12. $x' = 1 - xy$
    $y' = x - y^3$

13. $x' = 2y$
    $y' = -2x - y + y^4$

14. a) Convert the equation

$$x'' + ax' + bx + x^2 = 0, \qquad a, b > 0,$$

   into a system.

   b) Show that the origin is a stable focus and the point $(0, -b)$ is a saddle point.

   c) Conclude that the orbits have the form depicted in Fig. 12.20.

15. The *Van de Pol equation*

$$x'' + \varepsilon(x^2 - 1)x' + x = 0 \tag{12.43}$$

   arises in the study of a vacuum tube with three internal elements (triode). Show that if $\varepsilon < 0$, then the origin is asymptotically stable. [It can also be shown that if $\varepsilon > 0$, then (12.43) has a periodic solution that is approached by all other solutions as t tends to ∞. This phenomenon is called a *limit cycle.**]

16. Consider the following special case of the Lotka-Volterra equations (12.28):

$$x' = -x - 2x^2 + xy,$$

$$y' = -y + 7xy - 2y^2, \tag{12.44}$$

   where x and y are measured in hundreds of organisms.

   a) Show that the system (12.44) has four critical points but that only two of them have any biological meaning (negative populations are not permissible).

---

* See, for example, H. K. Wilson, *Ordinary Differential Equations*, Addison-Wesley, Reading, Mass., 1971.

Figure 12.20

b) Show that if both populations are small initially, then both species will become extinct.

c) Show that one biologically meaningful critical point is a saddle point, indicating that larger initial populations may lead to continued existence without the threat of extinction.

*12.4 LYAPUNOV'S METHOD

In the previous section we introduced a method for deriving the stability of nonlinear systems by comparing them with linear ones. This procedure, while often useful, has two obvious drawbacks. First, in the case of a center, no information can be obtained. Second, and much more serious, the method fails to yield information concerning the behavior of nonlinear systems that cannot easily be treated as perturbations of linear systems. Landmark work to resolve these difficulties was done by the Russian mathematician A. A. Lyapunov.* Lyapunov reasoned intuitively as follows. Suppose we have a system of differential equations which arises from a description of a physical system. If a critical point corresponds to a point of minimum potential energy of the system, and if the energy in the system

---

* Lyapunov's original paper on this subject appeared in 1892. His work was essentially unknown until a French translation appeared in 1907 under the title "Problème général de la stabilité du mouvement."

is constant or decreasing, then it is reasonable to "guess" that the critical point will be stable. On the other hand, if the critical point corresponds to a maximum of potential energy, then the point will be unstable.

Example 12.4.1 Consider the vibrations of a mass attached to a coiled spring, the upper end of which is securely fastened (see Section 3.9). If we make the natural assumption that damping forces (friction, air resistance, etc.) are present, then the equation of motion is of the form [from (3.56)]

$$x'' + \varepsilon x' + \mu x = 0. \tag{12.45}$$

For simplicity, we assume $\mu = 1$ so that the system (12.45) can be written as

$$
\begin{aligned}
x' &= y, \\
y' &= -x - \varepsilon y.
\end{aligned}
\tag{12.46}
$$

From the discussion in Section 12.2 it is obvious that the origin of this system is asymptotically stable. We shall use this simple problem to motivate the subsequent discussion.

Physically the potential energy of the system is proportional to the square of x, the distance from the mass to the origin, while the kinetic energy is proportional to the square of the velocity x'. This is so because potential energy is usually given as an integral of position. Here the position is represented by x, so that the potential energy is proportional to $\int x \, dx = x^2/2$. Kinetic energy is given by the formula

$$KE = mv^2 = m \left(\frac{dx}{dt} \right)^2 = my^2.$$

Thus the function $V(x,y)$, defined by

$$V(x,y) = \tfrac{1}{2}(x^2 + y^2), \tag{12.47}$$

where $y = x'$, is representative of the total energy of the system. Now, at the origin potential energy is a minimum, so that the origin can be expected to be stable. The function $V(x,y)$ defined in (12.47) will shortly be used to prove that the origin is indeed stable.

We can generalize these remarks. Let us consider the autonomous system

$$
\begin{aligned}
x' &= f(x,y), \\
y' &= g(x,y),
\end{aligned}
\tag{12.48}
$$

where we assume that the origin is a critical point. Let $V(x,y)$ be a continuous real-valued function on the xy-plane with continuous first partial derivatives. Let D be a region containing the origin and suppose that $V(0,0) = 0$ and $V(x,y) > 0$ for all other points (x,y) in D. Then $V(x,y)$ is said to be *positive definite* in D. If $V(0,0) = 0$ and $V(x,y) < 0$ for all other points (x,y) in D, we say that $V(x,y)$ is *negative definite* in D. If $V(x,y) \geqslant 0$ [or $V(x,y) \leqslant 0$], then the function is said to be *positive semidefinite* [or *negative semidefinite*] in D. If $V(x,y)$ satisfies none of these conditions, then V is said to be *indefinite* in D.

Example 12.4.2 (a) The function $V(x,y)$ given in (12.47) is positive definite for all real values of x and y, since $x^2 + y^2$ is the square of the distance from (x,y) to the origin. (b) The function $-(x^2 + y^2)$ is negative definite. (c) The function $V(x,y) = x^2$ is positive semidefinite, since $V(0,a) = 0$ for any real value of a. (d) The function $V(x,y) = -y^2$ is negative semidefinite. (e) The function $V(x,y) = xy$ is indefinite, since $V(a,a) = a^2 > 0$ and $V(a,-a) = -a^2 < 0$ for all numbers $a > 0$.

Often the total energy of a system is a polynomial in x and y, in which case the following theorem is useful.

Theorem 12.4 The function

$$V(x,y) = x^2 + axy + by^2$$

is (a) positive definite if and only if $4b - a^2 > 0$, (b) positive semidefinite if and only if $4b - a^2 = 0$.

Proof. If $4b - a^2 > 0$, then completing the squares, we find that

$$x^2 + axy + by^2 = \left(x + \frac{a}{2}y\right)^2 + \left(b - \frac{a^2}{4}\right)y^2. \tag{12.49}$$

If $y \neq 0$, then $[b - (a^2/4)]y^2 > 0$ by assumption and $V(x,y)$ is positive definite. On the other hand, suppose that $[b - (a^2/4)] \leqslant 0$. Then choosing any nonzero y and $x = -(a/2)y$, we find the expression (12.49) to be less than or equal to zero, which is a contradiction, since V is positive definite. Thus $4b - a^2 > 0$, and the first part of the theorem is proved. The second part follows in the same way.

Note that $V(x,y)$ is negative definite if and only if $-V(x,y)$ is positive definite, so that Theorem 12.4 is also a theorem about negative definiteness (and semidefiniteness).

Example 12.4.3 (a) The function $x^2 - xy + 2y^2$ is positive definite, since $4b - a^2 = 7 > 0$. (b) The function $V(x,y) = -x^2 + 4xy - 4y^2$ is negative semidefinite, since $-V(x,y) = x^2 - 4xy + 4y^2$, $4b - a^2 = 0$, and $-V$ is positive semidefinite. (c) The function $x^2 + 4xy - 4y^2$ is indefinite, since neither V nor $-V$ belongs to any of the other four categories.

Now let $V(x,y)$ be a continuously differentiable, positive definite function. We define the *derivative of V along the orbits of the system* (12.48) by

$$V'(x,y) = \frac{\partial V}{\partial x}x' + \frac{\partial V}{\partial y}y' = \frac{\partial V}{\partial x}f(x,y) + \frac{\partial V}{\partial y}g(x,y). \tag{12.50}$$

When this association between $V(x,y)$ and the system (12.48) holds, $V(x,y)$ is called a *Lyapunov function* for the system. We emphasize that for it to be a Lyapunov function for the system (12.48), $V(x,y)$ must be continuously differentiable, positive definite, and have its derivative along the orbits defined by (12.50). We should also emphasize that $V(x,y)$ is a Lyapunov function for (12.48) only when $V'(x,y)$ is defined with respect to (12.48) according to (12.50). There will always be many choices for a Lyapunov function. Often, however, most of these functions will not be useful. The great importance of defining the derivative of V along the orbits of (12.48) is given in the next theorem.

Theorem 12.5 Let $V(x,y)$ be a Lyapunov function for the system (12.48). Then:
  i) if $V'(x,y)$ is negative semidefinite, the origin is stable;
 ii) if $V'(x,y)$ is negative definite, the origin is asymptotically stable;
iii) if $V'(x,y)$ is positive definite, the origin is unstable.

*Proof.** i) Let $\varepsilon > 0$ be given. We must show that there is a $\delta > 0$ such that if (x_0,y_0) is in D and $\sqrt{x_0^2 + y_0^2} < \delta$, then the solution pair $(x(t,x_0),y(t,y_0))$ (see Section 12.2) satisfies the inequality

$$\sqrt{[x(t,x_0)]^2 + [y(t,y_0)]^2} < \varepsilon$$

for all $t \geqslant 0$. We define

$$m = \min_{\sqrt{x^2+y^2}=\varepsilon} V(x,y).$$

This minimum exists since $V(x,y)$ is continuous on the circle $x^2 + y^2 = \varepsilon^2$. Furthermore, by the continuity of V and the fact that $V(0,0) = 0$, there exists a $\delta > 0$ such that $V(x,y) < m$ whenever $\sqrt{x_0^2 + y_0^2} < \delta$. (See Fig. 12.21.) Suppose that there is a number $t^* > 0$ for which $x(t^*,x_0)^2 + y(t^*,y_0)^2 = \varepsilon^2$. Since $dV/dt = V'(x,y) \leqslant 0$, we have $V(x(t,x_0),y(t,y_0)) \leqslant V(x_0,y_0)$ for all $t \geqslant 0$. Then

$$V(x(t^*,x_0),y(t^*,y_0)) \leqslant V(x_0,y_0) < m \leqslant V(x(t^*,x_0),y(t^*,y_0)),$$

since m is a minimum on the circle $x^2 + y^2 = \varepsilon^2$. This contradiction establishes (i).

ii) By hypothesis, $V(x,y)$ decreases as t increases. Since V is positive for all x and y and decreasing along the orbits,

$$\lim_{t \to \infty} V(x(t),y(t)) = \lambda \geqslant 0$$

exists. We must show that $\lambda = 0$ since, then, $(0,0)$ being the only point for which V is zero, we will necessarily have

$$\lim_{t \to \infty} x(t) = \lim_{t \to \infty} y(t) = 0.$$

---

* This proof is difficult and may be omitted without loss of continuity.

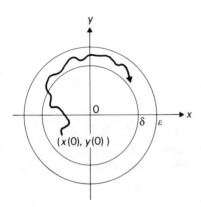

Figure 12.21

Suppose $\lambda > 0$. Then, as before, there is a $\eta > 0$ such that $V(x_0,y_0) < \lambda$ whenever $\sqrt{x_0^2 + y_0^2} < \eta$, meaning that the orbit $(x(t),y(t))$ never enters the circular region described by $x^2 + y^2 < \eta^2$. Let ε be as in part (i) and let

$$m_1 = \min_{\eta \leqslant \sqrt{x^2 + y^2} \leqslant \varepsilon} - V'(x,y).$$

Since $\eta > 0$ and V' is negative definite, we have $m_1 > 0$. Then, since $m_1 \leqslant -V'(x,y)$, we have $V'(x(t),y(t)) \leqslant -m_1$ for all $t \geqslant 0$. Therefore, we obtain upon integration

$$\int_0^t V'(x(s),y(s))\, ds \leqslant \int_0^t -m_1\, ds$$

or

$$V(x(t),y(t)) - V(x_0,y_0) \leqslant -m_1 t$$

or

$$V(x(t),y(t)) \leqslant V(x_0,y_0) - m_1 t.$$

But the right hand side of the last inequality becomes negative when t approaches ∞, thereby contradicting the positive definiteness of V unless $m_1 = 0$. But $m_1 = 0$ if and only if $\eta = 0$, which holds if and only if $\lambda = 0$. Therefore, (ii) is proved.

    iii) The proof of part (iii) is similar and therefore left as an exercise (see Exercise 6).

    The rest of this section indicates the wide range of applicability of Theorem 12.5.

Example 12.4.4 In Example 12.4.1, the function $V(x,y) = \frac{1}{2}(x^2 + y^2)$ is a Lyapunov function. In addition, by (12.46),

$$V'(x,y) = \frac{\partial V}{\partial x} x' + \frac{\partial V}{\partial y} y' = xy + y(-x - \varepsilon y) = -\varepsilon y^2,$$

which is negative semidefinite. Hence Theorem 12.5 implies that the origin is stable. Observe that Theorem 12.5 does not provide the best possible result since we know the origin to be asymptotically stable. Stronger results are known and can be found in many advanced differential equations books.*

Example 12.4.5 In our example of the movement of a mass attached to a frictionless spring, we assumed that the restoring force was proportional to the distance of the mass to the origin. More generally, we may assume the restoring force to be a nonlinear function $-f(x)$ of the distance to the origin. The equation of motion is then

$$x'' + f(x) = 0 \tag{12.51}$$

To analyze this equation, we make the assumptions that

$$f(0) = 0 \quad \text{and} \quad xf(x) > 0 \qquad \text{for } x \neq 0; \tag{12.52}$$

that is, $f(x)$ and x have the same sign. Equation (12.51), together with the conditions (12.52), is often called the equation of a *nonlinear spring*. We now write (12.51) as

$$x' = y,$$
$$y' = -f(x). \tag{12.53}$$

Clearly the origin is the only critical point of the system. The kinetic energy of this system is proportional to the square of the velocity $x'\,(=y)$, while the potential energy at any point x is given by

$$F(x) = \int_0^x f(x)\, dx.$$

Hence the total energy of the system is

$$V(x,y) = F(x) + \frac{y^2}{2}.$$

Since $f(x)$ and x have the same sign, $F \geq 0$ and $V(x,y)$ is positive definite. Moreover, $V'(x,y)$ is negative semidefinite since

$$V'(x,y) = F'(x)x' + yy' = f(x)y + y(-f(x)) = 0,$$

---

* See, for example, J. P. LaSalle and S. Lefshetz, *Stability by Lyapunov's Direct Method*, Academic Press, New York, 1961.

implying by Theorem 12.5 that the origin is stable. In this case, as in the linear case $x'' + x = 0$, it can be shown that the origin is a center.

Example 12.4.6 Consider the system

$$x' = -x - \frac{x^3}{3} - x \sin y,$$

$$y' = -y - \frac{y^3}{3}. \tag{12.54}$$

Here the origin is the only critical point. We define $V(x,y) = \frac{1}{2}(x^2 + y^2)$. Then

$$V'(x,y) = x\left(-x - \frac{x^3}{3} - x \sin y\right) + y\left(-y - \frac{y^3}{3}\right)$$

$$= -x^2 - \frac{x^4}{3} - y^2 - \frac{y^4}{3} - x^2 \sin y.$$

Now $|x^2 \sin y| \leqslant |x^2|$, so that $x^2 + x^2 \sin y \geqslant 0$. Hence

$$V'(x,y) = -\frac{x^4}{3} - y^2 - \frac{y^4}{3} - (x^2 + x^2 \sin y)$$

$$\leqslant -\frac{x^4}{3} - y^2 - \frac{y^4}{3}.$$

Therefore, V' is negative definite and the origin is asymptotically stable.

In the last example the Lyapunov function was pulled out of a hat. Actually, the problem of finding a Lyapunov function for a particular system is very difficult, especially if there is no way of estimating the total energy of the system. While a Lyapunov function cannot be found in every case, it is often worthwhile in the case of a system of two first order equations to try to find one of the form $V(x,y) = x^2 + axy + by^2$. By Theorem 12.4, such a Lyapunov function will be positive definite if $4b > a^2$, as illustrated by the next example.

Example 12.4.7 Consider the system

$$x' = xy^2 + x^2y + x^3,$$

$$y' = y^3 - x^3. \tag{12.55}$$

Assume that $V(x,y) = x^2 + axy + by^2$ and $4b > a^2$. Then by Theorem 12.4, V is positive definite and

$$V' = 2x(xy^2 + x^2y + x^3) + ax(y^3 - x^3)$$
$$+ ay(xy^2 + x^2y + x^3) + 2by(y^3 - x^3).$$

If we choose $a = 0$ and $b = 1$, then

$$V' = 2x^2y^2 + 2x^3y + 2x^4 + 2y^4 - 2yx^3$$
$$= 2(x^2y^2 + x^4 + y^4),$$

which is positive definite. By Theorem 12.5, the origin is unstable.

EXERCISES 12.4

1. Show that the zero solution of the system

$$x' = xy^2 - \frac{x^3}{2},$$

$$y' = -\frac{y^3}{2} + \frac{yx^2}{5}$$

is asymptotically stable. [*Hint*: Try a Lyapunov function of the form $V(x,y) = ax^2 + by^2$.]

2. Show that the zero solution of the system

$$x' = -6x^2y$$
$$y' = -3y^2 + 6x^3$$

is stable.

3. Consider the undamped pendulum of Example 12.1.2:

$$\theta'' + \omega^2 \sin\theta = 0, \qquad -\pi/2 \leqslant \theta \leqslant \pi/2.$$

Prove by means of an appropriate Lyapunov function that the origin is stable. [*Hint*: Use the result of Example 12.4.5.]

4. Consider the system

$$x' = y - xf(x,y),$$

$$y' = -x - yf(x,y),$$

where $f(0,0) = 0$ and $f(x,y)$ has a convergent power series expansion in a region D around the origin. Show that the zero solution is:

a) stable if $f(x,y) \geqslant 0$ in some region around $(0,0)$;
b) asymptotically stable if $f(x,y)$ is positive definite in some region around $(0,0)$; and
c) unstable if in every region around $(0,0)$ there are points (x,y) such that $f(x,y) < 0$.

5. Using the results of Exercise 4, examine the stability properties of the zero solutions of the following systems:

a) $x' = y - x(y^3 \sin^2 x)$
   $y' = -x - y(y^3 \sin^2 x)$
b) $x' = y - x(x^4 + y^6),$
   $y' = -x - y(x^4 + y^6)$

c) $x' = y - x(\sin^2 y)$
   $y' = -x - y(\sin^2 y)$

* 6. Prove part (iii) of Theorem 12.5.

* 7. Consider the equation

$$x'' + f(x,x') + g(x) = 0. \tag{12.56}$$

Assume that f and g possess continuous first derivatives, $f(0,0) = g(0) = 0$, and that $yf(x,y) > 0$ when $y \neq 0$ and $xg(x) > 0$ for $x \neq 0$. After writing (12.56) as a system, prove that the origin is stable.

8. Use the result of Exercise 7 to prove the stability of the solutions of the equation

$$x'' + x'^3 + x^5 = 0.$$

(Note that the associated linear system does us no good here since the zero solution of this system is a center.)

* 9. Consider the linear system

$$x' = a_{11}x + a_{12}y,$$
$$y' = a_{21}x + a_{22}y, \tag{12.57}$$

and assume that the zero solution is asymptotically stable (that is, the real parts of the roots of the auxiliary equation are negative). Let $(x_1(t),y_1(t))$ and $(x_2(t),y_2(t))$ be the solutions of (12.57) that satisfy the conditions

$$(x_1(0),y_1(0)) = (1,0) \quad \text{and} \quad (x_2(0),y_2(0)) = (0,1).$$

Define the function $V(x,y)$ by

$$V(x,y) = x^2 \int_0^\infty [x_1^2(t) + y_1^2(t)] \, dt + 2xy \int_0^\infty [x_1(t)x_2(t) + y_1(t)y_2(t)] \, dt$$

$$+ y^2 \int_0^\infty [x_2^2(t) + y_2^2(t)] \, dt. \tag{12.58}$$

   a) Show that $V(x,y)$ is positive definite.
   b) Show that $V'(x,y)$ is negative definite along the orbits of (12.57).

10. Use the result of Exercise 9 to find a Lyapunov function that could be used to prove the asymptotic stability of the following systems:

   a) $x' = -4x - y$ b) $x' = -5x - 3y$
      $y' = x - 2y$ $y' = 4x + 2y$

* 11. Consider the system

$$x' = a_{11}x + a_{12}y + f(x,y),$$
$$y' = a_{21}x + a_{22}y + g(x,y), \tag{12.59}$$

where f and g are "small" near the origin in the sense of property (12.25). Assume that the zero solution of the associated linear system is asymptotically stable. Prove that the zero solution of (12.59) is asymptotically stable. [*Hint*: Prove that there is a region D around $(0,0)$ such that the function $V(x,y)$ defined in Exercise 9 is a Lyapunov function for (12.59) and that $V'(x,y) < 0$ in D.]

INTEGRAL
TABLES

Appendix 1

1. $\displaystyle\int u\,dv = uv - \int v\,du$

2. $\displaystyle\int x^n\,dx = \frac{x^{n+1}}{n+1} + c, \quad n \neq -1$

3. $\displaystyle\int \frac{dx}{x} = \ln|x| + c$

4. $\displaystyle\int e^{ax}\,dx = \frac{e^{ax}}{a} + c$

5. $\displaystyle\int \ln x\,dx = x\ln x - x + c$

6. $\displaystyle\int \frac{dx}{a^2 + x^2} = \frac{1}{a}\tan^{-1}\frac{x}{a} + c$

7. $\displaystyle\int \frac{dx}{a^2 - x^2} = \frac{1}{2a}\ln\left|\frac{a+x}{a-x}\right| + c = \frac{1}{a}\tanh^{-1}\frac{x}{a} + c$

8. $\displaystyle\int \frac{dx}{x^2 - a^2} = \frac{1}{2a}\ln\left|\frac{x-a}{x+a}\right| + c = -\frac{1}{a}\coth^{-1}\frac{x}{a} + c$

9. $\displaystyle\int \frac{dx}{\sqrt{x^2 \pm a^2}} = \ln\left|x + \sqrt{x^2 \pm a^2}\right| + c$

10. $\displaystyle\int \frac{dx}{\sqrt{a^2 - x^2}} = \sin^{-1}\frac{x}{a} + c$

11. $\displaystyle\int \frac{x\,dx}{\sqrt{x^2 \pm a^2}} = \sqrt{x^2 \pm a^2} + c$

12. $\displaystyle\int \frac{x\,dx}{\sqrt{a^2 - x^2}} = -\sqrt{a^2 - x^2} + c$

13. $\displaystyle\int \sqrt{x^2 \pm a^2}\,dx = \frac{x}{2}\sqrt{x^2 \pm a^2} \pm \frac{a^2}{2}\ln\left|x + \sqrt{x^2 \pm a^2}\right| + c$

14. $\displaystyle\int \sqrt{a^2 - x^2}\,dx = \frac{x}{2}\sqrt{a^2 - x^2} + \frac{a^2}{2}\sin^{-1}\frac{x}{a} + c$

15. $\displaystyle\int \frac{\sqrt{x^2 - a^2}}{x}\,dx = \sqrt{x^2 - a^2} - |a|\sec^{-1}\frac{x}{a} + c.$

16. $\displaystyle\int \frac{\sqrt{a^2 \pm x^2}}{x}\,dx = \sqrt{a^2 \pm x^2} - a\ln\left|\frac{a + \sqrt{a^2 \pm x^2}}{x}\right| + c$

17. $\displaystyle\int \sin x\,dx = -\cos x + c$

18. $\displaystyle\int \sin^n x\,dx = -\frac{1}{n}\sin^{n-1} x\cos x + \frac{n-1}{n}\int \sin^{n-2} x\,dx$

19. $\int \sin mx \sin nx \, dx = \dfrac{\sin (m - n)x}{2(m - n)} - \dfrac{\sin (m + n)x}{2(m + n)} + c, \quad m^2 \neq n^2$

20. $\int \dfrac{dx}{a + b \sin x} = \dfrac{2}{\sqrt{a^2 - b^2}} \tan^{-1} \dfrac{a \tan x/2 + b}{\sqrt{a^2 - b^2}} + c, \quad a^2 > b^2$

21. $\int x^n \sin x \, dx = -x^n \cos x + n \int x^{n-1} \cos x \, dx$

22. $\int \cos x \, dx = \sin x + c$

23. $\int \cos^n x \, dx = \dfrac{1}{n} \cos^{n-1} x \sin x + \dfrac{n - 1}{n} \int \cos^{n-2} x \, dx$

24. $\int \cos mx \cos nx \, dx = \dfrac{\sin (m - n)x}{2(m - n)} + \dfrac{\sin (m + n)x}{2(m + n)} + c, \quad m^2 \neq n^2$

25. $\int \dfrac{dx}{a + b \cos x} = \dfrac{2}{\sqrt{a^2 - b^2}} \tan^{-1} \dfrac{\sqrt{a^2 - b^2} \tan (x/2)}{a + b} + c, \quad a^2 > b^2$

26. $\int x^n \cos x \, dx = x^n \sin x - n \int x^{n-1} \sin x \, dx$

27. $\int \sin mx \cos nx \, dx = \dfrac{\cos (n - m)x}{2(n - m)} - \dfrac{\cos (n + m)x}{2(n + m)} + c, \quad m^2 \neq n^2$

28. $\int \sin x \cos x \, dx = \frac{1}{2} \sin^2 x + c$

29. $\int \tan x \, dx = -\ln |\cos x| + c$

30. $\int \tan^n x \, dx = \dfrac{\tan^{n-1} x}{n - 1} - \int \tan^{n-2} x \, dx, \quad n \neq 1$

31. $\int \cot x \, dx = \ln |\sin x| + c$

32. $\int \cot^n x \, dx = -\dfrac{\cot^{n-1} x}{n - 1} - \int \cot^{n-2} x \, dx, \quad n \neq 1$

33. $\int \sec x \, dx = \ln |\sec x + \tan x| + c$

34. $\int \sec^2 x \, dx = \tan x + c$

35. $\int \sec^n x \, dx = \dfrac{1}{n - 1} \tan x \sec^{n-2} x + \dfrac{n - 2}{n - 1} \int \sec^{n-2} x \, dx$

36. $\int \csc x \, dx = \ln |\csc x - \cot x| + c$

37. $\int \csc^2 x \, dx = -\cot x + c$

38. $\int \csc^n x \, dx = -\dfrac{1}{n - 1} \cot x \csc^{n-2} x + \dfrac{n - 2}{n - 1} \int \csc^{n-2} x \, dx$

39. $\displaystyle\int \sin^{-1}\frac{x}{a}\,dx = x\sin^{-1}\frac{x}{a} + \sqrt{a^2 - x^2} + c$

40. $\displaystyle\int \cos^{-1}\frac{x}{a}\,dx = x\cos^{-1}\frac{x}{a} - \sqrt{a^2 - x^2} + c$

41. $\displaystyle\int \tan^{-1}\frac{x}{a}\,dx = x\tan^{-1}\frac{x}{a} - \frac{a}{2}\ln(a^2 + x^2) + c$

42. $\displaystyle\int \cot^{-1}\frac{x}{a}\,dx = x\cot^{-1}\frac{x}{a} + \frac{a}{2}\ln(a^2 + x^2) + c$

43. $\displaystyle\int \sec^{-1}\frac{x}{a}\,dx = x\sec^{-1}\frac{x}{a} - a\ln\left|x + \sqrt{x^2 - a^2}\right| + c$

44. $\displaystyle\int \csc^{-1}\frac{x}{a}\,dx = x\csc^{-1}\frac{x}{a} + a\ln\left|x + \sqrt{x^2 - a^2}\right| + c$

45. $\displaystyle\int x^n \ln(ax)\,dx = \frac{x^{n+1}}{n+1}\left[\ln(ax) - \frac{1}{n+1}\right] + c, \quad n \neq -1$

46. $\displaystyle\int (\ln x)^n\,dx = x(\ln x)^n - n\int (\ln x)^{n-1}\,dx, \quad n \neq -1$

47. $\displaystyle\int \frac{(\ln x)^n}{x}\,dx = \frac{(\ln x)^{n+1}}{n+1} + c$

48. $\displaystyle\int \frac{dx}{x\ln x} = \ln(\ln x) + c$

49. $\displaystyle\int x^n e^{ax}\,dx = \frac{x^n e^{ax}}{a} - \frac{n}{a}\int x^{n-1}e^{ax}\,dx$

50. $\displaystyle\int e^{ax}\sin bx\,dx = \frac{e^{ax}}{a^2 + b^2}(a\sin bx - b\cos bx) + c$

51. $\displaystyle\int e^{ax}\cos bx\,dx = \frac{e^{ax}}{a^2 + b^2}(a\cos bx + b\sin bx) + c$

52. $\displaystyle\int \sinh x\,dx = \cosh x + c$

53. $\displaystyle\int \cosh x\,dx = \sinh x + c$

54. $\displaystyle\int \tanh x\,dx = \ln(\cosh x) + c$

55. $\displaystyle\int_0^\infty x^{n-1}e^{-x}\,dx = \Gamma(n) = (n-1)\Gamma(n-1)$

56. $\displaystyle\int_0^\infty \frac{e^{-x}}{\sqrt{x}}\,dx = \Gamma(\tfrac{1}{2}) = \sqrt{\pi}$

57. $\displaystyle\int_0^{\pi/2} \sin^n x \, dx = \int_0^{\pi/2} \cos^n x \, dx = \begin{cases} \dfrac{\pi}{2^{n+1}} \dfrac{n!}{(n/2)!^2}, & n \text{ even} \\[2ex] \dfrac{2^{n-1}[(n-1)/2]!^2}{n!}, & n \text{ odd} \end{cases}$

58. $\displaystyle\int_0^\infty e^{-a^2 x^2} \, dx = \dfrac{1}{2a} \Gamma(\tfrac{1}{2}) = \dfrac{\sqrt{\pi}}{2a}, \quad a > 0$

$$\mu \frac{d\mu}{dx} = \frac{\partial M/\partial y - \partial N/\partial x}{N} \quad \text{dep on } x$$

$$\mu \frac{d\mu}{dy} = \frac{\partial M/\partial y - \partial N/\partial x}{-M} \quad \text{dep on } y$$

LAPLACE
TRANSFORMS

Appendix 2

For a more extensive list of Laplace transforms and their inverses see A. Erdelyi, et. al., *Tables of Integral Transforms* (2 vols.), McGraw-Hill, New York, 1954.

$F(s) = \mathcal{L}\{f(t)\}$	$f(t)$
1. $\dfrac{1}{s^r}, \quad r > 0$	$\dfrac{t^{r-1}}{\Gamma(r)}, \quad \Gamma(n+1) = n!$
2. $\dfrac{1}{(s-a)^r}, \quad r > 0$	$\dfrac{t^{r-1}e^{at}}{\Gamma(r)}$
3. $\dfrac{1}{(s-a)(s-b)}, \quad a \neq b$	$\dfrac{e^{at} - e^{bt}}{a - b}$
4. $\dfrac{s}{(s-a)(s-b)}, \quad a \neq b$	$\dfrac{ae^{at} - be^{bt}}{a - b}$
5. $\dfrac{1}{s^2 + k^2}$	$\dfrac{1}{k} \sin kt$
6. $\dfrac{s}{s^2 + k^2}$	$\cos kt$
7. $\dfrac{1}{s^2 - k^2}$	$\dfrac{1}{k} \sinh kt$
8. $\dfrac{s}{s^2 - k^2}$	$\cosh kt$
9. $\dfrac{1}{(s-a)^2 + k^2}$	$\dfrac{1}{k} e^{at} \sin kt$
10. $\dfrac{s-a}{(s-a)^2 + k^2}$	$e^{at} \cos kt$
11. $\dfrac{1}{(s-a)^2 - k^2}$	$\dfrac{1}{k} e^{at} \sinh kt$
12. $\dfrac{s-a}{(s-a)^2 - k^2}$	$e^{at} \cosh kt$
13. $\dfrac{1}{s(s^2 + k^2)}$	$\dfrac{1}{k^2} (1 - \cos kt)$
14. $\dfrac{1}{s^2(s^2 + k^2)}$	$\dfrac{1}{k^3} (kt - \sin kt)$
15. $\dfrac{1}{(s^2 + k^2)^2}$	$\dfrac{1}{2k^3} (\sin kt - kt \cos kt)$
16. $\dfrac{s}{(s^2 + k^2)^2}$	$\dfrac{t}{2k} \sin kt$

$F(s) = \mathcal{L}\{f(t)\}$	$f(t)$
17. $\dfrac{s^2}{(s^2 + k^2)^2}$	$\dfrac{1}{2k}(\sin kt + kt \cos kt)$
18. $\dfrac{1}{(s^2 + a^2)(s^2 + b^2)}, \quad a^2 \neq b^2$	$\dfrac{a \sin bt - b \sin at}{ab(a^2 - b^2)}$
19. $\dfrac{s}{(s^2 + a^2)(s^2 + b^2)}, \quad a^2 \neq b^2$	$\dfrac{\cos bt - \cos at}{a^2 - b^2}$
20. $\dfrac{1}{s^4 - k^4}$	$\dfrac{1}{2k^3}(\sinh kt - \sin kt)$
21. $\dfrac{s}{s^4 - k^4}$	$\dfrac{1}{2k^2}(\cosh kt - \cos kt)$
22. $\dfrac{1}{s^4 + 4k^4}$	$\dfrac{1}{4k^3}(\sin kt \cosh kt - \cos kt \sinh kt)$
23. $\dfrac{s}{s^4 + 4k^4}$	$\dfrac{1}{2k^2} \sin kt \sinh kt$
24. $\sqrt{s - a} - \sqrt{s - b}$	$\dfrac{e^{bt} - e^{at}}{2\sqrt{\pi t^3}}$
25. $\dfrac{s}{(s - a)^{3/2}}$	$\dfrac{e^{at}}{\sqrt{\pi t}}(1 - 2at)$
26. $\dfrac{1}{\sqrt{s + a}\sqrt{s + b}}$	$e^{-(a+b)t/2} I_0\left(\dfrac{a - b}{2}t\right)$
27. $\dfrac{(\sqrt{s^2 + k^2} - s)^r}{\sqrt{s^2 + k^2}}, \quad r > -1$	$k^r J_r(kt)$
28. $\dfrac{1}{(s^2 + k^2)^r}, \quad r > 0$	$\dfrac{\sqrt{\pi}}{\Gamma(r)}\left(\dfrac{t}{2k}\right)^{r-1/2} J_{r-1/2}(kt)$
29. $(\sqrt{s^2 + k^2} - s)^r, \quad r > 0$	$\dfrac{rk^r}{t} J_r(kt)$
30. $\dfrac{(s - \sqrt{s^2 - k^2})^r}{\sqrt{s^2 - k^2}}, \quad r > -1$	$k^r I_r(kt)$
31. $\dfrac{1}{(s^2 - k^2)^r}, \quad r > 0$	$\dfrac{\sqrt{\pi}}{\Gamma(r)}\left(\dfrac{t}{2k}\right)^{r-1/2} I_{r-1/2}(kt)$
32. $\dfrac{e^{-k/s}}{s^r}, \quad r > 0$	$\left(\dfrac{t}{k}\right)^{(r-1)/2} J_{r-1}(2\sqrt{kt})$
33. $\dfrac{e^{-k/s}}{\sqrt{s}}$	$\dfrac{1}{\sqrt{\pi t}} \cos 2\sqrt{kt}$

$F(s) = \mathscr{L}\{f(t)\}$	$f(t)$

34. $\dfrac{e^{k/s}}{s^r}, \quad r > 0$ $\qquad$ $\left(\dfrac{t}{k}\right)^{(r-1)/2} I_{r-1}(2\sqrt{kt})$

35. $\dfrac{e^{k/s}}{\sqrt{s}}$ $\qquad$ $\dfrac{1}{\sqrt{\pi t}} \cosh 2\sqrt{kt}$

36. $-\dfrac{1}{s} \ln s$ $\qquad$ $-\ln t - \gamma, \quad \gamma \approx 0.5772$

37. $\ln \dfrac{s-a}{s-b}$ $\qquad$ $\dfrac{e^{bt} - e^{at}}{t}$

38. $\ln\left(1 + \dfrac{k^2}{s^2}\right)$ $\qquad$ $\dfrac{2}{t}(1 - \cos kt)$

39. $\ln\left(1 - \dfrac{k^2}{s^2}\right)$ $\qquad$ $\dfrac{2}{t}(1 - \cosh kt)$

40. $\arctan\left(\dfrac{k}{s}\right)$ $\qquad$ $\dfrac{\sin kt}{t}$

41. $\dfrac{1}{s} \arctan\left(\dfrac{k}{s}\right)$ $\qquad$ $\mathrm{Si}\,(kt) = \displaystyle\int_0^{kt} \dfrac{\sin u}{u}\, du$

42. $e^{-r\sqrt{s}}, \quad r > 0$ $\qquad$ $\dfrac{r}{2\sqrt{\pi t^3}} \exp\left(-\dfrac{r^2}{4t}\right)$

43. $\dfrac{e^{-r\sqrt{s}}}{s}, \quad r \geqslant 0$ $\qquad$ $1 - \mathrm{erf}\left(\dfrac{r}{2\sqrt{t}}\right) = 1 - \dfrac{2}{\sqrt{\pi}} \displaystyle\int_0^{r/2\sqrt{t}} e^{-u^2}\, du$

44. $e^{r^2 s^2}(1 - \mathrm{erf}(rs)), \quad r > 0$ $\qquad$ $\dfrac{1}{r\sqrt{\pi}} \exp\left(-\dfrac{t^2}{4r^2}\right)$

45. $\dfrac{1}{s} e^{r^2 s^2}(1 - \mathrm{erf}(rs)), \quad r > 0$ $\qquad$ $\mathrm{erf}\left(\dfrac{t}{2r}\right)$

46. $\mathrm{erf}\left(\dfrac{r}{\sqrt{s}}\right)$ $\qquad$ $\dfrac{1}{\pi t} \sin (2k\sqrt{t})$

47. $e^{rs}(1 - \mathrm{erf}\sqrt{rs}), \quad r > 0$ $\qquad$ $\dfrac{\sqrt{r}}{\pi\sqrt{t(t+r)}}$

48. $\dfrac{1}{\sqrt{s}} e^{rs}(1 - \mathrm{erf}\sqrt{rs}), \quad r > 0$ $\qquad$ $\dfrac{1}{\sqrt{\pi(t+r)}}$

FORTRAN-BASIC
PRIMER

Appendix 3

This primer is designed to provide the minimum necessary vocabulary needed to write simple programs in two computer languages: FORTRAN and BASIC. If the reader is unfamiliar with computer languages, we recommend that he start with BASIC, since it is easier to learn. Later, if he desires, a knowledge of FORTRAN will allow him greater flexibility in the preparation and transmittal of information with a computer.

Every computer has an interval vocabulary for the elementary operations, such as adding, shifting, and storing, that it is designed to perform. This vocabulary is called the *machine language* for that computer. Machine languages change from computer to computer, since each computer has different capabilities. In addition, many routine procedures, such as input or output of data, require that complicated (and often different) sets of machine language instructions be performed. The FORTRAN (FORmula TRANslation) language was developed to avoid both of these problems by

i) providing a universal language which can be used on any computer, and

ii) performing certain routine procedures with a single statement.

It accomplishes this task by providing the user with a vocabulary of mathematical-like statements which must be used in accordance with precisely formulated rules. The user writes the sequence of FORTRAN statements he wishes the computer to perform, and transmits this list to the computer's FORTRAN compiler. This sequence of statements is called a *FORTRAN program*. The compiler then translates these statements into a sequence of machine language instructions, called a *machine language program*. Finally, the machine language program is executed by the computer. Although FORTRAN programs can be used on all computers having a FORTRAN compiler, the resulting machine language programs are generally not interchangeable.

The BASIC language works in very much the same way. The main difference between the two languages is that BASIC allows for easier handling of the computer, especially if the computer has a time-sharing capability.

The following steps must be followed in order to execute a computer program successfully.

1. *Gain access to the computer's monitor.* The monitor is a system that directs the flow of communications between the user and the various compilers and other systems available to the user.

2. *Gain access to the BASIC or FORTRAN compiler.* This is done by giving a specific command to the monitor.

3. *Create a BASIC or FORTRAN program* by giving a sequence of BASIC or FORTRAN instructions to the computer.

4. *Translate the above program into a machine program.* The compilers will produce a machine program only when no violations of the rules of the machine are encountered. If any rule is violated, the nature of the mistake will be communicated by the compiler to the user, who must then amend his BASIC or FORTRAN program.

5. *Execute the machine program.*

6. *Relinquish control of the computer.*

In this primer we shall discuss only Step 3, since the other steps may differ from one computer system to another. Generally, with most computer systems, one or two commands are all that is necessary to perform Steps 1, 2, 4, 5, and 6. (Most computer users will be delighted to provide you with this information.) In addition, we shall assume that the user is gaining access to the compiler from a teletype terminal, since this is a common practice.

Each line of type constitutes a *statement* in each of the languages. It is important that these statements be prepared exactly as described, since failure to do so will probably result in a *compiler error*. If an error is made in the process of typing a statement, you can correct it by pressing the RUBOUT key once for each character, going backward until the error is erased. Then correct this character and continue typing.

The following symbols have the same meaning in both languages:

Arithmetic Operations	Example	Meaning
+	A + B	add B to A
−	A − B	subtract B from A
*	A * B	multiply B by A
/	A / B	divide A by B
**	A ** B	raise A to the power B

Symbols	Example	Meaning
( )	(A + B) * C	multiply C by $A + B$
=	A = B	let A equal B

Functions	Meaning
SIN (A)	sine of the angle A (in radians)
COS (A)	cosine of the angle A (in radians)
EXP (A)	e^A
SQRT (A)	$\sqrt{A}$
ABS (A)	$|A|$

We suggest that in writing formulas a liberal use of parentheses be made to prevent misinterpretation. For example, the ambiguous expression $A + B * C$ might be interpreted algebraically as either $(A + B)C$ or $A + BC$. The computer resolves such difficulties in a set way that need not coincide with the user's. The algebraic equation

$$x = \frac{-B + \sqrt{B^2 - 4AC}}{2A}$$

could be typed as

```
X = (-B + SQRT ((B ** 2.) - (4. * A * C)))/(2. * A)
```

to avoid possible ambiguities.

BASIC

Constants In BASIC language numbers may be positive or negative and contain up to eight digits in decimal form (for example, $0, 2, -3.654, -987.65432$). Numbers which can not be written in this form may be produced by means of the arithmetic operations, for example,

```
1.23 * (10 ** 7), 3.456/(10 ** 8), 1/3
```

Variables Variables in BASIC are denoted by either a single letter or a letter followed by a single digit (for example, A, B7, X0).

Functions In addition to the previously mentioned functions common to both BASIC and FORTRAN, the following are also available in BASIC.

Function	Meaning
LN (A)	$\log_e A = \ln A$
ATN (A)	arctangent of A in radians
INT (A)	greatest integer less than or equal to A

Statements Each line of a BASIC program starts with a line number (of 1 to 5 digits) that serves to indicate the order in which this statement is to be performed. It is good programming practice to number lines by tens, to allow for possible insertions of other statements at a later time. Note that the compiler orders the statements according to their line numbers, which need not be the order in which they are typed. After the line number, skip a space and then type the command you wish the computer to obey (a list of these commands will follow). A typical statement is

```
100 PRINT X, SIN (X)
```

which will cause the computer to type on the teletype the present values of x and sin (x).

BASIC Commands

PRINT [list of variables, formulas, and "expressions"]
Types the present values of the specified variables and reproduces any expression in quotes. For example,

$$53 \text{ PRINT A, A**2, B7, "END"}$$

will type the current values of A, A^2, $B7$, and the word END. If closer spacing of the output is desired, use semicolons instead of commas between the entries in the list.

INPUT [list of variables]
Causes the teletype to type a ? to the user and waits for the user to supply the values of the specified variables. The user must type in these numbers, separated by commas, and must press the RETURN key RET before BASIC will continue the program. For example,

$$74 \text{ INPUT A5, B}$$

might require the user to reply

$$? \ 15.3, \ -9$$

LET [variable] = [formula]
Assigns to the given variable the result obtained by performing the operations of the formula with the current values of all the variables in the formula. *It is not necessary to type the word* LET. For example,

$$49 \text{ X = (-B + SQRT ((B*B) - (4*A*C)))/(2*A)}$$

will use the current values of the variables A, B, and C to calculate $(-B + \sqrt{B^2 - 4AC})/2A$ and assign the resulting value to the variable X.

GO TO [line number]
Transfers control from this line to the line with the specified line number and continues execution from that point. This instruction allows the program to jump from one point to another. Thus

$$87 \text{ GO TO } 42$$

sends the program back to line 42.

IF [formula] relation [formula] THEN [line number]
If the stated relationship between the two formulas is true, then control is transferred to the specified line; if not, then the next statement is executed. The

most commonly used relations are:

Relation	Example	Meaning
=	A = B	A equals B
<	A < B	A is less than B
<=	A <= B	A is less than or equal to B
<>	A <> B	A is not equal to B

A typical example of this command is furnished by the statement

$$124 \ \text{IF} \ A < A**3 \ \text{THEN} \ 515$$

Whenever the value of A is less than that of A^3 the program will transfer to line 515; otherwise, the program will go on to the next statement.

DEF FN [letter] ([variables]) = [formula]
This statement *defines* a *function* specified by three letters, the first two of which are FN. For example,

$$300 \ \text{DEF} \ \text{FNG}(X,Y) = \text{EXP} \ (-X)*\text{SIN} \ (Y)$$

tells the computer how to calculate FNG(X, Y), namely, whenever the function is encountered, the recipe given in line 300 is used in its calculation. For example, the statement

$$92 \ X4 = \text{SQRT} \ (\text{FNG}(X4,9.2))$$

will cause the computer to use 9.2 and the current value of the variable X4 to calculate

$$\text{FNG} \ (X4,9.2) = e^{x4} \sin (9.2),$$

take the square root of this number, and replace the value of X4 by the new number so obtained.

The main advantage of using such a statement is to reduce the job of typing a program in which a given function is used repeatedly. Also it may be employed to write programs which will work for many different functions, which can be adapted to each such function by a change in only one line of the program.

FOR [variable] = [formula$_1$] TO [formula$_2$] STEP [formula$_3$]
NEXT [variable]
These two commands go together and are used to loop repetitively through a series of steps. The variable, which is the same for both statements, is initially set by the value of formula$_1$. The intervening statements are performed until the NEXT statement is reached at which time the variable is increased by the value of formula$_3$. (The portion of the statement "STEP [formula$_3$]" can be

omitted, in which case the variable is increased by $+1$.) The resulting value is compared to formula$_2$. If less than or equal to formula$_2$, control is sent back to the statement following the FOR statement; otherwise, control is sent to the statement following the NEXT statement. For example, the statements

```
10 FOR A = 7 TO 9 STEP 1.2
20 PRINT A
30 NEXT A
40 PRINT A**2
```

would cause the program to set $A = 7$, print 7, increment A by 1.2, obtaining $A = 8.2$, which does not exceed 9. Then return to line 20, print A, and increment, yielding $A = 9.4$. Since 9.4 exceeds 9, the computer would exit from the loop and print $(9.4)^2$.

Loops can be nested as follows:

```
10 FOR A = 1 TO 10
20 FOR B = 3 TO 9 STEP 3
30 PRINT A*B
40 NEXT B
50 NEXT A
```

These statements would cause the computer to type sequentially the numbers 3, 6, 9, 6, 12, 18, 9, 18, 27, 12, 24, 36, . . . , 30, 60, 90.

END

This statement must be present and have the highest line number in every program. It signals to the computer that the job is complete.

Example 1 The following is a simple program for finding the roots of any quadratic equation

$$Ax^2 + Bx + C = 0.$$

The reader should study it carefully to understand why it will perform the indicated task.

```
10 INPUT A, B, C
20 D = (B*B) - (4*A*C)
30 IF D < 0 THEN 80
40 X1 = (-B + SQRT (D))/(2*A)
50 X2 = (-B - SQRT (D))/(2*A)
60 PRINT X1, X2
70 GO TO 110
80 R = -B/(2*A)
90 I = SQRT (-D)/(2*A)
100 PRINT R; "+I"; I, R; "-I"; I
110 END
```

After the values of A, B, C have been entered and the RETURN key has been pressed, the computer will calculate the discriminant D. If D is nonnegative, the roots are real and will be determined by lines 40 and 50 and typed out by line 60. If D is negative, the roots are complex. Then lines 80 and 90 will yield the real and imaginary parts, and the complex conjugate roots will be typed out by line 100. The program ends on line 110.

Example 2 Suppose that we wish to obtain a table of values of the function

$$f(x) = \left|\tan^{-1}\left(e^{x^{1.35}}\right)\right|$$

for 20 values of x between 0.1 and 0.3. The following program will accomplish the job:

```
10 DEF FNF(X) = ABS(ATN(EXP(X**1.35)))
20 FOR X = 0.1 TO 0.3 STEP 0.01
30 PRINT X, FNF(X)
40 NEXT X
50 END
```

The loop at line 20 will cause the values of x and $f(x)$ to be typed as a table. Values are typed for each 0.01 between 0.1 and 0.3.

This short list of BASIC commands is but a small fraction of those available. In fact, in several instances we did not describe all the capabilities of these commands. The interested reader should consult a BASIC handbook for additional information.

FORTRAN

Constants We will define two of the seven types of numbers which are allowed in FORTRAN. *Integer constants* may be positive or negative and consist of from one to eleven digits written *without* a decimal point (e.g., 0, 3, +5, −2, 8127). *Real constants* are positive or negative strings of up to nine digits which include a decimal point. They may be written in scientific notation by appending an E followed by a signed integer constant, indicating the power of ten involved. Some examples of real constants are 72., 0.0, .0359 ($= 3.59E - 2$), 1.23456789E+8. The power of ten should lie between 10^{-37} and 10^{37}.

Variables Variables in FORTRAN consist of one to six alphanumeric characters, the first of which must be alphabetic. If this first letter is an I, J, K, L, M, or N, then we have an *integer variable*. Any other first letter indicates a *real variable*.

Functions In addition to functions common to both BASIC and FORTRAN, the following are also available in FORTRAN

Function	Meaning
ALOG (A)	$\log_e A = \ln A$
ATAN (A)	arctangent of A in radians
ASIN (A)	arcsine of A in radians
ACOS (A)	arccosine of A in radians

(FORTRAN provides the principal branch of each of the last three functions.)

Statements Each line of a FORTRAN program consists of three fields: a statement number field, a line continuation field, and a statement field.

A statement number consists of one to five digits in the first five spaces of the line, all blanks being ignored. Any statement referred by another statement must have a statement number, but not all statements must be numbered. Statement numbers may be in any order, since they serve only to identify the statement. For this reason two statements must not have the same number.

Occasionally a FORTRAN statement is so long that it will not fit on one line. When the teletype returns you to a new line before finishing a statement, put a 1 in the sixth space of the next line, and continue typing the rest of the statement.

The FORTRAN commands are typed in the statement field, which consists of all spaces on the line from the seventh space on. Some simple commands are described below.

A typical FORTRAN statement is:

$$\begin{array}{c|c|c} 104 & & \text{GO TO 98} \end{array}$$
$$\text{in space:} \quad \begin{array}{c|c|c} 12345 & 6 & 7 \end{array}$$

It transfers control to the statement numbered 98.

FORTRAN Commands

GO TO [statement number]
   Transfers control to the statement whose number is specified. This command allows the program to jump from one point to another. An example of such a FORTRAN statement was just given.

[variable] = [formula]   (arithmetic statements)
   Assigns to the given variable the result obtained by calculating the formula with the current values of all variables in the formula. It is important not to mix the two different types of constants and variables in the formula. All constants in the formula must be of only one of the two types of constants. If the constants in the formula are integer constants, all variables in the formula must be integer variables. Similarly, if the constants are all real, then the variables must also be real. The variable on the left-hand side of the equal sign need not be of the same type as the variables and constants in the formula. For example

$$I = 9*(J**2)$$

will square the present value of J, multiply the result by 9, and store the answer as an integer constant in I. On the other hand,

$$K19B = (A**B) + 19.4$$

will raise the current value of A to the B power, add 19.4 to the result, and store the largest integer less than or equal to the result as an integer constant in K19B. Finally,

$$C = A**.5$$

will store the square root of A as a real constant in C.

IF ([formula]) n_1, n_2, n_3

This command will obtain the calculation of the current value of the formula and transfer control to the statement numbered n_1, n_2, n_3 if the value is negative, zero, or positive, respectively. For example

$$IF (X**2 - 7) 11, 12, 13$$

will calculate $X^2 - 7$ and transfer control to statement number 13 if $X^2 - 7 > 0$, to statement number 12 if $X^2 - 7 = 0$, and to statement number 11 otherwise.

DO n $i = m_1, m_2, m_3$

Here n is a statement number, i is an integer variable, and m_1, m_2, m_3 are positive integer constants or variables. If m_3 is not specified, it is assumed to be $+1$. This command will cause the statements that follow the DO statement, up to and including the statement numbered n, to be executed repeatedly. The number of times the loop will be performed depends on the values of m_1, m_2, and m_3. The integer variable i will be initially assigned the value m_1 and the statements including that numbered n will be performed. i will then be incremented by the amount m_3, and this new value of i will be compared to m_2. If $i \leq m_2$, the loop will be repeated; otherwise, control will pass to the statement immediately following statement number n. For example, the command

$$DO 17 I5 = 7, 19, 2$$

will set $I5 = 7$, perform all statements until the one numbered 17 is done, reset $I5 = 9$, and repeat the procedure until $I5 = 21$, at which time control will pass to the statement following the one numbered 17. A similar loop will be caused by

$$DO 93 J = I, K**2$$

or

$$DO 6 NM = I, J, K$$

Loops can be nested as shown below:

$$DO 7 N = 1, 5$$
$$DO 6 M = 1, 4$$

```
         DO 6 I = 1, 9
   6     X = X + I*M*N
   7     X = X*N
```

These three loops will initially set $N = M = I = 1$. The variable I will run through the values 1 through 9 before M changes to 2; and N will not become 2 until M has run through the values 1 through 4. When M changes to 2, I will be reset at 1 and must run through 9 before M changes.

Although it is permissible to transfer out of a loop, it is not possible to transfer into a loop.

      ACCEPT   *n*, [variables]

n   FORMAT (description)

These two commands go together and are used to input values in the program. The ACCEPT statement indicates the names of the variables that will be given values at this point in the program. As soon as the execution phase of the program begins, the user types in the values of the variables separated by commas. It is important to type in integer constants for integer variables, and real constants (which may involve powers of ten as specified previously) for real variables. The FORMAT statement numbered *n* tells the computer what type of and how many constants to expect. For the two types of constants which we shall be using, it suffices to let the description read kG where k is the number of variables being entered. As an example, consider the statements

```
         ACCEPT 17, A, B12, J, K17
   17    FORMAT (4G)
```

When this part of the program is reached, the computer will wait till the values are supplied and the RETURN button is pushed.

      TYPE *n*, [variables]

n   FORMAT (description)

These commands also go together and are used to output data on the user's teletype. The output is typed out according to the specifications in the FORMAT description. Each entry in the description corresponding to a variable that should be of the form

          Gw   (integer constant),

or

          Gw.d   (real constant),

where w specifies how many spaces you are setting aside for the answer, and d is the number of significant digits desired. The field w should always be larger than the expected answer, since otherwise leading digits will be lost. For real constants, no problems will result if $w \geqslant d + 8$. The computer can also be made to type out any desired expression by including it with a space in front within single quotes. All entries in the description must be separated by commas.

Thus

```
        Y = SIN (X)
        TYPE 82, X, Y
82      FORMAT (' SIN(', G13.5, ') =', G13.5)
```

will cause the following output for x = 1.04719 radians ($\approx 60°$);

```
    SIN ( 1.04719      ) = 0.50000
```

and

```
        TYPE 13, A, B, I, J
13      FORMAT (2G13.5, 2G5)
```

will cause the output

```
    -1.02345      0.12346E+12      18    -9
```

if A = −1.02345, B = 1.234567E+11, I = 18, and J = −9.

STOP

This statement terminates the program and returns control to the monitor. Every program must have such a statement.

END

This must be the last statement of the FORTRAN program, since it tells the FORTRAN compiler that the end of the program has been reached.

Example 3 The following program will yield the roots of any quadratic equation

$$Ax^2 + Bx + C = 0.$$

The reader should study it carefully to understand why it works.

```
        ACCEPT 10, A, B, C
10      FORMAT (3G)
        D = (B*B) - (4.*A*C)
15      IF (D) 40, 20, 20
20      X1 = (-B + SQRT (D))/(2.*A)
        X2 = (-B - SQRT (D))/(2.*A)
        TYPE 30, X1, X2
30      FORMAT (2G13.5)
        GO TO 60
40      XREAL = -B/(2.*A)
        XIMAG = SQRT (-D)/(2.*A)
        TYPE 50, XREAL, XIMAG, XREAL, XIMAG
50      FORMAT (G13.5,' +I', 2G13.5,' -I', G13.5)
60      STOP
        END
```

After the programmer has typed in the values A, B, C and pressed the RETURN key, the computer will calculate the discriminant D. If $D < 0$, statement 15 will transfer control to statement 40; otherwise, control will pass to statement 20. In the first case complex roots will be the output, while in the second case real roots will be typed out. The program exits at statement 60.

Example 4 To obtain a table of values for the sine function for every 0.1 radian between 0 and 3 radians, one could use the following program:

```
        TYPE 10
   10   FORMAT ('      X              SIN (X)')
        X = 0.0
   15   DO 30, I = 0, 30
        Y = SIN (X)
        TYPE 20, X, Y
   20   FORMAT (2G13.5)
   30   X = X + .1
   40   STOP
        END
```

This program will set up a loop between statements 15 and 30 inclusive. After 31 iterations, the process will exit at statement 40. Statement 10 provides headings for the output.

In this short list we have not described all the capabilities of the above commands. For more information on these and many more commands, the reader should consult a FORTRAN handbook.

DETERMINANTS

Appendix 4

In many parts of this book we made use of determinants. In this appendix we shall illustrate how determinants arise and discuss their uses.

We begin by considering the system of two linear equations in two unknowns

$$a_{11}x_1 + a_{12}x_2 = b_1,$$
$$a_{21}x_1 + a_{22}x_2 = b_2. \tag{A4.1}$$

For simplicity we assume that the constants a_{11}, a_{12}, a_{21}, and a_{22} are all nonzero (otherwise, the system can be solved directly). To solve the system (A4.1), we multiply the first equation by a_{22} and the second by a_{12} to obtain

$$a_{11}a_{22}x_1 + a_{22}a_{12}x_2 = a_{22}b_1,$$
$$a_{12}a_{21}x_1 + a_{22}a_{12}x_2 = a_{12}b_2. \tag{A4.2}$$

Then subtracting the second equation from the first, we have

$$(a_{11}a_{22} - a_{12}a_{21})x_1 = a_{22}b_1 - a_{12}b_2. \tag{A4.3}$$

Now we define the quantity

$$D = a_{11}a_{22} - a_{12}a_{21}. \tag{A4.4}$$

If $D \neq 0$, then (A4.3) yields

$$x_1 = \frac{a_{21}b_1 - a_{12}b_2}{D}, \tag{A4.5}$$

and x_2 may be obtained by substituting this value of x_1 into either of the equations of (A4.1). Thus if $D \neq 0$, *the system* (A4.1) *has a unique solution.*

On the other hand, suppose that $D = 0$. Then $a_{11}a_{22} = a_{12}a_{21}$, and if we subtract the second equation of (A4.2) from the first, we have

$$0 = a_{21}b_1 - a_{12}b_2.$$

Either this equation is true or it is false. If $a_{21}b_1 - a_{12}b_2 \neq 0$, then the system (A4.1) has *no* solution. If $a_{21}b_1 - a_{12}b_2 = 0$, then the second equation of (A4.2) is a multiple of the first and (A4.1) consists essentially of only one equation. Thus we may choose x_1 arbitrarily and calculate the corresponding value of x_2. In this case there are an *infinite* number of solutions. In sum, we have shown that *if* $D = 0$, *then the system* (A4.1) *has either no solution or an infinite number of solutions.*

These facts are easily visualized geometrically by noting that (A4.1) consists of the equations of two straight lines. A solution of the system is a point of intersection of the two lines. It is easy to show (see Exercise 5) that $D = 0$ if and only if the slopes of the two lines are the same. If the slopes are different, then $D \neq 0$ and the two lines intersect at a single point, which is the unique solution. If $D = 0$, we either have two parallel lines and no solution, since the lines never intersect,

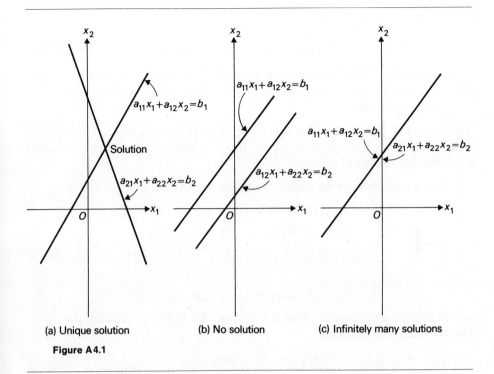

(a) Unique solution   (b) No solution   (c) Infinitely many solutions

Figure A4.1

or both equations yield the same line and every point on this line is a solution. This is illustrated in Fig. A4.1.

Example A4.1 Consider the following systems of equations:

i) $2x_1 + 3x_2 = 12,$ ii) $x_1 + 3x_2 = 3,$ iii) $x_1 + 3x_2 = 3,$
   $x_1 + x_2 = 5.$ $3x_1 + 9x_2 = 8.$ $3x_1 + 9x_2 = 9.$

In system (i), $D = 2 \cdot 1 - 3 \cdot 1 = -1 \neq 0$, so there is a unique solution, which is easily found to be $x_1 = 3$, $x_2 = 2$. In system (ii), $D = 1 \cdot 9 - 3 \cdot 3 = 0$. Multiplying the first equation by 3 and then subtracting this from the second equation, we obtain the equation $0 = 1$, which is impossible. Thus there is no solution. In (iii), $D = 1 \cdot 9 - 3 \cdot 3 = 0$. But now the second equation is simply three times the first equation. If x_2 is arbitrary, then $x_1 = 3 - 3x_2$, and there are an infinite number of solutions.

Returning again to the system (A4.1), we define the *determinant of the system* as

$$D = a_{11}a_{22} - a_{12}a_{21}. \qquad (A4.6)$$

For convenience of notation we denote the determinant by writing the coefficients of the system in a square array:

$$D = \begin{vmatrix} a_{11} & a_{12} \\ a_{21} & a_{22} \end{vmatrix} = a_{11}a_{22} - a_{12}a_{21}. \tag{A4.7}$$

Therefore, a 2×2 determinant is the product of the two elements in the upper-left-to-lower-right diagonal minus the product of the other two elements.

We have proved:

Theorem A4.1 For the 2×2 system (A4.1) there is a unique solution if and only if the determinant D is not equal to zero. If $D = 0$, then there is either no solution or an infinite number of solutions.

Let us now consider the general system of n equations in n unknowns:

$$a_{11}x_1 + a_{12}x_2 + \cdots + a_{1n}x_n = b_1,$$
$$a_{21}x_1 + a_{22}x_2 + \cdots + a_{2n}x_n = b_2,$$
$$\vdots \tag{A4.8}$$
$$a_{n1}x_1 + a_{n2}x_2 + \cdots + a_{nn}x_n = b_n,$$

and define the determinant of such a system. We begin by defining the determinant of a 3×3 system:

$$D = \begin{vmatrix} a_{11} & a_{12} & a_{13} \\ a_{21} & a_{22} & a_{23} \\ a_{31} & a_{32} & a_{33} \end{vmatrix} = a_{11} \begin{vmatrix} a_{22} & a_{23} \\ a_{32} & a_{33} \end{vmatrix} - a_{12} \begin{vmatrix} a_{21} & a_{23} \\ a_{31} & a_{33} \end{vmatrix} + a_{13} \begin{vmatrix} a_{21} & a_{22} \\ a_{31} & a_{32} \end{vmatrix}. \tag{A4.9}$$

Hence to calculate a 3×3 determinant, it is necessary to calculate three 2×2 determinants.

Example A4.2

$$\begin{vmatrix} 3 & 5 & 2 \\ 4 & 2 & 3 \\ -1 & 2 & 4 \end{vmatrix} = 3 \begin{vmatrix} 2 & 3 \\ 2 & 4 \end{vmatrix} - 5 \begin{vmatrix} 4 & 3 \\ -1 & 4 \end{vmatrix} + 2 \begin{vmatrix} 4 & 2 \\ -1 & 2 \end{vmatrix}$$
$$= 3 \cdot 2 - 5 \cdot 19 + 2 \cdot 10 = -69.$$

The general definition of the determinant of the $n \times n$ system of equations (A4.8) is simply an extension of this procedure:

$$D = \begin{vmatrix} a_{11} & a_{12} & \cdots & a_{1n} \\ a_{21} & a_{22} & \cdots & a_{2n} \\ \vdots & \vdots & & \vdots \\ \vdots & \vdots & & \vdots \\ a_{n1} & a_{n2} & \cdots & a_{nn} \end{vmatrix} = a_{11}A_{11} - a_{12}A_{12} + \cdots + (-1)^{n+1}a_{1n}A_{1n}, \tag{A4.10}$$

where A_{1j} is the $(n-1) \times (n-1)$ determinant obtained by crossing out the first row and jth column of the original $n \times n$ determinant. Thus an $n \times n$ determinant can be obtained by calculating $n(n-1) \times (n-1)$ determinants. [Note that in definition (A4.10), the signs alternate.]

Example A4.3

$$
\begin{vmatrix} 1 & 3 & 5 & 2 \\ 0 & -1 & 3 & 4 \\ 2 & 1 & 9 & 6 \\ 3 & 2 & 4 & 8 \end{vmatrix} = 1 \begin{vmatrix} -1 & 3 & 4 \\ 1 & 9 & 6 \\ 2 & 4 & 8 \end{vmatrix} - 3 \begin{vmatrix} 0 & 3 & 4 \\ 2 & 9 & 6 \\ 3 & 4 & 8 \end{vmatrix} + 5 \begin{vmatrix} 0 & -1 & 4 \\ 2 & 1 & 6 \\ 3 & 2 & 8 \end{vmatrix} - 2 \begin{vmatrix} 0 & -1 & 3 \\ 2 & 1 & 9 \\ 3 & 2 & 4 \end{vmatrix}
$$

$$
= 1(-92) - 3(-70) + 5(2) - 2(-16) = 160.
$$

(The values in parentheses are obtained by calculating the four 3×3 determinants.)

It is clear that calculating determinants by this method can be a tedious procedure, especially if $n \geqslant 5$. For this reason, techniques are available for significantly simplifying these calculations. We shall not discuss these methods here, since they can be found in most books on matrix theory.

The reason for considering determinants of systems of n equations in n unknowns is that Theorem A4.1 also holds for these systems (although this fact will not be proven here).

Theorem A4.2 For the system (A4.8) there is a unique solution if and only if the determinant D, defined by (A4.10), is not zero. If $D = 0$, then there is either no solutions or an infinite number of solutions.

There is a particular case of system (A4.8) which merits special consideration (for example, in checking whether or not solutions to a differential equation are linearly independent). This is the *homogeneous system*

$$
\begin{aligned}
a_{11}x_1 + a_{12}x_2 + \cdots + a_{1n}x_n &= 0, \\
a_{21}x_1 + a_{22}x_2 + \cdots + a_{2n}x_n &= 0,
\end{aligned}
$$

$$
\text{(A4.11)}
$$

$$
a_{n1}x_1 + a_{n2}x_2 + \cdots + a_{nn}x_n = 0,
$$

which occurs when all the values b_i, $i = 1, \ldots, n$ in (A4.8) are zero. Clearly, $x_1 = x_2 = \cdots = x_n = 0$ is a solution of (A4.11). Thus for homogeneous systems we have:

Theorem A4.3 If the determinant $D \neq 0$, the zero solution $x_1 = x_2 = \cdots = x_n = 0$ is the only solution of the homogeneous system (A4.11). If $D = 0$, then there are an infinite number of solutions of (A4.11).

We shall conclude this appendix by introducing a method for obtaining solutions of the system (A4.8). We define the determinants

$$
D_1 = \begin{vmatrix}
b_1 & a_{12} & \cdots & a_{1n} \\
b_2 & a_{22} & \cdots & a_{2n} \\
\cdot & \cdot & & \cdot \\
\cdot & \cdot & & \cdot \\
\cdot & \cdot & & \cdot \\
b_n & a_{n2} & \cdots & a_{nn}
\end{vmatrix},
$$

$$
D_2 = \begin{vmatrix}
a_{11} & b_1 & a_{13} & \cdots & a_{1n} \\
a_{21} & b_2 & a_{23} & \cdots & a_{2n} \\
\cdot & \cdot & \cdot & & \cdot \\
\cdot & \cdot & \cdot & & \cdot \\
\cdot & \cdot & \cdot & & \cdot \\
a_{n1} & b_n & a_{n3} & \cdots & a_{nn}
\end{vmatrix}, \quad \ldots,
$$

$$
D_k = \begin{vmatrix}
a_{11} & a_{12} & \cdots & a_{1,k-1} & b_1 & a_{1,k+1} & \cdots & a_{1n} \\
a_{21} & a_{22} & \cdots & a_{2,k-1} & b_2 & a_{2,k+1} & \cdots & a_{2n} \\
\cdot & \cdot & & \cdot & \cdot & \cdot & & \cdot \\
\cdot & \cdot & & \cdot & \cdot & \cdot & & \cdot \\
\cdot & \cdot & & \cdot & \cdot & \cdot & & \cdot \\
a_{n1} & a_{n2} & \cdots & a_{n,k-1} & b_k & a_{n,k+1} & \cdots & a_{nn}
\end{vmatrix}, \quad \ldots,
$$

$$
D_n = \begin{vmatrix}
a_{11} & a_{12} & \cdots & a_{1,n-1} & b_1 \\
a_{21} & a_{22} & \cdots & a_{2,n-1} & b_2 \\
\cdot & \cdot & & \cdot & \cdot \\
\cdot & \cdot & & \cdot & \cdot \\
\cdot & \cdot & & \cdot & \cdot \\
a_{n1} & a_{n2} & \cdots & a_{n,n-1} & b_n
\end{vmatrix}, \tag{A4.12}
$$

obtained by replacing the kth column of D by the column

$$
\begin{bmatrix}
b_1 \\
b_2 \\
\cdot \\
\cdot \\
\cdot \\
b_n
\end{bmatrix}.
$$

Then we have the following theorem, known as *Cramer's rule.*

Theorem A4.4. (Cramer's rule) Let D and D_k, $k = 1, 2, \ldots, n$, be given as in (A4.10) and (A4.12). If $D \neq 0$, then the unique solution to the system (A4.8) is given by the values

$$x_1 = \frac{D_1}{D}, \quad x_2 = \frac{D_2}{D}, \quad \ldots, \quad x_n = \frac{D_n}{D}. \tag{A4.13}$$

Example A4.4 Consider the system

$$2x_1 + 4x_2 - x_3 = -5,$$
$$-4x_1 + 3x_2 + 5x_3 = 14,$$
$$6x_1 + 3x_2 - 2x_3 = 5.$$

We have

$$D = \begin{vmatrix} 2 & 4 & -1 \\ -4 & 3 & 5 \\ 6 & -3 & -2 \end{vmatrix} = 112, \quad D_1 = \begin{vmatrix} -5 & 4 & -1 \\ 14 & 3 & 5 \\ 5 & -3 & -2 \end{vmatrix} = 224,$$

$$D_2 = \begin{vmatrix} 2 & -5 & -1 \\ -4 & 14 & 5 \\ 6 & 5 & -2 \end{vmatrix} = -112, \quad D_3 = \begin{vmatrix} 2 & 4 & -5 \\ -4 & 3 & 14 \\ 6 & -3 & 5 \end{vmatrix} = 560.$$

Therefore

$$x_1 = \frac{D_1}{D} = 2, \quad x_2 = \frac{D_2}{D} = -1, \quad x_3 = \frac{D_3}{D} = 5.$$

EXERCISES A4

1. For each of the following 2×2 systems, calculate the determinant D. If $D \neq 0$, find the unique solution. If $D = 0$, determine whether there is no solution or an infinite number of solutions.

   a) $2x_1 + 4x_2 = 6$
   $\quad x_1 + x_2 = 3$

   b) $2x_1 + 4x_2 = 6$
   $\quad x_1 + 2x_2 = 5$

   c) $2x_1 + 4x_2 = 6$
   $\quad x_1 + 2x_2 = 3$

   d) $\quad 6x_1 - 3x_2 = 3$
   $\quad -2x_1 + x_2 = -1$

   e) $\quad 6x_1 - 3x_2 = 3$
   $\quad -2x_1 + x_2 = 1$

   f) $\quad 6x_1 - 3x_2 = 3$
   $\quad -2x_1 + 2x_2 = -1$

2. Calculate the following determinants:

   a) $\begin{vmatrix} 1 & 2 & 3 \\ 6 & -1 & 4 \\ 2 & 0 & 6 \end{vmatrix}$

   b) $\begin{vmatrix} 4 & -1 & 0 \\ 2 & 1 & 7 \\ -2 & 3 & 4 \end{vmatrix}$

   c) $\begin{vmatrix} 7 & 2 & 3 \\ 0 & 4 & 1 \\ 0 & 0 & 5 \end{vmatrix}$

   d) $\begin{vmatrix} 1 & 0 & 0 \\ 0 & -3 & 0 \\ 0 & 0 & 7 \end{vmatrix}$

   e) $\begin{vmatrix} 1 & 7 & 2 & 3 \\ 3 & 4 & 1 & 6 \\ 2 & 0 & 5 & -1 \\ -1 & 2 & 0 & 4 \end{vmatrix}$

   f) $\begin{vmatrix} 2 & -1 & 3 & 5 \\ 0 & 3 & 1 & 6 \\ 0 & 0 & -2 & 4 \\ 0 & 0 & 0 & 5 \end{vmatrix}$

$$g) \quad \begin{vmatrix} 1 & 0 & -2 & 3 & 5 \\ 0 & 4 & 2 & 3 & 6 \\ 0 & 0 & 5 & 7 & -8 \\ 0 & 0 & 0 & -1 & 73 \\ 0 & 0 & 0 & 0 & 2 \end{vmatrix}$$

3. Determine whether each of the following homogeneous systems has a solution other than the zero solution.

a) $\begin{aligned} 2x_1 - 3x_2 + 4x_3 &= 0 \\ -x_1 + x_2 - 6x_3 &= 0 \\ 4x_1 - 5x_2 + 16x_3 &= 0 \end{aligned}$ b) $\begin{aligned} x_1 + 6x_2 - x_3 &= 0 \\ -2x_1 + 7x_2 + x_3 &= 0 \\ 3x_1 - x_2 + 5x_3 &= 0 \end{aligned}$

4. Solve each of the following systems by Cramer's rule:

a) $\begin{aligned} 3x_1 - x_2 &= 13 \\ -4x_1 + 6x_2 &= -8 \end{aligned}$ b) $\begin{aligned} 2x_1 + 6x_2 + 3x_3 &= 9 \\ -3x_1 - 17x_2 - x_3 &= 4 \\ 4x_1 + 3x_2 + x_3 &= -7 \end{aligned}$

c) $\begin{aligned} 2x_1 \qquad + x_3 &= 0 \\ 3x_1 - 2x_2 + 2x_3 &= -4 \\ 4x_1 - 5x_2 \qquad &= 3 \end{aligned}$ d) $\begin{aligned} x_1 + 2x_2 - x_3 - 4x_4 &= 1 \\ -x_1 \qquad + 2x_3 + 6x_4 &= 5 \\ -4x_2 - 2x_3 - 8x_4 &= -8 \\ 3x_1 - 2x_2 \qquad + 5x_4 &= 3 \end{aligned}$

5. Show that the two lines in system (A4.1) have the same slope if and only if the determinant of the system is zero.

SOLUTIONS
TO
ODD-NUMBERED
EXERCISES

CHAPTER 1

Section 1.1

1. a) $y = \dfrac{x^2}{2} + 3x + 2$

   b) $y = \ln\left(\dfrac{x^2 + 5}{6}\right) + 4$

   c) $y = \tan x - x + 1$

   d) $y = \dfrac{x^4}{12} - \dfrac{9x^2}{2} + 3x + 1$

   e) $y = \cos x - \sin x + 2x + (1 - \pi)$

3. 13.2 ft/sec^2

5. $x(t) = [v_0 \cos \theta]t + x_0;\ y(t) = -\frac{1}{2}gt^2 + [v_0 \sin \theta]t + y_0$

Section 1.2

1. a) First   b) Second   c) Third   d) Fifth
   e) Second   f) First   g) Third

5. $y_1(x) = e^{4x},\ y_2(x) = e^{-x}$ 7. $\varphi(x) = x$

11. $\varphi'(1) = 5,\ \varphi''(1) = 22,\ \varphi'''(1) = 140$

Section 1.3

1. a) First   b) Third   c) Second   d) First
   e) Fifth   f) First   g) Fourth

5. a) $P_{n+2} = 1.3P_n$ b) Second   c) $P_2 = 1300,\ P_4 = 1690$

7. $y_n^{(1)} = 4^n,\ y_n^{(2)} = (-1)^n$

CHAPTER 2

Section 2.1

1. $y = \sqrt{e^x + c}$

3. $y = \frac{1}{2}\ln(x^2 + 1) + c$

5. $z = \tan\left(\dfrac{r^3}{3} + c\right)$

7. $P = ce^{\sin Q - \cos Q},\ c > 0$

9. $s = e^{(t^3/3) - 2t}$

11. $y = \dfrac{2301}{(1 + x)^3}$

13. $y = c \cos x - 3$

15. $y = \sin(c - \sin^{-1} x)$

17. $y = \dfrac{3x}{4x - 3}$

19. $\dfrac{1}{1 - e^x} = \dfrac{e^t}{1 - e}$

23. a) $x(t) = \beta + (x(0) - \beta)e^{-\alpha t}$

25. $\dfrac{2916\frac{2}{3}}{\sqrt{2g}} \sec$

27. The king.

Section 2.2

1. $x = ce^{2\sqrt{y/x}}$

3. $\ln x + e^{-y/x} = 1$

5. $t = \sqrt{st} + c$

7. $xy(y - x) = 2$

9. $y + \sqrt{x^2 + y^2} = cx^2$

11. $x = e^{(y - x)/y}$

13. a) $\frac{1}{2}\tan^{-1}\left(\dfrac{x + y}{2}\right) = x + c$

   b) $\frac{3}{2}\ln\dfrac{2 + x + y}{4 - x - y} = x + c$

c) $\dfrac{y}{x} + \ln\left(\dfrac{y}{x} - 1\right) = 3x + c$

15. $n = -\tfrac{1}{2};\ xy^2 = \ln x + c$

Section 2.3

1. $x = e^{3t}$

3. $x = 2e^t - 1$

5. $x = e^{-y}y^y(e^y + c)$

7. $y = x^4 + 3x^3$

9. $x = ce^{2t} + \dfrac{t^3 e^{2t}}{3}$

11. $s = ce^{-u} + \dfrac{u^2}{2}e^{-u} + 1$

13. $x = e^{-y}(c + y)$

15. a) $\dfrac{1}{k}\ln 2$

b) $\dfrac{2\ln 2}{k}$

17. $3\ln 10 \approx 6.9$ yr.

Section 2.4

1. a) $z'' - z = 0$

b) $z'' + z = 0$

c) $z'' + z' - z = 0$

3. $y^2(6 + ce^{-x}) = x$

5. $e^{2x} = y^2(x^2 + c)$

7. $5x^2y^3 = x^5 + 4$

9. $x^3 = 3\sin t + \dfrac{9\cos t}{t} - \dfrac{18\sin t}{t^2} - \dfrac{18\cos t}{t^3} + \dfrac{c}{t^3}$

11. $y^2 = x^2(c - 2\ln|x|)$

13. $y = cx - \tfrac{1}{2}(c - 4)^2,\ y = \tfrac{1}{2}x^2 + 4x$

15. $y = cx + \dfrac{1}{c},\ y^2 = 4x$

17. $y = cx - \sqrt{c},\ 4xy + 1 = 0$

Section 2.5

1. $x^2y + y = c$

3. $x^4y^3 + \ln\left|\dfrac{x}{y}\right| = e^3 - 1$

5. $x^2 - 2y\sin x - \dfrac{\pi^2}{4} + 2 = 0$

7. $e^{xy} + 4xy^3 - y^2 + 3 = 0$

9. $x^2y + xe^y = c$

11. $\tan^{-1}\dfrac{y}{x} + \ln\left|\dfrac{x}{y}\right| = c$

13. $2x + \ln(x^2 + y^2) = c$, and $x = y = 0$

15. $x^2\ln|x| - y - cx^2 = 0$, and $x = 0$

17. $2x^2y^2 + 2y^4 + 4y^2 = c$, and $y = 0$

19. $y = \dfrac{c}{x^2}$

Section 2.6

1. $y_n = y_0 + 2\left(1 - \dfrac{1}{2^n}\right)$

3. $y_n = (-1)^n y_0$

5. $y_n = n + 1$

7. $y_n = \dfrac{5^{n+1}}{2} - \dfrac{1}{2}$

9. $y_n = e^{-n^2 + n}\left(y_0 + \dfrac{e^n - 1}{e - 1}\right)$

11. $P_n = \tfrac{2}{3} + \tfrac{1}{3}(-\tfrac{1}{2})^n$

13. $P_n = \dfrac{1}{n!} P_0$

17. a) $(n + 8)x_{n+2} + 2x_{n+1} - x_n = 0$

   b) $2 \cdot 3^n(3^{n+1} - 2)x_{n+2} - 2(3^n - 1)x_{n+1} + x_n = 0$

Section 2.7

1. 1.8171   3. 0.5671   5. 1.1656

7. $x_{n+1} = \dfrac{x_{n-1}F(x_n) - x_n F(x_{n-1})}{F(x_n) - F(x_{n-1})}$

Section 2.8

1. $I(0.1) = 0.01 + (600\pi(1 - e^{-10})/[(100)^2 + (120\pi)^2]$

3. $Q(60) = \dfrac{1}{2 \cdot 10^3}[1 - e^{-(10^6 t + 10^5 t^2)}]\Big|_{t=60}$

5. $I(t) = \dfrac{E_0 C}{1 + (\omega RC)^2}\left[RC\omega^2 \cos \omega t - \omega \sin \omega t + \dfrac{1}{RC}e^{-t/RC}\right];$

   $Q(t) = \dfrac{E_0 C}{1 + (\omega RC)^2}[(\cos \omega t - e^{-t/RC}) + \omega RC \sin \omega t]$

Section 2.9

3. $\dfrac{y + \sqrt{x^2 + y^2}}{x} = \left(\dfrac{x}{b}\right)^{v/w}$ or $y = \dfrac{x}{2}\left[\left(\dfrac{x}{b}\right)^{v/w} - \left(\dfrac{b}{x}\right)^{v/w}\right]$

5. Yes, at $\left(b, -\dfrac{bv}{w}\right)$ 7. $r = e^{\theta/\sqrt{3}}; r = 2e^{(\theta - \pi)/\sqrt{3}}$

9. a units

Section 2.10

1. 2000-year-old wood has $(\frac{1}{2})^{20/57} \approx 0.784$ the concentration of new wood.

3. $I(t) = \dfrac{1 + N}{1 + Ne^{-k(1+N)t}}$ 5. $t = 25 \ln 2 \approx 17.3$ min

7. $47\frac{43}{91}$ g/liter   9. $P(t) = 400 - \dfrac{4}{\pi} - \dfrac{t}{6} - \dfrac{4}{\pi}\sin\dfrac{\pi}{12}(t - 6)$

11. $x(t) = \dfrac{A}{k}(1 - e^{-kt}) + x(0)e^{-kt} + \dfrac{B}{k^2 + \omega^2}[k \sin \omega t + \omega(e^{-kt} - \cos \omega t)]$

CHAPTER 3

Section 3.1

1. Linear, homogeneous, variable coefficients   3. Nonlinear

5. Linear, nonhomogeneous, constant coefficients

7. Linear, nonhomogeneous, variable coefficients

9. Linear, homogeneous, constant coefficients

Section 3.2

7. $-\dfrac{2}{x}$

11. $y = -\frac{3}{2}\cos x - \frac{5}{2}\sin x + \frac{5}{2}e^x$

Section 3.3

1. $y_2(x) = xe^x$

3. $y_2(x) = \dfrac{3x^2 - 1}{4}\ln\left|\dfrac{1 + x}{1 - x}\right| - \dfrac{3x}{2}$

5. $y_2(x) = \dfrac{1}{x^2}$

7. $y_2(x) = J_0(x)\displaystyle\int \dfrac{dx}{x[J_0(x)]^2}$

9. b) $y_3 = \dfrac{1}{x}$

Section 3.4

1. $y = c_1 e^{2x} + c_2 e^{-2x}$ 3. $y = c_1 e^x + c_2 e^{2x}$ 5. $x = (1 + \frac{9}{2}t)e^{-5t/2}$

7. $x = -\frac{1}{5}(e^{3t} + 4e^{-2t})$ 9. $y = c_1 + c_2 e^{5x}$

11. $y = e^{-\pi(x-1)}\left(\left[1 - \dfrac{1}{\pi} - \pi\right] + \left[\dfrac{1}{\pi} + \pi\right]x\right)$

13. b) $y = c_1 e^{\lambda_1 x} + c_2 e^{\lambda_2 x} + c_3 e^{\lambda_3 x}$ c) $y = c_1 e^{\lambda_1 x} + (c_2 + c_3 x)e^{\lambda_2 x}$
    d) $y = (c_1 + c_2 x + c_3 x^2)e^{\lambda_1 x}$

15. $y = (1 + x)e^x$ 17. $y = \frac{7}{9} + \frac{1}{9}(e^{3x} + e^{-3x})$

19. $y = \dfrac{c_1 e^x - c_2 e^{-x}}{c_1 e^x + c_2 e^{-x}}$

21. $y = \dfrac{z'}{az}$, where $z = c_1 e^{\lambda_1 t} + c_2 e^{\lambda_2 t}$ and $\left.\begin{matrix}\lambda_1 \\ \lambda_2\end{matrix}\right\} = \dfrac{-b \pm \sqrt{b^2 - 4ac}}{2}$

Section 3.5

1. $y = e^{-x}(c_1 \cos x + c_2 \sin x)$ 3. $x = e^{-t/2}\left(c_1 \cos\dfrac{3\sqrt{3}}{2}t + c_2 \sin\dfrac{3\sqrt{3}}{2}t\right)$

5. $x = -\frac{3}{2}\cos 2\theta + \sin 2\theta$ 7. $y = 2e^x - \cos x$

Section 3.6

1. $y = c_1 \sin 2x + c_2 \cos 2x + \sin x$ 3. $y = c_1 e^x + c_2 e^{2x} + 3e^{3x}$

5. $y = e^x(c_1 + c_2 x - 2x^2)$ 7. $y = 3e^{5x} - 10e^{2x} + 10x + 7$

9. $y = c_1 + c_2 e^{-x} + \dfrac{x^4}{4} - \frac{4}{3}x^3 + 4x^2 - 8x$

11. $y = c_1 e^{2x}\cos x + c_2 e^{2x}\sin x + 5xe^{2x}\sin x + \dfrac{e^{-2x}}{4}(2\cos x - \sin x)$

13. $y = e^{-3x}(c_1 + c_2 x + 5x^2)$

15. $y = c_1 e^{3x} + c_2 e^{-x} + \frac{20}{27} - \frac{7}{9}x + \frac{1}{3}x^2 - \frac{1}{4}e^x$

17. $y = (c_1 + c_2 x)e^{-2x} + (\frac{1}{9}x - \frac{2}{27})e^x + \frac{3}{25}\sin x - \frac{4}{25}\cos x$

Section 3.7

1. $y = c_1 \cos x + c_2 \sin x - \frac{1}{3} \cos 2x$

3. $y = c_1 \cos 2x + c_2 \sin 2x + \frac{1}{2}x \sin 2x + \frac{1}{4} \cos 2x \ln |\cos 2x|$

5. $y = (c_1 + c_2 x)e^x - e^x \ln |1 - x|$

7. $y = c_1 + c_2 e^{-x} + \dfrac{e^x}{4} - \dfrac{xe^{-x}}{2}$

9. $y = \left(c_1 + c_2 x + \dfrac{x^2}{2}\right)e^{2x}$

11. $y = \dfrac{1}{2} - \dfrac{x}{4} + \left(\dfrac{x}{2} - \dfrac{1}{2x}\right) \ln |x + 1|$

15. $y = \dfrac{e^{2x}}{17}$

Section 3.8

1. $y = \frac{4}{3}x^2 + \dfrac{5}{3x}$

3. $y = \dfrac{1}{x}(c_1 \cos \ln |x| + c_2 \sin \ln |x|)$

5. $y = \dfrac{1}{x^2}(1 + 5 \ln |x|)$

7. $y = c_1 x + c_2 x^2 + c_3 x^{-4}$

9. $y = c_1 e^{3t} + c_2 e^{-4t} = c_1 x^3 + \dfrac{c_2}{x^4}$

11. $y = \dfrac{c_1}{x^5} + \dfrac{c_2}{x} + \frac{1}{12}x$

13. $y = c_1 x^{\sqrt{2}} + \dfrac{c_2}{x^{\sqrt{2}}} - \frac{1}{2} \ln x + \frac{1}{4}, x > 0$

Section 3.9

1. $\dfrac{\sqrt{g}}{50}$ m/sec

3. $\dfrac{2\pi}{\sqrt{5\pi g}} \sec; x(t) = \left(\dfrac{1}{5\pi} - 1\right) \cos \sqrt{5\pi g t}$

5. $\lambda = \dfrac{100\pi^2 + 1}{160}$

7. $x(t) = e^{-\sqrt{17}t}(c_1 \cos 9t + c_2 \sin 9t) + (47 \sin 2t - 2\sqrt{17} \cos 2t)/22770$

Section 3.10

1. $I(t) = \dfrac{80\sqrt{87}}{29} e^{-25t/2} \sin \dfrac{25\sqrt{87}}{6} t;$

$Q(t) = \dfrac{3}{5} - \dfrac{87}{145} e^{-25t/2} \cos \dfrac{25\sqrt{87}}{6} t - \dfrac{3\sqrt{87}}{145} e^{-25t/2} \sin \dfrac{25\sqrt{87}}{6} t$

3. $Q(t) = -\dfrac{e^{-600t}}{116,000}(9 \cos 800t + \dfrac{51}{4} \sin 800t) + \dfrac{1}{116,000}(9 \cos 600t + 8 \sin 600t);$ differentiating and evaluating at $t = 0.001$ shows that the transient current is larger.

5. $Q = c_1 \cos \dfrac{t}{\sqrt{LC}} + c_2 \sin \dfrac{t}{\sqrt{LC}} + (CE_0 + LC\omega^2) \sin \omega t;$ if $\omega = \dfrac{1}{\sqrt{LC}}$ then

$Q = \left(c_1 - \dfrac{E_0 t}{2\omega L}\right) \cos \omega t + c_2 \sin \omega t$

7. b) $\omega = \dfrac{\sqrt{(4L/C) - R^2}}{2L}$

9. b) No ω produces resonance

Section 3.11

1. $t = \dfrac{25}{2} \ln 3 \approx 13.7$ min

5. 780.9 counts per minute

Section 3.12

7. $y = c_1 e^x + c_2 x e^x + c_3 x^2 e^x$

9. $y = c_1 e^{2x} + c_2 x e^{2x} + c_3 e^{-3x} + c_4 x e^{-3x}$

11. $y = c_1 + c_2 \cos x + c_3 \sin x + c_4 x \cos x + c_5 x \sin x$

13. $y = e^{\sqrt{2}x}(c_1 \cos \sqrt{2}x + c_2 \sin \sqrt{2}x) + e^{-\sqrt{2}x}(c_3 \cos \sqrt{2}x + c_4 \sin \sqrt{2}x)$

15. $y = \frac{11}{30}e^{-2x} + \frac{9}{5}e^{3x} - \frac{7}{6}e^{4x}$

17. $y_p = \dfrac{x}{24} + \dfrac{1}{288}$

19. $y_p = -\dfrac{e^{-2x}}{4}$

21. $y_p = \dfrac{x}{24} + \dfrac{37}{288}$

23. $y_p = \dfrac{x^5}{120} + \dfrac{x^6}{360} + \dfrac{x^8}{1680}$

25. $y_p = \frac{1}{4}(\sin x + \cos x)$

27. $y_p = x - \frac{1}{4}e^x$

29. $y_p = -\frac{45}{8}x + \frac{21}{8}x^2 - \frac{3}{4}x^3 + \frac{1}{8}x^4 - \frac{1}{6}e^x$

CHAPTER 4

Section 4.1

1. Linear, nonhomogeneous, constant coefficients

3. Nonlinear

5. Linear, homogeneous, constant coefficients

7. Nonlinear

9. Linear, nonhomogeneous, constant coefficients

Section 4.2

1. $C_n = -2^n$

3. $C_n = 1\left(\text{since } \sin (n + 1)\dfrac{\pi}{2} = \cos \dfrac{\pi}{2} \text{ and } \cos (n + 1)\dfrac{\pi}{2} = \sin \dfrac{n\pi}{2}\right)$

5. $C_n = 12^n$

Section 4.3

1. $y_n = n$

3. $y_n = \displaystyle\sum_{k=0}^{n-1} \dfrac{(-1)^{k-1}}{(k + 1)!}$

5. $y_n = 2^n$

7. $n! \displaystyle\sum_{k=0}^{n-1} \dfrac{(-1)^k}{k+1}$

Section 4.4

1. $y_n = c_1 4^n + c_2(-1)^n$

3. $y_n = (\tfrac{1}{3})^{n-1} - (\tfrac{1}{2})^{n-1}$

5. $y_n = c_1(-1)^n + c_2 n(-1)^n$

7. $y_n = (\tfrac{1}{3})^n(c_1 + nc_2)$

9. $y_n = c_1 \left(\dfrac{1 + \sqrt{41}}{20}\right)^n + c_2 \left(\dfrac{1 - \sqrt{41}}{20}\right)^n$

11. $P_n = \dfrac{\sqrt{5}}{2}\left[\left(\dfrac{1 + 1/\sqrt{5}}{2}\right)^n - \left(\dfrac{1 - 1/\sqrt{5}}{2}\right)^n\right]$; $P_{11} = 0.088$

13. $P_n = 1000(1.6^n + 1.4^n)$

15. $y_n = c_1 + c_2 2^n + c_3 3^n$

17. $y_n = c_1 + c_2 n + c_3(-2)^n$

19. $y_n = c_1 + c_2 n + c_3 n^2$

Section 4.5

1. $y_n = c_1 \cos\dfrac{n\pi}{4} + c_2 \sin\dfrac{n\pi}{4}$

3. $y_n = 2^{3n/2}\left(c_1 \cos\dfrac{n\pi}{2} + c_2 \sin\dfrac{n\pi}{2}\right)$

9. $y_n = \dfrac{x_n}{x_{n+1}} + 1$, where $x_n = c_1\left(\dfrac{1}{\sqrt{2}}\right)^n + c_2\left(-\dfrac{1}{\sqrt{2}}\right)^n$

11. $y_n = \dfrac{x_n}{x_{n+1}} + \dfrac{3}{4}$, where $x_n = c_1\left(\dfrac{-2}{3}\right)^n + c_2\left(\dfrac{-4}{3}\right)^n$

13. $y_n = c_1 2^n + c_2 \cos\dfrac{n\pi}{2} + c_3 \sin\dfrac{n\pi}{2}$

15. $y_n = c_1 + c_2 \cos\dfrac{n\pi}{2} + c_3 \sin\dfrac{n\pi}{2}$

17. $y_n = c_1(-6)^n + 2^n\left(c_2 \cos\dfrac{n\pi}{3} + c_3 \sin\dfrac{n\pi}{3}\right)$

Section 4.6

1. $y_n = \tfrac{1}{6} \cdot 5^{n+1}$

3. $y_n = 2^n\left(\dfrac{n^3}{24} - \dfrac{n^2}{8}\right)$

5. $y_n = \dfrac{1}{4}\sin\dfrac{n\pi}{2}$

11. $y_n = n \cdot 2^{n-1} + \dfrac{2^{2-n}}{3} + c_1 + c_2 2^n$

13. $y_r = \dfrac{\sin n + \sin(n-2)}{2(1 + \cos 2)} + c_1 \cos\dfrac{n\pi}{2} + c_2 \sin\dfrac{n\pi}{2}$

15. $y_n = c_1(-1)^n + 3^{n/2}\left[c_2 \cos\dfrac{n\pi}{6} + c_3 \sin\dfrac{n\pi}{6}\right] + \dfrac{5}{26}\cos\dfrac{n\pi}{2} - \dfrac{1}{26}\sin\dfrac{n\pi}{2}$

17. $y_n = -6 - 2n - n^2 + \left(\dfrac{3}{2}\right)^{n/2}[c_1 \cos n\theta + c_2 \sin n\theta]$, where $\tan\theta = \sqrt{5}$

Section 4.7

3. Natural frequencies are $\omega_N = \left[2\sqrt{CL} \sin \dfrac{N\pi}{2(n+1)} \right]^{-1}$. Cut-off frequency is $\omega_n \approx \dfrac{1}{2\sqrt{CL}}$.

Section 4.8

1. \$9 to have a better than even chance; Player B will *never* have better than a 55% chance to win all of A's money no matter how much he starts with.

3. $P_{n+3} + \dfrac{p}{q} P_{n+2} - \dfrac{1}{q} P_{n+1} + \dfrac{r}{q} P_n = 0, P_0 = 0, P_{2N} = P_{2N+1} = 1$

5. The general solution is $P_n = c_1 + c_2(-1 + \sqrt{2})^n + c_3(-1 - \sqrt{2})^n$ and use $P_0 = 0$, $P_{2N} = P_{2N+1} = 1$ to evaluate c_1, c_2, c_3.

7. a) $P_n = \dfrac{3n + (-2)^{n-2N} - (-2)^{-2N}}{6N + 1 - (-2)^{-2N}}$; no; $P_{10} \approx \frac{30}{61}$

   b) $P_n = \dfrac{3n + (-2)^{n-30} - (-2)^{-30}}{91 - (-2)^{-30}}$; $P_{10} \approx \frac{30}{91}$

CHAPTER 5

Section 5.1

11. $R = 1$ 13. $R = 0$ 15. $R = 1/4$

Section 5.2

1. $y = 4e^x$ 3. $y = e^x + x + 1$ 5. $y = c_0 e^{x^2/2}$

7. $y = c_1 x e^x$ 9. $y = c_n x^n$ 11. $y = c_0 \cos 2x + c_1 \sin 2x$

13. $y = c_0 \cosh x + c_1 \sinh x$ 15. $y = c_0(1 + x \tan^{-1} x) + a_1 x$

17. $y = c_0 \cos x + c_1 \sin x + \dfrac{e^{2x}}{5}$ 19. $y = c_2 x^2 + Ax^{-2}$

21. a) No   b) No

Section 5.3

1. $y_0 = c_0 \cos \sqrt{x} + c_1 \sin \sqrt{x}$ 3. $y = c_0 \dfrac{\sinh x}{x^3} + c_1 \dfrac{\cosh x}{x^3}$

5. $y = c_0 \dfrac{\sin x^2}{x^2} + c_1 \dfrac{\cos x^2}{x^2}$ 7. $y = c_0 x + \dfrac{c_1}{x^2}$

9. $y = c_0 \dfrac{\sin x}{x} + c_1 \dfrac{\cos x}{x}$ 11. $y = c_0 e^x + c_1 e^x \ln|x|$

13. $y = c_0 \sqrt{x} e^x + c_1 \sqrt{x} e^x \ln|x|$ 15. $y = \dfrac{c_0}{1-x} + c_1 \dfrac{\ln|x|}{1-x}$

17. $y = (c_0 + c_1 \ln|x|)\left(1 + \dfrac{x^2}{2^2} + \dfrac{x^4}{(2 \cdot 4)^2} + \cdots\right) - c_1\left(\dfrac{x^2}{4} + \dfrac{3x^4}{8 \cdot 16} + \cdots\right)$

19. $y = c_0\sqrt{x}\left(1 - \dfrac{7}{6}x + \dfrac{21}{40}x^2 + \cdots\right) + c_1(1 - 3x + 2x^2 + \cdots)$

21. $y = c_0x + c_1\left(x \ln|x| + \displaystyle\sum_{n=1}^{\infty} \dfrac{(-1)^n}{n!n} x^{n+1}\right)$

25. b) $y_2 = xe^{-1/x}$

Section 5.4

1. $\left(\dfrac{384}{x^4} - \dfrac{72}{x^2} + 1\right) J_1(x) - \left(\dfrac{192}{x^3} - \dfrac{12}{x}\right) J_0(x)$ 13. Use Eq. (5.117).

15. Set $z = \sqrt{x}$, then $y = AJ_p(\sqrt{x}) + BY_p(\sqrt{x})$.

17. Set $y = xu$, then $y = x[AJ_1(x) + BY_1(x)]$.

19. Set $y = ux^{-k}$, then $y = x^{-k}(AJ_k(x) + BY_k(x))$.

21. Set $y = \sqrt{x}u$, $z = kx^3/3$, then $y = \sqrt{x}(AJ_{1/6}(kx^3/3) + BY_{1/6}(kx^3/3))$.

Section 5.5

1. $P_5 = \dfrac{63x^5 - 70x^3 + 15x}{8}, P_6 = \dfrac{231x^6 - 315x^4 + 105x^2 - 5}{16}$,

$P_7 = \dfrac{429x^7 - 693x^5 + 315x^3 - 35x}{16}$,

$P_8 = \dfrac{429(15x^8 - 28x^6) + 630(11x^4 - 2x^2) + 35}{128}$

9. d) $H_0(x) = 1, H_1(x) = x, H_2(x) = x^2 - 1, H_3(x) = x^3 - 3x, H_4(x) = x^4 - 6x^2 + 3$

CHAPTER 6

Section 6.1

1. $\dfrac{5}{s^2} + \dfrac{2}{s}$ 3. $\dfrac{2}{s^3} + \dfrac{a}{s^2} + \dfrac{b}{s}$ 5. $\dfrac{e^2}{s - 5}$

7. $(\cos b)\dfrac{s}{s^2 + a^2} - (\sin b)\dfrac{a}{s^2 + a^2}$ 9. $(\cosh b)\dfrac{s}{s^2 - a^2} + (\sinh b)\dfrac{a}{s^2 - a^2}$

11. $\dfrac{1}{(s + 1)^2 + 1}$ 15. $7t$

17. $a_1 + a_2t + \dfrac{a_3}{2!}t^2 + \cdots + a_{n+1}\dfrac{t^n}{n!}$ 19. $7e^{3t}$

21. $\cos\sqrt{3}t - 2\sin\sqrt{3}t$

Section 6.2

9. $\dfrac{s(s^2 - a^2 - b^2)}{(s^2 - (a + b)^2)(s^2 - (a - b)^2)}$ 11. $\dfrac{b(s^2 + a^2 + b^2)}{(s^2 + a^2 + b^2) - 4a^2s^2}$

13. $\dfrac{2abs}{(s^2 + a^2 + b^2)^2 - 4a^2s^2}$

19. $\dfrac{b}{s(e^{-as} + 1)}$

21. $\dfrac{s^2 + 2a^2}{s(s^2 + 4a^2)}$

23. $\dfrac{2a(3s^2 - a^2)}{(s^2 + a^2)^3}$

25. $\dfrac{1 - e^{-as}}{s^2(1 + e^{-2as})}$

27. $\dfrac{(1 - e^{-as})^2}{s^2(1 + e^{-2as})}$

29. $\dfrac{2(s^2 - 3a^2)}{(s^2 + a^2)^3}$

31. $\dfrac{s^2 + 2a^2}{s^2(s^2 + 4a^2)}$

33. $e^t - 1$

35. $\dfrac{1 - \cos 2t}{2}$

37. $\dfrac{(1 + a)(1 - e^{-at})}{a} - at$

39. $\dfrac{1 + \sin 2t - \cos 2t}{2}$

Section 6.3

1. $\cos t$

3. $A \cosh at + \dfrac{B}{a} \sinh at$

5. $e^{-t}(\cos 2t + \sin 2t)$

7. $\frac{1}{3} - e^t + \frac{5}{3}e^{3t}$

9. $-\dfrac{4}{27} - \dfrac{t}{9} + \frac{23}{27}e^{3t} + \frac{4}{27}e^{-3t}$

11. e^{-t}

13. $\frac{1}{2}(e^t - e^{-t}) = \sinh t$

15. $\left(1 + \dfrac{t}{2k}\right) \sin kt$

17. $\left(a + \dfrac{1}{2a^2}\right) \sin at + \left(a - \dfrac{t}{2a}\right) \cos at$

19. $\frac{5}{4} \cosh t - \frac{1}{4}(\cos t + t \sin t)$

21. $\dfrac{s^2 - k^2}{(s^2 + k^2)^2}$

23. $\dfrac{2ks}{(s^2 + k^2)^2}$

25. $\dfrac{s^2 + 1}{(s^2 - 1)^2}$

27. $\dfrac{2s}{(s^2 - 1)^2}$

29. $\arctan \dfrac{k}{s}$

31. $\dfrac{1}{2} \ln \dfrac{s^2 - a^2}{s^2}$

33. $\dfrac{1}{2s} \ln \dfrac{s^2 + a^2}{s^2}$

35. $\dfrac{e^{s^2/4}}{s}\left(1 - \operatorname{erf}\left(\dfrac{s}{2}\right)\right)$

37. $\dfrac{2}{t}(1 - \cos at)$

39. $\dfrac{\sin t}{t}$

Section 6.4

1. $\dfrac{3!}{s^4(s^2 + 1)}$

3. $\dfrac{3!5!}{s^{10}}$

5. $\dfrac{19!}{s^{20}(s - 17)}$

7. $\dfrac{at - \sin at}{a^2}$

9. $\dfrac{1}{a^2}(1 - \cos at)$

11. $e^{-t}(1 - t)^2$

15. $e^{3t/2}\left[\cosh \dfrac{\sqrt{5}t}{2} + \dfrac{1}{\sqrt{5}} \sinh \dfrac{\sqrt{5}t}{2}\right]$

Section 6.5

1. $\frac{1}{5}((-1)^n - (-6)^n)$

3. $3^n(5 - n)$

5. $2^{n/2}\left(\cos \dfrac{n\pi}{4} + \sin \dfrac{n\pi}{4}\right)$

7. $2^{n-2}\left[\dfrac{n^2(n+1)}{2} - \displaystyle\sum_{k=2}^{n}(k^2 - k)\right]$

9. $\frac{1}{2}(4^n - 2^n)$

17. $-20 + 120(2)^n$

19. $\frac{1}{4} - n - \frac{2}{3}(2)^n + \dfrac{5^{n+1}}{12}$

CHAPTER 7

Section 7.1

1. $x' = -c_1 x,$
$y' = c_1 x - c_2 y$

3. When $t = \dfrac{100}{2\sqrt{3}}\ln\dfrac{3+\sqrt{3}}{3-\sqrt{3}} \approx 38$ min; ≈ 39 lb salt

5. $m_1 x_1'' = -\lambda_1 x_1 + \lambda_2(x_2 - x_1),$
$m_2 x_2'' = -\lambda_2(x_2 - x_1) + \lambda_3(x_3 - x_2),$
$m_3 x_2'' = -\lambda_3(x_3 - x_2)$

7. a) $x_1 = \dfrac{\alpha x_1(0)x_2(0)e^{(\alpha x_1(0)-\beta)t} + \alpha x_1^2(0) - \beta x_1(0) - \alpha x_1(0)x_2(0)}{\alpha x_1(0) - \beta},$

$x_2 = x_2(0)e^{(\alpha x_1(0)-\beta)t},$

$x_3 = \dfrac{\beta x_2(0)e^{(\alpha x_1(0)-\beta)t} + \alpha x_3(0)x_1(0) - x_3(0)\beta - x_2(0)\beta}{\alpha x_1(0) - \beta}$

c) An epidemic occurs.

Section 7.2

1. $x_1' = x_2,$
$x_2' = -3x_1 - 2x_2$

3. $x_1' = x_2,$
$x_2' = x_3,$
$x_3' = x_1^3 - x_2^2 + x_3 + t$

5. $x_1' = x_2,$
$x_2' = x_3,$
$x_3' = x_1^4 x_2 - x_1 x_3 + \sin t$

7. $x_1' = x_2,$
$x_2' = x_3,$
$x_3' = x_1 - 4x_2 + 3x_3$

9. a) $x_1(t) = c_1 e^t;\ x_2(t) = c_2 e^t;\ c_1, c_2$ constant

Section 7.3

1. $x = c_1 e^{-t} + c_2 e^{4t},$
$y = -c_1 e^{-t} + \frac{3}{2}c_2 e^{4t}$

3. $x = c_1 e^{-3t} + c_2(1 - t)e^{-3t},$
$y = -c_1 e^{-3t} + c_2 t e^{-3t}$

5. $x = c_1 e^{10t} + c_2 t e^{10t},$
$y = -(2c_1 + c_2)e^{10t} - 2c_2 t e^{10t}$

7. $x = \frac{1}{8}(-2t^2 - 18t - 6) + c_1 - 3c_2 e^{2t},$
$y = \frac{1}{8}(2t^2 + 14t) - c_1 + c_2 e^{2t}$

9. $x = e^{4t}(17c_1 \cos 2t + 17c_2 \sin 2t),$
$y = e^{4t}[(8c_1 - 2c_2)\cos 2t + (8c_2 + 2c_1)\sin 2t]$

11. $x_1 = 2c_1 e^t,$
$x_2 = -3c_1 e^t + c_2 e^t \cos 2t + c_3 e^t \sin 2t,$
$x_3 = 2c_1 e^t + c_2 e^t \sin 2t - c_3 e^t \cos 2t$

Section 7.4

1. b) $W = e^{-6t}$
   c) $\{c_1 e^{-3t} + c_2(1 - t)e^{-3t}, -c_1 e^{-3t} + c_2 t e^{-3t}\}$

Section 7.5

1. $\{e^t, e^t\}, \{3e^{-t}, 5e^{-t}\}$

3. $\{e^t \cos t, e^t(2 \cos t - \sin t)\}, \{e^t \sin t, e^t(2 \sin t + \cos t)\}$

5. $\{e^{-3t}, -e^{-3t}\}, \{(t - 1)e^{-3t}, -te^{-3t}\}$ 7. $\{3, 4\}, \{e^{-2t}, 2e^{-2t}\}$

11. $\{\frac{1}{8}(-2t^2 - 18t - 6), \frac{1}{8}(2t^2 + 14t)\}$ 13. $\{\sin t - \cos t, 2 \sin t - \cos t\}$

15. a) $x' = -0.134x + 0.02y$,
       $y' = 0.036x - 0.02y$
    b) $\{10c_1 e^{-0.14t} + c_2 e^{-0.014t}, -3c_1 e^{-0.14t} + 6c_2 e^{-0.014t}\}$

Section 7.6

1. $\{1 - 1.025e^{-0.05}, 1 - 1.05e^{-0.05}\}$

3. $\{100(k_1 e^{\lambda_1 t} + k_2 e^{\lambda_2 t}) + 1, -8(k_1 \lambda_1 e^{\lambda_1 t} + k_2 \lambda_2 e^{\lambda_2 t}) + 1\}$
   where $\begin{Bmatrix} k_1 \\ k_2 \end{Bmatrix} = \dfrac{\pm 3\sqrt{2} - 4}{800}$, $\begin{Bmatrix} \lambda_1 \\ \lambda_2 \end{Bmatrix} = -50 \pm 25\sqrt{2}$

5. The general solution $\{I_L, I_R\} = \{I_L, I_R\}_h + \{I_L, I_R\}_p$ is given by

   $$\{I_L, I_R\}_p = \{A \sin \omega t + B \cos \omega t, C \sin \omega t + D \cos \omega t\},$$

   where $\omega = 60\pi$, $\Delta = \omega^2 + 2500$, $A = (2500/\Delta)^2$, $B = -25\omega(\omega^2 + 7500)/\Delta^2$, $C = -2500(\omega^2 - 2500)/\Delta^2$, $D = -250000\omega/\Delta^2$, and

   a) $\{I_L, I_R\}_h = \left\{ \dfrac{e^{-50t}}{2}[k_1 + k_2(t + \frac{1}{25})], e^{-50t}[k_1 + k_2(t + \frac{1}{25})] \right\}$

   b) $\{I_L, I_R\}_h = \{e^{-50t}[k_1 \cos 50\sqrt{3}t + k_2 \sin 50\sqrt{3}t], e^{-50t}[k_3 \cos 50\sqrt{3}t + k_4 \sin 50\sqrt{3}t]\}$

   c) $\{I_L, I_R\}_h = \{e^{-50t}(k_1 e^{25\sqrt{2}t} + k_2 e^{-25\sqrt{2}t}),$
      $e^{-50t}[(2 - \sqrt{2})k_1 e^{25\sqrt{2}t} + (2 + \sqrt{2})k_2 e^{-25\sqrt{2}t}]\}$

7. $I_1 = \frac{1}{20}(8 \sin t - 6 \cos t + 5e^{-t} + e^{-3t})$;
   $I_2 = \frac{1}{20}(2 \sin t - 4 \cos t + 5e^{-t} - e^{-3t})$

Section 7.7

1. a) $\dfrac{x''}{x} - \left(\dfrac{x'}{x}\right)^2 + b_1 x' = \left(a_1 - b_1 x - \dfrac{x'}{x}\right)\left[-a_2 + c_2 x + \dfrac{b_2}{c_1}\left(a_1 - b_1 x - \dfrac{x'}{x}\right)\right]$

   b) $(0, a_2/b_2), (a_1/b_1, 0), \left(\dfrac{a_1 b_2 - a_2 c_1}{b_1 b_2 - c_1 c_2}, \dfrac{a_2 b_1 - a_1 c_2}{b_1 b_2 - c_1 c_2}\right)$ c) $y = cx$

   d) $\dfrac{dy}{dx} = \left(\dfrac{y}{x}\right)\left(\dfrac{b_2(y/x) + c_2}{b_1 + c_1(y/x)}\right)$; starvation of both species

3. a) $(n\pi, 0), n = 0, 1, 2, 3, \ldots$. These points represent positions where the pendulum is at the bottom or top of its trajectory.

c) $\theta = \theta_0 \cos \sqrt{\dfrac{g}{l}}\, t; \psi = -\sqrt{\dfrac{g}{l}}\, \theta_0 \sin \sqrt{\dfrac{g}{l}}\, t, T \approx 2\pi \sqrt{\dfrac{l}{g}}$

Section 7.8

1. $x = \cosh t; y = \sinh t$

3. $x = e^{3t}(2 \cos 3t + 2 \sin 3t); y = e^{3t}(-2 \cos 3t + 4 \sin 3t)$

5. $x = \frac{3}{2} \cos t + \frac{1}{2} \sin t + \frac{7}{2} e^t - 5e^{-t} - 4t;$
   $y = \cos t + \frac{7}{2} e^t - \frac{5}{2} e^{-t} - 1 - 3t$

7. $x = \dfrac{e^{3t}}{4(13)^3}(6887 \cos 2t + 2637 \sin 2t) + \dfrac{e^t}{4} - \dfrac{5t^2}{13} - \dfrac{34t}{(13)^2} - \dfrac{74}{(13)^3};$

   $y = \dfrac{e^{3t}}{4(13)^3}(-9524 \cos 2t + 4250 \sin 2t) + \dfrac{4}{13}t^2 + \dfrac{48t}{(13)^2} + \dfrac{184}{(13)^3}$

Section 7.9

3. $x_{n+1} = y_n,$
   $y_{n+1} = -n^2 x_n + n y_n + 3^n$

5. $x_{n+1} = y_n,$
   $y_{n+1} = -x_n y_n$

7. $x_n = c_1 2^n + c_2(-3)^n,$
   $y_n = c_1 2^{n+2} - c_2(-3)^n$

9. $x_n = \begin{cases} -3, n = 1 \\ 0, n \neq 1 \end{cases} \quad y_n = \begin{cases} 1, n = 0, 1 \\ 0, n = 2, 3, \ldots \end{cases}$

11. $x_n = -\frac{1}{4}(3^n - 5) + \dfrac{n}{2}$

   $y_n = \frac{1}{4}(3^n - 1) + \dfrac{n}{2}$

13. $51 \times 10^{-7}, 52.5 \times 10^{-7}, 52 \times 10^{-7}$ g/m^3 at $t = 1$ hr

Section 7.10

1. $\mathbf{x}'(t) = A\mathbf{x}(t); A = \begin{pmatrix} 2 & 3 \\ 4 & -6 \end{pmatrix}$

3. $\mathbf{x}'(t) = A(t)\mathbf{x}(t) + \mathbf{f}(t); A(t) = \begin{pmatrix} 0 & 1 & 0 \\ 0 & 0 & 0 \\ 1 & 4t & -2 \end{pmatrix}; \mathbf{f}(t) = \begin{pmatrix} 0 \\ 0 \\ \sin t \end{pmatrix}$

5. $\mathbf{x}'(t) = A(t)\mathbf{x}(t) + \mathbf{f}(t); A(t) = \begin{pmatrix} 2t & -3t^2 & \sin t \\ 2 & 0 & -4 \\ 0 & 17 & 4t \end{pmatrix}; \mathbf{f}(t) = \begin{pmatrix} 0 \\ -\sin t \\ e^t \end{pmatrix}$

17. a) $\mathbf{x}'(t) = (1, \cos t); \displaystyle\int \mathbf{x}(t)\, dt = \left(\dfrac{t^2}{2}, -\cos t \right)$

   b) $\mathbf{y}'(t) = \begin{pmatrix} e^t \\ -\sin t \\ \sec^2 t \end{pmatrix}; \displaystyle\int \mathbf{y}(t)\, dt = \begin{pmatrix} e^t \\ \sin t \\ -\ln|\cos t| \end{pmatrix}$

c) $A'(t) = \begin{pmatrix} \dfrac{1}{2\sqrt{t}} & 2t \\ 2e^{2t} & 2\cos 2t \end{pmatrix}$; $\displaystyle\int A(t)\,dt = \begin{pmatrix} \dfrac{2}{3}t^{3/2} & \dfrac{t^3}{3} \\ \dfrac{e^{2t}}{2} & -\dfrac{\cos 2t}{2} \end{pmatrix}$

d) $B'(t) = \begin{pmatrix} \dfrac{1}{t} & e^t(\sin t + \cos t) & e^t(-\sin t + \cos t) \\ \frac{5}{2}t^{3/2} & \sin t & -\cos t \\ -\dfrac{1}{t^2} & e^{t^2}(1 + 2t^2) & e^{t^3}(2t + 3t^4) \end{pmatrix}$;

$\displaystyle\int B(t)\,dt = \begin{pmatrix} t\ln|t| - t & \frac{1}{2}e^t(\sin t - \cos t) & \frac{1}{2}e^t(\sin t + \cos t) \\ \frac{2}{7}t^{7/2} & -\sin t & \cos t \\ \ln|t| & \dfrac{e^{t^2}}{2} & \dfrac{e^{t^3}}{3} \end{pmatrix}$

Section 7.11

1. Fundamental
3. Fundamental
5. Not fundamental

7. $C = \begin{pmatrix} 1 & -2 \\ -1 & 1 \end{pmatrix}$

9. $C = \begin{pmatrix} 1 & -1 & 1 \\ 0 & 1 & 1 \\ 1 & 0 & 0 \end{pmatrix}$

13. $\Psi(t) = e^{6t}\begin{pmatrix} \cos 2t - \sin t & \frac{1}{2}\sin t \\ -4\sin t & \cos 2t + \sin 2t \end{pmatrix}$

15. $\Psi(t) = \dfrac{1}{6}\begin{pmatrix} -e^{-t} + 9e^t - 2e^{2t} & -2e^{-t} + 2e^{2t} & 7e^{-t} - 9e^t + 2e^{2t} \\ 6e^t - 6e^{2t} & 6e^{2t} & -6e^t + 6e^{2t} \\ -e^{-t} + 3e^t - 2e^{2t} & -2e^{-t} + 2e^{2t} & 7e^{-t} - 3e^t + 2e^{2t} \end{pmatrix}$

17. a) $e^t\begin{pmatrix} 1 - 2t \\ 2 - 2t \end{pmatrix}$

b) $e^t\begin{pmatrix} -10t - 2 \\ -10t + 3 \end{pmatrix}$

c) $e^{t-1}\begin{pmatrix} 2 - 2t \\ 3 - 2t \end{pmatrix}$

d) $e^{t+1}\begin{pmatrix} 4 + 2t \\ 3 + 2t \end{pmatrix}$

e) $e^{t-3}\begin{pmatrix} 3 \\ 3 \end{pmatrix}$

f) $e^{t-a}\begin{pmatrix} 2(c - b)(a - t) + b \\ 2(c - b)(a - t) + c \end{pmatrix}$

19. a) $\mathbf{x}' = A(t)\mathbf{x}$, where $A(t) = \begin{pmatrix} 0 & 1 \\ -b(t) & -a(t) \end{pmatrix}$

21. $\varphi_2(t) = e^{t^2}\displaystyle\int e^{-t^2}\,dt$

23. $\Psi(t) = \begin{pmatrix} \frac{5}{2}e^t - \frac{3}{2}e^{-t} & -\frac{3}{2}e^t + \frac{3}{2}e^{-t} \\ \frac{5}{2}e^t - \frac{5}{2}e^{-t} & -\frac{3}{2}e^t + \frac{5}{2}e^{-t} \end{pmatrix}$

25. $\Psi(t) = e^t\begin{pmatrix} \cos t - 2\sin t & \sin t \\ -5\sin t & 2\sin t + \cos t \end{pmatrix}$

27. $\Psi(t) = e^{-3t}\begin{pmatrix} 1 - t & -t \\ t & 1 + t \end{pmatrix}$

29. $\Psi(t) = \begin{pmatrix} 3 - 2e^{-2t} & -\frac{3}{2} + \frac{3}{2}e^{-2t} \\ 4 - 4e^{-2t} & -2 + 3e^{-2t} \end{pmatrix}$

Section 7.12

1. $\varphi_p(t) = \begin{pmatrix} \dfrac{e^{2t}}{3} \\ \dfrac{e^t}{2} - \frac{2}{3}e^{2t} \end{pmatrix}$

3. $\varphi_p(t) = \begin{pmatrix} -t\cos t \\ -t\sin t - 2t\cos t + \cos t + \sin t \end{pmatrix}$

5. $\varphi_p(t) = \begin{pmatrix} -\frac{3}{2}e^{-t} + \frac{7}{4}e^t + \frac{3}{2}te^t + \frac{1}{3}te^{2t} - \frac{1}{9}e^{2t} - \frac{1}{8}e^{3t} \\ 2e^t + te^t + te^{2t} + \frac{1}{2}e^{3t} \\ -\frac{3}{2}e^{-t} + \frac{3}{4}e^t + \frac{1}{2}te^t - \frac{1}{9}e^{2t} + \frac{1}{3}te^{2t} + \frac{3}{8}e^{3t} \end{pmatrix}$

7. $\varphi_p(t) = \begin{pmatrix} t \\ t^2 \\ t^4 \end{pmatrix}$

9. $\varphi_p(t) = \begin{pmatrix} -5\sin t \ln|\csc t + \cot t| \\ 1 - (2\sin t - \cos t)\ln|\csc t + \cot t| \end{pmatrix}$

11. $\Phi(t) = \begin{pmatrix} 1/t & \ln t/t \\ -1/t^2 & (1 - \ln t)/t^2 \end{pmatrix}$

Section 7.13

1. $x(t) = -\frac{1}{5}\begin{pmatrix} 12 + 11i \\ 6 + 2i \end{pmatrix} e^{it}$

3. $x(t) = -\frac{1}{754}\begin{pmatrix} 353 + 205i \\ -197 - 595i \end{pmatrix} e^{-5it}$

9. a) Let $x_1 = x$, $x_2 = x'$, then $\begin{pmatrix} x_1 \\ x_2 \end{pmatrix}' = \begin{pmatrix} 0 & 1 \\ -\lambda/m & 0 \end{pmatrix}\begin{pmatrix} x_1 \\ x_2 \end{pmatrix} + \begin{pmatrix} 0 \\ k/m \end{pmatrix} e^{i\omega t}$

   b) $x(t) = \dfrac{k}{\lambda - \omega^2 m}\begin{pmatrix} 1 \\ i\omega \end{pmatrix} e^{i\omega t}$

   c) $\Psi(t) = \begin{pmatrix} \cos\sqrt{\lambda/m}\,t & \sqrt{\lambda/m}\sin\sqrt{\lambda/m}\,t \\ -\sqrt{\lambda/m}\sin\sqrt{\lambda/m}\,t & \cos\sqrt{\lambda/m}\,t \end{pmatrix}$. Resonance

CHAPTER 8

Section 8.1

1. a) $y(1) \approx 2.98$ $(y(1) = 2(e - 1) \approx 3.44)$

   b) $y(3) \approx 2.16$ $\left(y(3) = 2\left(1 + \dfrac{1}{e^2}\right) \approx 2.27\right)$

   c) $y(1) \approx 0.72$ $(y(1) = \sqrt{3} - 1 \approx 0.73)$

3. No method will work since the solution to the differential is the hyperbola

$$x^2 - 2xy - y^2 = 4,$$

which isn't defined if $x = 1$.

Section 8.2

1. $|e_n| \leqslant 0.859h$ b) $h = 0.1, |e_n| \leqslant 0.086; h = 0.2, |e_n| \leqslant 0.172$

3. $|e_n| \leqslant 6.06h$ a) $1{,}212{,}000$ b) $12{,}120{,}000$

Section 8.3

1. $a = \frac{1}{4}, b = 0, c = \frac{3}{4}, d = 0$ 3. $n = p = \frac{2}{3}$

5. Because the determinant of the resulting 4×4 system is zero. There are an infinite number of solutions; no.

Section 8.4

9. $y(1) \approx 3.333$ $(y(1) = 3.4365)$

Section 8.6

1. $x(1) \approx -17.43; y(1) \approx 7.77$ 9. Maximum occurs at $y(\frac{1}{2}) = \sqrt{2y'(0) - 1}$

CHAPTER 9

Section 9.1

1. Eigenvalues: $\left(\dfrac{n\pi}{T}\right)^2$, eigenfunctions: $A \sin \dfrac{n\pi t}{T}, n = 1, 2, 3, \ldots$

3. Eigenvalues: $-\left(\dfrac{n\pi}{T}\right)^2$, eigenfunctions: $A \sin \dfrac{n\pi t}{T}, n = 1, 2, 3, \ldots$

5. Eigenvalues: $\left(\dfrac{n\pi}{2T}\right)^2$, eigenfunctions: $A \sin \dfrac{n\pi(t + T)}{2T}, n = 1, 2, 3, \ldots$

Section 9.2

1. $\dfrac{1}{\sqrt{2\pi}}, \dfrac{\cos x}{\sqrt{\pi}}, \dfrac{\cos 2x}{\sqrt{\pi}}, \ldots$ 3. $\dfrac{1}{\sqrt{T}}, \sqrt{\dfrac{2}{T}} \cos \dfrac{\pi x}{T}, \sqrt{\dfrac{2}{T}} \cos \dfrac{2\pi x}{T}, \ldots$

5. $\sqrt{\dfrac{1}{\pi}}, \sqrt{\dfrac{2}{\pi}} \cos 2x, \sqrt{\dfrac{2}{\pi}} \cos 4x, \ldots$

7. a) $\sinh (1)P_0(x) + \dfrac{3}{e} P_1(x) + \dfrac{5}{2}\left(e - \dfrac{7}{e}\right) P_2(x)$

   b) $3 \sin (1)P_1(x) + 15 \sin (1)P_2(x)$

   c) $\sin (1)P_0(x) + [15 \cos (1) - 10 \sin (1)]P_2(x)$

   d) $3[\cosh (1) - \sinh (1)]P_1(x)$

   e) $\sinh (1)P_0(x) + [20 \sinh (1) - 15 \cosh (1)]P_2(x)$

   f) $\dfrac{\pi}{4} P_0(x) - \dfrac{5}{2}(3 - \pi)P_2(x)$

Section 9.3

1. $\pi, 2, T/n, 1/k$ 3. $2bT/a$

5. $2(\sin x - \frac{1}{2}\sin 2x + \frac{1}{3}\sin 3x - \frac{1}{4}\sin 4x + \cdots)$

7. $\dfrac{\pi^2}{3} - 4(\cos x - \frac{1}{4}\cos 2x + \frac{1}{9}\cos 3x - \frac{1}{16}\cos 4x + \cdots)$

9. $\dfrac{\pi}{2} - \dfrac{4}{\pi}\left[\cos x + \dfrac{1}{3^2}\cos 3x + \dfrac{1}{5^2}\cos 5x + \cdots\right]$

11. $\dfrac{2}{\pi}[(1 + \sin(\pi - 1))\sin x - \frac{1}{2}(1 + \frac{1}{2}\sin 2(\pi - 1))\sin 2x + \frac{1}{3}(1 + \frac{1}{3}\sin 3(\pi - 1))\sin 3x - \cdots]$

13. $\dfrac{\sinh \pi}{\pi}\left[1 + 2\displaystyle\sum_{n=1}^{\infty}\dfrac{(-1)^n}{1 + n^2}\cos nx - 2\displaystyle\sum_{n=1}^{\infty}\dfrac{n(-1)^n}{1 + n^2}\sin nx\right]$

15. $1 - \dfrac{2}{\pi}\displaystyle\sum_{n=1}^{\infty}\dfrac{\sin n\pi x}{n}$ 17. $\dfrac{1}{3} + \dfrac{4}{\pi^2}\displaystyle\sum_{n=1}^{\infty}\dfrac{(-1)^n}{n^2}\cos n\pi x$

19. $\dfrac{1}{2} - \dfrac{2}{\pi}\left(\sin \pi x + \dfrac{1}{3}\sin 3\pi x + \dfrac{1}{5}\sin 5\pi x + \cdots\right)$

21. $\dfrac{3}{4} - \dfrac{1}{\pi}\displaystyle\sum_{n=1}^{\infty}\dfrac{1}{n}\sin n\pi x - \dfrac{2}{\pi^2}\left(\cos \pi x + \dfrac{1}{3^2}\cos 3\pi x + \dfrac{1}{5^2}\cos 5\pi x + \cdots\right)$

Section 9.4

1. $\lambda_k = k^2$, $y_k = A\sin kx$, $k = 1, 2, 3, \ldots$

3. $\lambda_k = k^2$, $y_k = A\sin k\left(x + \dfrac{\pi}{2}\right)$, $k = 1, 2, 3, \ldots$

5. $\lambda_k = \dfrac{k^2\pi^2}{4}$, $y_k = A\sin\left(\dfrac{k\pi \ln x}{2}\right)$, $k = 1, 2, 3, \ldots$

7. $\lambda_k = k^2\pi^2$, $y_k = A\sin\dfrac{k\pi}{x}$, $k = 1, 2, 3, \ldots$

Section 9.5

1. $\lambda_k = k^2$ are real and simple, $y_k = A\sin kx$, $k = 1, 2, 3, \ldots$, are orthogonal.

3. $\lambda_k = k^2$ are real and simple, $y_k = A\sin k\left(x + \dfrac{\pi}{2}\right)$, $k = 1, 2, 3, \ldots$, are orthogonal.

5. $\lambda_k = k^2\pi^2$ are real and simple, $y_k = A\sin\dfrac{k\pi}{x}$, $k = 1, 2, 3, \ldots$, are orthogonal with respect to $1/x^2$ on $1/2 \leqslant x \leqslant 1$.

9. $\lambda \neq 1$, since $y = Ax + Bx \ln x$ cannot satisfy the boundary condition. Use Theorem 3.9 with $\lambda = \lambda_1 + i\lambda_2$. Then $\lambda_1 = 1$ and $\lambda_2 = \pm\dfrac{2k\pi}{\ln 2}, k = 1, 2, 3, \ldots$.

Section 9.6

5. Yes, trivially

7. Infinitely many

Section 9.7

1. $\displaystyle\sum_{k=1}^{\infty} \frac{2(-1)^k}{k(k^2 - 1)} \sin kx$

3. $\displaystyle 4\sum_{k=1}^{\infty} \frac{\left[1 + (-1)^k\left(\dfrac{2k-1}{2}\pi\right)\right]}{\left(\dfrac{2k-1}{2}\pi\right)^3\left[\left(\dfrac{2k-1}{4}\pi\right)^2 - 1\right]} \sin\left(\frac{2k-1}{2}\right)\pi x$

5. $\displaystyle 2(\ln 2)^2 \sin(\ln 2)\pi \sum_{k=1}^{\infty} \frac{(-1)^k k}{[3(\ln 2)^2 - (k\pi)^2][(\ln 2)^2 - (k\pi)^2]} \sin\left(\frac{k\pi}{\ln 2} \ln x\right)$

13. $K(x,t) = \begin{cases} -\dfrac{\sinh(1-t)\sinh x}{\sinh(1)}, & x < t, \\[3mm] -\dfrac{\sinh(1-x)\sinh t}{\sinh(1)}, & t < x \end{cases}$

17. $K(x,t) = \begin{cases} -\dfrac{\sin 2x \cos 2(1-t)}{2\cos 2}, & x < t, \\[3mm] -\dfrac{\cos 2(1-x)\sin 2t}{2\cos 2}, & t < x \end{cases}$

CHAPTER 10

Section 10.1

1. $W(x,y) = \frac{1}{2} + e^{x + (5/4)y} - \frac{7}{2}e^{y/2}$

3. $W(x,y) = ye^x$

Section 10.2

1. -0.0075 ft

3. 0.0 ft

5. 0.1 ft

9. $-\dfrac{0.045}{\pi}$ ft

11. $\dfrac{0.03}{\pi}\left[\dfrac{\pi}{6} - \dfrac{1}{4}\right]$ ft

Section 10.3

1. $\displaystyle T(x,t) = \frac{4}{\pi}\sum_{n=1}^{\infty} \frac{[1 - (-1)^n]}{n^3}(\sin nx)e^{-n^2\delta t}$

3. $\displaystyle T(x,t) = 4\sum_{n=1}^{\infty}\left(\frac{2(-1)^{n+1} - 1}{n^3}\right)(\sin nx)e^{-n^2\delta t}$

5. $\displaystyle T(x,t) = \frac{2}{\pi}\sum_{n=1}^{\infty}\frac{1}{n^2}\left[\sin\frac{n\pi}{2} - \frac{n\pi}{2}\cos\left(\frac{n\pi}{2}\right)\right](\sin nx)e^{-n^2\delta t}$

7. $[(L - x)T_1 + xT_2]/L$

9. $f(x) = \dfrac{c_0^2}{T_1} \cos cx + c_0 c_2 \sin cx$, where $c_2 = c_0(1 - \cos cL)/T \sin cL$. If cL is a

multiple of π, then c_2 is arbitrary.

11. $y(x,t) = e^{kx + (t/k)}$

13. $y(x,t) = e^{x - kt}(c_1 e^{\sqrt{1 + kx}} + c_2 e^{-\sqrt{1 + kx}})$

15. $y(x,t) = \sqrt{x}(c_1 e^{\sqrt{kt}} + c_2 e^{-\sqrt{kt}})(b_1 x^{\sqrt{1 + 4k/2}} + b_2 x^{-\sqrt{1 + 4k/2}})$

Section 10.4

1. $T(x,y) = \dfrac{100}{\pi} \displaystyle\sum_{n=1}^{\infty} \left(\dfrac{1 - (-1)^n}{n \sinh n\pi} \right) \sin \dfrac{n\pi x}{L} \sinh \dfrac{n\pi y}{L}$, where we evaluate at $\left(\dfrac{L}{2}, \dfrac{L}{2} \right)$ and

$\left(\dfrac{L}{4}, \dfrac{L}{4} \right)$.

3. $T(x,y) = \dfrac{100}{\sinh \pi} \sin \dfrac{\pi x}{L} \sinh \dfrac{\pi y}{L}$

5. a) $A_0 = 1$, $A_2 = -\frac{1}{2}$, all other coefficients are zero, so that $T(r,\theta) = \frac{1}{2}(1 - r^2 \cos 2\theta)$.
   Hence $T(\frac{1}{2},0) = \frac{3}{8}°$.

   b) $T(\frac{1}{2},0) = \dfrac{3}{8\pi} \displaystyle\int_{-\pi}^{\pi} \dfrac{\sin^2 \varphi}{5/4 - \cos \varphi} \, d\varphi = \frac{3}{8}°$. This last equality can be proved using the
   calculus of residues.

7. a) $T(r,\theta) = -\dfrac{\pi^2}{3} - 4 \displaystyle\sum_{n=1}^{\infty} (-r)^n \left(\dfrac{\cos n\theta}{n^2} + \dfrac{\pi \sin n\theta}{n} \right)$

   b) $T(r,\theta) = \dfrac{1 - r^2}{2\pi} \displaystyle\int_{-\pi}^{\pi} \dfrac{\varphi(2\pi - \varphi) \, d\varphi}{r^2 + 1 - 2r \cos(\theta - \varphi)}$

Section 10.5

3. $y(x,t) = H(t - 4x)\{\sin(t - 4x) - (t - 4x)\} + t$

5. $T(x,t) = T_0\left(1 - \text{erf } \dfrac{x}{2\sqrt{\delta t}}\right) + e^{-(x + \delta t)} - \displaystyle\int_0^t e^{-\delta(t - u)} \dfrac{x}{2\sqrt{\pi \delta u^3}} e^{-x^2/4\delta u} \, du$

7. $T(x,t) = T_0 e^{-(\mu x/2\delta) - (\mu^2/4\delta)} \displaystyle\int_0^t e^{\mu^2(t - u)/4\delta} \dfrac{x}{2\sqrt{\pi \delta u^3}} e^{-x^2/4\delta u} \, du$

CHAPTER 11

Section 11.1

1. Yes; $\delta = \frac{1}{216}$ (better: $\delta = \frac{1}{50}$)

3. Yes, but $a = b = 1$ is not possible. You need $a + b < 1$. If $a = b = \frac{1}{3}$, then $\delta = \frac{1}{4}$.

5. Yes; $\delta = 1$

7. Yes, but you need $b < 1$. If $b = \frac{1}{2}$, then $\delta = 1$.

9. Yes; $\delta = \frac{1}{2}$ (better: $\delta = +\infty$; there is a unique solution defined for $-\infty < t < \infty$).

11. a) $k = 1$ b) $k = 1$

     c) $k = 3 \left(\text{a smaller constant is } \sup\limits_{|x| \leqslant 1} \left| 2x \sin \dfrac{1}{x} - \cos \dfrac{1}{x} \right| \right)$

     d) $972\sqrt{3} \approx 1683$

13. $x_n(t) = 3 \sum\limits_{k=0}^{n} \dfrac{(t-3)^k}{k!}$

Section 11.2

9. a) $\begin{pmatrix} x_1 \\ x_2 \end{pmatrix}' = \begin{pmatrix} tx_1 x_2 + t^2 \\ x_1^2 + x_2^2 + t \end{pmatrix}; x(0) = \begin{pmatrix} 1 \\ 2 \end{pmatrix}$

   b) $k = 10$ c) $\delta = \dfrac{1}{7\sqrt{5}} \approx 0.064$

11. a) $\begin{pmatrix} x_1 \\ x_2 \\ x_3 \end{pmatrix}' = \begin{pmatrix} x_1^2 x_2 \\ x_3 + t \\ x_3^2 \end{pmatrix}; x(0) = c$

   b) $k \leqslant \max\limits_{\substack{1 \leqslant i \leqslant 3 \\ (t,x) \in D}} \left| \dfrac{\partial f}{\partial x_i} \right| \leqslant \max\{1 + a, 2(b + |c|)^2\}$

   c) $\delta = \min\left\{ a, \dfrac{b}{M} \right\}$ where $M \leqslant ((b + |c|)^6 + (b + |c| + a)^2 + (b + |c|)^4)$

CHAPTER 12

Section 12.1

1. a) $x' = x$ b) $x = 0, x = 1$ c) $x(t) = \dfrac{x(0)e^t}{x(0)e^t + 1 - x(0)}$

3. a) $x' = 2x$ b) $x = 0, x = -\frac{2}{3}$ c) $x(t) = \dfrac{2x(0)e^{2t}}{2 + 3x(0)(1 - e^{2t})}$

5. a) $x' = 2x$ b) $x = 0, x = 1, x = 2$

   c) $\dfrac{x^2 - 2x}{(x-1)^2} = ce^{2t}$ where $c = \dfrac{x(0)^2 - 2x(0)}{(x(0) - 1)^2}$

7. a) The orbits are the circles $x^2 + y^2 = a^2 + b^2$.

9. The orbits are $x = \exp\left[\frac{1}{2}(\ln^2 y - 2t_0 \ln y)\right]$.

Section 12.2

1. Saddle point   3. Unstable focus

5. Stable focus   7. Unstable focus

15. a) $x = c_1, y = c_1 + c_2 e^t$

   c) (c,c) is a critical point for any real number c.

Section 12.3

1. $(0,0)$ is a saddle point (unstable). 3. $(0,0)$ is unstable.

5. $(0,0)$ is an unstable node.

7. $(0,0)$ is the only critical point and is an unstable node or focus.

9. $(0,0)$ is an unstable node or focus, $(0,\frac{1}{3})$ and $(1,0)$ are saddle points and $(\frac{8}{7},\frac{1}{7})$ is a stable focus.

11. $(0,0)$, $(3,-9)$ and $(-2,-4)$ are saddle points; $(1,-1)$ is a stable focus and $(-1,-1)$ is an unstable focus.

13. $(0,0)$ is the only critical point and is a stable focus.

Section 12.4

1. Let $V(x,y) = x^2 + 2y^2$. 3. Let $V(x,y) = \dfrac{y^2}{2} + \omega^2(1 - \cos x)$.

5. a) Unstable b) Asymptotically stable c) Stable

7. Let $V(x,y) = \dfrac{y^2}{2} + \displaystyle\int_0^x g(x)\, dx$.

APPENDIX A4

1. a) $D = -2; x_1 = 3, x_2 = 0$ b) $D = 0$; no solutions

   c) $D = 0$; infinite number of solutions of the form $x_1 = c$, $x_2 = \dfrac{3 - c}{2}$, where c is a constant

   d) $D = 0$; infinite number of solutions of the form $x_1 = c$, $x_2 = 2c - 1$

   e) $D = 0$; no solutions f) $D = 6; x_1 = \frac{1}{2}, x_2 = 0$

3. a) Yes b) No

INDEX